W9-AAC-888

ENVIRONMENTAL ANALYSIS USING CHROMATOGRAPHY INTERFACED WITH ATOMIC SPECTROSCOPY

ELLIS HORWOOD SERIES IN ANALYTICAL CHEMISTRY

Series Editors: Dr R. A. CHALMERS and Dr MARY MASSON, University of Aberdeen
Consultant Editor: Prof. J. N. MILLER, Loughborough University of Technology

S. Alegret — **Developments in Solvent Extraction**
S. Allenmark — **Chromatographic Enantioseparation—Methods and Applications**
G.E. Baiulescu, P. Dumitrescu & P.Gh. Zugravescu — **Sampling**
H. Barańska, A. Łabudzińska & J. Terpiński — **Laser Raman Spectrometry**
G.I. Bekov & V.S. Letokhov — **Laser Resonant Photoionization Spectroscopy for Trace Analysis**
K. Beyermann — **Organic Trace Analysis**
O. Budevsky — **Foundations of Chemical Analysis**
J. Buffle — **Complexation Reactions in Aquatic Systems: An Analytical Approach**
D.T. Burns, A. Townshend & A.G. Catchpole — **Inorganic Reaction Chemistry Vol. 1: Systematic Chemical Separation**
D.T. Burns, A. Townshend & A.H. Carter — **Inorganic Reaction Chemistry: Vol. 2: Reactions of the Elements and their Compounds: Part A: Alkali Metals to Nitrogen, Part B: Osmium to Zirconium**
J. Churáček — **New Trends in the Theory & Instrumentation of Selected Analytical Methods**
E. Constantin, A. Schnell & A. Pape — **Mass Spectrometry**
R. Czoch & A. Francik — **Instrumental Effects in Homodyne Electron Paramagnetic Resonance Spectrometers**
T.E. Edmonds — **Interfacing Analytical Instrumentation with Microcomputers**
Z. Galus — **Fundamentals of Electrochemical Analysis, Second Edition**
S. Görög — **Steroid Analysis in the Pharmaceutical Industry**
T. S. Harrison — **Handbook of Analytical Control of Iron and Steel Production**
J.P. Hart — **Electroanalysis of Biologically Important Compounds**
T.F. Hartley — **Computerized Quality Control: Programs for the Analytical Laboratory**
Saad S.M. Hassan — **Organic Analysis using Atomic Absorption Spectrometry**
M.H. Ho — **Analytical Methods in Forensic Chemistry**
Z. Holzbecher, L. Diviš, M. Král, L. Šůcha & F. Vláčil — **Handbook of Organic Reagents in Inorganic Chemistry**
A. Hulanicki — **Reactions of Acids and Bases in Analytical Chemistry**
David Huskins — **Electrical and Magnetic Methods in On-line Process Analysis**
David Huskins — **Optical Methods in On-line Process Analysis**
David Huskins — **Quality Measuring Instruments in On-line Process Analysis**
J. Inczédy — **Analytical Applications of Complex Equilibria**
M. Kaljurand & E. Küllik — **Computerized Multiple Input Chromatography**
S. Kotrlý & L. Šůcha — **Handbook of Chemical Equilibria in Analytical Chemistry**
J. Kragten — **Atlas of Metal-ligand Equilibria in Aqueous Solution**
A.M. Krstulović — **Quantitative Analysis of Catecholamines and Related Compounds**
F.J. Krug & E.A.G. Zagotto — **Flow Injection Analysis in Agriculture and Environmental Science**
V. Linek, V. Vacek, J. Sinkule & P. Beneš — **Measurement of Oxygen by Membrane-covered Probes**
C. Liteanu, E. Hopîrtean & R. A. Chalmers — **Titrimetric Analytical Chemistry**
C. Liteanu & I. Rîcă — **Statistical Theory and Methodology of Trace Analysis**
Z. Marczenko — **Separation and Spectrophotometric Determination of Elements**
M. Meloun, J. Havel & E. Högfeldt — **Computation of Solution Equilibria**
M. Meloun, J. Militky & M. Forina — **Chemometrics in Instrumental Analysis: Solved Problems for the IBM PC**
O. Mikeš — **Laboratory Handbook of Chromatographic and Allied Methods**
J.C. Miller & J.N. Miller — **Statistics for Analytical Chemistry, Second Edition**
J.N. Miller — **Fluorescence Spectroscopy**
J.N. Miller — **Modern Analytical Chemistry**
J. Minczewski, J. Chwastowska & R. Dybczyński — **Separation and Preconcentration Methods in Inorganic Trace Analysis**
T.T. Orlovsky — **Chromatographic Adsorption Analysis**
D. Pérez-Bendito & M. Silva — **Kinetic Methods in Analytical Chemistry**
B. Ravindranath — **Principles and Practice of Chromotography**
V. Sedivec & J. Flek — **Handbook of Analysis of Organic Solvents**
O. Shpigun & Yu. A. Zolotov — **Ion Chromatography in Water Analysis**
R.M. Smith — **Derivatization for High Pressure Liquid Chromatography**
R.V. Smith — **Handbook of Biopharmaceutic Analysis**
K.R. Spurny — **Physical and Chemical Characterization of Individual Airborne Particles**
K. Štulík & V. Pacáková — **Electroanalytical Measurements in Flowing Liquids**
J. Tölgyessy & E.H. Klehr — **Nuclear Environmental Chemical Analysis**
J. Tölgyessy & M. Kyrš — **Radioanalytical Chemistry, Volumes I & II**
J. Urbanski, *et al.* — **Handbook of Analysis of Synthetic Polymers and Plastics**
M. Valcárcel & M.D. Luque de Castro — **Flow-Injection Analysis: Principles and Applications**
C. Vandecasteele — **Activation Analysis with Charged Particles**
F. Vydra, K. Štulík & E. Juláková — **Electrochemical Stripping Analysis**
N. G. West — **Practical Environmental Analysis using X-ray Fluorescence Spectrometry**
J. Zupan — **Computer-supported Spectroscopic Databases**
J. Zýka — **Instrumentation in Analytical Chemistry**

ENVIRONMENTAL ANALYSIS USING CHROMATOGRAPHY INTERFACED WITH ATOMIC SPECTROSCOPY

Edited by
ROY M. HARRISON
Department of Chemistry
University of Essex

and

SPYRIDON RAPSOMANIKIS
Chemical Engineering Department
Aristotelian University of Thessaloniki
Thessaloniki
Greece

ELLIS HORWOOD LIMITED
Publishers · Chichester

Halsted Press: a division of
JOHN WILEY & SONS
New York · Chichester · Brisbane · Toronto

First published in 1989 by
ELLIS HORWOOD LIMITED
Market Cross House, Cooper Street,
Chichester, West Sussex, PO19 1EB, England
The publisher's colophon is reproduced from James Gillison's drawing of the ancient Market Cross, Chichester.

Distributors:
Australia and New Zealand:
JACARANDA WILEY LIMITED
GPO Box 859, Brisbane, Queensland 4001, Australia
Canada:
JOHN WILEY & SONS CANADA LIMITED
22 Worcester Road, Rexdale, Ontario, Canada
Europe and Africa:
JOHN WILEY & SONS LIMITED
Baffins Lane, Chichester, West Sussex, England
North and South America and the rest of the world:
Halsted Press: a division of
JOHN WILEY & SONS
605 Third Avenue, New York, NY 10158, USA
South-East Asia
JOHN WILEY & SONS (SEA) PTE LIMITED
37 Jalan Pemimpin # 05–04
Block B, Union Industrial Building, Singapore 2057
Indian Subcontinent
WILEY EASTERN LIMITED
4835/24 Ansari Road
Daryaganj, New Delhi 110002, India

British Library Cataloguing in Publication Data
Harrison, Roy, M., *1948–*
Environmental analysis using chromotography interfaced with atomic spectroscopy.
1. Environment. Chemical constituents. Chemical analysis
I. Title II. Rapsomanikis, Spyridon
543

Library of Congress Card No. 88–34705

ISBN 0–85312–979–7 (Ellis Horwood Limited)
ISBN 0–470–21407–4 (Halsted Press)

Typeset in Times by Ellis Horwood Limited
Printed in Great Britain by The Camelot Press, Southampton

Table of contents

Chapter 7 Tin and germanium

O. F. X. Donard
Group d'Océanographie Physico-Chimique, Laboratoire de Chimie Physique A, Université de Bordeaux I, 33405 Talence, France
R. Pinel
Laboratoire de Chimie Analytique, Faculté des Sciences et Techniques, 64000 Pau, France

Chapter 8 Lead

M. Radojević
Department of Chemistry, University of Manchester,
Institute of Science and Technology, PO Box 88, Manchester M60 1DQ, England.

Chapter 9 Arsenic and antimony

S. C. Apte,
Water Research Centre, Medmenham Laboratory, Henley Road, Marlow, Buckinghamshire, SL7 2HD, England.
A. G. Howard and A. T. Campbell,
The University, Southampton, Hampshire, SO9 5NH, England.

Preface

The subject of this book lies at the boundary of two rapidly advancing areas of science: analytical chemistry and environmental chemistry. Much of the driving force behind development of more specific and sensitive analytical procedures has come from the need of environmental chemists to analyse trace quantities of pollutants with a view to quantification of human exposure and elucidation of environmental pathways. This has been especially important in the field of organometallic species in the environment, for which analytical methods involving interfaced chromatography/atomic spectroscopy methods are extremely valuable.

We have not restricted our remit rigidly to atomic spectroscopy. Where closely related molecular spectroscopic techniques exist, particularly those involving flame photometric detectors, we have felt it logical and useful to include them. Thus well tested methods such as gas chromatography/flame photometric detection for sulphur compounds are included. These are readily commercially available. At the other extreme, the development of interfaces between gas chromatography or liquid chromatography and atomic-absorption, atomic-fluorescence or plasma emission detectors is a rapidly advancing field of research which has yet to lead significantly to commercially available interfaces.

The book divides into two sections. The first section, Chapters 1–6, covers the basic principles of chromatographic separations and spectroscopic detection methods. The second part of the book, Chapters 7–12, deals with the practical application of these techniques in environmental analysis and is treated element by element, rather than by technique. Thus both the analytical chemist and the environmental chemist should find material to meet their needs.

No one author is expert across this whole field, and an edited book appeared the best approach. We have been extremely fortunate to attract leading scientists in the field as authors of the individual chapters and we are delighted with the fine work which they have done. We hope that the reader will find their chapters as illuminating as we have done.

Colchester
March 1988

Roy M. Harrison
Spyros Rapsomanikis

1

Basic principles of chromatography and atomic spectroscopy

S. J. de Mora
Department of Chemistry, University of Auckland, Private Bag, Auckland, New Zealand

Both chromatography and atomic-absorption spectrometry are major branches of analytical chemistry. Chromatography is the science of separations. As such it can be considered to be basically qualitative in nature. The types of separations feasible are extremely diverse. Different phases may be involved and resolution of analytes may be according to chemical behaviour or size differentiation. Quantification is possible only with the incorporation of a detector. Atomic spectroscopy has recently gained widespread usage as a detector in chromatographic techniques. As with chromatography, atomic spectroscopy encompasses a wide variety of methodologies. The common factor in these techniques is that quantification is realized by measuring the amount of energy associated with electronic transitions in atoms. The nature of the electronic transition considered defines the specific technique.

From a theoretical point of view, chromatography and atomic spectroscopy make a clumsy marriage. They constitute quite different disciplines. The theoretical principles governing these two branches are completely independent. Both areas have undergone accelerated growth in the past two decades and have vast literatures associated with them. Limited space allows only the fundamental principles of the theory to be presented here. A brief outline of practical considerations in chromatography is included. For the most part citations are made of monographs which deal at great length with the different types of techniques involved in both chromatography and atomic spectroscopy.

1.1 THE THEORY OF CHROMATOGRAPHY

1.1.1 Introduction

The term chromatography was introduced about the turn of the century by the

Russian botanist Mikhail Tswett, working in Warsaw. He separated pigments in a plant extract by passing the mixture through a glass column packed with calcium carbonate. Coloured bands were seen on the column, and gave rise to the name of the technique. Such techniques are not limited to coloured compounds and so the term is now a misnomer.

Chromatography is a generic term describing a major branch of separation science, incorporating several types of experimental procedures and applicable to a wide range of components. In chromatography practice, several classification schemes prevail, which depend on the principles or the different phases utilized to achieve separation [1,2]. The common factor in the separation techniques is the requirement for two phases, a stationary phase and a mobile phase (called the eluent), and that during elution solutes migrate through the system at different rates depending on their degree of interaction with the different phases. As illustrated in Table 1.1, the mechanism responsible for the retardation of solutes varies. Clearly

Table 1.1 — Classification of chromatographic techniques

Name	Mobile phase	Stationary phase	Principle of separation
Gas–liquid	Gas	Liquid	Partition
Gas–solid	Gas	Solid	Adsorption of diffusion
Liquid–liquid	Liquid	Liquid	Partition
Liquid–solid	Liquid	Solid	Adsorption
Thin-layer	Liquid	Liquid/solid	Adsorption and partition
Paper	Liquid	Liquid/solid	Adsorption and partition
Size-exclusion	Liquid	Solid	Diffusion
Ion-exchange	Liquid	Solid	Electrostatic interaction
Affinity	Liquid	Solid	Specific adsorption

not all the chromatographic techniques noted in Table 1.1 are readily amenable to interfacing with element-specific detectors. The discussion here is limited to gas and liquid chromatographic techniques, although size exclusion chromatography is included in Section 1.2.3.

In both gas and liquid chromatography, the distribution of solutes between the stationary and mobile phases is an equilibrium process and so solutes are resolved according to their differing affinities towards the two phases, which results in differential migration rates through the system. Solutes exhibiting a preference for the mobile phase are eluted relatively quickly while those with a high affinity for the stationary phase are retarded. Qualitative analysis of solutes is based on their migration rates in an operationally defined system. The migration rate of a solute is characterized either by the retention time, t_R, i.e. the total time elapsed from the injection of a sample into the system until detection of the maximum of the elution peak of the solute, or by the retention volume, V_R, i.e. the total volume of eluent that flows through the system during the time t_R. Thus

$$V_R = t_R F_C \tag{1.1}$$

where F_C is the volume flow-rate of the mobile phase, usually expressed in ml/min, at the column temperature. Because the sample aliquot is injected into the flowing mobile phase, t_R includes the time taken for the mobile phase to travel from the point of injection to the detector. This is known as the hold-up time (or dead-time), t_M, and is equivalent to the time taken for a completely unretained solute to pass through the system. By definition, V_R includes the hold-up volume (or dead-volume), V_M, corresponding to the total volume of the system, including that of the injector and detector. In some instances it is useful to consider the adjusted retention time, t'_R:

$$t'_R = t_R - t_M \tag{1.2}$$

or the adjusted retention volume, V'_R:

$$V'_R = t'_R F_C \tag{1.3}$$

Retention times and volumes can also be corrected for temperature or pressure, thereby allowing for the compressibility of the fluid. This may not be necessary in liquid chromatography, but in gas chromatography a significant pressure gradient may exist between the inlet and outlet of the system. The specific retention volume, V_G, incorporates these corrections and constitutes the net volume of eluent at 0°C necessary to transfer one-half of the solute through a system normalized to 1 g of stationary phase and exhibiting no pressure drop:

$$V_G = t'_R F_C j(273/T_C w_s) \tag{1.4}$$

where T_C is the temperature of the column in Kelvin, w_s is the weight in g of the stationary phase and j is the pressure correction given by:

$$j = \frac{3[(P_i/P_o)^2 - 1]}{2[(P_i/P_o)^3 - 1]} \tag{1.5}$$

where P_i and P_o are the inlet and outlet pressures, respectively.

The various retention times and retention volumes outlined above and depicted in Fig. 1.1 are diagnostic of the migration rate of a solute through an operationally defined system. Further nomenclature will be introduced later, but it should be noted that Grob [2] provides a comprehensive glossary of chromatographic definitions and terms, and takes care to point out the preferred IUPAC notation and usage.

It is important to appreciate that though solutes exhibit differential migration rates through a system, broadening of the solute band also occurs. Such zone broadening is unavoidable but generally occurs at a slower rate than zone separation. Nonetheless, the resolution of solutes will depend upon both processes and peak broadening may be quite significant when a complete separation between two solutes is necessary. Theoretical treatments of chromatography must be able to account for both migration rates and zone broadening effects.

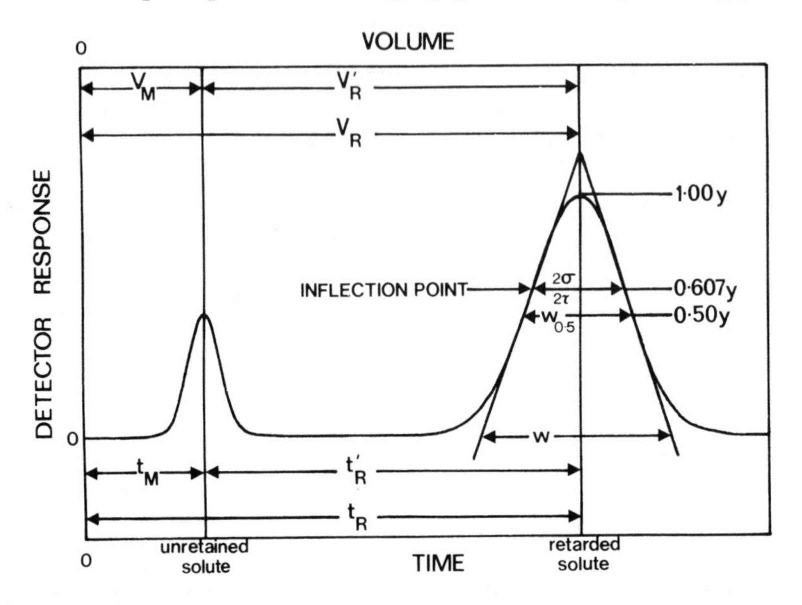

Fig. 1.1 — A linear chromatogram illustrating some features of a Gaussian curve and some chromatographic parameters.

1.1.2 Equilibrium considerations and plate theory

Chromatographic separations rely upon solutes attaining equilibrium between a stationary and mobile phase. For a system at constant temperature and pressure, the equilibrium condition is:

$$(\Delta G)_{T,P} = 0 \tag{1.6}$$

If we consider a solute distributed between a stationary and mobile phase, then the chemical potential, μ, of the solute in each phase will be equal at equilibrium:

$$\mu_{\mathrm{s}} = \mu_{\mathrm{m}} \tag{1.7}$$

where the subscripts s and m refer to the stationary and mobile phases, respectively. This can also be stated in terms of the standard chemical potential, μ_i^{o}, and activity, a_i such that:

$$\mu_{\mathrm{s}}^{\mathrm{o}} + RT\ln a_{\mathrm{s}} = \mu_{\mathrm{m}}^{\mathrm{o}} + RT\ln a_{\mathrm{m}} \tag{1.8}$$

Rearranging gives:

$$\ln(a_{\mathrm{s}}/a_{\mathrm{m}}) = (\mu_{\mathrm{m}}^{\mathrm{o}} - \mu_{\mathrm{s}}^{\mathrm{o}})/RT \tag{1.9}$$

In many systems the activity coefficients are not known, but in chromatographic techniques such a small amount of solute is present that the solution tends towards infinite dilution and so activity coefficients can be assumed to be unity. This gives:

$$C_s/C_m = \exp(-\Delta\mu^o/RT) \tag{1.10}$$

where C_s and C_m are the concentrations of the solute in the stationary and mobile phases, respectively, and $\Delta\mu^o = \mu_s^o - \mu_m^o$. This concentration ratio is an equilibrium constant, also known as a distribution constant:

$$K = C_s/C_m \tag{1.11}$$

At constant temperature and pressure, a plot of C_s *vs.* C_m should be linear with a slope of K. This type of chromatography is referred to as linear chromatography. Under these circumstances, the migration rate through the system is independent of solute concentration and the peak shape is a Gaussian curve. That is to say, the peak is symmetrical, with a constant retention time. Figure 1.1 shows a linear chromatogram and several useful chromatographic parameters. Some of the descriptions of the peak shape rely upon the characteristic properties of a Gaussian peak. Consider a Gaussian peak normalized to a height of y and with standard deviation σ (using volume units) or τ (using time units). The inflection points occur at $0.607y$ with a peak width $W_i = 2\sigma$. Tangents to these points define a triangle which intersects the x-axis to give a peak width, W, equal to 4σ. The area of this triangle accounts for approximately 96% of the area under the peak. The peak width at half height is symbolized as $W_{0.5}$ and equals 2.354σ. It should be noted that all peak widths may be alternatively expressed in time units, e.g. $W = 4\tau$.

Non-linear chromatography occurs when the C_s/C_m distribution is not linear at a constant temperature. Elution peak profiles are then asymmetric, exhibiting either a leading or a trailing edge. Positive deviations from linearity (i.e. C_s *vs.* C_m is concave) results in a trailing edge and retention times increase with solute concentration. These features are symptomatic of overloading the stationary phase and may be rectified by injecting a smaller sample aliquot. Conversely, negative deviations from linear behaviour (i.e. convex curve for C_s *vs.* C_m) give a sharp leading edge followed by an extended tail. Retention times then decrease with solute concentration. Trailing edges are indicative of adsorption effects and often the remedy is to replace the column, possibly changing the stationary phase.

The separation of solutes relies upon the differential retardation of solutes during their transport through the system. Differing values of K reflect the differences in affinity of a solute towards the stationary phase. Chromatography is dynamic in nature and certainly more than one equilibration process is involved. The sample aliquot is injected into the mobile phase, and on contacting the stationary phase solute is transferred from one phase to another as the system attempts to attain equilibrium. The mobile phase continues to flow, encountering stationary phase containing none of the solute, so further transfer occurs; in this way the process continues, thereby establishing the leading edge of the solute peak. At the trailing edge of the peak there is a net transfer of solute from the stationary phase into the mobile phase. Symmetric profiles arise only when there are no constraints on the system's approach to equilibrium. This conceptual framework is analogous to the Craig countercurrent system of multi-step solvent extraction. Clearly peak broadening will occur as a result of each additional equilibrium step. This approach to interpreting chromatographic processes was first suggested by Martin and Synge [3]. In their plate theory, the chromatographic column was considered to consist of a series of discrete but continuous narrow bands of equal volume in which a perfect equilibrium distribution of the solute between the mobile and stationary phases was achieved. These bands were termed theoretical plates, acknowledging that the plates did not exist as physical entities in the column. Martin and Synge showed that if the solute distribution was assumed to be controlled exclusively by equilibrium considerations (i.e. no solute–matrix interactions, no diffusion between plates), then the number of theoretical plates can be determined from the extent to which zone broadening occurs for the peak concerned. The theoretical plate number, n, is defined in terms of retention time, as

$$n = (t_R/\tau)^2 \tag{1.12}$$

or of retention volume, as

$$n = (V_R/\sigma)^2 \tag{1.13}$$

From consideration of the characteristics of Gaussian curves, these equations (shown below only for t_R) may be expressed as:

$$n = 5.54\ (t_R/W_{0.5})^2 \tag{1.14}$$

or

$$n = 16(t_R/W)^2 \tag{1.15}$$

The theoretical plate number is defined by the retention time and so includes a time period (the dead-time) during which no equilibration is possible because there is no sample present in the mobile phase that is in contact with the stationary phase between the injection point and the detector. This difficulty is overcome by using the effective theoretical plate number, N, which is based on the adjusted retention time:

$$N = 5.54\ (t'_R/\tau)^2 \tag{1.16}$$

Obviously the number of theoretical and effective plates increases as the length of the column increases. Zone spreading will increase with the number of equilibration steps involved. For the comparison of different stationary phases it is convenient to consider the height equivalent to a theoretical plate (HETP), h, defined as:

$$h = L/n \tag{1.17}$$

where L is the length of the column; or the height equivalent to an effective plate, H, which is

$$H = L/N \qquad (1.18)$$

Alternatively, h can be expressed as a measure of zone spreading:

$$h = \sigma^2/L \qquad (1.19)$$

where σ^2 is the variance of the retention volume.

The plates lack physical reality and a number of assumptions have been made in order to relate the number of equilibration steps to zone broadening effects. First, that perfect equilibration occurs at each step. This means that the distribution constant for a solute must be the same at each plate, therefore being invariant with solute concentration or matrix composition. Secondly, that the mobile phase flows in a stepwise manner from plate to plate, with equilibrium being achieved rapidly at each stage. Plate boundaries prevent axial diffusion between adjacent plates. Not all of these assumptions are valid and zone broadening does not occur exclusively because of multi-step equilibria. Nonetheless, the concepts of the theoretical plate and height equivalent to a theoretical plate are useful criteria for the comparison of the separation efficiency in different chromatographic systems.

1.1.3 Rate theory

The plate theory does not adequately take into account the mechanisms causing peak broadening. Some of the experimental parameters which affect broadening can be influenced by the analyst. The plate theory gives no indication as to how such variables can be adjusted in order to achieve a low value of h. The rate theory overcomes these inadequacies by considering carrier-gas flow-rates, hydrodynamics of the mobile phase, various diffusional processes and the rate of solute transfer between phases. Thus, retention times and widths of elution bands are explained by the rate theory.

The rate theory considers the chromatographic process in terms of partitioning dynamics rather than multi-step equilibrium. The migration involves the repeated transfer of a solute between the stationary and mobile phases. The solute is transported through the system only during residence in the mobile phase, at the linear velocity, u, of that phase. Thus, for any individual particle, the migration through the column is irregular. For a multitude of particles, the time spent in the mobile phase relative to that spent in the stationary phase will vary. The mechanisms responsible are considered below, but the manifestation of this effect is a symmetric distribution of migration rates about some mean value. The broadening is a kinetic effect and so zone breadth increases during solute transfer through the system.

The rate theory is best described in terms of a random walk model [2,4]. Various types of diffusion are considered to be the zone broadening mechanisms. These act as random processes causing a general back and forth motion of the solute along the axis of eluent flow. The resultant concentration profile is Gaussian in shape. The zone broadening is described by the standard deviation, σ. In terms of the random walk model, σ is given by the number of steps n, each of length l:

$$\sigma = ln^{\frac{1}{2}} \tag{1.20}$$

Variances, but not standard deviations, are additive and so the overall variance of peak spreading is handled as a cumulative set of the variances, σ_i^2, due to individual effects. Therefore, the overall effect is:

$$\sigma^2 = \Sigma\sigma_i^2 \tag{1.21}$$

The individual variances are summarized below; derivations for individual terms can be found elsewhere [2,4,5].

Eddy diffusion. Zone broadening due to eddy diffusion results from the tortuosity of the path that individual solute molecules follow during transport through the column. Packing of the solid support in the column leaves void spaces between individual particles. Under ideal conditions such void spaces would be of uniform size. However, variations in the diameter, d_p, of the solid support particles, and inhomogeneities in the packing density, will cause variations in the void volume down the column. Solute molecules will follow paths of different length during migration through the column. The resulting variance in zone broadening is given by:

$$\sigma_A^2 = 2\lambda d_p L \tag{1.22}$$

where λ is the coefficient of eddy diffusion.

Longitudinal diffusion in the mobile phase. This is also known as ordinary or axial diffusion and is a process associated with any concentration gradient. Diffusion occurs from a zone of high concentration to a zone of low concentration. Longitudinal diffusion occurs at a molecular level, owing to the collision of molecules. When such diffusion occurs in the mobile phase, the resulting variance is:

$$\sigma_{Bm}^2 = 2\gamma_m D_m L/u \tag{1.23}$$

where γ_m is an obstruction factor for longitudinal diffusion of the solute in the mobile phase (and is usually less than unity), D_m is the coefficient of molecular diffusion for the solute in the mobile phase, and u is the linear velocity of the mobile phase.

Longitudinal diffusion in the stationary phase. As in the case of the mobile phase, the diffusion of solute will occur whenever a concentration gradient is established in the stationary phase. The corresponding relationship for the variance is:

$$\sigma_{Bs}^2 = 2\gamma_s D_s L(1 - R)/Ru \tag{1.24}$$

where γ_s is the obstruction factor for longitudinal diffusion of the solute in the stationary phase, D_s is the coefficient of molecular diffusion of the solute in the stationary phase and R is the fraction of solute molecules in the mobile phase (therefore $1-R$ is the fraction of solute molecules in the stationary phase).

Resistance to mass transfer in the stationary phase. Chromatographic bands also spread because the mobile phase moves too fast for equilibrium with the stationary phase to be attained. Non-equilibrium conditions occur at both the leading edge of the solute peak where the net transfer of solute is from the mobile phase into the stationary phase and at the trailing edge where the reverse process is in operation. The effect of non-equilibrium transfer is directly related to the flow-rate of the mobile phase in that slower flow-rates allow a closer approach to equilibrium to be attained. The zone spreading due to this effect is described by the term:

$$\sigma^2_{\mathrm{Cs}} = (8/\pi^2)[k/(1+k)^2](d_f^2/D_s)Lu \tag{1.25}$$

where k is the retention factor for the column [i.e. $k=(1-R)/R$], d_f is the effective film thickness of the stationary phase, and D_s, as given above, is the diffusion coefficient of the solute in the stationary phase.

Resistance to mass transfer in the mobile phase. As noted by Novák [6] zone spreading may also arise from non-equilibrium conditions in the mobile phase. He distinguishes two cases: first, for the mobile phase present in the spaces between the particles of the solid support the effect is characterized by

$$\sigma^2_{\mathrm{Cm}} = \omega d_p^2 Lu/2D_m \tag{1.26}$$

where ω is a factor characterizing the geometrical structure of the chromatographic packing and d_p is the particle diameter; secondly, for non-equilibrium in the mobile phase present within microporous solid support particles the resulting variance of zone speading is

$$\sigma^2_{\mathrm{C^*m}} = [(1-\phi_m R)^2/30\gamma'_m(1-\phi_m)]d_p^2 Lu/D_m \tag{1.27}$$

where ϕ_m is the fraction of the mobile phase present in the intraparticle space and γ'_m is an obstruction factor for lateral diffusion in the intraparticle mobile phase.

As noted previously, Martin and Synge [3] defined the height equivalent to a theoretical plate in terms of the total variance of the zone broadening (i.e. $h=\sigma^2/L$). From Eq. (1.21) we get:

$$h = \Sigma\sigma_i^2/L \tag{1.28}$$

or, in terms of the various effects causing zone spreading, we have:

$$h = \sigma_A^2/L + \sigma_{Bm}^2/L + \sigma_{Bs}^2/L + \sigma_{Cs}^2/L + \sigma_{Cm}^2/L + \sigma_{C^*m}^2/L, \quad (1.29)$$

Summarizing, three types of diffusion mechanism cause zone spreading. Characterized in the terms above by the subscripts A, B and C, respectively, these effects are contributions from non-equal paths, longitudinal diffusion and non-equilibrium conditions. The relative importance of the different terms depends on the type of chromatographic technique considered.

The generalized relationship [5] is known as the van Deemter equation:

$$h = A + B/u + Cu \quad (1.30)$$

In gas-liquid chromatography, zone spreading is predominantly due to eddy diffusion, longitudinal diffusion in the mobile phase and mass transfer resistance in the liquid phase. In this case the constants in the van Deemter equation are $A = 2\lambda d_p$, $B = 2\lambda D_m$ and $C = (8/\pi^2)[k/(1+k)^2](d_f^2/D_s)$. The influence of the various terms is evident in the van Deemter plot of h *vs.* u, as illustrated in Fig. 1.2. The performance

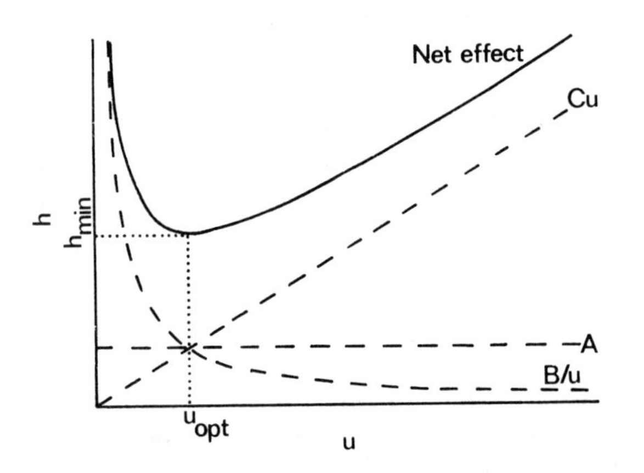

Fig. 1.2 — A van Deemter plot of h *vs.* u, indicating the relative importance of the A, B and C terms in the equation.

of the column is optimal at the minimum value for the theoretical plate height, so

$$h_{min} = A + 2(BC)^{\frac{1}{2}} \quad (1.31)$$

and

$$u_{opt} = (B/C)^{\frac{1}{2}} \quad (1.32)$$

In the case of liquid chromatography, the van Deemter equation must also include consideration of longitudinal diffusion in the stationary phase and mass transfer resistances in the mobile phase. Diffusion coefficients are 3–4 orders of magnitude lower in liquids than in gases. Accordingly, the B term in the equation becomes less significant than in gas–liquid chromatography. The value of h becomes minimal at a very low flow-rate with the consequence that liquid chromatography is conducted at velocities greater than u_{opt}. As the mass transfer resistances dominate the theoretical plate height, the particle size of the solid support must be kept as small as possible.

1.1.4 Characterization of column performance

The resolution of two components is ultimately a function of the difference in their equilibrium constants in different chromatographic systems. The choice of chromatographic column has a great influence on the extent to which solutes can be separated. A number of parameters have been used to compare the performance of columns [7].

Theoretical plate number. As noted previously, the number of theoretical plates, n, can be used as a guide to column performance. This number is operationally defined and dependent on the solute considered.

Retention factor. This is a measure of the time a solute spends in the mobile phase relative to the time spent in the stationary phase. The retention factor, k, is:

$$k = V'_R / V_M \tag{1.33}$$

or

$$k = t'_R / t_M \tag{1.34}$$

The retention factor can be expressed in terms of the ratio of the amounts of solute in the stationary and mobile phases:

$$k = (1 - R)/R \tag{1.35}$$

Separation factor. The separation factor $\alpha_{A/B}$ is the ratio of the distribution constants for solutes A and B measured under identical conditions. The relationship is

$$\alpha_{A/B} = K_A / K_B \tag{1.36}$$

By convention, component A is the solute more strongly retained by the stationary phase so $\alpha_{A/B}$ is always greater than unity. From the definition of K:

$$K_i = C_s / C_m \tag{1.37}$$

we get:

$$K_i = [(1 - R)/V_s]/(R/V_m) \quad (1.38)$$

where V_m is the volume of the mobile phase and V_s is the volume of the stationary phase and therefore:

$$K_i = k_i V_s/V_m \quad (1.39)$$

The separation factor can be expressed in terms of retention factors as

$$\alpha_{A/B} = k_A/k_B \quad (1.40)$$

or in terms of the adjusted retention times, which can be readily obtained from a chromatogram:

$$\alpha_{A/B} = (t'_R)_A/(t'_R)_B \quad (1.41)$$

Resolution. Column performance can be judged from the resolution, R_s, of solutes A and B. This is the degree of separation between a pair of peaks and defined as:

$$R_s = 2\Delta t_R/(W_A + W_B) \quad (1.42)$$

Several expressions relating the resolution to chromatographic parameters have been developed. The general resolution equation:

$$R_s = (N^{\frac{1}{2}}/4)[(\alpha_{A/B} - 1)/(\alpha_{A/B})][k_A/(k_A + 1)] \quad (1.43)$$

relates the resolution between two peaks to the number of theoretical plates, the selectivity of the column and the retention factor. For a derivation of the general resolution expression see, for example, Parris [8].

A criterion which may be important in selecting a chromatographic system is the analysis time. From the resolution, the time required to elute the more strongly bound solute is:

$$(t_R)_A = 16R_s^2 h/u[\alpha_{A/B}/(\alpha_{A/B} - 1)]^2[(1 + k_A)^3/k_A^2] \quad (1.44)$$

Trennzahl value. The 'Trennzahl value', also known as the separation number, is defined as the resolution between two solutes, A and B, which are consecutive members of the n-paraffin homologous series:

$$TZ = \{[(t_R)_A - (t_R)_B]/[(W_{0.5})_A + (W_{0.5})_B]\} - 1 \quad (1.45)$$

The Trennzahl value is the number of peaks that could be placed between the two solute peaks and is predominantly used in characterizing capillary columns.

Retention index. For an operationally defined system (i.e. specified column at a given temperature), the logarithm of the adjusted retention time is a linear function of the carbon number for a homologous series of compounds. Kováts [9] defined a retention index as 100 times the hypothetical carbon number for an n-alkane having an adjusted retention time equivalent to that of the analyte of interest. Mathematically, the retention index, I, is

$$I = 100Z + 100\left[\frac{\log(t'_R)_i - \log(t'_R)_Z}{\log(t'_R)_{Z+1} - \log(t'_R)_Z}\right] \tag{1.46}$$

where i is the analyte of interest, Z is the carbon number of the n-alkane that would be eluted just prior to i and $Z+1$ is the carbon number of the n-alkane that would be eluted just after i. As noted above, the column temperature and stationary phase must be defined and so are usually designated as subscript and superscript to I, respectively.

1.2 PRACTICAL ASPECTS OF CHROMATOGRAPHY

The theoretical principles of chromatography have been outlined in the previous section. Chromatography in practice is more an empirical art than a theoretical science. Accordingly, a few practical aspects of chromatography are introduced here, regarding the choice of stationary phase and procedural strategies for improving chromatographic performance. Also, the basic components of chromatographic instruments are described.

1.2.1 Gas chromatography

Gas chromatography encompasses both gas–liquid and gas–solid categories, the major characteristic thus being the use of a gaseous mobile phase. Only a limited discussion can be presented here and readers should refer elsewhere for more complete descriptions of gas chromatographic procedures [2,7,10].

The basic components of a gas chromatograph are:

(1) carrier gas system
(2) sample introduction system
(3) column oven
(4) column
(5) detection system
(6) data acquisition/reduction system.

The first four components will be discussed here, and detectors are dealt with in subsequent chapters.

Carrier gas system. The carrier gas must be inert, not reacting with the components of the analyte mixture or with the column materials. A high degree of purity is required and usually impurities such as air or water vapour are removed by passing the carrier gas through a molecular sieve. The most commonly utilized carrier gases are Ar, CO_2, He, H_2, N_2 and O_2. The choice of gas may be limited by the detector system. In the absence of such constraints, separations requiring low mobile phase velocities should be done with gases with small diffusion coefficients (Ar, N_2, CO_2). Conversely, gases with larger diffusion coefficients (H_2, He) are preferred when high flow-rates are utilized.

The carrier gas flow-rate determines the resolution and analysis time for a given column. Flow-rates can be affected by the column temperature as both the viscosity and density of gases are temperature-dependent. Fluctuations in flow-rate may affect detector response in some cases and always influence the retention times of solutes. Thus, good control of the flow-rate is required for both quantitative and qualitative analyses. Flow-rate fluctuations are kept below 0.2% by either controlling the inlet pressure of the carrier gas or using a flow controller (i.e. a needle valve which provides an orifice with a variable size dependent upon the downstream pressure).

Sample introduction systems. The sample is introduced into the flowing mobile phase in the sample inlet system. The design of this system depends on the type of sample (especially its phase) and the type of column. Ideal characteristics of the inlet system are a small volume chamber with no local stagnation of the carrier gas. The sample chamber can be heated to ensure flash vaporization of the sample mixture introduced (i.e. solute and solvent). These features ensure that the sample is introduced into the column with the minimum amount of zone broadening.

Gases can be introduced into the mobile phase either by means of a gas sampling loop or by injection with a gas syringe through a self-sealing silicone septum. Liquids are commonly injected in microlitre volumes by syringe. The sample can be injected into a flash vaporization chamber or in some cases directly into the end of the column (on-column injection).

Capillary columns have very low capacities, taking only 1–500 nl of sample and requiring special injection procedures. Introduction systems may involve split injection, whereby a relatively large volume is injected into a heated port; the sample is instantly vaporized but only 0.1–10% of it is introduced into the column, the rest being vented. Splitless injection can be utilized whereby a trace analyte is diluted in an appropriate solvent. The boiling point of this solvent must be high relative to the column temperature but lower than the boiling point of the solutes to be determined. The sample is injected into the vaporization chamber and is carried by the mobile phase to the column. The column temperature is sufficiently low for the solvent to condense and thereby trap the solutes. Thereafter, a portion of the carrier gas purges the vaporization chamber and a temperature ramp is applied to the column. Finally, on-column injection can be utilized. The amount of sample introduced can be carefully controlled. The injection is made rapidly, with up to 2 μl of solvent. The column temperature must be below the boiling point of the solvent during the injection.

Column ovens. The oven must provide a uniform temperature throughout the whole space in which the columns are placed. Thus, no spatial temperature gradients should be present at any time. Oven temperatures during operation in both the isothermal and programming modes must be reproducible.

The temperature of the system is an important parameter that the analyst can adjust. During isothermal separations, volatile components (i.e. early peaks) may not be well resolved, whereas solutes with higher boiling point are eluted slowly and can undergo severe zone spreading. Temperature programming allows a low initial temperature to be maintained so that the volatile solutes can be resolved. Thereafter, the temperature is increased in a linear or stepwise manner as necessary. Hence, analysis times can be decreased while resolution, detection limits and quantification are improved.

Temperatures utilized in column ovens may range from -50°C up to 400°C. The upper temperature limit may be set by the temperature tolerance of the materials used in the gas chromatograph. For example, Teflon transfer lines and column connectors should not be used at temperatures greater than 220–250°C. Alternatively, thermal decomposition or evaporation of the stationary phase may define the maximum possible temperature.

Columns. The choice of stationary phase provides the greatest flexibility for influencing chromatographic performance. A few physicochemical criteria must be met by any potential liquid phase. These requirements are:

(i) chemical inertness with respect to the mobile phase and constituents of the sample mixture
(ii) thermal stability and a wide temperature range for existence in the liquid state
(iii) low vapour pressure and low viscosity in the desired temperature range
(iv) ability to wet the solid support or inner surface of a capillary column.

With respect to a particular analysis, the component of interest in a sample mixture must have a high solubility in the stationary phase. There must also be a suitable range of distribution constants of the analytes of interest between the mobile and stationary phases if adequate resolution is to be achieved. The empirical approach to selecting a liquid phase for a particular separation is to consider the polarity of the stationary phase relative to that of the analytes to be resolved. Similarity in polarity would lead to maximum retention. Commonly utilized phases vary in polarity from the non-polar hydrocarbon squalane (2,6,10,15,19,23-hexamethyltetracosane) to the very polar phases such as TCEP [1,2,3-tris(2-cyanoethoxy)propane].

The concept of polarity is rather subjective. Stationary phases cannot be assigned a universal rank order of polarity as the polarity of a column is dependent on both the stationary phase and the solutes being eluted. The apparent polarity of the liquid phase may also be temperature-dependent. In an attempt to produce a polarity scale, chromatographers have developed solute mixtures and compared their retention indices on a particular column with those on a column with a reference stationary phase. This approach was first adopted by Rohrschneider [11] with benzene, ethanol, butanone, nitromethane and pyridine as probe solutes at a column temperature of 100°C. Squalane was taken as reference stationary phase. The difference in retention index for a given solute, i, is given by

$$\Delta I_i = I_{b,i} - I_{a,i} \tag{1.47}$$

where $I_{b,i}$ is the retention index for the solute i on phase b and $I_{a,i}$ that on a squalane column. Thus, a comparison of the ΔI_i values for two phases provides an indication of their relative polarities with respect to solute i in that the higher the value of ΔI, the more polar is the stationary phase. Phases can be ranked overall in order of polarity on the basis of $\Sigma \Delta I_i$ for the five test solutes.

McReynolds [12] extended this treatment by using a test mixture of ten solutes. These solutes, specified in Table 1.2, were meant to be representative of a wide range

Table 1.2 — Ten test solutes proposed by McReynolds as representative of different organic functional groups. (Reproduced by permission from W. O. McReynolds, *J. Chromatog. Sci.*, 1970, **8**, 685. Copyright 1970, Preston Publications, Inc.)

Symbol	Probe	Functional group
X′	benzene	aromatics alkenes
Y′	1-butanol	alcohols nitriles acids
Z′	2-pentanone	ketones ethers aldehydes esters epoxides dimethylamino-derivatives
U′	1-nitropropane	nitro-derivatives nitrilo-derivatives
S′	pyridine	bases aromatic *N*-heterocycles
H′	2-methyl-2-pentanol	branched chain compounds, especially alcohols
J′	1-iodobutane	halogenated compounds
K′	2-octyne	alkynes
L′	1,4-dioxan	ethers bases
M′	*cis*-hydrindane	non-polar steroids terpenes naphthenic structures

of compounds. He employed a slightly higher column temperature, namely 120°C. Additionally, constants were defined which provide information on the efficiency of the stationary phase in resolving members of a homologous series. The constant b is the slope of the line defined by the two points obtained when the logarithms of the net retention times of decane and dodecane are plotted against their retention indices (i.e. 1000 and 1200, respectively). The constant *r* is the square root of the ratio of the net retention time of dodecane to that of decane. In comparing phases, that with the higher values of *b* and *r* will give better resolution of the constituents of a homologous series.

As a result of their utility and widespread acceptance, the ΔI values for the test solutes specified by McReynolds are now known as McReynolds constants. McReynolds constants for selected stationary phases are tabulated in Table 1.3. These data aid in selecting an appropriate phase for a separation. The high X′ value (benzene) exhibited by TCEP indicates good separation would be achieved between aliphatic and aromatic hydrocarbons with the same carbon number. The Y′ and Z′ values for tricresyl phosphate and QF-1 indicate that ketones and ethers would be eluted before alcohols from a tricresyl phosphate column. This elution order would be reversed for QF-1. Also it should be noted that while these two phases have a similar overall polarity, the larger *b* and *r* constants for tricresyl phosphate indicate that it would better resolve a homologous series.

The discussion above has emphasized gas–liquid chromatography. Media are available which require no liquid phase yet can resolve a wide range of organic materials. The Porapak and Chromosorb Century Series of stationary phases consist of cross-linked styrene–divinylbenzene copolymers. The polymerization mixture can include compounds with functional groups of varying polarity. As with liquid phases, the polarity of different stationary phases can be characterized by using McReynolds constants.

1.2.2 Liquid chromatography

Liquid chromatography comprises separations based on the liquid–liquid, liquid–solid, size-exclusion and ion-exchange techniques. Only the first procedure listed will be considered in this section, which is intended as only a brief introduction to liquid chromatography. Further information can be obtained from [8,10,13,14].

Most often liquid chromatography refers to liquid–liquid techniques, also known as bonded-phase chromatography. The separation mechanism is principally the partitioning of the solute between two phases, but adsorption on the solid support may play a role. Usually the stationary phase is more polar than the mobile phase. Reversed-phase chromatography involves a non-polar stationary phase that is chemically bonded to the solid support. The mobile phase is the more polar phase and usually consists of an aqueous solution containing a water-miscible organic solvent.

Liquid chromatography differs from gas chromatography in one important way. Because of the low diffusion coefficients of solutes in liquids relative to those in gases, the mass transfer resistance terms dominate the van Deemter equation (p. 24). Thus, the column efficiency is influenced to a great extent by the particle size of the column packing, improving as the particle size decreases. The size of available packing material was a technological barrier to enhanced performance in liquid chromatography. Packings in the 3–10 μm size range are now commonly available. Classical liquid chromatography relied on the hydrostatic pressure of the solvent reservoir, placed above the column, to force the mobile phase through the column. Flow-rates were low and packing material was limited to particles 150–200 μm in diameter. Reasonable resolution required long columns and separations were time-

Table 1.3 — McReynolds constants for selected stationary phases. (Reproduced by permission, from W. O. McReynolds, *J. Chromatog. Sci.*, 1970, **8,** 685. Copyright 1970, Preston Publications, Inc.)

Stationary phase	X′	Y′	Z′	U′	S′	H′	J′	K′	L′	M′	*b*	*r*
Squalane	0	0	0	0	0	0	0	0	0	0	0.2891	1.945
Liquid paraffin	11	6	2	7	13	2	12	2	9	9	0.2887	1.944
OV-101	17	57	45	67	43	33	4	23	46	−2	0.2484	1.771
Tricresyl phosphate	176	321	250	374	299	242	169	131	254	76	0.2630	1.832
QF-1	144	233	355	463	305	203	136	53	280	59	0.2094	1.619
XE-60	204	381	340	493	367	289	203	120	327	94	0.2237	1.674
Triton X-305	262	467	314	488	430	336	229	183	366	113	0.2404	1.739
Carbowax 20-M	322	536	368	572	510	387	282	221	434	148	0.2235	1.673
FFAP	340	580	397	602	627	423	298	228	473	161	0.2204	1.661
Diglycerol	371	826	560	676	854	608	245	141	724	36	0.2568	1.806
TCEP	593	857	752	1028	915	672	503	375	853	267	0.1789	1.509

consuming. Modern chromatographic practice, using a dense packing of small particles, requires pressurized systems to ensure relatively high flow-rates. This type of technique is termed high-pressure (or high-performance) liquid chromatography (HPLC).

The main components of an HPLC instrument are:

(1) eluent reservoir
(2) pumping system
(3) sample introduction system
(4) column
(5) detection system
(6) data acquisition/reduction system
(7) fraction collection.

Only the first four components are within the scope of this section. It should be noted that in contrast to gas chromatography, recovery of the fractionated sample mixture may be possible, depending on the detector utilized.

Eluent reservoir. A wide range of mobile phases may be utilized in HPLC. The eluent may be a single solvent but more often a solvent mixture. Analysis in which a single eluent, whether a single solvent or a mixture, is utilized is termed isocratic elution. Gradient elution refers to the case in which the eluent composition is varied during the separation by mixing two or more eluents. In liquid chromatography, gradient elution is analogous to temperature programming in gas chromatography and so serves to speed up analyses.

It is important that the eluents used in HPLC be deaerated. The formation of bubbles in the column or detector system by a decrease in pressure causes zone spreading. A degassing facility may be incorporated into the HPLC instrument in the form of a vacuum pumping system. Alternatively, before use eluents may be filtered through a membrane by suction, which will tend to remove gases and solid residues.

Finally, it should be noted that owing to the high resolution of HPLC columns and the sensitivity of many detector systems, the solvents used must be of high purity. HPLC-grade solvents are commonly available from many chemical manufacturers and chromatographic equipment suppliers.

Pumps. HPLC requires a pump capable of delivering an accurate and reproducible flow-rate of 0.5–3.0 ml/min against a column back-pressure of up to 5000 psi (34.5 MPa). Short-term fluctuations in the eluent flow-rate as a result of pumping operations (e.g. piston movement, valve function) must be kept to a minimum. Gradient elution requires at least two pumps, although eluents may be mixed before their introduction into the high-pressure part of the system.

Two types of high-pressure pumps are available: constant pressure and constant volume. Constant-pressure pumping systems may involve a gas displacement pump.

This is not a true pump; in it fluid flows from a closed reservoir as a result of gas pressure (He or N_2) applied through a diaphragm. This system delivers a pulse-free flow but may have a limited reservoir volume. Pneumatic amplifying pumps are a second type of constant-pressure system. In this case, a piston is operated by applying air pressure at one end to drive the piston forward. Non-return valves prevent back-flow into the eluent reservoir and so the fluid is forced into the column. The surface area exposed to the air pressure is larger than that in the eluent, thereby giving a constant compression ratio. At the end of the power stroke the gas is vented and the piston cylinder filled with fresh eluent. One type of constant-volume pump involves a mechanically driven syringe. This gives a pulse-free delivery but again suffers from having a reservoir of limited volume. The most commonly utilized pump in HPLC is the reciprocating or metering pump. The single-piston version has a low volume chamber (30–1000 μl). Again, non-return valves regulate the direction of eluent flow into and out of the piston chamber. The flow is initially non-uniform, but flow pulsations can be eliminated by using pulse dampers or multi-head reciprocating pumps.

Sample introduction system. Samples may be injected in a fashion analogous to that used in gas chromatography, by means of a syringe and a self-sealing septum. The eluent is, of course, under high pressure. Potential problems are sample leakage during injection, ghost peaks due to the eluent leaching material from the septum, and poor reproducibility (variation generally $>2\%$). These difficulties are overcome by using a microvolume valve; 4-port and 6-port varieties are utilized. Precise volumes can be delivered with very little change in pressure in the system.

Columns. Usually two columns are used in HPLC. The first is known as the precolumn or guard column. The precolumn contains the same packing as the second (analytical) column except that the particles are larger to prevent a significant pressure drop. This column protects the analytical column by removing impurities from the eluent and saturating it with the stationary phase.

The analytical column is made of stainless steel or glass [for operating pressures <600 psi (4.1 MPa)]. The solid support consists of small (3–10 μm) particles (of uniform size) of silica gel, alumina or diatomaceous earth. The silica(te) surface is treated to form silicate esters or stable silicone polymers. The surface functional groups may subsequently be modified. The separation techniques are known as bonded-phase (or normal-phase) chromatography when the stationary phase is more polar than the mobile phase. Typical functional groups in this case are shown in Table 1.4. Alternatively, the silicate esters may be modified to be hydrophobic in nature. The surface is then non-polar and the procedures are classified as reversed-phase chromatography. The functional group in this case typically consists of a straight-chain hydrocarbon with carbon number >8. Several manufacturers produce a reversed-phase medium incorporating octadecyl-silica.

While temperature programming is an important feature in gas chromatography, most liquid chromatography is conducted at ambient temperature. Columns can be mounted inside ovens when temperature control is required. Gradient elution is used

Table 1.4 — Typical functional groups used as stationary phases for reversed-phase chromatography

Group	Structure
Amino-	$-(CH_2)_3NH_2$
Cyano-	$-(CH_2)_3CN$
Diamino-	$-(CH_2)_2NH(CH_2)_2NH_2$
Dimethylamino-	$-(CH_2)_3N(CH_3)_2$
Diol	$-(CH_2)_3OCH_2CH(OH)CH_2OH$

to speed the transfer of solutes through the system. The solubility of analytes in the mobile phase is altered by varying the composition (especially the polarity) of the eluent.

1.2.3 Size-exclusion chromatography

Size-exclusion chromatography, also known as gel chromatography, allows compounds to be separated according to their size and shape. The principles have been discussed, for example, by Kremmar and Boross [15], Tomášek [16] and Dawkins [17]. Applications to trace metals in natural waters have been reviewed by de Mora and Harrison [18].

The stationary phase consists of microporous beads. These may consist of an insoluble, cross-linked polymeric matrix or controlled pore-size glass. The beads are saturated with the mobile phase, usually an aqueous solution, and then packed vertically in a column. When a sample is applied, size differentiation is achieved by the extent to which solutes diffuse into and out of the pores of the stationary phase, which is determined by the size and shape of the molecules. One extreme is represented by large molecules, which are completely excluded from the interstitial cavities of the stationary phase, and thus pass through the column at the same flow-rate as the eluent and have a retention volume equivalent to the void volume of the column. Thereafter molecules are eluted in decreasing order of size. The lower limit for size fractionation is that set by small molecules that can freely diffuse into the pores of the stationary phase. They have a retention volume that approximates the total bed volume of the column.

Ideally the mechanism of this type of separation is diffusion. Adsorption processes can be pronounced in some instances and lead to an underestimation of the molecular size. Adsorption is manifested as asymmetric peaks with obvious downstream tailing. Severe adsorption can give peaks with retention volumes greater than the total bed volume of the column. Molecular weights then cannot be assigned and quantification is unreliable, owing to the possibility of incomplete solute recovery.

Components in various size ranges can be investigated by using size-exclusion media with pores of different dimensions. The appropriate operating ranges are usually designated on a molecular weight basis, rather than size, with calibration by use of a set of globular proteins. Operating ranges for different media are noted in Table 1.5. The matrix of the stationary phase may be borosilicate glass (controlled pore glass), or polymers of dextran, acrylic esters or polystyrene–divinylbenzene.

Table 1.5 — The effective size-differentiation range of selected stationary phases

Size-exclusion medium	Operating range for molecular weight differentiation
CPG-40	1×10^3–8×10^3
CPG-100	1×10^3–3×10^4
CPG-240	2.5×10^3–1.2×10^5
CPG-500	1.1×10^4–3.5×10^5
Bio-Beads S-X12	$\leqslant 4.0 \times 10^2$
Bio-Beads S-X4	$\leqslant 1.4 \times 10^3$
Bio-Beads S-X2	1×10^2–2.7×10^3
Bio-Beads S-X1	6×10^2–1.4×10^4
Sephadex G-10	$\leqslant 7.0 \times 10^2$
Sephadex G-25	1×10^3–5×10^3
Sephadex G-50	1.5×10^3–3×10^4
Sephadex G-200	5×10^3–6×10^5

Polymeric media are disadvantaged relative to controlled pore glass in that they are gels, and bed volumes may change with the composition of the eluent.

1.3 THE THEORY OF ATOMIC SPECTROSCOPY

1.3.1 Atomic spectra

Atomic spectroscopy is a broad classification of analytical techniques which depend on the interpretation of atomic spectra and quantification of the interactions between atoms and energy which cause electronic transitions. Electrons within an atom can occupy only discrete energy levels. Electronic transitions between such levels give rise to atomic spectra. The quantized nature of the energy levels ensures that atomic spectra exist as a series of discrete lines, or more properly narrow bands. Furthermore, a set of selection rules governs which electronic transitions are permitted. An understanding of the origin of atomic spectra therefore requires a review of the basic principles of quantum mechanics dealing with electron energy levels and electronic transitions in atoms. More extensive treatment can be found elsewhere [19–21].

1.3.1.1 Origin of atomic lines

The energy level of an electron in an atom can be described by four quantum numbers. The principal quantum number, n, where $n = 1,2,3. . .$, defines the principal energy level of the electron with respect to the nucleus. The levels are known as K-, L-, M-, etc., shells for $n = 1,2,3$, etc. The angular momentum (or azimuthal) quantum number, l, where $l = 0,1,2 . . . (n-1)$, defines subshells or sublevels within a principal energy level. For $l = 0,1,2,3$ the subshells are specified as s, p, d and f orbitals. The magnetic quantum number, m, where m ranges from $-l$ to $+l$, characterizes the spatial orientation of the orbital about the nucleus. In the absence of electric or magnetic fields, orbitals with the same values for n and l are energetically equivalent (i.e. degenerate). The fourth quantum number is the so-called magnetic spin quantum number, m_s, which denotes the notional spin orientation of the electron about its own axis. Only two values are possible and so $m_s = \pm s$ where s, the spin quantum number, is $\frac{1}{2}$. Two criteria must be met in order to assign the electronic configuration of any atom. First, orbitals are filled in sequence of ascending energy, starting with the one of minimum energy. Secondly, Pauli's exclusion principle states that no two electrons in the atom may have the same four quantum numbers. For example, the electronic configuration for Na and Hg atoms can be denoted as:

Na: $1s^2, 2s^2, 2p^6, 3s^1$
Hg: $1s^2, 2s^2, 2p^6, 3s^2, 3p^6, 3d^{10}, 4s^2, 4p^6, 4d^{10}, 4f^{14}, 5s^2, 5p^6, 5d^{10}, 6s^2$

where the prefix is the principal quantum number and the superscript the number of electrons in the orbital(s) designated.

The quantum numbers outlined above do not entirely characterize the energy state of an atom. Various interactions between electrons, and also spin–orbit coupling, must be taken into account. The electron is conventionally regarded as spinning about its own axis as well as describing a path around the nucleus. The orientation of the spin relative to the orbiting motion will affect the overall angular momentum. Such interactions can be taken into account by vector addition of the orbital angular momentum vector, $\boldsymbol{l}$, and the spin momentum vector, $\boldsymbol{s}$. Two coupling schemes can be used. In the first, known as Russell–Saunders or L–S coupling, interactions between different electrons predominate and the $\boldsymbol{l}_i$ vectors for all the electrons are added to give a resultant orbital angular momentum: $\boldsymbol{L} = \sum_i \boldsymbol{l}_i$ and similarly the $\boldsymbol{s}_i$ vectors are combined to give a resultant spin angular momentum: $\boldsymbol{S} = \sum_i \boldsymbol{s}_i$. The total electronic angular momentum, $\boldsymbol{J}$, is the vector sum of $\boldsymbol{L}$ and $\boldsymbol{S}$. The alternative approach, used especially for heavy atoms, is called j–j coupling. Spin–orbital coupling is then more important than coupling between different electrons, so $\boldsymbol{l}_i$ and $\boldsymbol{s}_i$ for each electron are added to give a resultant angular momentum $\boldsymbol{j}_i$ and the total electronic angular momentum for the atom as a whole is the vector sum of all values of $\boldsymbol{j}$. Thus, the two approaches are distinguished as:

$$\text{L–S coupling: } \boldsymbol{J}_i = \sum_i \boldsymbol{l}_i + \sum \boldsymbol{s}_i \qquad (1.48)$$

$$j\text{–}j \text{ coupling: } \boldsymbol{J}_i = \sum(\boldsymbol{l}_i + \boldsymbol{s}_i) \qquad (1.49)$$

The quantum numbers that describe the magnitude of the different vector sums are: L, the integral angular momentum quantum number; S, the integral spin quantum number and J, the total electronic angular momentum quantum number. Atomic states for $L = 0, 1, 2, 3$ are termed S-, P-, D-, F-states. For any given atomic state, J can have values between $|L - S|$ and $|L + S|$, so each J value distinguishes an energy level. Such levels vary only slightly in energy, and the number of different levels, known as the multiplicity, is defined by $2S + 1$. Atomic states characterized by the quantum numbers n, L, S and J are symbolized by the notation: $n^{2S+1}L_J$. For example, the valence electrons for Na and Hg atoms, respectively, have the configuration $3s^1$ and $6s^2$. The atomic states are designed as $3^2S_{1/2}$ for Na ($n=3$, $L=0, S=1/2, J=1/2$) and 6^1S_0 for Hg ($n=6, L=0, S=0, J=0$). As is evident from the example of Hg above, any filled shell or subshell will have $L=S=0$ because the symmetrical nature of the orbitals ensures that the vector sums for $\boldsymbol{l}_i$ and $\boldsymbol{s}_i$ will both be zero. When the *j–j* coupling scheme is utilized, the analogous condition is that $J=0$.

Recapitulating, the energy of electron orbitals in atoms is given by the quantum numbers n and l. Atomic states and their associated energy levels due to spin–orbital coupling are further defined by quantum numbers L, S and J. For any element, the electron orbitals and atomic states can be graphically presented in a Grotrian or energy-level diagram. Grotrian diagrams for Na and Hg are shown in Fig. 1.3. Each energy level is designated by a horizontal line. The lowest energy level is assigned an energy of 0 and the others are then arranged in ascending order of associated energy. The upper limit is defined by the first ionization potential of the atom. Units can be eV or, less commonly, wavenumbers. For convenience the energy levels are spread out along the x-axis according to their atomic state.

The atomic states $3^2S_{1/2}$ for Na and 6^1S_0 for Hg represent the condition wherein the outer electrons are at the lowest possible energy level. These atoms are termed ground-state atoms. If an electron absorbs the appropriate quantum of energy it is promoted to a vacant orbital of corresponding higher energy, giving rise to an excited state. Alternatively, electrons in excited states can lose energy by the emission of photons and so fall to fill vacancies in orbitals of lower energy. Transitions involving the ground state are known as resonance transitions. Resonance transitions, especially those involving the first excited state, give rise to intense, stable atomic lines which are usually the most sensitive analytical lines.

The most important electronic transitions for Na and Hg atoms are depicted on the Grotrian diagrams (Fig. 1.3). Usually only one electron changes its energy level and a set of selection rules determines which transitions are permitted.

(1) There is no restriction on the change for the principal quantum number, n.
(2) The change in l is restricted to $\Delta l = \pm 1$.
(3) The change in J is restricted to $\Delta J = \pm 1$ or 0, except that transitions $J=0 \rightarrow J=0$ are forbidden.
(4) Where Russell–Saunders coupling prevails, $\Delta L = \pm 1$ or 0 and $\Delta S = 0$.
(5) Where *j–j* coupling occurs, $\Delta j = \pm 1$ or 0.

Rules 1 and 2 have the effect that for Na, various series of spectral lines result from transitions between S and P levels, between P and D levels, etc. Atomic spectra will also contain fine structure due to the presence of doublet and triplet excited states. Again with Na as example, the electron transfer from a $3s$ to a $3p$ orbital gives excited states with the quantum numbers $n=3$, $L=1$, $S=1/2$ and $J=1/2$ or $3/2$. This state is designated by $3^2P_{1/2}$ or $3^2P_{3/2}$. The two energy levels arise from Russell–Saunders coupling and are slightly different in energy. The transitions $3^2S_{1/2} \rightarrow 3^2P_{1/2}$ and $3^2S_{1/2} \rightarrow 3^2P_{3/2}$ give rise to the doublet observed in the Na spectrum. When an atom possesses two valence electrons then the spin angular momentum vectors can result in $\boldsymbol{S}=0$ or $\boldsymbol{S}=1$. For Ca atoms, the presence of an electron in a $4p$ orbital would then give rise to the following atomic states: the singlet state 4^1P_1 (quantum

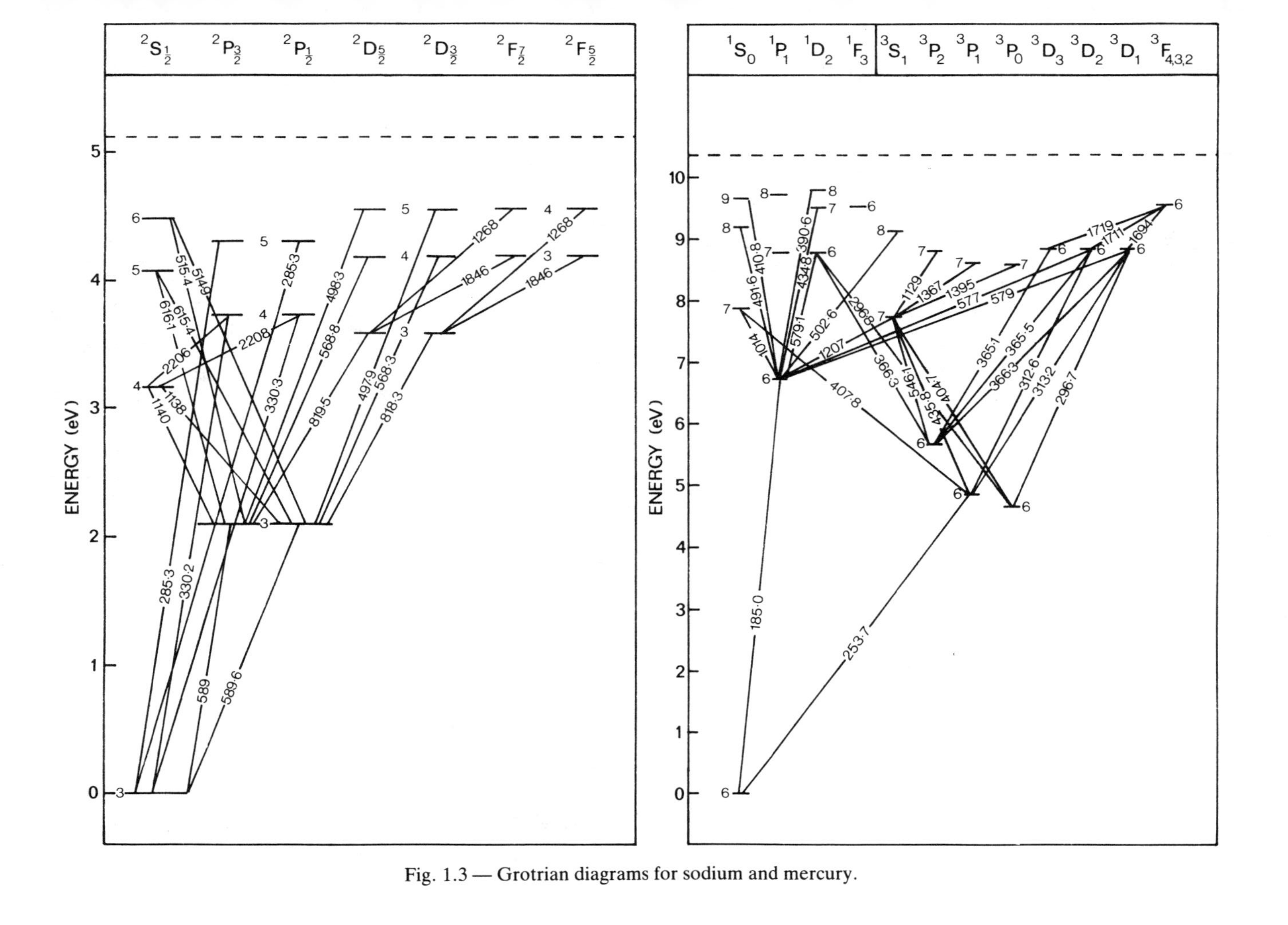

Fig. 1.3 — Grotrian diagrams for sodium and mercury.

numbers $n=4$, $L=1$, $S=0$, $J=1$) and the triplet state $4^3P_{0,1,2}$ (quantum numbers: $n=4$, $L=1$, $S=1$, $J=0$, 1 or 2). As Ca atoms experience Russell–Saunders coupling, the condition $\Delta S=0$ in rule 4 prohibits the transition $4^1S_0 \rightarrow 4^3P_{0,1,2}$ from the ground state. Thus, the singlet and triplet states exhibit distinct spectra. For Hg atoms, *j–j* coupling occurs instead of Russell–Saunders coupling and this allows singlet to triplet transitions (i.e. $6^1S_0 \rightarrow 6^3P_1$). However, the $6^1S \rightarrow 6^3P$ transition has no fine structure in this case, because rule 3 restricts the transitions $6^1S_0 \rightarrow 6^3P_0$ and $6^1S_0 \rightarrow 6^3P_2$. Electrons in these energy levels cannot lose energy by falling to levels of lower energy and so are said to exist in a metastable state.

Atomic spectra can also be influenced by the nucleus of the atom. Interactions between the total electronic angular momentum and the nuclear spin momentum cause hyperfine structure. The mass of the nucleus also influences the energy levels in atomic states and so isotopic effects may be exhibited. Insofar as analytical atomic spectroscopy is concerned, such nuclear influences are manifest as line-broadening effects.

Finally, it should be noted that atomic lines can exhibit further structure if the atoms experience an applied magnetic field. Degeneracy is lost for electrons that are in the same atomic state but with different magnetic quantum numbers. The lines are split into Zeeman components — those polarized parallel to the magnetic field (termed π-components) and those polarized perpendicular to the magnetic field (known as σ-components). This phenomenon can be utilized in atomic-absorption spectroscopy to correct for spectral interferences and background effects [22,23].

1.3.1.2 Energy associated with electronic transitions

Electron energy levels in atoms are quantized and atomic spectra result from permitted electron transitions between different levels. Such transitions are accompanied by the absorption or emission of photons. The radiation characteristic of a given transition is defined by:

$$\Delta E = h\nu \tag{1.50}$$

where h is Planck's constant (6.626×10^{-38} J.sec) and ν is the frequency of the radiation. In terms of wavelength, the expression becomes:

$$\Delta E = hc/\lambda \tag{1.51}$$

where c is the velocity of light (2.998×10^8 m/sec) and λ is the wavelength of the radiation. The energy involved in electronic transitions within atoms corresponds to radiation in the ultraviolet and visible regions of the spectrum. In the case of sodium, as shown in Fig. 1.3, $3s \rightarrow 3p$ transitions require the absorption of radiation of wavelengths 589.00 and 589.59 nm, depending on the atomic state involved. Alternatively, radiation of these wavelengths is emitted for $3p \rightarrow 3s$ transitions, giving rise to the characteristic orange doublet of sodium.

For the most part, the energy required for electronic transitions in different atoms is sufficiently different for the corresponding wavelength to be diagnostic for the analyte, thus allowing qualitative and selective analyses. There are only a few instances of spectral coincidence [24]. Quantitative analyses are achieved by measuring the amount of monochromatic radiation that has been either absorbed (atomic-absorption spectrophotometry) or emitted (atomic-emission spectrometry). A special category of atomic-emission spectrometry is atomic-fluorescence spectrometry in which the emission of photons is measured following excitation of the atoms and some internal conversion of energy.

1.3.2 Atomic-absorption spectrometry (AAS)

The relationships between concentration, optical path-length and absorption of radiation were derived by Bouguer, Lambert, Beer and Bernard in the 18th and 19th centuries. Theoretical treatments have been presented elsewhere [19,21,25,26], and are only outlined here. When a beam of parallel and continuous incident radiation of intensity I_0 and frequency ν passes through a homogeneous atomic vapour of thickness b, the intensity of the transmitted radiation, I_t, is given by:

$$I_t = I_0 \, e^{-k_\nu b} \tag{1.52}$$

where k_ν is the atomic absorption coefficient. This relationship can be expressed in terms of absorbance, A, as:

$$A = \log(I_0/I_t) = 2.303 k_\nu b \tag{1.53}$$

There is a frequency dependence of the absorbance, with the integrated absorption given as:

$$\int k_\nu \mathrm{d}_\nu = (\pi e^2/mc)Nf \tag{1.54}$$

where $\mathrm{d}\nu$ is the frequency interval considered, e is the charge on an electron, m is the mass of an electron, c is the velocity of light, N is the number of atoms per unit volume capable of absorbing in the frequency range from ν to $\nu + \mathrm{d}\nu$, and f is the oscillator strength (i.e. the average number of electrons per atom capable of absorbing the incident radiation). The oscillator strength is dimensionless and typically of the order of unity for strong atomic lines. Alkemade *et al.* [21] have tabulated f-values for some metal atoms.

For any given transition, an equilibrium will be established between the number, N_0, of atoms at the lower energy level and the number, N_i, of atoms at the higher energy level. The thermodynamic equilibrium is described by the Boltzmann distribution:

$$N_i = N_0(g_i/g_0)e^{-E_i/kT} \tag{1.55}$$

where g_i and g_0 are the statistical weights of the excited and ground states (calculated by using the expression $g = 2J + 1$), E_i is the energy difference between the states, k is Boltzmann's constant (1.381×10^{-23} J/K) and T is the absolute temperature. The relative number of atoms in the excited state (i.e. N_i/N_0) increases exponentially with an increase in temperature. Some values for N_i/N_0 are given in Table 1.6 [27]. At

Table 1.6 — Temperature dependence of N_i/N_0 for different atomic lines. (Reprinted with permission, from A. Walsh, *Spectrochim. Acta*, 1955, **7**, 108. Copyright 1955, Pergamon Press.)

Element	Atomic line (nm)	E_i (eV)	g_i/g_0	N_i/N_0			
				2000 K	3000 K	4000 K	5000 K
Zn	213.9	5.8	3	7.3×10^{-15}	5.6×10^{-10}	1.5×10^{-7}	4.3×10^{-6}
Ca	422.7	2.9	3	1.2×10^{-7}	3.7×10^{-5}	6.0×10^{-4}	3.3×10^{-3}
Na	589.0	2.1	2	9.9×10^{-6}	5.9×10^{-4}	4.4×10^{-3}	1.5×10^{-2}
Cs	852.1	1.5	2	4.4×10^{-4}	7.2×10^{-3}	3.0×10^{-2}	6.8×10^{-2}

temperatures typically used in atomic-absorption spectrometry, i.e. below 3000 K, N_i is small relative to N_0 and virtually represents the total number of atoms present. This accounts for the intensity of the resonance lines in atomic-absorption spectra.

From the discussion above, it is clear that the magnitude of the absorbance of an atom depends upon the observational path-length through the atom cloud, and the number of atoms present, in accordance with the observations of Lambert and Beer, respectively, which resulted in the empirical relationship governing analytical measurements:

$$A = abc \tag{1.56}$$

where A is the absorbance, a is a constant for a given system, b is the length of the absorption cell and c is the analyte concentration. This expression, known as Beer's law or sometimes the Beer–Lambert law, states that the absorbance exhibited by an absorbing medium of constant path-length is directly proportional to the concentration of the absorbing species. Thus, it is predicted that calibration curves should be linear.

The principles outlined above ideally relate to the absorption of monochromatic radiation. As shown in Fig. 1.4, absorption and emission profiles for metal atoms are peaks rather than lines. Natural line-widths are characterized by the half-width (i.e. the frequency or wavelength interval at half the peak height) and are of the order of 10 fm. Line broadening can occur as a result of self-absorption, the movement of atoms (Doppler effect), collisions with similar atoms (Holtsmark effect), collisions with different atoms (Lorentz effect), the influence of magnetic fields (Stark–Zeeman effect). Several of these line-broadening effects are utilized in atomic spectrometry. For atomic-absorption spectrometry in particular, line broadening is the underlying principle in the use of hollow-cathode lamps. Incident radiation is provided from an emission source consisting of the analyte element encapsulated under very low pressure. The radiation is then absorbed by ground-state atoms of the analyte in the gas phase in a flame or graphite furnace. The Lorentz effect is much smaller in the emission source than in the absorption cell (the analyte atom cloud). As a result, the emission line-widths are much smaller than the absorption line-

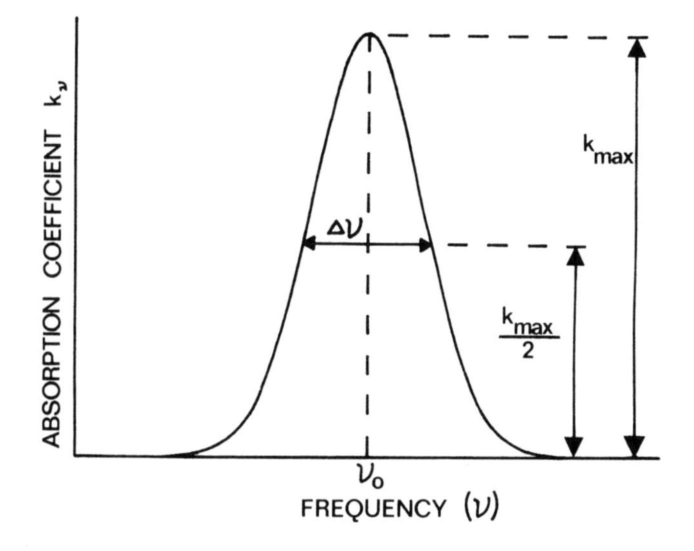

Fig. 1.4 — Profile of an atomic-absorption line.

widths and therefore absorption by an analyte is measured only at the centre of the absorption line. Pulsed hollow-cathode lamps used for background correction make use of self-absorption and enhanced Lorentz effects [28,29]. It should be remembered that the Lorentz effect produces an asymmetric red-shift of the line profile, so the absorbance maximum shifts from the original line centre.

Beer's law predicts that calibration curves should be linear and pass through the origin. This is often the case, but departures from this ideal behaviour can occur and are exemplified in Fig. 1.5. Curve I is linear but has an intercept on the ordinate,

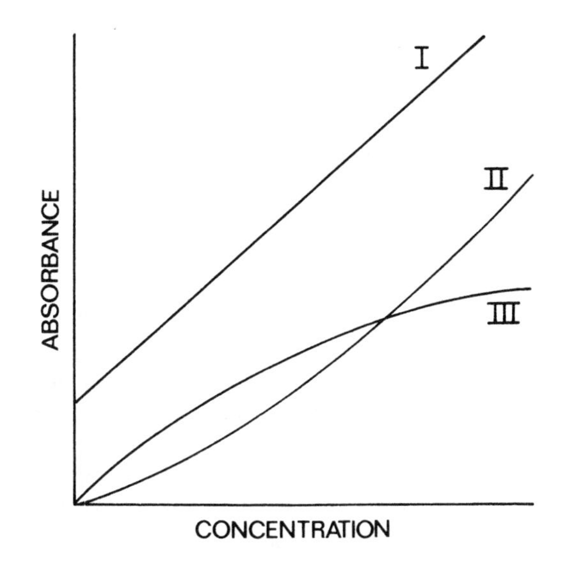

Fig. 1.5 — Non-ideal calibration curves observed in atomic-absorption spectrometry.

owing to non-specific interference. This is also known as background absorption or broad-band absorption and can be caused by the occurrence of a molecular absorption band overlapping with the atomic line, or by light-scattering effects. Several procedures can be utilized to correct for large background absorption due to the matrix. These are discussed in Section 1.4.3.

Positive deviations from linearity, Curve II in Fig. 1.5, are due to ionization interference. When atoms are easily ionized, the ratio of ions to atoms increases with decrease in analyte concentration, causing the initial curvature in the plot.

Negative deviations from linearity, as shown in Curve III, may be due to either spectral or chemical interferences. Two types of spectral interference have been recognized [24]. Inadequacies in the monochromator-slit system may allow several spectral lines to reach the detector. Curvature results, as the lines will have different absorption coefficients. Similarly, if the line-width of the emission source is greater than the line-width of the absorption band then the integrated absorption of the line is measured rather than the absorption at the centre of the line. This case is thus analogous to several closely-spaced lines reaching the detector and curvature results from the frequency-dependence of the absorption coefficient. Chemical interferences are quite diverse and depend upon the analyte considered and the atomization process in use.

1.3.3 Atomic-emission spectrometry (AES)

Radiation is emitted from electrons when they fall from one energy level to a lower one. The amount of radiation emitted is a function of the analyte concentration. The relationships governing the concentration dependence of the emission are complex and dependent upon the excitation process utilized. Several excitation processes have been used, giving rise to a wide variety of techniques under the general classification of atomic-emission spectrometry. In atomic-fluorescence spectrometry the analyte atoms are excited by ultraviolet/visible radiation. The atomization is generally achieved by means of a flame. This technique is distinguished from flame-emission spectrometry, also known as flame photometry, in which the atoms undergo thermal excitation. Thermal excitation also occurs in an inductively coupled plasma [30]. Flame temperatures in an argon torch vary in the range 6000–10000 K, compared to 2000–3000 K for air–acetylene or nitrous oxide–acetylene flames. Atoms may also be excited in a high-voltage spark or an electric arc. Arc/spark spectrometry is also known as optical-emission spectrometry and is generally used in the analysis of solids.

Despite the variations in excitation methodology, some general considerations apply. Analyte information is obtained from the atoms in an excited state. As discussed previously, the ratio N_i/N_o is temperature-dependent (Boltzmann distribution). Only a small fraction of the total number of atoms will be present in an excited state, and whereas temperature fluctuations have little influence in absorbance measurements (because $N_0 \ll N_i$), in atomic-emission spectrometry they can cause significant variations in N_i and hence the emitted radiation intensity. The strongest resonance lines in emission spectra do not necessarily correspond to the strongest resonance lines in absorption spectra. L'vov [25] has demonstrated that the intensity does depend on the wavelength corresponding to the transition. As evident from Table 1.6, the ratio N_i/N_0 depends on the excitation energy. Emission intensities fall with decrease in wavelength, so the most sensitive lines for emission analysis may be at longer wavelengths than those used for absorbance measurements. This is exemplified by the most sensitive lines for Al:

AAS 309.3 nm $3^2P_{3/2} \rightarrow 3^2D_{5/2,1/2}$
AES 396.1 nm $3^2P_{3/2} \rightarrow 4^2S_{1/2}$

Calibration in atomic-emission spectrometry is conducted on an empirical basis. The intensity of the emission is proportional to the analyte concentration, and in the case of atomic-fluorescence spectrometry is also directly proportional to the intensity of the excitation radiation source. Sychra *et al.* [31] provide a comprehensive coverage of the theory and practice of atomic-fluorescence spectrometry.

Atomic-emission spectrometry is subject to spectral interferences to a greater extent than atomic-absorption spectrometry is. This arises partly because the emission spectra are rich in lines, but also because spectral resolution is dependent on the quality of the monochromator of the instrument. In contrast, atomic lines in absorption spectra are sufficiently narrow to ensure a high degree of specificity. However, chemical and ionization interferences may occur and self-absorption can cause negative deviations from linearity for calibration curves in emission techniques.

1.4 SOME INSTRUMENTAL CONSIDERATIONS IN ATOMIC SPECTROSCOPY

Many monographs on atomic spectrometry describe the development of instrumentation and applications. The purpose of this section is to describe briefly some practical and instrumental features of atomic spectrometry that are not dealt with elsewhere in this text. Atomic-absorption spectrometry is primarily considered, but many aspects of the discussion of radiation sources are pertinent to atomic-fluorescence spectrometry. Similarly, the atomization processes and interferences described are common to many atomic-absorption and atomic-emission techniques.

1.4.1 Radiation sources

Atomic-absorption spectrometry is dependent on measuring the absorbance of monochromatic radiation. Such radiation can be obtained from either a continuum source or an atomic-line source. A continuum source produces emission over a considerable range of the spectrum. Hydrogen, deuterium, xenon and tungsten–iodine lamps have been used. They generate a stable light signal, suitable for multi-element analysis. O'Haver and Messman [32] note several advantages of continuum over atomic line sources. Background correction can be applied in both AAS and AES modes. Spectrum scanning allows the measurement of molecular absorption fine structure and can be diagnostic for background and spectral interferences. The dynamic range is 5 orders of magnitude for several elements.

A number of disadvantages of continuum source techniques should be recognized, however. First, no commercial instrument is yet available. Secondly, absorption occurs within a narrow frequency range and even intense continuous sources would have only a limited emission intensity over a bandwidth typical of atomic-absorption profiles, about 10 fm. Thirdly, the use of a continuous emission source requires a monochromator of very high resolution. Inadequate spectral resolution causes a decrease in sensitivity, and increased risk of spectral interferences and non-linear calibration curves. Finally, detection limits are poorer in continuum-source AAS than in atomic-line source AAS, particularly at wavelengths shorter than 250 nm.

The alternative radiation source would be a discrete emitter or atomic-line source. In this case the emission takes the form of a series of intense, narrow, separated lines, usually characteristic of only one element. For any analyte, the emission lines will be narrow relative to the absorption lines, so the need for a high-resolution monochromator is eliminated. However, each analyte will require a suitable (usually individual) radiation source.

Lamps and sources have been reviewed by Sullivan [33] and Alkemade *et al.* [21]. The more commonly used lamps are described below, but it is worth considering the preferred characteristics of an atomic-line source. Small spectral band-widths (i.e. narrow lines) ensure good sensitivity, while a stable light output (i.e. low drift or long-term fluctuations) gives improved accuracy in single-beam instruments. A high-intensity source and low noise level (i.e. flicker or short term fluctuations) improves precision and favours a high signal to noise ratio, thereby allowing a lower detection limit. Economic considerations are short warm-up times (for single-beam instruments) and long lamp lifetimes.

The most common radiation source used in atomic-absorption spectrometry is the hollow-cathode lamp. The resonance radiation emitted from such lamps has a narrow band-width, usually < 10 pm for most elements. A sealed tube is evacuated and then filled to 2–10 mmHg pressure with a noble gas, typically neon or argon. The lamp contains an anode made of W, Ni or Zr, and a cathode either made of or coated with the analyte metal. The radiation is emitted through a quartz window mounted opposite the cathode.

A small current of the order of a few mA is passed between the anode and the cathode, causing the noble gas to be ionized. These ions subsequently bombard the cathode, causing metal atoms to be sputtered. The resulting atomic vapour is in turn excited by collisions with the noble gas and emits its characteristic atomic resonance spectrum. Ideally the emission spectrum would consist only of the atomic lines of the analyte element, but in practice will include the spectrum of the filler gas. In some cases, then, the choice of the filler gas may be critical if spectral interferences are to be avoided. In the absence of spectral coincidence, other considerations are that argon is more efficient than neon for sputtering, but neon has a higher ionization potential than argon.

The conditions for operating the lamps are critical. A stabilized current source is required in order to ensure stable emission. Increasing the current between the anode and the cathode decreases the lamp's lifetime but increases the emission intensity. However, the benefits gained by increasing the intensity in this manner are limited. The vapour pressure of the atomic vapour must be kept relatively low in order to prevent self-reversal or self-absorption, which would otherwise lead to line-broadening. A design feature which overcomes this problem is the incorporation of a second pair of electrodes in the lamp. High-intensity or boosted-output hollow-cathode lamps thus have two electrode pairs. As in a conventional hollow-cathode lamp, the first pair has a limited current passed through it to produce an atomic cloud with low vapour pressure. The second electrode pair ensures efficient excitation of the atom cloud. Increasing the current through this secondary discharge increases the lamp intensity, while the vapour pressure of the analyte element remains constant.

Hollow-cathode lamp characteristics such as radiation intensity, stability and lifetime are generally poor for elements such as As, Pb, Se, Sn and Te. Electrodeless discharge lamps are available for these analytes. The design involves placing a small amount of the analyte, either as pure metal or as a metal salt, usually the iodide, in a sealed silica tube filled with a noble gas at a few mmHg pressure. This sealed tube is placed in the centre of a radiofrequency coil, typically of 27 MHz, frequency. For ease of handling, these components are then housed in a tube of dimensions similar to those of hollow-cathode lamps.

Electrodeless discharge lamps may have up to 100 times the emission intensity of hollow-cathode lamps. However, they exhibit line-broadening effects, which may limit their suitability for atomic-absorption spectrometry. The very great increase in intensity renders such lamps suitable as a source for atomic-fluorescence spectrometry, however. In atomic absorption, the enhanced emission improves the signal to noise ratio. Improved detection limits and linearity in calibration curves can result.

1.4.2 Atomization processes

Atomic-absorption spectrometry as a technique is dependent upon the ability to produce a volatile cloud of ground-state atoms in an optical pathway. For most elements, sample vaporization and atomization of the analyte are achieved by means of thermal energy. Temperature control is important to ensure adequate and reproducible atomization efficiency. Atomization efficiency is relatively low at low temperatures but excessive temperatures may lead to the formation of ions as well as atoms. Atomization systems in AAS usually involve either flames or an electrically heated graphite furnace. Aspects of the instrumentation and atomization processes are outlined below for both of these general types.

Flame atomic-absorption spectrometry. The analyte in flame AAS must be either in gaseous form or in solution. In the latter case, the solution is aspirated and nebulized by the venturi action of the oxidant gas flowing past an inlet nozzle. The nebulization can be improved by impinging the spray on a glass bead. A fuel gas is introduced and the sample spray and gases pass through a premix chamber. This chamber usually contains baffles to ensure gas mixing and to collect the large droplets (i.e. up to 90% of the total volume), which then drain to waste. The fine mist/gas mixture leaves the burner head and is burned. Gaseous analytes can be introduced directly into the premix chamber.

The flame is inhomogeneous in composition and exhibits a marked temperature gradient. The primary combustion zone is characterized by intense emission and reducing conditions. Oxidation of carbon monoxide predominates in the secondary combustion zone. Between these two extremes is the interconal zone where thermodynamic equilibrium is attained and most atomic-absorption measurements are made. Several reactions occur within the flame. The analyte undergoes desolvation and the resulting aerosol becomes vaporized, forming discrete molecules, often oxides of the analyte. These molecules undergo thermal dissociation to produce a cloud of atoms which can then become excited by absorbing incident radiation.

Maximum sensitivity is achieved by measuring the absorption through that part of the flame in which the number of atoms is greatest. Atomization processes are controlled by the thermodynamics and kinetics of the system. In practice, the maximum sensitivity for an analyte can be determined by measuring the atomic absorbance in various volume elements (i.e. at varous heights) in the flame. The optimum height of observation in a given flame type depends on the nature of the size distribution and composition of the solid aerosol particles formed by desolvation [34,35]. The flame composition can be altered by adjusting the oxidant to fuel ratio. Various oxidant/fuel mixtures have been used, but the most common are air/acetylene and nitrous oxide/acetylene. These flames generate temperatures in the ranges of 2100–2400 and 2600–2800°C, respectively.

Graphite-furnace atomic-absorption spectrometry. In GFAAS the sample, either solid or liquid, is placed in a graphite container aligned in or just beneath the optical pathway. The graphite furnace is heated electrically (hence the often used term electrothermal atomization), usually in a series of four steps.

(i) Drying stage — the temperature is increased to just below the boiling point of the solvent. The solvent is evaporated, care being taken to prevent sputtering.
(ii) Charring stage — the temperature is increased to some intermediate value at which components of the matrix can be volatilized without loss of analyte.
(iii) Atomization stage — the temperature is increased, generally as fast as possible, to that at which the analyte is atomized.
(iv) Decontamination (cleaning) stage — the maximum temperature, about 2700°C, is maintained for a brief time to ensure there is no accumulation of analyte or matrix which might otherwise cause memory effects.

In many cases, the graphite furnace is a tube. The flow-rate of a purge gas through the tube can be continuously controlled to remove volatiles during the drying and charring stages, but slowed or stopped during atomization, giving the atom cloud a longer residence time in the optical pathway. GFAAS is more prone than the flame techniques to interference difficulties but the sensitivity can be 3 orders of magnitude better, owing to the enhanced atomization efficiencies and longer residence times of atoms in the optical pathway.

Atom formation processes are not well resolved for GFAAS. Likely mechanisms include the reduction of analyte compounds by graphite, or the thermal dissociation of analyte compounds. Such compounds may be oxides or in some cases carbides.

1.4.3 Interferences and background correction techniques

Several types of interferences are possible in atomic-absorption spectrometry. They may not necessarily be predictable and, depending on the sample or analyte, individual analytical procedures may require investigation in order for the analysis to be optimized by minimization or preferably elimination of interferences. Four types of interferences can be broadly classified: chemical, physical, spectral and non-specific. These are briefly outlined below, but it is beyond the scope of this chapter to review adequately the procedures for dealing with all interferences. For general purposes reference may be made to Van Loon [24] or Welz [26]. Frech *et al.* [36] have dealt with interferences in graphite-furnace atomic-absorption spectrometry.

Chemical interferences. Such interferences are matrix-dependent and arise through chemical processes affecting the free atom population of the analyte element. In many instances, chemical interferences can be controlled by modifying the matrix composition or by altering the atomization procedure. Interference may result from reactions during atomization which result in refractory compounds (generally metal oxides but carbides may be formed in graphite furnaces) or comparatively large particles that are relatively slow or difficult to dissociate, as exemplified by sulphate and phosphate interference in calcium determination [34]. Ionization interference can occur for atoms with low ionization potentials. Volatilization interference may be observed in a graphite furnace, whereby some vaporization of the analyte occurs prior to the atomization step.

Physical interferences. Physical processes that can affect the free atom population of an analyte in the optical pathway give rise to physical interferences. Liquid samples are generally introduced into flames and plasmas by an aspiration/nebulization system. Aspiration rates are affected by the visocity of the solution. Similarly, viscosity together with density and surface tension influence the nebulizer efficiency and the resulting drop-size distribution. Thus, sensitivities can be enhanced in flame analyses when the analyte is present in an organic solvent instead of in water. For example, there is a 5-fold increase in sensitivity for manganese when MIBK (methyl isobutyl ketone; 4-methylpentan-2-one) is used as the solvent instead of water. The higher vapour pressure, lower viscosity and lower surface tension result in a decrease in mean droplet diameter.

Spectral interferences. Three principal mechanisms can lead to spectral interference. First, the radiation source may emit intense radiation at wavelengths close to the analyte emission line used for the absorbance measurement. Such emission may be due to impurities in the cathode, atomic lines of the filler gas or non-absorbed lines of the analyte. Secondly, the atomization system (flame or graphite furnace) may emit radiation at the atomic-line wavelength at which the absorbance is measured. Though this can be overcome by modulating the source emission and tuning the detector, the noise is increased and, in severe cases, the photomultiplier may be saturated. Finally, incident radiation may be absorbed by elements other than the analyte at wavelengths in close proximity to that being measured. Such spectral overlap can be easily predicted, and avoided by decreasing the slit-width of the detector or by using a different atomic line for the analyte.

Non-specific interferences. Non-specific interference is also known as non-atomic absorption, broad-band absorption or background absorption. Such interference has two principal causes. First, molecular species may absorb radiation of the wavelength used for determining the analyte. Molecular absorption, sometimes classified as spectral interference, has a broad profile relative to atomic absorption, owing to changes in vibrational levels concurrent with the electronic transition. Fine structure due to narrow rotational lines may be superimposed on the molecular absorption profile and can be difficult to compensate when coincident with the atomic line. Molecular absorption is most severe at wavelengths shorter than 250 nm. Secondly, light-scattering by solid particles in the optical pathway contributes to the background absorption. This effect is particularly prevalent in the graphite-furnace procedures where 'smoke' forms during the atomization step. Light-scattering may be considered as physical or spectral interference, and is also most pronounced at wavelengths <250 nm.

There are many strategies for reducing the background absorption; one is modification of the chemical matrix of the analyte solution. However, the severity and commonness of the problem has resulted in the development of a number of spectral procedures by which the background absorbance can be measured. Subtraction of the background absorbance from the total absorbance gives the absorbance due to the analyte.

The non-absorbing line method relies on measuring the absorbance at a wavelength close to that of the atomic line but at which the analyte itself does not absorb radiation. Also, there should be no spectral overlap. The radiation source may be the hollow-cathode lamp for the analyte (i.e. an atomic line of the filler gas or a non-absorbed emission line of the cathode) or the hollow-cathode lamp for a different element. The background absorption profile may not be uniform and most instruments cannot make simultaneous total and background absorbance measurements. Examples of this method are: for Pb, the non-absorbed line at 282.0 nm and the atomic line at 283.3 nm, and for Mn the non-absorbed line at 281.7 nm and the atomic line at 279.5 nm.

In the continuum method, a deuterium arc lamp provides a continuous emission spectrum with an intensity maximum at about 250 nm. The light beams from the D_2 lamp and the analyte hollow-cathode lamp are optically aligned through the atom cloud and adjusted to equal intensity. Alternate rapid pulsing of the D_2 lamp and the hollow-cathode lamp allows the 'simultaneous' measurement of the background and total absorbances, respectively. Owing to the emission spectrum, poor background correction performance is observed at wavelengths longer than 350 nm. Generally, background absorbances >1 cannot be adequately corrected for and may lead to spurious background-corrected measurements.

Background correction by the pulsed-lamp method [28,29] relies on operating a hollow-cathode lamp at high and low applied current alternately. A high current through the hollow-cathode lamp causes a broadening of the emission spectrum, but at the same time self-reversal or self-absorbance causes a pronounced valley centred on the atomic line. The background-corrected signal is therefore the difference between that recorded under low and high current conditions. Corrections can be made for very high background absorbances, up to an absorbance of about 3.

Finally, the Zeeman method can be used to correct for non-specific absorption [22,23]. This requires rather expensive instrumentation and is based on the observation that fine structure can be imparted to atomic lines by the presence of a magnetic field. The atomic line is split into a central π-component which is at the original wavelength and polarized parallel to the applied magnetic field, and σ^+/σ^--components which are shifted to shorter and longer wavelengths respectively as well as being polarized perpendicularly to the magnetic field. The extent of separation of the components is a function of the strength of the applied magnetic field. The π-component is coincident with the atomic line and so is the sum of the atomic and background absorption. Ideally the $\sigma^{\pm}$ components do not overlap with the atomic-absorption profile and hence represent only the non-specific absorption. The independent absorption of the π- and $\sigma^{\pm}$-components can be resolved by using a polarizing filter, and the difference between the two signals gives the background-corrected signal. Zeeman correction can be applied to background absorbance of up to 2, and resolves some instances of spectral interference, but is not universally applicable and can lead to double-valued calibration curves.

1.5 SUMMARY

Chromatography is fundamentally a qualitative technique dealing with the analytical separation of solutes in a sample mixture. Two main theories have been developed to describe the transfer of solutes through a column. Plate theory explains chromatography in terms of multi-step equilibria. Rate theory considers solute transport in terms of kinetics and hydrodynamics. The latter theory accounts for zone spreading as the manifestation of diffusion and non-equilibrium mechanisms.

The van Deemter equation provides a theoretical framework for optimizing chromatographic performance. However, the empirical approach to chromatography should not be underestimated. Solute retardation is often explained in a practical sense in terms of polarity. McReynolds constants define an operational scale of polarity for stationary phases used in gas chromatography. The analyst using HPLC can readily control the polarity of the stationary phase, but also, and more importantly, can alter the eluent polarity during the elution. This technique, known as gradient elution, is analogous to temperature programming in gas chromatography and can significantly reduce analysis times.

Atomic spectrometry is quantitative. Atomic spectra result from electronic transitions within atoms. Such transitions are governed by a set of selection rules. Radiation is absorbed by electrons during excitation to higher energy states. Beer's law relates the energy absorbed to analyte concentration, thereby enabling quantification in atomic-absorption spectrometry. In atomic-emission spectrometry, analyte concentrations are determined from the amount of radiation emitted when electrons in excited states decrease in energy. With the exception of some well known instances of spectral overlap, the wavelength associated with the electronic transition is diagnostic of the element.

Hollow-cathode lamps are generally used as a radiation source in atomic-absorption spectrometry. Atomization is achieved either by combustion in a flame or by electrothermal atomization in a graphite furnace. Several types of interference (chemical, physical, spectral and non-specific) are encountered, especially in graphite furnaces. The background absorbance can be compensated for by measuring the absorbance at an adjacent non-absorbing line, by using a continuum source as well as an atomic-line source, by operating a hollow-cathode lamp at high and low current alternately or by making use of the Zeeman effect, whereby atomic lines are split in the presence of an applied magnetic field.

Acknowledgements — I wish to thank Dorothy Chaffe for typing the manuscript and Richard Barton for preparing the figures.

REFERENCES

[1] O. Mikeš (ed.), *Laboratory Handbook of Chromatographic and Allied Methods*, Horwood, Chichester, 1979.
[2] R. Grob, *Modern Pracice of Gas Chromatography*, 2nd Ed., Wiley, New York, 1985.
[3] A. J. P. Martin and R. L. M. Synge, *Biochem. J.*, 1941, **35,** 1358.
[4] J. C. Giddings, *Dynamics of Chromatography: Part I, Principles and Theory*, Dekker, New York, 1965.
[5] J. J. van Deemter, F. J. Zuiderweg and A. Klinkenberg, *Chem. Eng. Sci.*, 1956, **5,** 271.
[6] J. Novák, in *Laboratory Handbook of Chromatographic and Allied Methods*, O. Mikeš (ed.), Chapter 2, Horwood, Chichester, 1979.
[7] D. T. Burns, *Pure Appl. Chem.*, 1986, **58,** 1291.
[8] N. A. Parris, *Instrumental Liquid Chromatography.*, Elsevier, Amsterdam, 1976.
[9] E. Kováts, *Helv. Chim. Acta*, 1958, **41,** 1915.
[10] C. F. Poole and S. A. Schuette, *Contemporary Practices of Chromatography*, Elsevier, Amsterdam, 1984.
[11] L. Rohrschneider, *J. Chromatog.*, 1965, **17,** 1; 1966, **22,** 6.
[12] W. O. McReynolds, *J. Chromatog. Sci.*, 1970, **8,** 685.
[13] C. F. Simpson, *Techniques in Liquid Chromatography*, Wiley, Chichester, 1982.
[14] L. R. Snyder and J. J. Kirkland, *Introduction to Modern Liquid Chromatography*, 2nd Ed., Wiley, New York, 1979.
[15] T. Kremmer and L. Boross, *Gel Chromatography*, Wiley, Chichester, 1979.
[16] V. Tomášek, in *Laboratory Handbook of Chromatographic and Allied Methods*, O. Mikeš (ed.), Chapter 6, Horwood, Chichester, 1979.
[17] J. V. Dawkins, in *Techniques in Liquid Chromatography*, C. F. Simpson (ed.) Chapter X, Wiley, Chichester, 1982.
[18] S. J. de Mora and R. M. Harrison, *Water Res.*, 1983, **17,** 723.
[19] A. C. G. Mitchell and M. W. Zemansky, *Resonance Radiation and Excited Atoms*, Cambridge University Press, Cambridge, 1934.
[20] A. Corney, *Atomic and Laser Spectroscopy*, Clarendon Press, Oxford, 1977.
[21] C. T. J. Alkemade, T. Hollander, W. Snelleman and P. J. T. Zeegers, *Metal Vapours in Flames*, Pergamon Press, Oxford, 1982.
[22] K. Yasuda, H. Koizumi, K. Ohishi and T. Noda, *Prog. Anal. Atom. Spectrosc.*, 1980, **3,** 299.
[23] M. T. C. de Loos-Vollebregt and L. de Galan, *Prog. Anal. Atom. Spectrosc.*, 1985, **8,** 47.
[24] J. C. Van Loon, *Analytical Atomic Absorption Spectroscopy*, Academic Press, New York, 1980.
[25] B. V. L'vov, *Atomic Absorption Spectrochemical Analysis*, Hilger, London, 1970.
[26] B. Welz, *Atomic Absorption Spectroscopy*, Verlag Chemie, Weinheim, 1976.
[27] A. Walsh, *Spectrochim. Acta*, 1955, **7,** 108.
[28] D. Siemer, *Appl. Spectrosc.*, 1983, **37,** 552.
[29] S. B. Smith and G. M. Hieftje, *Appl. Spectrosc.*, 1983, **37,** 419.
[30] M. Thompson and J. N. Walsh, *A Handbook of Inductively Coupled Plasma Spectrometry*, Blackie, Glasgow, 1983.
[31] V. Sychra, V. Svoboda and I. Rubeška, *Atomic Fluorescence Spectroscopy*, Van Nostrand–Reinhold, London, 1975.
[32] T. C. O'Haver and J. D. Messman, *Prog. Anal. Atom. Spectrosc.*, 1986, **9,** 483.
[33] J. V. Sullivan, *Progr. Anal. Atom. Spectrosc.*, 1981, **4,** 311.
[34] B. Smets, *Analyst*, 1980, **105,** 482.
[35] J. Aggett and G. O'Brien, *Analyst*, 1981, **106,** 497.
[36] W. Frech, E. Lundberg and A. Cedergren, *Progr. Anal. Atom. Spectrosc.*, 1985, **8,** 257.

2

Atomic-absorption detectors

C. N. Hewitt
Institute of Environmental and Biological Sciences, University of Lancaster, Lancaster, LA1 4YQ, England

The successful analysis of organometallic compounds in environmental media is dependent on three requirements being met. First, the compounds of interest must be separable from one another to allow their individual determination. Secondly, their analysis must be specific and not subject to interference from any other element or compound present in the sample, and thirdly the method of detection must be sufficiently sensitive to allow their determination at trace and ultra-trace concentrations. The interfacing of a chromatographic separation technique (either gas or liquid) with an atomic-absorption spectrophotometer (GC–AAS or LC–AAS) is therefore an ideal analytical tool for organometallics in environmental samples, satisfying as it does these three considerations. Since 1966, when Kolb *et al.* [1] first demonstrated the capabilities of AAS as a detector for gas chromatography, much development work on this method has been undertaken.

This chapter is devoted to the description of the ways in which atomic-absorption spectrophotometers have been used as detectors for chromatographic separation methods. Inevitably some consideration is given to the methods used for sample collection, preconcentration and, where used, derivatization, as well as the separation methods used. However, particular emphasis is given to the design of the interface between the chromatographic column and the detector, to the ways in which the sample is introduced into the atom cell, and to the atom cell itself. The optimization of the various components of the system is also described and numerous examples of applications are cited.

2.1 THEORETICAL ASPECTS OF GC–AAS DESIGN

The basic theory of chromatography and atomic spectroscopy, as described in Chapter 1, indicates that consideration must be given to several different aspects of the design of a chromatography–atomic-absorption interfaced system. First, the sample must be transferred from the chromatographic column to the detector with minimal sample loss, with minimal peak broadening, and with no interfering chemical reactions taking place, hence care must be taken in the design of the interface tube itself. Secondly, the sample must be introduced into the detector atom cell without loss. This also has design implications. Thirdly, and probably most critically, the atom cell itself must be designed to have maximum sensitivity and the lowest possible detection limits for the analytes of interest. The most important factors here are the efficiency of atomization of the element in the atom cell, the residence time of the analytes in the light-beam and the intensity of the light-source.

Other factors which must be considered are the carrier gas and any other ancillary gas flows into the atom cell (since these will affect the residence times of the analytes in the cell), the light-source, the optical system and the signal processing and display systems. These various components are listed in Table 2.1 and are considered in detail below.

Table 2.1 — Breakdown of detector components in GC/LC–AAS

Component	Notes
Interface	material/temperature
Sample introduction	direct or indirect to atom cell
Atom cell	flame/flameless/silica/graphite
Carrier gas	flow-rate affects chromatography and sensitivity
Ancillary gas	flow-rate affects sensitivity
Light source	hollow-cathode/electrodeless-discharge
Optical system	single/double beam
Background correction	essential for some matrices
Signal processing	integration/peak height

2.2 INTERFACE DESIGN

Ideally, the eluent from the chromatographic column should be introduced directly into the detector atom cell without any intervening tubing or connectors. In this way sample loss, peak broadening or peak tailing are absolutely minimized, as the analytes are detected immediately they leave the column. In practice, however, this is not usually possible and the column is connected to the detector by means of a short length of tubing.

Various materials have been employed for this purpose but no consensus has been reached on the most suitable, nor on whether or not it is necessary to heat the tube above ambient temperature. Stainless-steel tubing of 2 mm i.d. has been used for the transfer of tetra-alkyllead compounds, heated to 150°C [2] or to 80°C [3]. Unheated glass-lined stainless-steel tubing of 1.5 mm o.d. was also successfully used in the first of these two systems without noticeable peak broadening or sample loss by condensation [4]. Teflon-lined aluminium tubing at 80°C [5], Pyrex capillary at 100°C [6], glass tubing at 200°C [7], 0.5 mm i.d. nickel tubing at 140°C [8] and tungsten tubing [9] have also been used with tetra-alkyllead compounds.

In the last of these studies the length of transfer tube used was only 10 cm but no indication was given as to whether or not sample loss occurred with a longer tube. This problem has been investigated [10] by varying the length of 1.5 mm o.d. stainless-steel tubing used for the interface from 0.5 to 1.5 m. No difference in tetra-alkyllead signal was found. Peak tailing was also investigated, with different interface temperatures. No change in the chromatographic separation was found, whether the interface was heated or not. However, it may be prudent to use an elevated temperature during routine work in order to avoid condensation and progressive accumulation of solvent and other volatile components in the tube, which might eventually lead to a loss of sensitivity, owing to consequent lossses of analyte on the tube walls.

The elimination of a separate interface line may be possible by use of a capillary GC column, the downstream end of which could be uncoiled and led from the oven directly to the detector. This could be advantageous for the analysis of complex mixtures of analytes, where the determination of all the compounds present is critically dependent upon their separation. Although this technique has not yet been reported in the literature, Forsyth and Marshall [11] have partially developed the method by interfacing a 1.8 m, 6 mm diameter, glass GC column packed with 10% OV-101 on 80–100 mesh Supelcoport to a Zeiss FMD-3 AAS by means of a 1 m length of 0.327 mm OV-101 fused silica capillary column surrounded by 6.25 mm copper tubing and maintained at 250°C with heating tape. This method of interfacing has been reported in the context of atomic-fluorescence detection [12].

2.3 SAMPLE INTRODUCTION

2.3.1 From a gas chromatograph

In the first description of a coupled GC–AAS system [1] the column eluent was introduced into the nebulizer of an unmodified air–acetylene flame atomic-absorption spectrometer. The same procedure was used by other workers [13,14] for the determination of tetra-alkyllead compounds, and also for the determination of some silylated aliphatic alcohols in the presence of non-silylated species, by use of an oxygen–acetylene or nitrous oxide–acetylene flame [15]. This method of sample introduction is generally unsatisfactory as excessive condensation of volatile species may occur in the unheated nebulizer and there will be appreciable dilution of the sample by mixing in the burner chamber [16–18].

Instead of introducing the column eluent into the nebulizer, Coker [19] introduced the sample into the base of the burner by means of a gas union threaded into a small hole made in the base of the burner neck assembly. The gas flow from the column was therefore mixed into the fuel and oxidant gases at this point, thus by-passing the nebulizer chamber. In this way detection limits of 10–20 ng (as lead) could be obtained for each tetra-alkyllead compound. The same method of coupling has been used in the analysis of chromium compounds [20]. The samples were digested with sulphuric acid–hydrogen peroxide mixture, then heated with trifluoro-acetyl-acetone. This produced a volatile chromium diketonate complex which was extractable into hexane. This was injected into a PTFE column packed with 10% SE-30 on Chromosorb WHP heated to 180°C and so swept into the AAS burner assembly for chromium measurement. The limit of detection was 1 ng (as chromium).

The next stage in the development of a more efficient method of sample introduction was to by-pass all the nebulizer and burner assembly and to pass the column eluent directly into the atom cell. As will be discussed in detail below, there are broadly two types of atom cell currently used in GC/LC–AAS systems, those employing a silica furnace tube heated either by a flame or electrically, and those employing a graphite tube heated electrically. In the case of the silica furnace (often called a quartz furnace) the column eluent may be introduced through a small hole in its side (6 mm diameter in one case [2], shown in Fig. 2.1a) or by a more sophisticated

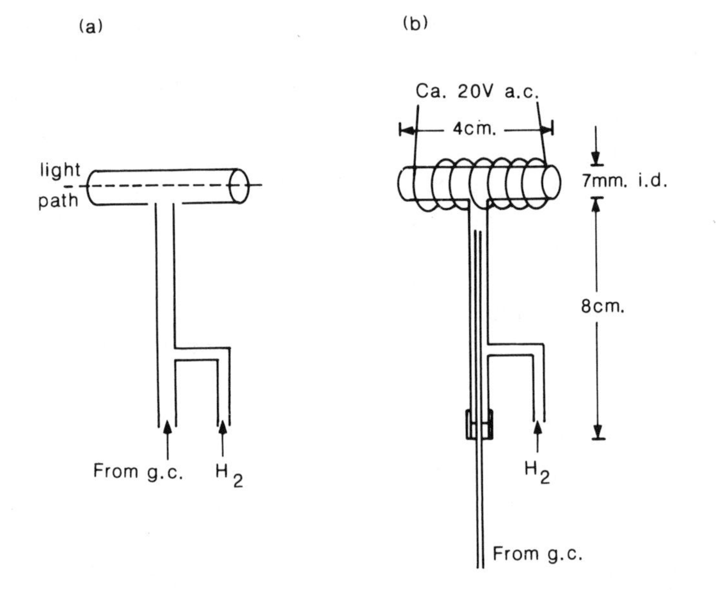

Fig. 2.1 — Sample introduction into a silica furnace (a) through an open entry port, (b) through a T-piece side arm.
[(b) Reproduced with permission, from Y. K. Chau, P. T. S. Wong and P. D. Goulden, *Anal. Chim. Acta*, 1976, **85**, 421. Copyright 1976, Elsevier Science Publishers.]

T-piece side-arm arrangement, e.g. [21]. The latter method, shown in Fig. 1.1b, has the advantage of eliminating the entry of ambient air through the sample port, which would dilute the analytes inside the atom cell. It can be made even more effective by using gas-tight fittings at the T-piece. When sensitivity was not a critical consideration and detection limits of the order of micrograms of tetra-alkyllead were sufficient, these workers connected their interface line to the nebulizer of an unmodified, conventional, AAS system. They found that a glass liner was necessary inside the nebulizer to prevent the surface adsorption of the analytes.

The connection of a capillary column transfer line to the side-arm of a silica furnace has been achieved with two reducing unions and a short length of alumina tubing [11]. The capillary column was positioned inside an 8.5 cm length of 0.32 cm o.d. alumina tube by means of a modified stainless-steel reducing union. The capillary column–alumina tubing assembly was then positioned inside the furnace side-arm by means of another reducing union. Both reducing unions were modified to accept separate 0.16 cm stainless-steel tubes and so acted as T-pieces as well as reducing unions. These steel tubes served as entry ports for the make-up gases to the furnace and allowed for the flow of separate gases between the capillary tubing and the alumina support tube and between the support tube and the inner wall of the silica side-arm. This arrangement is shown in Fig. 2.2.

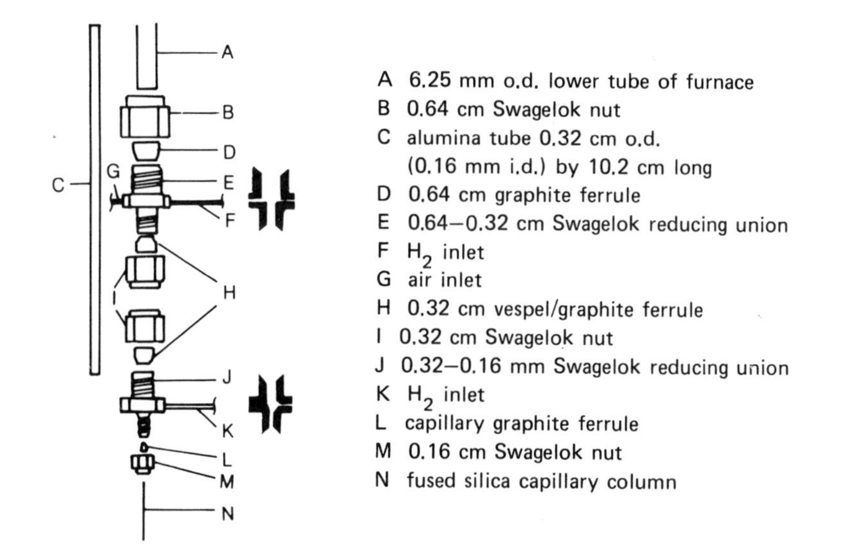

Fig. 2.2 — Silica T-tube–capillary interface.
(Reproduced with permission, from D. S. Forsyth and W. D. Marshall, *Anal. Chem.*, 1985, **57**, 1299. Copyright 1985, American Chemical Society.)

In the case of the graphite furnace atomizer the sample introduction has been accomplished in a variety of ways. Figure 2.3 shows an arrangement whereby a Pyrex capillary interface line heated to 110°C terminates inside the base of a 'hollow-T' carbon atomization chamber [6]. For tetramethyllead the detection limit was about 0.04 ng. Alternatively an alumina tube placed tangentially near the end of the graphite tube has been used [22], as shown in Fig. 2.4. The alumina tube was connected to the gas chromatograph by a stainless-steel tube heated to ~130°C. The tangential inlet gave a lower background absorption than a radial inlet did and a detection limit of 0.12 ng (as lead) for tetramethyllead and 1.1 ng (as lead) for tetraethyllead. Segar [9] and Radziuk *et al.* [5] both used conventional graphite tubes but with somewhat enlarged side-entry sample ports with the eluent impinging directly onto the heated graphite surface. In the latter case this was achieved by machining a tantalum connector from a 6.4 mm diameter rod to receive the Teflon-lined aluminium transfer tube. This connector then fitted into the entry port in the

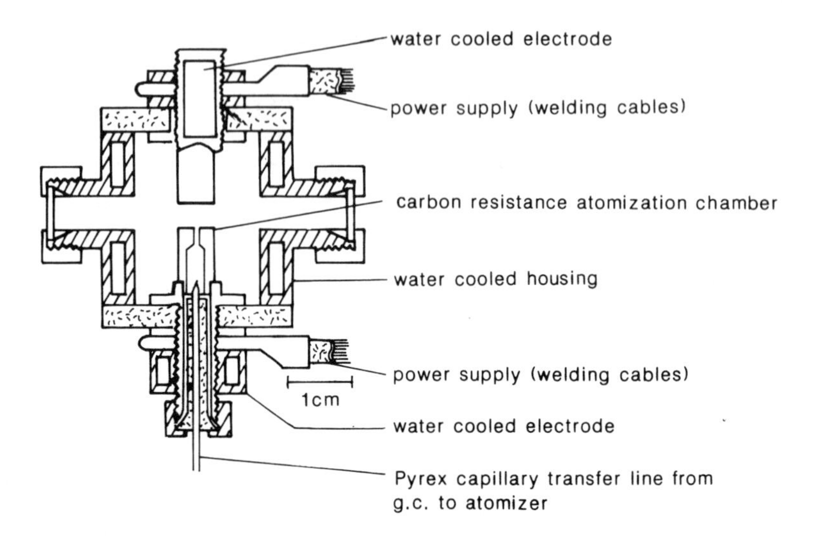

Fig. 2.3 — Schematic diagram of capillary transfer line into a graphite furnace. (Reproduced with permission, from J. W. Robinson, E. L. Kiesel, J. P. Goodbread, R. Bliss and R. Marshall, *Anal. Chim. Acta*, 1977, **92**, 321. Copyright 1977, Elsevier Science Publishers.)

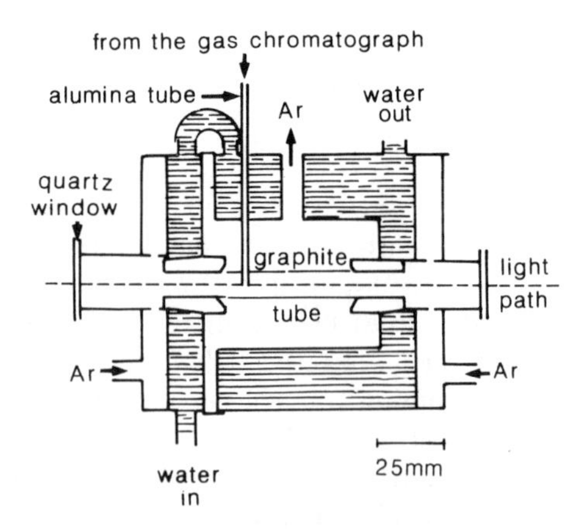

Fig. 2.4 — Schematic diagram of tangential sample introduction into a graphite furnace. (Reproduced with permission, from R. Bye, P. E. Paus, R. Solberg and Y. Thomassen, *At. Abs. Newsl.*, 1978, **17**, 131. Copyright 1978, Perkin-Elmer Corp. Inc.)

graphite tube, as shown in Fig. 2.5, allowing higher atomization temperatures than could otherwise be used. However, it may be that some premature sample decomposition or, if mixtures of compounds are present, disproportionation, may take place at the heated tantalum surface.

A simple and elegant solution to the problem of introducing the sample gases into a graphite furnace has been proposed [7]. A silica T-piece was used to connect a glass transfer tube of 6.2 mm o.d. and 0.5 mm i.d. to both the inner gas-flow entrances of a Perkin-Elmer HGA-74 graphite furnace. In this way the column eluent was forced to follow the flow path that would normally be taken by the purge gas inside the furnace, i.e. a symmetrical path from both ends of the tube and escape through the normal sample port in the side of the tube. This arrangement is shown in Fig. 2.6. The advantage of this method is that it allows an unmodified furnace assembly to be quickly and simply connected to the chromatograph. A glass interface line about 1 m in length was used. In order to avoid breakage four short pieces were joined by shrinkable Teflon tubing, giving a flexible assembly. A temperature of 200°C was maintained by wrapping the tube with heating wire and insulating tape. These workers believe that if the analyte is introduced into the graphite tube from both sides symmetrically, most of the solvent molecules are decomposed, ensuring that deuterium background correction is sufficient to correct for any background absorption.

2.3.2 From a liquid chromatograph

The methods used for introducing the eluate from a liquid chromatograph into an AAS detector are not yet as well developed as those for the gaseous eluate from a GC. However, since the eluent flow-rate from an LC is typically similar to the uptake rate of a flame AAS nebulizer (~2–6 ml/min) it is possible to couple the two instruments directly by an interface tube. To overcome flow-rate differences between the column and the nebulizer it may be necessary to use an auxiliary solvent flow into the nebulizer [23], to allow excess of column eluate to overflow from a cup from which liquid is drawn by the nebulizer, or to collect discrete aliquots of the eluate in Teflon cups, followed by their introduction into the detector [24]. Ebdon *et al.* [25] allowed the nebulizer of a modified flame atomic-absorption spectrometer (flame AAS) to draw air and aliquots of the sample, in order to balance the flows. The sample uptake capillary of the nebulizer was connected to a wider bore tube which in turn was connected to the analytical column. A side-arm was fitted to the wide tube in such a way that a liquid trap was not formed, thus avoiding sample cross-contamination. When no sample was being passed, the nebulizer drew air through the side-arm, allowing the spray chamber to run dry. This arrangement is shown in Fig. 2.7. The damping of the spectrometer was adjusted to 0.5 sec in order to remove the pulsed response which was otherwise observed.

Similar introduction of LC eluate into a flame AAS has been used by other workers for the determination of tetra-alkyllead [26]. Removal of the internal flow spoiler from the nebulizer allowed about 80% of the eluate uptake to reach the flame, permitting the determination of 0.25–50 μg of tetramethyllead and tetraethyllead in a water–methanol (3:2 v/v) mobile phase at a flow-rate of 1 ml/min. Similarly the eluate from a reversed-phase μ Bondapak C_{18} column (with acetonitrile–water mobile phase) was introduced into the aspiration uptake capillary of the nebulizer of

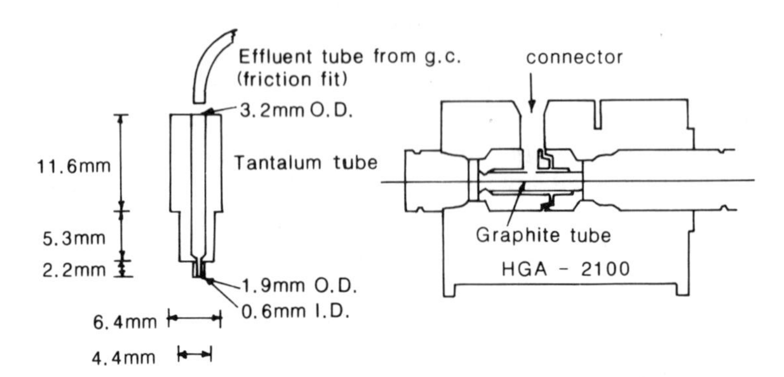

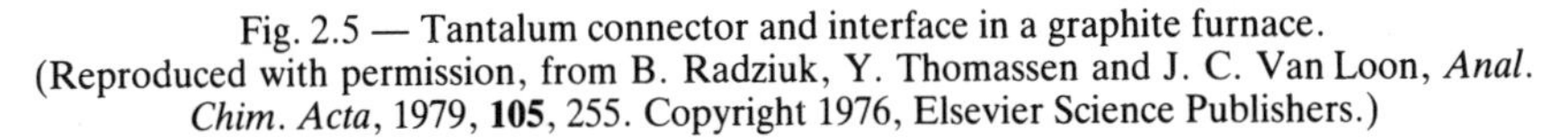

Fig. 2.5 — Tantalum connector and interface in a graphite furnace.
(Reproduced with permission, from B. Radziuk, Y. Thomassen and J. C. Van Loon, *Anal. Chim. Acta*, 1979, **105**, 255. Copyright 1976, Elsevier Science Publishers.)

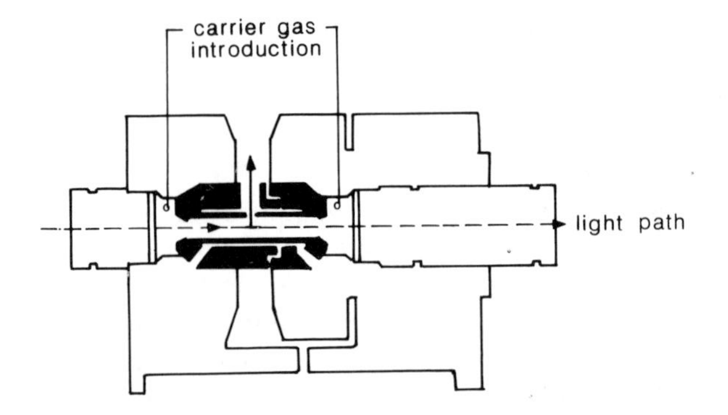

Fig. 2.6 — Cross-section of a graphite furnace, showing gas paths.
(Reproduced with permission, from W. R. A. De Jonghe, D. Chakraborti and F. Adams, *Anal. Chim. Acta*, 1980, **115**, 89. Copyright 1980, Elsevier Science Publishers.)

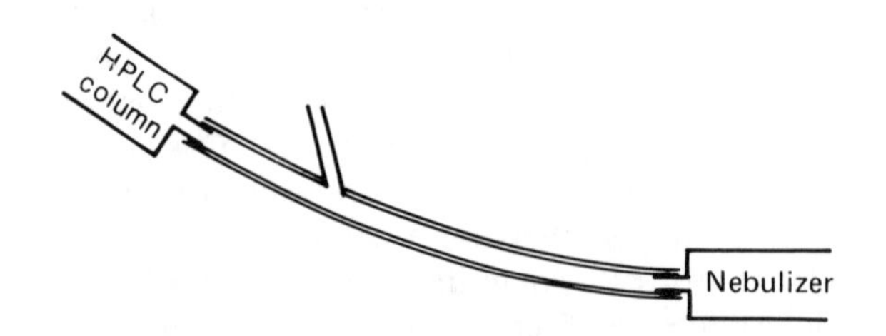

Fig. 2.7 — Schematic diagram of HPLC–AAS nebulizer interface.
(Reproduced with permission, from L. Ebdon, S. J. Hill and P. Jones, *Analyst*, 1985, **110**, 515. Copyright 1985, Royal Society of Chemistry.)

an air–acetylene flame AAS, allowing all five tetra-alkyllead compounds (Me_4Pb, Me_3EtPb, Me_2Et_2Pb, $MeEt_3Pb$ and Et_4Pb) to be separately detected at 10 ng level (as Pb) for each compound [27].

The direct coupling of an LC to a flameless AAS is not as straightforward as to a flame instrument, owing to the different flow characteristics of the systems. In LC (including HPLC) and in flame AAS the liquid flows are continuous, both out of the chromatography column and into the detector's nebulizer. In electrothermal AAS, however, discrete volumes of the sample are introduced into the graphite-furnace atomizer, and drying, charring, atomization and purging then take place sequentially. This problem has been overcome by using multiport sampling and injection valves whereby the column eluate is stopped at fixed time intervals and collected in sampling cups. Aliquots of these fractions are then injected in sequence into the atomizer [28].

A conventional graphite-furnace carousel auto-sampler has also been used for the injection of aliquots of discrete samples of HPLC eluate [29]. The carousel revolves normally at a rate dependent on the eluate flow-rate, thus segmenting the eluate stream into a number of samples. The AAS auto-sampler is then used to introduce these into the graphite furnace. Alternatively the carousel may be kept stationary and the column eluate introduced into the bottom of a specially constructed Teflon well-sampler. Excess of liquid is withdrawn at the top by suction so that a continuously changing stream of eluate is available in the well. The auto-sampler pipette is programmed for periodical withdrawal of a preselected volume of sample from the well and its injection into the graphite furnace. These arrangements were both used for the analysis of some organic arsenic, lead, tin and mercury compounds.

A simple interface between an LC and an AAS has been described, consisting of a sampling valve, dispenser, timing circuit and analyte-addition facility [30] and used for the determination of selenium compounds with a Zeeman graphite-tube analyser. A graphite connector has also been used to link an interface tube from an LC to a graphite-furnace AAS [31] and used for the determination of tetra-alkyllead compounds in gasoline. Alkyllead and alkyltin compounds have also been separated by HPLC and introduced directly into a nitrous oxide–acetylene flame [32].

Alkyllead and alkyltin compounds have also been determined sequentially off-line following separation by LC. The various eluate fractions were collected in capillary tubes and stored during the chromatography run, with subsequent analysis by AAS, either directly in the case of the tin compounds or following digestion with methanolic iodine in the case of the lead compounds [33].

The dispersion characteristics of an LC–flame AAS system with three different interfaces have been examined [34]. It was found that the detector produced a significant degree of peak dispersion, of which about 80% was due to dispersion in the burner system following sample introduction into the nebulizer. The remaining 20% was attributed to the response time of the AAS electronics. The other major source of peak dispersion identified in this study was that taking place within the interface tube. In the case of a 0.58 mm i.d. polyethylene tube a 25 cm length resulted in the same amount of peak dispersion as that caused by the AAS. In practice, however, an interface tube of twice this length was required.

The other tubes investigated as interfaces in this study were a serpentine tube,

0.25 mm i.d. and 49 cm long, the ends of which were soldered into 1.5 cm lengths of 1.27 mm i.d. straight tube so that standard LC connectors could be used, and a straight tube of 0.25 mm i.d. and 59 cm length. The form of the serpentine tube is shown in Fig. 2.8. It consists of a series of semi-circular bends in a cylindrical tube,

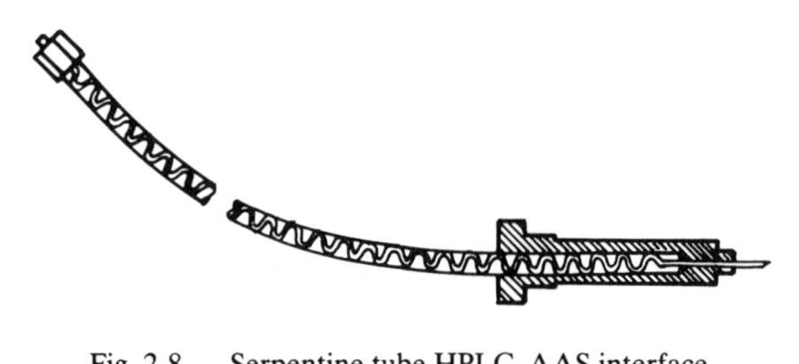

Fig. 2.8 — Serpentine tube HPLC–AAS interface.
(Reproduced with permission, from E. D. Katz and R. P. W. Scott, *Analyst*, 1985, **110**, 253. Copyright 1985, Royal Society of Chemistry.)

the bends being as tight as possible without causing any kinking of the tube and consequently without there being any straight portions. In any straight portions of tube the velocity profile of the moving liquid would again become parabolic (as it is in a straight tube as a result of friction at the tube walls), whereas in a completely serpentine tube the flow of the liquid is continually changing in direction, producing secondary radial flow inside the tube and disrupting the parabolic velocity profile. This allows the whole band of liquid to move at approximately the same velocity down the tube with no static layer of liquid at the tube walls. In this way band spreading, and consequently peak dispersion, is significantly reduced. These workers are of the opinion that the use of present-day AAS systems (with their inherently poor sample dispersion characteristics) as detectors for LCs prevents the high performance of modern chromatographic columns from being realized. They point out that should future spectrophotometers be designed to provide low dispersion, then the major contribution to solute dispersion will reside in the interface, making the use of low-dispersion serpentine tubing essential.

2.4 GENERAL ASPECTS OF DETECTOR DESIGN

In an AAS detector it is first necessary to produce free metal atoms from the analyte. This can be simply done by using a flame as the atom cell, as in conventional flame atomic-absorption spectrometry. The sample is nebulized to form very fine aerosol droplets and introduced into the flame where there is rapid evaporation of the solvent, volatilization of the resultant solid particles and dissociation of the vaporized analyte to yield free metal atoms which absorb the incident radiation. Whilst this utilizes simple equipment with low cost and great versatility in terms of the number of elements which can be excited, the efficiency of atomization in flames is very low. Because of the high flow-rate of fuel and oxidant used to support the flame and to transport the sample the atomic concentration in the flame is severely limited by dilution effects. Absorption takes place by only one atom in about every 10^8 atoms that are fed into the flame [35]. For this reason other types of atom cell have been developed which allow a much longer residence time of the atoms in the light-beam path, with a commensurate increase in sensitivity. As well as this, the ability to use an inert or reducing gas inside the atom cell is a further advantage, as this enhances the decomposition of many metal species and also allows atomic-absorption measurements in the vacuum ultraviolet region, by excluding oxygen. The first description of a tubular atom cell was that by L'vov, who built an atomizer which used a carbon rod heated with an electrical discharge [36–38] and from this has developed the modern electrothermal graphite furnace used in commercial AAS instruments.

Because of the sensitivity generally required for the analysis of environmental samples (which is the main application of GC/LC–AAS as an analytical tool) the simple flame is invariably not used as the atom cell in such systems. As mentioned above, either flame or electrically heated silica furnace tubes are generally used or else an electrothermal graphite furnace is modified to accept the GC eluate whilst being continuously heated. This continuous heating is required both because of the continuous nature of the flow from a chromatograph and because the volatile organometallic species would otherwise escape from the atom cell before atomization and absorption can take place. Some organometallic compounds, in particular those of lead, present problems in their AAS detection, as similar amounts of different compounds of the same metal can given different atomic-absorption responses. For example, tetramethyllead may give a larger absorbance than does the same amount of tetraethyllead [39]. For this reason it is necessary to ensure that organometallic compounds are dissociated as completely as possible in the atom cell. This will minimize the different responses for a given amount of metal when present in different species. In particular it is necessary to ensure that the gas temperature inside the furnace is as close to the wall temperature as possible. Whilst most GC–AAS furnaces have been developed with this in mind but without specifically making high gas temperature a design requirement, others have been developed with this aspect as a prime requirement [32,40].

The primary variable in detector design is the size and shape of the atom cell itself. To a certain extent these parameters are controlled by the configuration of the spectrometer to be used in the system. For example, when a flame AAS is being modified to accept a silica furnace the internal diameter of the furnace will be dictated by the width of the light-beam. With too narrow a tube an annulus of radiation will be excluded from the atom cell, while use of too wide a tube will result in an unnecessarily large volume in the atom cell and consequntly greater dilution of the analyte atoms. Both situations will lead to an avoidable decrease in sensitivity. Other variables, such as furnace temperature and carrier gas and ancillary gas flow-rates, are more or less easily controlled, and some method of optimization of equipment design and operating conditions is required. The degree of sophistication of the optimization procedure used has varied tremendously, from virtually none, through laborious univariate stepwise methods to highly developed variable step-size simplex methods, e.g. [41,42].

2.5 SILICA FURNACE DETECTORS FOLLOWING GC SEPARATION

2.5.1 Electrothermally heated

In 1976 Chau and co-workers described the application of a silica furnace atom cell to the detection of tetra-alkyllead compounds with an enhancement in sensitivity of three orders of magnitude over that achieved with conventional flame AAS [21]. The furnace was constructed of silica tubing (7 mm i.d., 6 mm long) with open ends and a side-arm through which the GC eluate and hydrogen were introduced. The furnace was heated by a winding of 26-gauge chromel wire, resistance 5 ohm, with an applied voltage of about 20 V a.c. With a hydrogen flow-rate of 135 ml/min a temperature of about 1000°C was maintained. The detection limits achieved for the various lead compounds with this system were about 100 pg (of Pb).

This system has been used for the determination of the ionic dialkyllead and trialkyllead compounds, following their extraction from water [43]. A chelating agent, sodium diethyldithiocarbamate, was used to assist their extraction into benzene, following which the species were butylated with a Grignard reagent to their tetra-alkyl forms. Any lead(II) ions present are also extracted and derivatized by this procedure and so the whole spectrum of alkyllead compounds and their degradation products is analysable. Detection limits of about 0.1 μg/l. could be achieved by using a 1-litre water sample, 5 ml of solvent and a 5-μl sample injection into the GC–AAS, with absolute detection limits of $<$100 pg.

The same system was also applied to the determination of methylated arsenic compounds from cultures of lake sediments and from pure bacterial cultures to which inorganic arsenic had been added [44]. The detection limit of the method was about 100 pg (of As) for each compound.

Arsenic species were also determined in natural waters by a similar method of analysis after gas stripping and collection in a cold trap [45]. The detection limit for AsH_3 was about 50 pg (as As) with the 193.7 nm line, corresponding to 0.5 ng/l. for a 100-ml water sample. For the methylated arsines detection limits of several ng/l. were found. The silica atom cell AAS used in this study was also used for the detection of selenium species in natural waters, with sensitivities and precision similar to those found for arsenic [46].

A 10 mm i.d., 14 mm o.d. and 140 mm long silica tube, heated to 1000°C by nichrome wire has been used as the atom cell in an AAS for the determination of triethyllead, diethyllead and inorganic lead in urine [47]. The samples were pre-treated with buffering reagents before the addition of sodium borohydride (2 ml of 10% $NaBH_4$ solution). The volatile lead hydrides produced were swept by helium carrier gas into a trap cooled with liquid nitrogen and maintained there until the reaction was complete. When the trap was warmed to room temperature the analytes were swept individually into the detector, in the order of increasing boiling points. Detection limits of $\sim$5 ng were found for Et_3Pb^+ and Et_2Pb^{2+} and $\sim$100 ng for Pb^{2+}, making the method suitable for the routine screening of occupationally-exposed lead workers.

The silica furnace AAS described above [21] was also applied to the determination of methyltin and inorganic tin(IV) species in water. Instead of formation of the volatile hydrides of the analytes to allow their separation and detection, the compounds were further alkylated to the tetra-alkyltin form by use of a butyl Grignard reagent. With furnace gas flows of 85 ml/min hydrogen and 20 ml/min air and a furnace temperature of 850–900°C, detection limits of about 0.1 ng for each of the compounds were obtained, corresponding to $\sim$40 ng/l. in water samples.

An automated GC–AAS system utilizing an electrothermally heated silica tube has been used to study the decomposition–atomization process for alkyllead compounds [11]. A 10 cm long, 7 mm i.d. silica tube with a 6 cm, 4 mm i.d. side-arm was heated with 3.6 m of 22-gauge Chromel 875 wire of resistance 4.53 ohm/m. The furnace was encased in shaped firebrick and held in place by a hinged aluminium tube. This assembly had insulated entry ports for electrical connections and two thermocouples placed 1–2 mm from the quartz surface. It was held aligned in the light-beam of a Zeiss FMD-3 AAS by an aluminium cradle. Deuterium background correction was used to eliminate interference from the solvent used. The interface between the GC column and the detector atom cell has been described in detail above. The furnace temperature was maintained at 900±2°C by means of a controller circuit connected to one of the thermocouples. The second thermocouple was connected to a digital read-out display.

One problem associated with the use of silica atom cells which has been noted by several workers, e.g. [48], is that over time a carbonaceous deposit of incompletely combusted solvent can build up on the cell walls. In this automated system [11] the cell was periodically purged with air, which effectively removed the wall deposit. This was achieved by the use of a timing circuit which activated two solenoid valves which in turn controlled the flow of hydrogen and air into the cell.

The operating conditions of the detector were optimized by a univariate procedure. The system response was recorded as a function of furnace temperature, position of the transfer line within the base of the furnace assembly, the detector hydrogen flow-rate, and the column flow-rate. Of these parameters the flow-rate of the hydrogen make-up gas was found to influence the detector response considerably. It was concluded that a certain quantity of hydrogen must be present in the furnace for maximum atomization of the analyte, but that at higher hydrogen flow-rates the analytes are removed more rapidly from the optical beam, to the detriment of the detector response. The role of hydrogen in the atomization process for the alkyllead molecules was investigated by using various sources of hydrogen atoms, including NH_2 and CH_3 radicals. It was concluded that the hydrogen was atomized within the silica furnace and that the resulting hydrogen radicals act as scavengers which revolatilize metallic lead from the tube surface. The resultant lead hydride is then atomized by reaction with two further hydrogen radicals and absorption of the incident light-beam then occurs.

The sensitivity achieved with an electrically heated silica furnace as the atom cell in a GC–AAS system has been compared with that achieved with a flame photometric detector (FPD) in a GC–FPD system, for butyltin compounds as the analytes [49]. With hydrogen and air as the combustion gases in the silica furnace, introduced through a T-piece, a temperature of 900°C and a tin electrodeless-discharge lamp, the minimum amount of tributylpentyltin detectable at 224.6 mm was found to be 150 pg (as Sn) with the GC–AAS system. With the GC–FPD the detection limit was the same for the tributyl compound, and slightly lower than the corresponding amounts for the GC–AAS system for the Bu_2Sn^{2+}, $BuSn^{3+}$ and Sn^{4+} species. However, these workers considered that the ease of operation of the GC–AAS system, the ease of replacement of contaminated or broken silica furnaces and the potential application of the technique to the detection of other volatile organometallic compounds, made it preferable to the GC–FPD method.

Substantial improvements have been made in the design, optimization and application of a low-cost electrothermal silica furnace for the detection of alkyllead species in environmental samples [50]. With a 16 cm long, 12 mm i.d., silica tube wrapped with 22-gauge nickel–chromium wire and insulated with glass string and porcelain sleeving, a detection limit for tetramethyllead as low as 10 pg has been achieved. By use of a derivatization procedure utilizing solvent extraction and propylation with propyl magnesium chloride, the ionic alkyllead species are detectable in aqueous samples at the sub-ng/l. level.

2.5.2 Flame heated

Substantial fundamental improvements in the design of the silica furnace atom cell have been made by Ebdon and co-workers [41]. Four different atom cells were designed and the performance of each was optimized with respect to the detection of tetra-alkyllead. In each case flame, rather than electrothermal, heating of the detector cell was used. A Pye Unicam SP192 AAS with a deuterium-arc background corrector was used as the detector and each atom cell was assessed in turn. In the simplest design the GC eluate was delivered directly to the air–acetylene flame, the interface tube from the GC terminating at one end of the burner slot. A fivefold increase in sensitivity was immediately achieved by suspending a ceramic tube in the light-beam, with the interface tube terminating just below a hole in its lower side. A further refinement was to pass the interface tube through the burner so that it terminated just above the burner slot. The eluate was then more efficiently carried into the ceramic tube through the hole in its lower side. One of the disadvantages of this arrangement was that some of the combustion gases entered the atom cell and so reduced the residence time of the analyte atoms in the light-path. In the fourth design the sample was introduced independently of the air–acetylene flame by terminating the interface tube at a hole in the side of the ceramic tube. A small hydrogen diffusion flame was used to atomize the alkyllead species, the hydrogen being supplied by means of a glass-lined stainless-steel T-piece in the interface tube, and the air–acetylene flame heating the ceramic tube and so keeping the diffusion flame alight.

To optimize these four atom cells a variable step-size simplex method was used with each parameter (nitrogen gas flow, air flow, fuel gas flow, hydrogen gas flow, GC column temperature and position of the burner head relative to the light-beam) being varied to give maximum peak height for a given injection of alkyllead. The optimum conditions predicted by the simplex method were then confirmed by using a univariate search procedure in which each parameter was varied in turn, all the others being held constant. Of all the variables the acetylene flow-rate proved to be the most critical, with a narrow optimum range. The peak height responses to varying air and acetylene flow-rates are shown in Fig. 2.9. After optimization, detection limits [defined as twice the standard deviation (σ) of the blank signal] of 17 pg (as Pb) were found for both TML and TEL with the most sensitive atom cell. The linear working ranges and detection limits for each of the four atom cells investigated are shown in Table 2.2.

The silica furnace atom cell heated in an air–acetylene flame has been used by several other groups investigating the occurrence and chemistry of organometallics, especially organolead compounds, in the environment. For example a GC–AAS system has been interfaced with a two-stage thermal desorber to allow the determination of tetra-alkyllead (TAL) compounds collected from the atmosphere on a

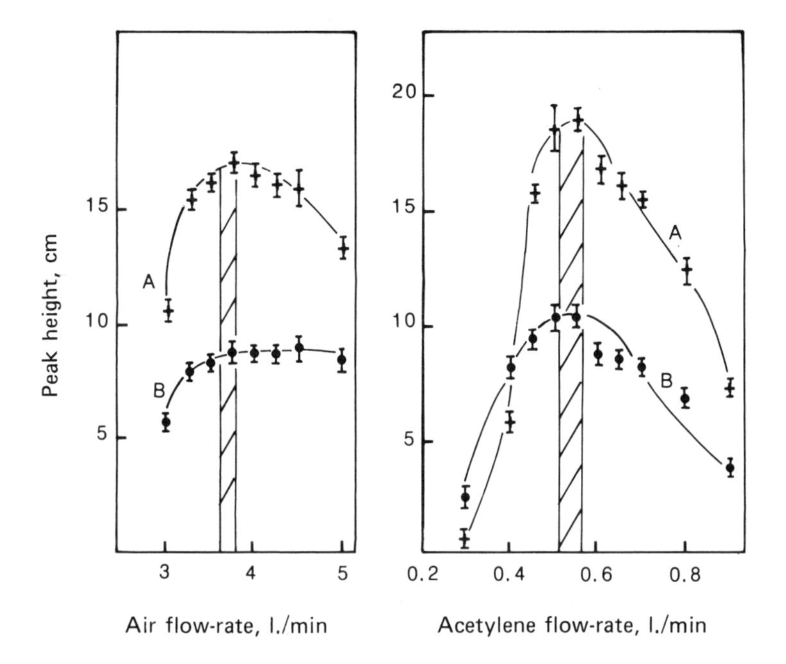

Fig. 2.9 — Peak height response at various air and acetylene flow-rates. (Reproduced with permission, from L. Ebdon, R. W. Ward and D. A. Leathard, *Analyst*, 1982, **107**, 129. Copyright 1982, Royal Society of Chemistry.)

Table 2.2 — Linear working range and detection limits (as Pb) for four different atom cells. (Reproduced with permission, from L. Ebdon, R. W. Ward and D. A. Leathard, *Analyst*, 1982, **107**, 129. Copyright 1982, Royal Society of Chemistry)

Atom cell	Linear range (ng)	Detection limits (pg)	
		tetramethyllead	tetraethyllead
I	10–300	1000	2000
II	1.0–50	58	75
III	0.8–20	48	71
IV	0.1–15	17*	17*

* Equivalent to 82 fmole of compound.

porous polymer adsorbent [2,48]. In this system a 14 mm o.d., 9 mm i.d. ceramic tube of recrystallized alumina, of length 150 mm, was suspended in the air–acetylene flame of an Evans Electroselenium Model 240 Mk2 AAS. A 6 mm diameter hole in its side was used as a sample introduction port with hydrogen at a flow-rate of 80 ml/

min being mixed into the GC eluate just prior to the atom cell. TAL compounds were collected from the air by adsorption on 0.5 g of Porapak-Q polymer (Waters Associates) with iron(II) sulphate crystals used as a prefilter to remove ozone from the air stream during sampling. A two-stage GN concentrator (GN Instrumentation Consultancy) with a variable-temperature oven and a glass-lined stainless-steel U-tube cold trap was used to desorb the sample into the GC. Initial desorption from the sampling tubes into the cold trap was done at 150°C with subsequent second-stage desorption by flash heating of the cold trap from −196°C to 90°C. Chromatograms of a standard solution of the five TAL compounds in hexane [1 ng of each (as Pb)] and of a typical rural air sample are shown in Fig. 2.10. This method gave detection limits

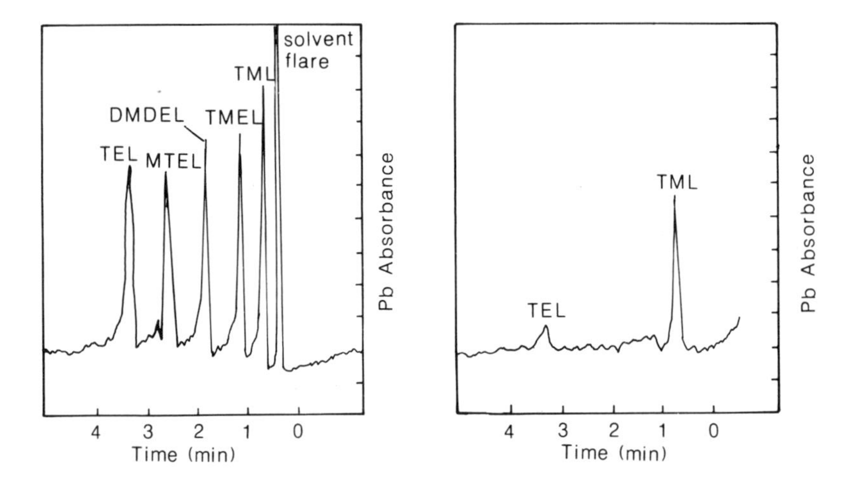

Fig. 2.10 — Chromatograms of a standard solution of five tetra-alkyllead compounds in hexane (1 ng Pb in each) and, right, a typical rural air sample.
(Reproduced with permission, from C. N. Hewitt and R. M. Harrison, *Anal. Chim. Acta*, 1985, **167**, 277. Copyright 1985, Elsevier Science Publishers.)

(3σ) of 20–30 pg Pb, or about 0.25 ng/m^3 for TML (as Pb) and 0.37 ng/m^3 for TEL (as Pb) in an air sample of 80 l. collected over a 3–24 hr period.

A flame-heated silica cell of a slightly different configuration has also been used for the determination of the ionic alkyllead compounds [49]. In this system the 1 cm i.d., 16.5 cm long, silica cell normally used with a Perkin-Elmer MHS-10 Mercury Hydride System was mounted on top of a three-slot burner and heated in an air–acetylene flame in the light-beam of a Perkin-Elmer 503 AAS. A hydrogen flow-rate of 110 ml/min and carrier gas flow-rate of 40 ml/min were used, with acetylene and air flow-rates as recommended by the manufacturers for the three-slot burner. An extraction, concentration and butylation procedure was used to render the ionic alkyllead species in aqueous solution analysable by this system, with absolute detection limits between 50 pg (as Pb) for the $PbMe_3Bu$ species and 100 pg (as Pb) for $PbEt_2Bu_2$. These correspond to 1.25 ng/l. for $PbMe_3^+$ and 2.5 ng/l. for $PbEt_2^{2+}$ (both calculated as Pb) in aqueous samples.

2.6 GRAPHITE FURNACE DETECTORS FOLLOWING GC SEPARATION

As an alternative to a silica tube furnace it is possible to use either a commercially available or a purpose-built graphite electrothermal furnace as the detector cell in an interfaced GC–AAS. For example, the atomizers shown in Fig. 2.3 [6] and Fig. 2.4 [22] were developed specifically for use in GC–AAS systems whereas those shown in Fig. 2.5 [5] and Fig. 2.6 [7] are commercially available components.

Under normal operating conditions the graphite furnace in an AAS undergoes heating in three or four stages, to allow evaporation of the solvent at a relatively low temperature, followed by ashing of the sample at an increased temperature, with atomization of the analyte at high temperature. When such a furnace is used in a GC–AAS system the high atomization temperature must be maintained for the time it takes for all of the sample to be eluted from the GC column. This may be tens of minutes and can cause premature deterioration of the carbon in some furnaces. Specially developed furnaces have the advantage that this potential problem can be tackled at the design stage. However, for most laboratories the ease of use of an existing graphite furnace in a GC–AAS system will be the major consideration. Several readily assembled/disassembled interfaces of this type have been described, e.g. [7], which allow the AAS to be used either for conventional graphite-furnace work or as the detector in a GC–AAS system.

Many of the chemical and physical considerations affecting the performance of the detector in GC–AAS systems have been investigated and described in an early major paper [51]. The inner surface of the atom cell in an HGA-2100 graphite-furnace atomizer fitted to a Perkin-Elmer 360 AAS was constructed of various materials in turn: fused silica, alumina, bare graphite and pyrolytic carbon. The effect of each of these on the detection of organic compounds of arsenic, selenium and tin was assessed, together with the effects of varying the atomization temperature and altering the argon/hydrogen ratio in the carrier gas. On balance, the graphite-walled furnace performed as well as, or better than, the others, and the system was optimized by using the bare graphite furnace running continuously at about 1800°C with a 90% Ar, 10% H_2 carrier gas mixture to give detection limits of 5 ng As, 7 ng Se and 12 ng Sn for trimethylarsine, dimethylselenium and tetramethyltin respectively.

Graphite-furnace detectors have been extensively used for the detection of alkyllead compounds in environmental media, e.g. [3,5–7,22], and these applications are described in detail in Chapter 8. Suffice it to say here that the environmental and analytical chemistry of the organic compounds of lead have received more study than those of any other metal and that the detection systems currently applied to alkyllead compounds may be readily adaptable for the study of other organometallics and metalloids. For example chromium compounds [52], organoindium and gallium compounds [53] and alkylmercury compounds [54] have been determined in this way and an adaptation of the method using high-temperature GC separation has been investigated for the speciation and determination of inorganic sodium, copper, manganese and magnesium compounds [40].

2.7 AA DETECTORS FOLLOWING LC SEPARATION

The coupling of a graphite-furnace AAS to an HPLC through an automated carousel sampler [29] has been described in detail above (see Section 2.3.2). A Perkin-Elmer 360 AAS with deuterium lamp background corrector and HGA-2100 graphite furnace and electrodeless-discharge lamps was used in this study for the detection of organoarsenic, lead, mercury and tin compounds with the signal produced being processed by a digital integrator. The relative sensitivity of the HPLC–AAS system was found to be mainly a function of the LC flow-rate and of the relative AA sensitivity for each element. Triphenylarsine was used for the evaluation of the complete system (with a detection limit of 4.8 ng) with the detection limits for the other analytes (tripropyl, triphenyl and tributyltin, methyl, ethyl, propyl and butylmercury and a hexylated lead compound) being calculated relative to this.

An HPLC–AAS system has been used for the separation and determination of the copper complexes of the complexones EDTA, NTA, EGTA and DCTA [55]. The chelates were retained on a weak-base anion-exchange resin column as a function of charge, and eluted in the order $Cu_2(EGTA)$, $Cu(NTA)^-$, $Cu(EDTA)^{2-}$ and $Cu(DCTA)^{2-}$ with absolute detection limits of 14, 16, 29 and 450 ng (as Cu) respectively. The method was considered applicable to analysis for individual chelating agents such as those used in industrial water treatment, food processing, metal cleaning and pharmaceuticals.

Rather than segmenting the entire LC eluate from a chromatographic run into discrete aliquots and then sampling each of these for AA analysis it is also possible to monitor the eluate as it leaves the LC column, for example with a UV detector, and to divert the eluate flow into a storage capillary once an increase in signal is detected. Once the peak has passed the eluate flow may be switched back to waste. In this way that part of the eluate containing the sample peak may be more intensively sampled and analysed by successive injections into the graphite furnace and more AA analyses per peak obtained than by using the 'pulsed' mode of sampling. Tetraphenyllead has been determined by both the peak-storage pulsed mode and the on/line pulsed mode and a significant improvement in sensitivity was found for the former method [56].

This last study utilized an AAS (Hitachi 170–70) which employed the Zeeman effect. This has the advantage that very large background absorption signals can be corrected for and has been successfully employed in other similar studies. For example tetraethyllead in gasoline has been determined by HPLC–Zeeman AAS, and the ability of the detector to suppress interference by $MgCl_2$, $CuCl_2$ and $CaCl_2$ was investigated [31]. In this system the furnace was designed to give a high (2800°C) gas temperature, so ensuring the complete atomization of the organometallic species. Arsenic species have also been separated and determined by HPLC–Zeeman AAS [57], the analytes being arsenobetaine bromide, arsenocholine bromide and inorganic arsenic (arsenate/arsenite). Arsenate, arsenite, methylarsonic acid and dimethylarsinic acid have also been determined by HPLC–AAS with both deuterium-lamp background correction and Zeeman-effect correction [58]. A similar system has been used for the speciation of di- and triorganotin compounds in synthetic and natural waters [59] and for the identification of inorganic arsenic and organoarsenic compounds in oil shale retort and process waters [60].

A Hitachi Zeeman graphite-furnace AAS has been used as a selenium-specific detector with a Dionex Model 16 ion chromatograph [61]. Selenite and selenate were separable from water samples and could be detected at levels of 20 ng (as Se) for each compound or at levels of 5 ng if a preconcentration technique was used. Each AAS determination took approximately 1.5 min so the peak of each selenium compound was quantified by three or four determinations. The interference caused by the presence of chloride, nitrite, nitrate, bromide and sulphate salts was found to be negligible although the selenate peak was broadened with increasing sulphate concentration.

The interfacing of a Pye Unicam Model SP9 flame AAS with an HPLC has been described in detail above [25]. In this system a slotted-tube atom cell accessory was fitted above a 5-cm slot burner and flow-rates of 4.0 l./min for air and 2.6 l./min for hydrogen were used. Deuterium-lamp background correction was used to avoid interference from other absorbing species, a tin hollow-cathode lamp was used at 6 mA and the 226.4 nm line was monitored with a band-pass of 0.5 nm. Tributyltin, tin(II) and tin(IV) compounds were separated on a Partisil 10 SCX cation-exchange analytical column within 8 min and could be determined with detection limits of ~200 ng (as Sn). Changing the air flow-rate to the burner did not greatly affect the AAS response, although at low flow-rates excessive noise was observed. A marked response to the hydrogen flow-rate was found for tributyltin chloride, but not for tin(IV). The concentration of the ammonium acetate buffer used was found to affect the AAS response to tribultyltin ions and also to affect their retention time significantly. Also the composition of the methanol–water mobile phase greatly affected the response, the peak height obtained for a given quantity of tributyltin doubling when the percentage of methanol in the mixture was increased from 60 to 70%. Natural water samples were analysed by this system after extraction of any organotin compounds present into chloroform, concentration by evaporation and redissolution in methanol. One harbour water sample was found to contain 0.47 μg/l. tributyltin (as Sn).

An HPLC–flame AAS system [34] has been used for the determination of iron in a blood sample. After optimization and investigation of the characteristics of the system with magnesium nitrate solution, its application to iron in blood was studied. The determination of iron and other metals in blood by conventional AAS presents analytical difficulties because of the presence of various protein molecules. After separation of the protein components with a cation-exchange LC packing, the iron content could be specifically determined by using the 248.3 nm absorption line.

2.8 SYSTEMS EMPLOYING HYDRIDE GENERATION

The analytical methods described above have all relied on the sample being introduced into the system in a form amenable to chromatographic separation. This is done by relying on the inherent volatility of the analyte (e.g. tetra-alkyllead species), or employing a derivatization technique prior to analysis (e.g. butylation of tri- and dialkyllead species with a Grignard reagent), or utilizing the different polarities of the analytes in a sample.

However, the use of hydride generation as a means of volatilizing some organometallic compounds from aqueous solution has also been shown to be a feasible and sensitive technique. This method has the advantages that it can be directly coupled to the GC/LC–AAS, so avoiding sample loss at the extraction or reaction stage (in the case of derivatization with a Grignard reagent), and that the resultant phase separation between the sample matrix and the hydride vapour reduces or eliminates chemical interferences from the matrix.

A continuously operated hydride generation system for the volatilization of ionic alkyltin compounds from aqueous solution, has been interfaced with a silica tube furnace AAS [62] and with a GC/HPLC–AAS [32]. A Perkin-Elmer 493 AAS fitted with deuterium-lamp background correction and an electrothermally heated silica cell of length 10 cm and i.d. 1 cm, was used as the detector [32]. The atom cell had a T-piece side-arm through which the column eluate and air (300 ml/min) were introduced and a cell temperature of 1000°C was maintained by using a Variac power supply, platinum–rhodium thermocouples and a temperature indicator. Detection limits of about 2 pg (as Sn) were found for Me_4Sn, Me_3SnCl and Me_2SnCl_2 after GC separation, and about 9 pg (as Sn) after HPLC separation.

A slightly different approach involves the chromatographic separation of the various hydride species produced from a sample after, rather than before, the hydride generation. In the case of alkyltin, acidification of an aqueous sample to pH 2, followed by addition of sodium borohydride solution, produced a large volume of hydrogen which helped flush the organotin hydrides from the sample [63]. The analytes were then trapped at −196°C on 2.5 g of Chromosorb G AW-DMCS coated with 3% SP-2100 chromatographic packing material. After 4 min the trap was removed from the liquid nitrogen bath and electrically heated to 200°C. Helium carrier gas flushed the trapped compounds sequentially into an electrothermal silica furnace suspended in the light-path of a Perkin-Elmer 503 AAS. A furnace temperature of 950°C was used, with addition of hydrogen and oxygen into the atom cell. A tin electrodeless-discharge lamp (EDL) was used as the radiation source with AA detection at 224.6 nm. Absolute detection limits (3σ) based on 100 ml samples ranged from 30 pg (as Sn) for the methyltin compounds (Me_4Sn, Me_3Sn^+, Me_2Sn^{2+} and $MeSn^{3+}$) to 200 pg (as Sn) for $(n\text{-}Bu)_3Sn^+$, with linear response up to 30 ng.

Three different furnace configurations were tested in this study, with a mixed inorganic tin and Me_3SnCl standard. A 110 mm × 12 mm i.d. cell was found to give a 50% greater response then either a 60 mm × 9 mm i.d. or a 110 mm × 15 mm i.d. cell. The residence time of the alkyltin hydrides in the best cell was ~0.3 sec. It was found that efficient mixing of the helium, oxygen and hydrogen before their introduction into the atom cell increased the signal-to-noise ratio, with a further increase by use of a blanketing stream of argon around the cell, from the AAS burner head. When the silica surface of the atom cell was in good condition, the sensitivity was greater than when deterioration had occurred. Soaking the inside surfaces with 10% hydrofluoric acid was used, with partial success, to restore full sensitivity.

The species antimony(III), antimony(V), methylstibonic acid and dimethylstibinic acid have been determined in natural waters by a similar method [64]. The apparatus used is shown in Fig. 2.11 and consisted of a reaction vessel in which sodium borohydride solution was added to the sample. The stibines produced were collected on a chromatographic packing material at −196°C, with subsequent removal by heating. AAS detection was achieved with either a graphite furnace or a

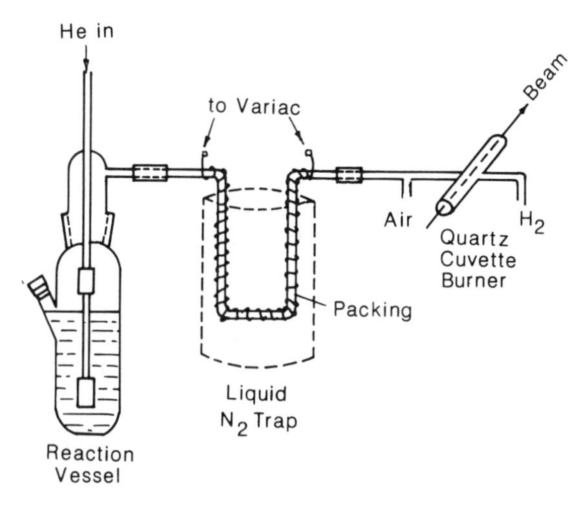

Fig. 2.11 — Hydride generation–purge and trap–AAS apparatus for the determination of antimony species in water.
(Reproduced with permission, from M. O. Andreae, J. F. Asmodé, P. Foster and L. Van't dack, *Anal. Chem.*, 1981, **53**, 1766. Copyright 1981, American Chemical Society.)

silica atom cell, the latter giving about twice the sensitivity of the former. Separation of the organic species from one another and from inorganic antimony was achieved chromatographically. However, this did not allow differentiation between Sb(III) and Sb(V). Since the efficiency of the hydride generation process depends strongly on the reaction pH it is possible to identify these two species by first operating under near-neutral conditions, Sb(III) but not Sb(V) being reduced, and then under highly acidic conditions, in which both Sb(III) and Sb(V) are reduced. Sb(V) is then determined by difference. Detection limits of 30–60 pg (as Sb), corresponding to 0.3–0.6 ng/l. for a 100 ml sample, were found.

A very similar method has been employed for the determination of methylgermanium species in natural waters [65]. After reduction to the gaseous germanes and methylgermanium hydrides, collection in a cold trap and separation by electrothermal heating, the species were atomized in a graphite furnace at 2700°C and detected by AAS. Detection limits in the range 75–175 pg (as Ge) were found. The method was applied to the determination of these compounds in a selection of natural waters: monomethylgermanium was the dominant species in sea-water.

Organic and inorganic arsenic species have been separated by ion chromatography, reduced to their arsine derivatives by reaction with sodium borohydride, and detected by electrothermal silica furnace AAS [66]. In this system a Perkin-Elmer 372 AAS was used with a silica cell held in the light-path and heated to 800°C. The arsenic absorbance was monitored at 193.7 nm, with an arsenic electrodeless-discharge lamp; the detection limits for arsenite, arsenate, monomethylarsonate, dimethylarsinate and *p*-aminophenylarsonate were <10 ng/ml (as As). By use of a sample preconcentration method employing anion-exchange columns with HPLC–hydride generation–AAS, the detection limits for these arsenic compounds have been reduced by an order of magnitude to <1 ng/ml (as As) [67]. A Pye-Unicam SP9 AAS fitted with an arsenic hollow-cathode lamp and a silica furnace heated by an air–acetylene flame was used. The method was used for the determination and speciation of arsenic in commercial bottled waters and in soil-pore waters [68]. Arsenic species have also been determined by hydride generation–AAS with separation of the gaseous compounds from the liquid phase by gas diffusion [69]. This method is claimed to give enhanced sensitivity and reduced interferences and although not yet used in speciation studies may prove to be a useful technique in hydride generation–GC–AAS systems.

The use of hydride generation with lead compounds is limited because of the instability of the alkyllead hydride and because of hydrogen–alkyl group exchange. A similar method, also based on reaction–purge and trap–desorption–AAS determination, has been developed for lead and methyllead compounds in water [42]. This uses the reaction of lead ions with sodium tetraethylborate to produce volatile tetra-alkyllead compounds which can be collected in a trap cooled by liquid nitrogen and then separated by electrothermal heating, with detection by AAS with an electrically heated silica furnace. Detection limits (3σ) of <1 pg/ml were found.

REFERENCES

[1] B. Kolb, G. Kemmner, F.H. Schleser and E. Wiedeking, *Z. Anal. Chem.*, 1966, **21**, 166.
[2] C. N. Hewitt and R. M. Harrison, *Anal. Chim. Acta*, 1985, **167**, 277.
[3] R. B. Cruz, C. Lorouso, S. George, Y. Thomassen, J. D. Kinrade, L. R. P. Butler, J. Lye and J. C. Van Loon, *Spectrochim. Acta*, 1980, **35B**, 775.
[4] S. J. De Mora, C. N. Hewitt and R. M. Harrison, *Anal. Proc.*, 1984, **21**, 415.
[5] B. Radziuk, Y. Thomassen and J. C. Van Loon, *Anal. Chim. Acta*, 1979, **105**, 255.
[6] J. W. Robinson, E. L. Kiesel, J. P. Goodbread, R. Bliss and R. Marshall, *Anal. Chim. Acta*, 1977, **92**, 321.
[7] W. R. A. De Jonghe, D. Chakraborti and F. Adams, *Anal. Chim. Acta*, 1980, **115**, 89.
[8] D. Chakraborti, W. R. A. De Jonghe, W. E. Van Mol, R. J. A. Van Cleuvenbergen and F. C. Adams, *Anal. Chem.*, 1984, **56**, 2692.
[9] D. A. Seger, *Anal. Lett.*, 1974, **7**, 89.
[10] R. M. Harrison and C. N. Hewitt, *Int. J. Environ. Anal. Chem.*, 1985, **21**, 89.
[11] D. S. Forsyth and W. D. Marshall, *Anal. Chem.*, 1985, **57**, 1299.
[12] A. D'Ulivo and P. Papoff, *Talanta*, 1985, **32**, 383.
[13] T. Katou and R. Nakagawa, *Bull. Inst. Environ. Sci. Technol.*, 1974, **1**, 19.
[14] Y. K. Chau, P. T. S. Wong and J. Saitoh, *J. Chromatog. Sci.*, 1976, **14**, 162.
[15] R. W. Morrow, J. A. Dean, W. D. Shults and M. R. Guerin, *J. Chromatog. Sci.*, 1969, **7**, 572.
[16] J. G. Gonzalez and R. T. Ross, *Anal. Lett.*, 1972, **5**, 683.
[17] J. E. Longbottom, *Anal. Chem.*, 1972, **44**, 111.
[18] N. K. Rudneveskii, D. A. Vyakhirev, V. T. Demarin, M. V. Zueva and A. I. Lukyanova, *Dokl. Akad. Nauk. SSSR*, 1975, **223**, 887.
[19] D. T. Coker, *Anal. Chem.*, 1975, **47**, 386.
[20] W. R. Wolf, *Anal. Chem.*, 1976, **48**, 1717.
[21] Y. K. Chau, P. T. S. Wong and P. D. Goulden, *Anal. Chim. Acta*, 1976, **85**, 421.
[22] R. Bye, P. E. Paus, R. Solberg and Y. Thomassen, *At. Abs. Newsl.*, 1978, **17**, 131.
[23] N. Yoza and S. Ohashi, *Anal. Lett.*, 1975, **6**, 595.
[24] J. C. Atwood, G. J. Schmidt and W. Slavin, *Pittsburgh Conference on Analytical Chemistry and Applied Spectroscopy*, Cleveland, 1979.
[25] L. Ebdon, S. J. Hill and P. Jones, *Analyst*, 1985, **110**, 515.
[26] C. Botre, F. Cacace and R. Cozzani, *Anal. Lett.*, 1976, **9**, 825.
[27] J. D. Messman and T. C. Rains, *Anal. Chem.*, 1981, **53**, 1632.
[28] A. Y. Cantillo and D. A. Segar, *Proc. Int. Conf. Heavy Metals in the Environment*, *Toronto*, 1975.
[29] F. E. Brinckman, W. R. Blair, K. L. Jewett and W. P. Iverson, *J. Chromatog. Sci.*, 1977, **15**, 493.
[30] T. M. Vickrey, M. S. Buren and H. E. Howell, *Anal. Lett.*, 1979, **11**, 1075.
[31] H. Koizumi, R. D. McLaughlin and T. Hadeishi, *Anal. Chem.*, 1979, **51**, 387.

[32] D. T. Burns, F. Glockling and M. Harriott, *Analyst*, 1981, **106**, 921.
[33] T. M. Vickery, H. E. Howell, G. V. Harrison and G. J. Ramelow, *Anal. Chem.*, 1980, **52**, 1743.
[34] E. D. Katz and R. P. W. Scott, *Analyst*, 1985, **110**, 253.
[35] S. S. M. Hassan, *Organic Analysis Using Atomic Absorption Spectrometry*, p. 384. Ellis Horwood, Chichester, 1984.
[36] B. L'vov, *J. Eng. Phys.*, 1959, **2**, 44.
[37] B. L'vov, *J. Eng. Phys.*, 1959, **2**, 56.
[38] B. L'vov, *Spectrochim. Acta*, 1961, **17**, 761.
[39] I. M. Kovina and N. K. Rozhkova, *Dokl. Akad. Nauk. Uzb. SSR*, 1968, **25**, No. 7, 25.
[40] K. Ohta, B. W. Smith and J. D. Winefordner, *Anal. Chem.*, 1982, **54**, 320.
[41] L. Ebdon, R. W. Ward and D. A. Leathard, *Analyst*, 1982, **107**, 129.
[42] S. Rapsomanikis, O. F. X. Donard and J. H. Weber, *Anal. Chem.*, 1986, **58**, 35.
[43] Y. K. Chau, P. T. S. Wong and O. Kramar, *Anal. Chim. Acta*, 1983, **146**, 211.
[44] P. T. S. Wong, Y. K. Chau, L. Luxon and G. A. Bengert, *Proc. Symp. Trace Substances in Environmental Health — XI*, Columbia, Missouri, 1977.
[45] M. O. Andreae, *Anal. Chem.*, 1977, **49**, 820.
[46] G. A. Cutter, *Anal. Chim. Acta*, 1978, **98**, 59.
[47] H. Yamaguchi, F. Arai and Y. Yamamura, *Ind. Health*, 1981, **19**, 115.
[48] C. N. Hewitt and R. M. Harrison, *Proc. Int. Conf. Heavy Metals in the Environment, Athens*, pp. 171–173. CEP, Edinburgh, 1985.
[49] R. J. Maguire and R. J. Tkacz, *J. Chromatog.*, 1983, **268**, 99.
[50] M. Radojević, A. Allen, S. Rapsomanikis and R. M. Harrison, *Anal. Chem.*, 1986, **58**, 658.
[51] G. E. Parris, W. R. Blair and F. E. Brinckman, *Anal. Chem.*, 1977, **49**, 378.
[52] W. R. Wolf, *Anal. Chem.*, 1976, **48**, 1717.
[53] A. F. Shushunova, V. T. Demarin, G. I. Makin, L. V. Sklemina, N. K. Rudenskii and Yu. A. Aleksandrov, *Zh. Analit. Khim.*, 1980, **35**, 349.
[54] R. Bye and P. E. Paus, *Anal. Chim. Acta*, 1979, **107**, 169.
[55] D. R. Jones and S. E. Manahan, *Anal. Chem.*, 1976, **48**, 502.
[56] T. M. Vickrey, H. E. Howell and M. T. Paradise, *Anal. Chem.*, 1979, **51**, 1880.
[57] R. A. Stockton and K. J. Irgolic, *Intern. J. Environ. Anal. Chem.*, 1979, **6**, 313.
[58] F. E. Brinckman, K. L. Jewett, W. P. Iverson, K. J. Irgolic, K. C. Ehrhardt and R. A. Stockton, *J. Chromatog.*, 1980, **191**, 31.
[59] K. L. Jewett and F. E. Brinckman, *J. Chromatog. Sci.*, 1981, **19**, 583.
[60] R. H. Fish, F. E. Brinckman and K. L. Jewett, *Environ. Sci. Technol.*, 1982, **16**, 174.
[61] D. Chakraborti, D. C. J. Hillman, K. J. Irgolic and R. A. Zingaro, *J. Chromatog.*, 1982, **249**, 81.
[62] V. F. Hodge, S. L. Seidel and E. D. Goldberg, *Anal. Chem.*, 1979, **51**, 1256.
[63] O. F. X. Donard, S. Rapsomanikis and J. H. Weber, *Anal. Chem.*, 1986, **58**, 772.
[64] M. O. Andreae, J. F. Asmodé, P. Foster and L. Van't dack, *Anal. Chem.*, 1981, **53**, 1766.
[65] G. A. Hambrick, III, P. N. Froelich, M. O. Andreae and B. L. Lewis, *Anal. Chem.*, 1984, **56**, 421.
[66] G. R. Ricci, L. S. Shepard, G. Colovos and N. E. Hester, *Anal. Chem.*, 1981, **53**, 610.
[67] C. T. Tye, S. J. Haswell, P. O'Neill and K. C. C. Bancroft, *Anal. Chim. Acta*, 1985, **169**, 195.
[68] S. J. Haswell, P. O'Neill and K. C. C. Bancroft, *Talanta*, 1985, **32**, 69.
[69] G. E. Pacey, M. R. Straka and J. R. Gord, *Anal. Chem.*, 1986, **58**, 502.

3

Flame photometric detectors

S. Kapila, D. O. Duebelbeis, S. E. Manahan and T. E. Clevenger
Environmental Trace Substances Research Center, University of Missouri, Colombia, MO 65203, U.S.A.

3.1 INTRODUCTION

Chromatography continues to develop as a powerful tool for separation of a wide variety of analytes, including those of toxicological and environmental significance. Chromatograms of many environmental samples consist of a complex array of peaks from a variety of species, including hydrocarbons, organohalides, nitrogen-, sulphur- and phosphorus-containing organics, and organometallics (including organometalloids). The identification of these peaks is often a challenging task which has been greatly aided by the use of mass spectrometric detectors. The analytical power of chromatography in tandem with mass spectrometry is truly outstanding, particularly in its specificity for low-level analytes. The major drawbacks of these systems until recently have been the cost and complexity of operation and maintenance. These shortcomings have been overcome in the last few years by development of small computer-controlled quadrupole systems; however, it should be pointed out that true elemental specificity can still only be achieved with high-cost, high-resolution systems. The exception to these statements is provided by compounds of the elements with more than one abundant stable isotope, such as Cl, Br, Se and S, etc., which give readily identifiable ion clusters.

The potential of element-selective detectors in reducing the complexities of chromatographic analysis was recognized very early in the development of chromatographic systems, and such detectors continue to play a vital role in qualitative and quantitative analysis. A versatile element-selective detector is the flame photometric detector (FPD). It is based on emission in the near ultraviolet and visible regions, resulting from the formation of excited molecular species in a hydrogen-rich hydrogen/air or hydrogen/oxygen flame.

The purpose of the present chapter is to give the reader a concise description of the principles and salient design features of various flame photometric detectors used in monitoring organics containing sulphur, phosphorus and other hetero-atoms.

3.2 RESPONSE CHARACTERISTICS OF SULPHUR AND PHOSPHORUS COMPOUNDS

The flame photometric detector is the emission detector with the longest history of successful application in gas chromatography. The detector was first disclosed as a device for monitoring sulphur and phosphorus by Draegerwerk and Draeger in 1962 [1]. The device was adapted for monitoring gas chromatographic effluents by Brody and Chaney in 1966 [2] and the detector of their original design has since been marketed by Tracor, Inc. as the "Melpar flame photometric detector". Flame photometric detectors have also been marketed by other commercial manufacturers, including Varian, Shimadzu, Pye–Unicam, Carlo Erba and Perkin–Elmer. The FPDs from these commercial manufacturers differ little in their basic lay-out from the original detector introduced by Brody and Chaney. The reason for the success of the detector has been its sensitivity and selectivity, coupled with its relative stability and trouble-free operation.

Since its introduction, the FPD has been the subject of a large number of scientific publications, e.g. a recent computer search of a chemical database showed well over 250 citations, the largest number of which dealt with applications in monitoring sulphur compounds and with the idiosyncrasies of sulphur response. The impetus for development of the detector was provided by the needs of the pesticide residue chemists for monitoring sulphur and phosphorus compounds. The basic modus operandi of the detector is quite simple and based on the fact that phosphorus and sulphur compounds yield chemiluminescent species S_2^* and HPO^* in fuel-rich hydrogen/oxygen flames. These chemiluminescent species emit radiation in the blue and green regions of the visible range, characteristic for these elements.

In its original and most common form the FPD was coupled to a gas chromatograph using nitrogen as the carrier gas. The carrier gas was mixed with oxygen to achieve a ratio similar to that in air. An excess of hydrogen was brought directly into the detector chamber. The hydrogen and air were burned in a hollow tip which acted as a flame-shield for the photomultiplier tube used to measure the intensity of emission from the flame. A simplified schematic of the detector is shown in Fig. 3.1. The emission spectra obtained with the detector for sulphur and phosphorus compounds are shown in Fig. 3.2. Two emission bands, at 394 and 526 nm, are generally selected for monitoring the sulphur and phosphorus species, respectively.

A number of modifications to the original FPD design have been reported. These modifications can be divided into four categories. (1) Those devoted to enhancement of the sulphur and phosphorus response. (2) Those dedicated to monitoring of other elemental species. (3) Those involved with use of the FPD with other chromatographic techniques. (4) Those dedicated to use of the FPD in non-chromatographic applications. One of the first modifications was reported by Bowman and Beroza [3]. In this the focusing mirror of the Melpar detector was replaced by a mounting for an interference filter and photomultiplier tube, thus allowing simultaneous monitoring of sulphur and phosphorus compounds. These authors also demonstrated that the dual-channel arrangement can be used not only in confirming the presence of

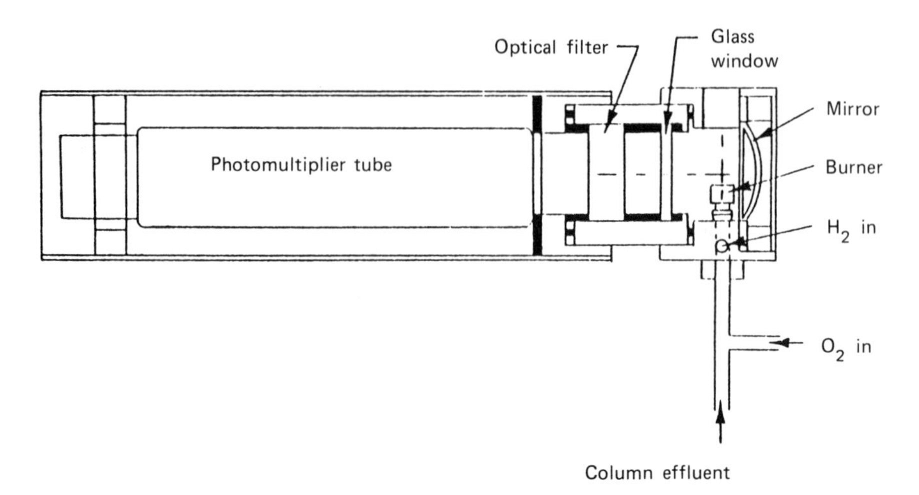

Fig. 3.1 — Schematic of the Melpar flame photometric detector. (Reproduced with permission, from S. S. Brody and J. E. Chaney, *J. Gas Chromatog.*, 1966, **4**, 42. Copyright 1966, Preston Publications Inc.)

phosphorus and sulphur analytes in a chromatogram but also in determining compounds with PS, PS_2 and PS_3 species, through the use of a response ratio — $R(\text{P})/\sqrt{R}(\text{S})$ [3]. Other modifications to improve the signal/noise characteristics of the detector have included a cooling system for the photomultiplier tube [4], installation of a thermal insulator between the photomultiplier and the detector housing, and use of light-pipes so that the photomultiplier can be mounted away from the detector [5,6].

One of the persistent problems of the original FPD design has been flame extinction resulting from a momentary starvation of the flame upon injection of a few μl of solvent. The modifications to alleviate this problem include incorporation of a solvent-venting system [7], mixing the column effluent with hydrogen instead of oxygen [8–11], premixing the hydrogen and oxygen [12], and use of an oxygen hyperventilated diffusion system which allows injection of up to 50 μl of the sample without extinction of the flame [13]. Flame-extinction by the solvent has also been overcome by use of dual-flame arrangements which permit automatic re-ignition of the emission flame [14].

The dual-flame detector designs have also been promoted for their ability to reduce emission quenching by hydrocarbons in the flame [15]. A schematic of a dual-flame photometric detector is shown in Fig. 3.3. In this detector the column effluent is introduced into a small hydrogen–oxygen/air flame; the decomposition products formed in this flame are then introduced into the analytical flame located downstream. The result of this arrangement is that all incoming sulphur compounds are decomposed to yield such reduced species as H_2S and S_2, and in addition, organic carbon is converted into CO_2 (a much weaker quencher), thus minimizing or eliminating the matrix effect. The emission results from the conversion of H_2S into

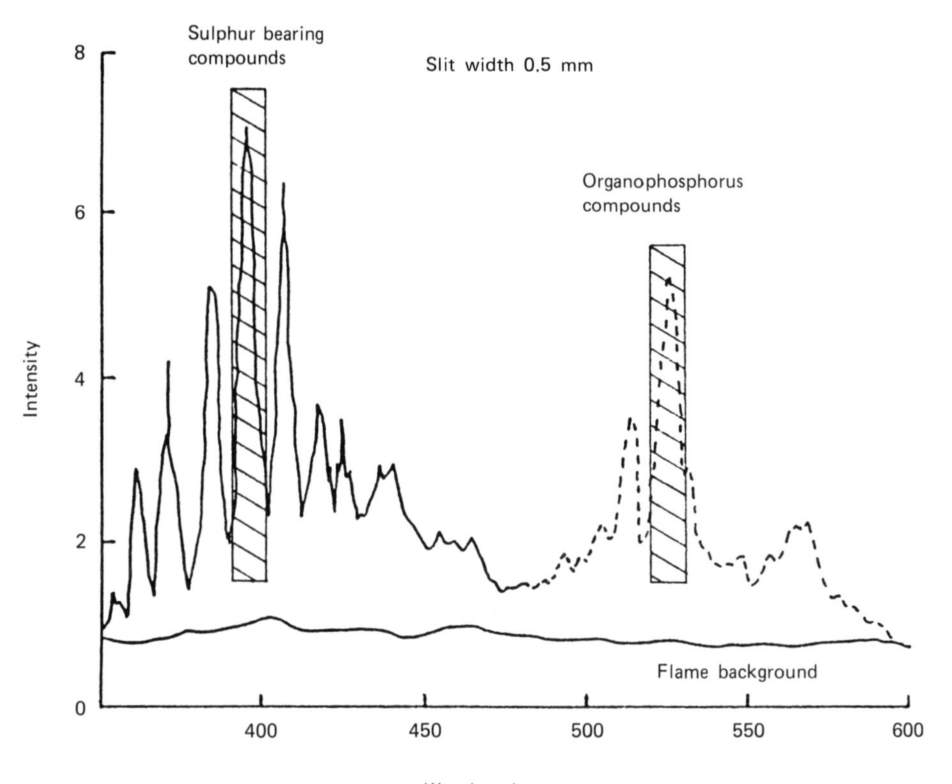

Fig. 3.2 — Emission bands obtained in FPD for sulphur and phosphorus organic compounds. Shaded areas represent commonly monitored regions. (Reproduced with permission, from S. S. Brody and J. E. Chaney, *J. Gas Chromatog.*, 1966, **4**, 42. Copyright 1966, Preston Publications Inc.)

excited S_2^* ($B^3\,\Sigma u$) which in return to its ground state ($B^3\,\Sigma g$) emits the characteristic bluish radiation.

$$\left.\begin{array}{l} \text{R–SH} \\ \text{R–S–R}_1 \\ \text{R–S–SR}_2 \end{array}\right. \xrightarrow{\text{Flame 1}} H_2S + S_2 + SO_2$$

$$2H_2S + 2H \xrightarrow{\text{Flame 2}} S_2^* + H_2$$

$$S_2 + 2H \xrightarrow{\text{Flame 2}} S_2^* + H_2$$

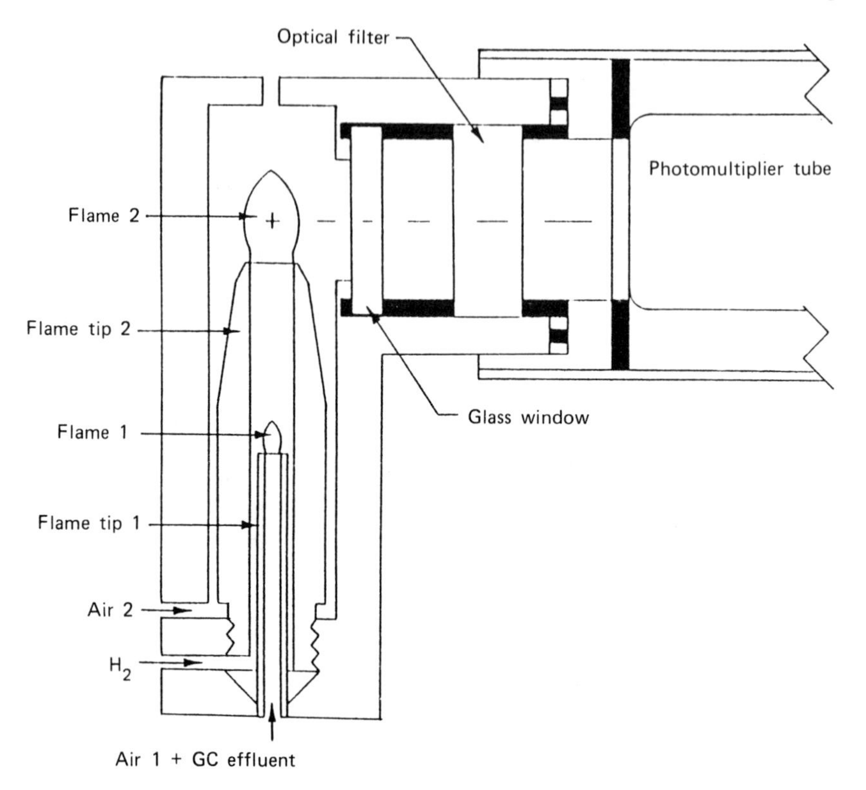

Fig. 3.3 — Schematic of the Varian dual-flame photometric detector. Adapted from [14].

The first flame in the detector with this geometry has aptly been described by Farwell and Barinaga as a "matrix-normalizing reactor" [16].

The response of FPD for sulphur compounds is generally considered to be proportional to the square of the sulphur concentration introduced into the detector. In an attempt to linearize this response, electronic circuits are routinely incorporated into the electrometers for FPD by the instrument manufacturers. These circuits give the square root of the detector signal as the output to strip-chart recorders or other data-collection devices. However, a number of researchers have shown that the FPD response for sulphur compounds frequently deviates from the square relationship and considerable error can be introduced by use of square-root linearizers [17, 18].

The anomalies associated with the response for sulphur compounds in FPD has been a subject of considerable interest since the introduction of this detector. However, only a limited discussion of the controversies and inconsistencies in this regard will be presented here. An exhaustive paper on the subject has recently been published by Farwell and Barinaga [16]. Drawing on the experimental data from laser fluorescence studies of flames by Muller and co-workers [19,20], and related thermodynamic and kinetic data [21,22], Farwell and Barinaga concluded that the discrepancies result from the fact that S_2 is not the only end-species formed on introduction of sulphur organics into the flame, and a number of other possible species ranging from H_2S, HS, S and SO to SO_2 are also formed. The relative concentrations of these sulphur species in the detector flame are determined by the kinetics of the reactions [16].

As discussed earlier, the simplest model for the sulphur response of the FPD is based on the assumption that most of the sulphur introduced into the detector is converted into S_2^* which is entirely responsible for the measured chemiluminesence emission intensity, *I*:

$$I \propto [S_2^*]$$

and further, if S_2^* is obtained by the reaction

$$S + S \longrightarrow S_2^*$$

then

$$I \propto [S]^2$$

However, it has been shown that in reality the intensity (*I*) is given by

$$I \propto [S]^n$$

Here *n* is the exponential proportionality constant or linearity factor [20–22], with values ranging from ~1 to 2. However, most commonly reported values fall between 1.8 and 2.2 [23–29]. The deviations from the square relationship have been attributed to a number of factors, including non-optimum flame conditions, incomplete or inconsistent compound decomposition, other competing flame reactions, quenching effects and non-Gaussian chromatographic peaks, sulphur background from impure gases and the detector housing, and the kinetics of various competing reactions occurring in the flame [16,30–33]. A number of methods for counteracting the non-linearity of the detector for sulphur detection have been proposed.

Attar *et al.* obtained a linearity factor of 2 for a number of sulphur compounds by using a two-parameter calibration based on peak area (*A*) and peak height (*H*). Other researchers have developed calibrations based on *H*/*W* or *A*/*H* where *W* is the width at half peak height [27,28, 34, 35]. Another procedure for calculation of *n* has been developed by Marriott and Cardwell [35a]:

$$n = \frac{\text{peak width determined with linear detector}}{\text{peak width obtained with non-linear detector}}$$

However, some ambiguities in calculating the value of *n* result from the uncertainties related to selecting the proper peak width in the non-linear detection.

A comprehensive quadratic response model for sulphur response has been developed by Ševčík [36] and by Farwell and Barinaga [16], a generalized form of which is

$$R = k_2[S]^2 + k_1[S] + k_0$$

In this expression, in addition to the contributions of classical S_2^* emission, contributions of other phenomena such as flame background emission, non-sulphur emission and/or quenching are taken into consideration. However, the effects of these contributions are difficult to assess in the routine operation of an FPD. Thus, in routine operation for sulphur determination, attention should be paid to optimizing the individual detector in terms of gas flow-rate, the most critical factor being the O_2/H_2 ratio. A ratio of 0.2–0.4 has often been suggested as a good starting range [37]; however, the gas flows and ratios should be determined and optimized experimentally. Farwell and Barinaga suggest that for best results a calibration must be made for each analyte of interest under set experimental conditions. When matrix conditions cannot be exactly matched, contributions from enhancement and quenching effects should be taken into consideration at low analyte levels [16]. Review of the available literature suggests that dual-flame detectors are somewhat more immune to changes in experimental conditions and provide more accurate results for sulphur-containing analytes.

The first observation of the green emission from phosphorus in the fuel-rich hydrogen/oxygen flame was reported by Salet about a hundred years ago [38,39]. The light-emitting species was identified as HPO* by Lam Thanh and Peyron in 1963 [40]. As shown earlier, in Fig. 3.2, the emission spectrum of HPO extends from 490 to 600 nm with maxima at 510, 526 and 560 nm. The relatively sharp band at 526 nm is most commonly used for phosphorus determination.

It is generally believed that introduction of phosphorus-containing compounds results in formation of PO, which according to Gilbert undergoes chemiluminescent reactions with or without a third body (M), leading to the formation of HPO* [41]:

$$PO + H + M \longrightarrow HPO^* + M$$

$$PO + OH + H_2 \longrightarrow HPO^* + H_2O$$

The detector exhibits very good sensitivity and linearity for phosphorus organics, as shown in Fig. 3.4. Detection limits (as a function of flow-rate) down to 0.1–0.5 pg/sec for phosphorus, with a linear dynamic range of approximately 10^5 have been reported [2]. A selectivity factor (with respect to carbon) of well over 4 orders of magnitude can easily be obtained. In this context an interesting observation made by Aue *et al.* [42] should be noted. These authors state that with shielded flames higher sensitivities with little or no loss of selectivity for phosphorus or sulphur relative to carbon can be obtained without the use of optical filters. This can be explained by the

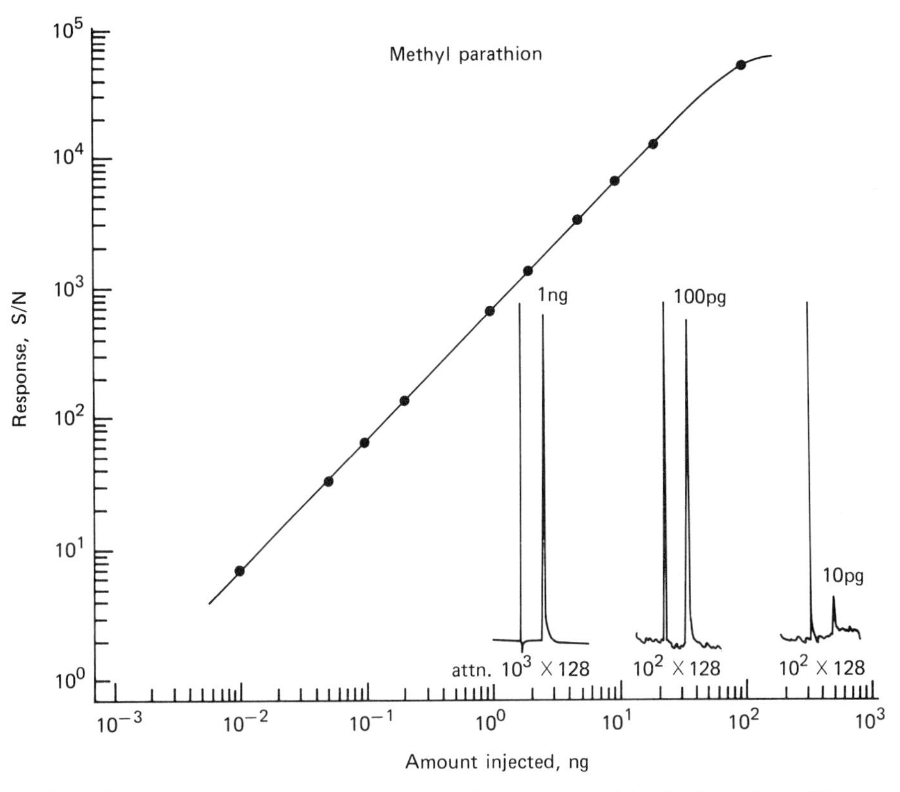

Fig. 3.4 — Calibration curve for methyl parathion, measured in the P mode at 525 nm. (Reproduced with permission, from S. Kapila and C. R. Vogt, *J. Chromatog. Sci.*, 1979, **17**, 327. Copyright 1979, Preston Publications Inc.)

fact that the carbon species formed in the fuel-rich H_2/O_2 flames are very weak emitters. However, interference filters are essential to distinguish between phosphorus, sulphur and other strongly emitting hetero-atom species. The selectivity of phosphorus relative to sulphur is variable and depends on the concentration of the sulphur compound. Owing to the extensive range of S_2^* emission bands, sulphur compounds at high concentrations can simulate lower concentrations of phosphorus compounds when the emission is monitored with a 526 nm filter. With large amounts ($>$200–400 ng) of sulphur as much as 25% of the emission seen at 526 nm may result from S_2^*.

The dependence of FPD response on the structure of organophosphorus compounds has been investigated by Sass and Parker. These authors concluded that the molar response of phosphorus was linear only within homologous series [43]. Similar results were obtained by Vogt and Kapila with a single-flame FPD; however, the phosphorus response of a dual-flame detector where primary degradation and emission regions were separated, was reported to be less dependent on compound type [44,45]. The results obtained are given in Fig. 3.5, which shows a set of calibration graphs obtained by using both a single- and a dual-flame FPD. In

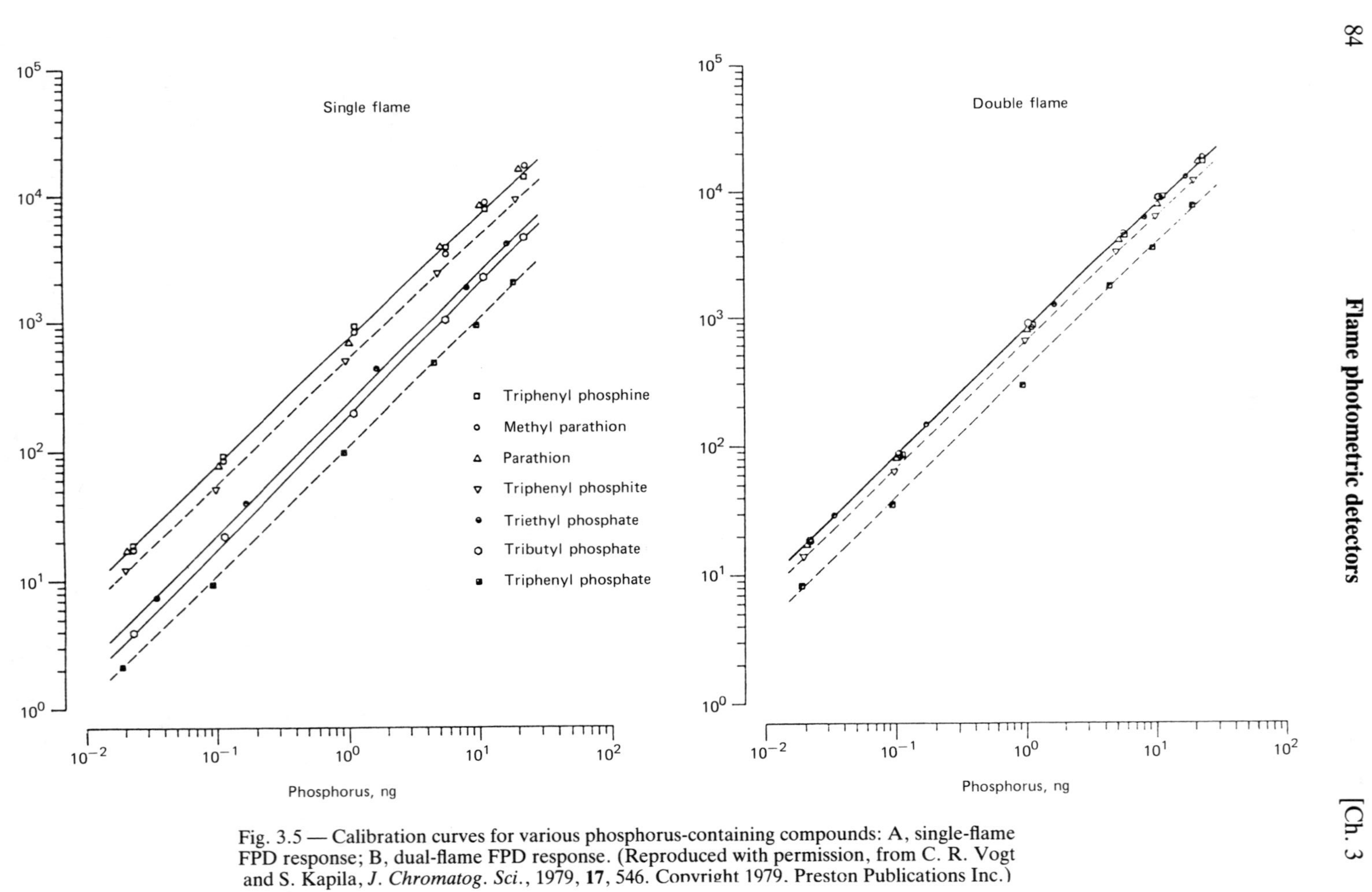

Fig. 3.5 — Calibration curves for various phosphorus-containing compounds: A, single-flame FPD response; B, dual-flame FPD response. (Reproduced with permission, from C. R. Vogt and S. Kapila, *J. Chromatog. Sci.*, 1979, **17**, 546. Copyright 1979. Preston Publications Inc.)

agreement with results reported by Patterson *et al.* [14], the phosphorus response of the dual-flame detector was less dependent on compound type than the response of the single-flame detector. The dependence of phosphorus response for phosphates, phosphites, phosphines and thiophosphates was attributed to the relative ease of converting these species into PO. Thus phosphates (P = O) (bond dissociation energies of 140 kcal/mole) are harder to degrade than phosphines (P − C) (bond energies of 63 kcal/mole) [46] and give a lower signal in single-flame detectors [15].

Overall, the FPD, with its good sensitivity and stability, has proved to be a very useful detector for trace determination of organophosphorus compounds and, contrary to opinions expressed by others, we believe that its overall performance in organophosphorus determination is superior to that of the more sensitive but temperamental thermionic detector.

3.3 RESPONSE CHARACTERISTICS FOR OTHER HETERO-ATOMS

3.3.1 Halogens

The determination of organochlorines continues to be of great environmental interest. Because of their high toxicities, lipophilicity and low degradation rates, organochlorine compounds have often topped the list of organics of environmental concern. The electron-capture detector (ECD), because of its exceptional sensitivity, continues to be the most commonly used detector for the determination of organochlorine compounds; however, its lack of specificity has plagued its use. Another successful detector for these compounds is the electrolytic conductivity detector. Its operation with capillary gas chromatography is, however, somewhat more cumbersome. Karmen and Haut demonstrated an interesting use of the FPD for detection of fluorine and chlorine compounds through the release of the sulphur and phosphorus species from CaS or $Ca(H_2PO_2)_2$ placed in proximity to the flame [47]. The applicability of FPD for determination of organochlorine compounds was explored by Tomkins and Frank, by an adaptation [48] of the classical Beilstein test. The burner assembly provided direct contact between the chlorine-containing molecules and hot metallic copper. The HCl formed by decomposition of the organochlorine compound interacted with copper surfaces to form CuCl, which on volatilization and subsequent decomposition introduced Cu atoms into the flame, resulting in the characteristic green emission (which is due to CuOH [48]). The concentration of copper introduced into the flame was reported to be directly dependent on the concentration of chlorine entering the flame.

An indium-coated copper burner assembly and air–hydrogen flame for chlorine determination was described by Gilbert [49], who reported that interaction of organochlorine compounds and indium in the burner leads to the release of InCl into the flame. The InCl emission at 360 nm can then be used for determination of the chlorine compound. A similar approach with an indium-sensitized chlorine-selective FPD has been reported by Wells [50]. A dual-flame geometry similar to the one described by Patterson *et al.* [14] was employed. The organochlorine compounds were decomposed in the first flame and the chlorine converted into InCl in the region between the flames. The InCl was then excited in the second flame and its emission monitored at around 360 nm with a 30-nm band-pass glass filter and a 1P28 photomultiplier tube. The distinctive feature of this design was that a porous ceramic cylinder with powdered indium dispersed in it served as the indium reservoir for the detector, a schematic of which is shown in Fig. 3.6. The use of a reservoir gave better

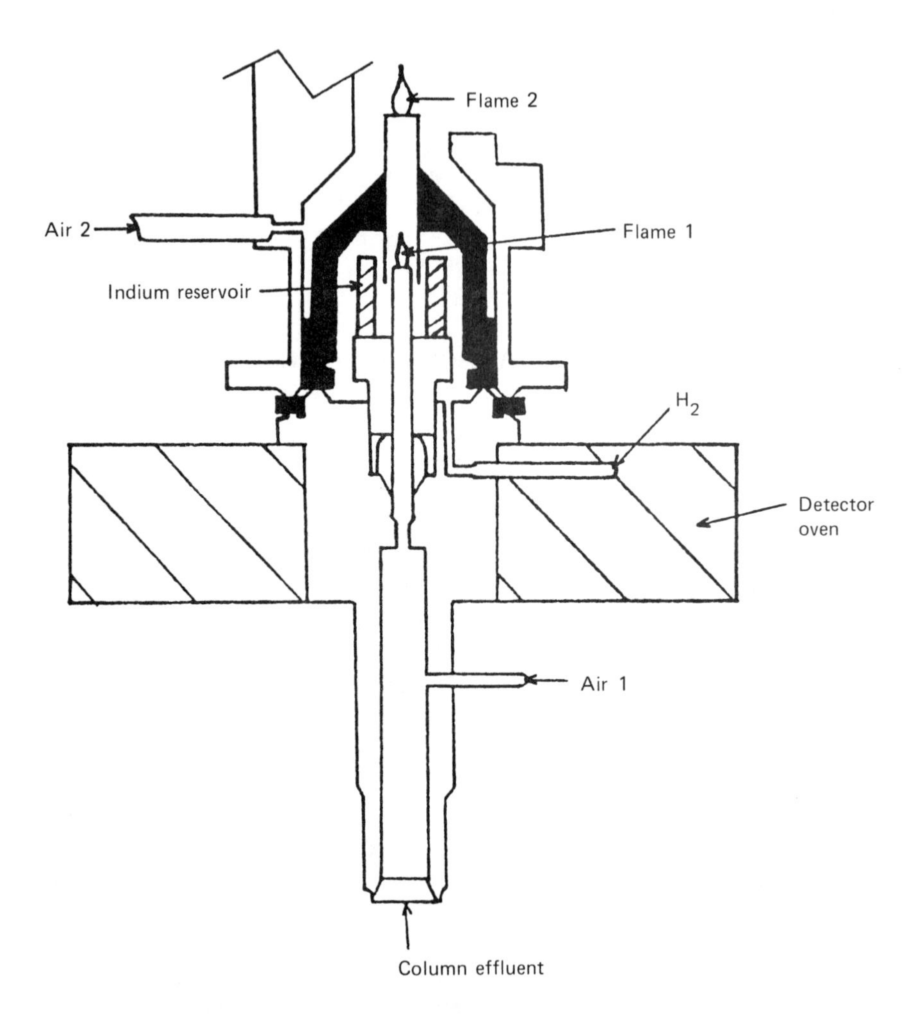

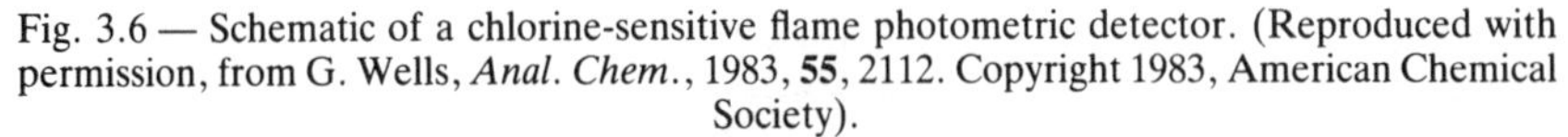
Fig. 3.6 — Schematic of a chlorine-sensitive flame photometric detector. (Reproduced with permission, from G. Wells, *Anal. Chem.*, 1983, **55**, 2112. Copyright 1983, American Chemical Society).

stability to the detector and a lifetime of approximately 2–3 months was reported for the reservoir. A lower detection limit (as a function of flow-rate through the detector) of 2.6 pg/sec was obtained for chlorine. The major sources of interferences and variability of response were phosphorus compounds, which were found to poison the catalytic surfaces required for facilitating the InCl formation. Recently an FPD design which incorporates catalytic conversion (with Pd or Ni) of organohalogen compounds into HX and subsequent reaction with indium and release of InX into the flame has been reported for determination of organic halides [51]. Despite these efforts, however, an FPD design with sensitivity and stability comparable to those of the electrolytic conductivity detector or the electron-capture detector has not been demonstrated.

3.3.2 Boron

The response for boron compounds in a Melpar FPD has been monitored by Sowinski and Suffet [52]. The green emission was monitored after passage through a 546 nm band-pass filter (half band-width 10 nm). BO* was identified as the emitting species. The optimum O_2/H_2 flow ratio was found to be 0.58; under the optimized conditions a lower detection limit of 0.72 ng for decaborane ($B_{10}H_{14}$) was reported. The potential interferences from phosphorus and/or sulphur compounds and selectivity relative to carbon were, however, not discussed. An innovative application of the FPD as an in-process monitor for phosphine (PH_3) and diborane (B_2H_6) in the semiconductor industry has been reported by Suzuki *et al.* [53].

3.3.3 Chromium

The use of the FPD for determination of Cr(III) in human urine by chromatographic separation as chromium(III) trifluoroacetylacetonate was reported by Ross and Shafik [54]. Emission from the chromium resonance line was monitored through a 425.4 nm filter. A signal to noise ratio of about 6 was obtained for 2.5 ng of chromium. Potential interferences from other elements and selectivity relative to carbon were not reported.

3.3.4 Tin and germanium

During the past few years the response characteristics of flame photometric detectors for tin, germanium, selenium and tellurium organic compounds has been explored by Aue and co-workers [42,54–61]. The application of the sharp SnH emission band in a cool hydrogen/air diffusion flame for the determination of inorganic tin was demonstrated by Dagnall *et al.*, and a minimum detectable limit of 1.5 ppm was obtained [62]. Braman and Tompkins reported detection limits down to 0.01 ng of Sn with an FPD of their own design [63].

The emission from organotin compounds in the small flame of a conventional flame photometric detector has been studied extensively by Aue and co-workers [54,56,59], who observed an intense blue emission in addition to the red emission observed by Dagnall *et al.* The blue emission is thought to result from surface reactions on clean silica surfaces, but the species responsible for it is unknown. The characteristics of tin emission in a flame photometric detector with and without silica flame enclosures have also been examined by Kapila and Vogt [64], who observed that the intensities of the blue and red emissions were directly dependent on the material of the flame enclosure used in the detector. The blue emission was observed when quartz and other silica surfaces were used. The differences in the observed emissions are shown in Fig. 3.7, which depicts the emission scan of tetra-n-butyltin with and without use of the quartz flame enclosures. These emission 'spectra' were obtained with a 0.25 m monochromator (band-pass 3 nm) by repeated injection of tetra-n-butyltin. As evident from these data, better absolute sensitivity can be obtained by monitoring the blue region, e.g. Aue and Flinn reported a minimum

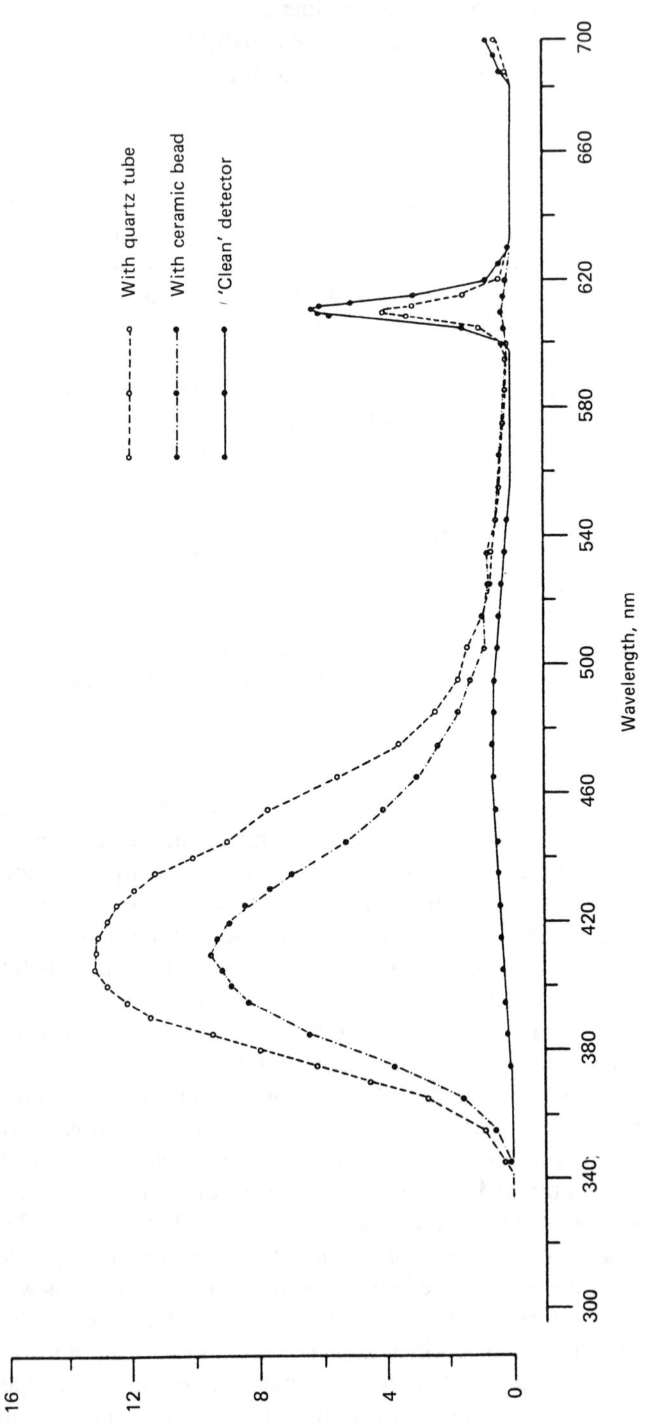

Fig. 3.7 — Emission spectra obtained by repeated introduction of organotin compounds into a flame photometric detector with different flame enclosures. (Reproduced with permission, from S. Kapila and C. R. Vogt, *J. Chromatog. Sci.*, 1980, **18**, 144. Copyright 1980, Preston Publications Inc.)

detection limit of 0.04 pg for tetrapropyltin, corresponding to 5 attomole of Sn per sec [61]. However, the dependence of this blue emission on the nature and history of the flame enclosure makes the detector response more temperamental than that for the red emission, centred at 610 nm. Excellent sensitivities down to 1 pg can be obtained with the use of red-sensitive multi-alkali metal photomultiplier tubes; furthermore, the distortion of peak shape is also minimal. The responses obtained in the two modes are shown in Fig. 3.8. The sensitivities compare very favourably with

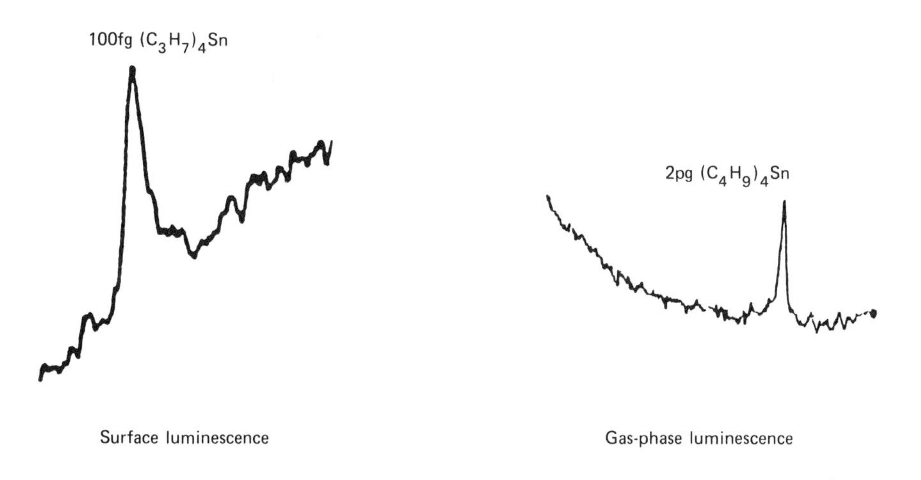

Fig. 3.8 — Response for organotin compounds, resulting from surface-mediated (A) and gas-phase (B) luminescence. (Reprinted by permission, from (A) [61], copyright 1980, American Chemical Society; (B) [64], copyright 1980, Preston Publications Inc.)

those obtained with techniques such as ICPES and AAS. The interferences in either mode result from sulphur, phosphorus and germanium species. Selectivities of $>10^3$ and $>10^2$ for Sn relative to S and P have been obtained. Application of a tin-selective FPD for the determination of triphenyltin hydroxide and its degradation products has been reported by Wright *et al.* [65].

On the basis of their experiences with organic tin compounds, Aue and Flinn investigated the flame photometric detection of a germanium compound. The emission characteristics for germanium compounds were found to be similar to those for tin compounds. The luminescence was observed in the gas phase and on a quartz surface in proximity to the flame. The features of the emission spectra obtained by repeated injections of tetra-n-butylgermanium in the two modes are given in Fig. 3.9. The surface-mediated blue emission was reported to be ten times more intense than the gas-phase red emission (at around 615 nm) attributed to GeH. Minimum detectable limits of 1.3 and 15 pg for tetra-n-butyl germanium were reported for surface and gas-phase emission respectively. However, the analytical applicability of the surface-mediated emission is severely limited by the strong interference from tin and sulphur compounds (at high concentrations) as well as the severe tailing imparted to the chromatographic peak by interactions with the quartz surface. It was

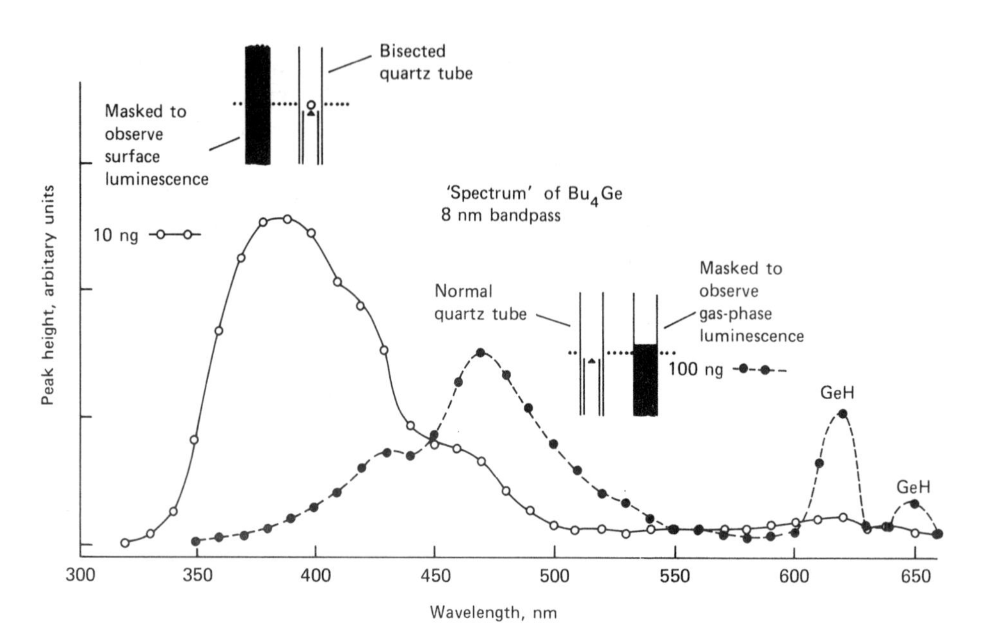

Fig. 3.9 — Emission 'spectra' of organogermanium compounds in FPD. Solid line represents luminescence obtained with bisected quartz enclosure. Dotted line represents luminescence in the gas phase, observed through a masked enclosure.

concluded by the authors that owing to these limitations the gas-phase GeH emission in the red region is better suited for analytical applications. A selectivity of better than two orders of magnitude was obtained by monitoring the red luminescence through a 650 nm band-pass filter and a red-sensitive photomultiplier tube. Phosphorus was found to be the major interfering element under these conditions.

3.3.5 Selenium and tellurium

The response characteristics of selenium and tellurium compounds with the flame photometric detector have been examined by Aue and Flinn [59,60]. The response for these two chalcogens was analogous to that for sulphur. Maximum intensity for Se was obtained with flow conditions similar to those required for sulphur, i.e. an O_2/H_2 ratio of 0.12–0.14. The emission scan obtained by repeated injections of piazselenol extended from 460 to 562 nm and is shown in Fig. 3.10. For analytical purposes a 484 nm filter with an 8.6 nm band-pass was employed. For all selenium compounds tested a quadratic response was obtained. This observation led the authors to conclude that by analogy with sulphur the emitting species is likely to be Se_2^*. A similar conclusion was drawn by Emeléus and Riley [66]. The selectivity for Se relative to carbon, and the quenching effect of carbon species, were both examined. It was reported that the Se emission is quenched by carbon compounds, but the effect is less severe than for sulphur, and under certain conditions can be used to differentiate between sulphur and selenium compounds. Under optimized conditions a minimum detectable amount of 2 pg/sec was obtained for Se. In an

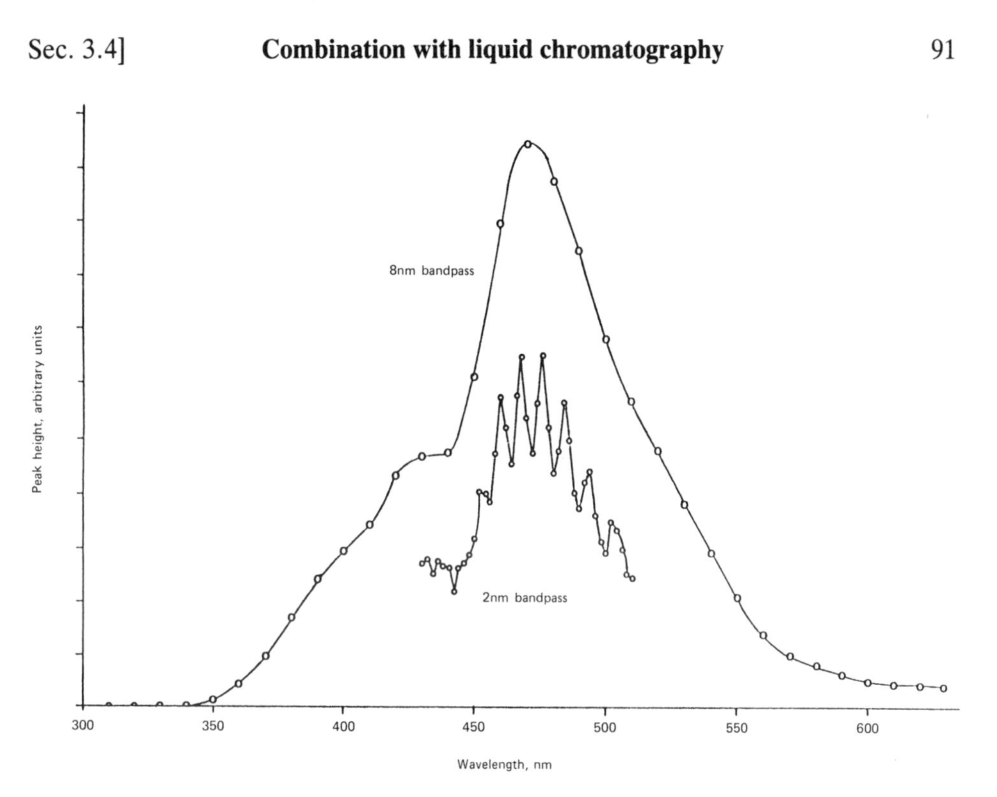

Fig. 3.10 — Emission 'spectra' obtained by repeated injection of an organoselenium compound.

interesting article, Aue and Flinn demonstrated that a linearized response can be calculated by means of the expression [60]:

$$R_s = k_s^2 S_a^2 + 2k_s' S_a k_s S_b$$

where R_s = total response, S_a = sulphur, selenium or tellurium analyte concentration, S_b = sulphur background concentration, k_s and k_s' are proportionality constants. It was speculated that the mutual enhancement resulted from the formation of mixed chalcogen species such as SeS and TeS. The relative sensitivities for S, Se and Te compounds with the FPD were reported to be 1, 0.1 and 0.01, respectively. The FPD sensitivity for selenium compounds, while appreciably lower than that of graphite-furnace AAS, compares well with that obtained with ICPES [67].

3.4 COMBINATION WITH LIQUID CHROMATOGRAPHY

The combination of flame emission spectrometers and ion-exchange chromatography has had an obvious appeal for the determination of such elements as Na, K, Ca, Sr and Ba [68]. Liquid chromatography offers an easier means of separating thermally labile, polar and ionic compounds. However, successful combinations of the traditional FPD (with its small flame) and liquid chromatography (with 1–2 ml/min liquid flow-rates) have been hard to achieve. Various approaches have been explored to alleviate the incompatibility problems arising from flame instability and the quenching effects of carbon species, alluded to earlier. Early approaches to achieving an LC–FPD combination were based on the use of mass transport systems such as moving wires, porous belts or rotating disks [69–71]. The other approaches have been based on the use of nebulizers. An application of dual-flame FPD and a nebulizer has been reported by Chester [72]. In this system, column effluent was introduced into the secondary flame of the FPD by a nebulizer. The combustion products and excess of hydrogen were then introduced into the emission flame. The response characteristics for a number of polar and ionic phosphorus compounds were investigated. A detection limit down to 0.01 mg/l. for phosphorus was obtained, with a linear dynamic range of 5×10^4. The detector design gave satisfactory performance with mixed aqueous phases, but the performance for normal-phase liquid chromatography with mobile phases such as hexane was quite erratic.

The introduction of microbore LC has reduced many of the problems associated with the LC–FPD combination. A low dead-volume FPD suitable for use with packed microbore LC was described by McGuffin and Novotny. The total column effluent at a rate of ~ 1 µl/min was aspirated into the diffusion flame of the FPD. Chromatographic separation and flame photometric determination of thiophosphate pesticides and dimethylthiophosphate derivatives was illustrated. A detection limit of 0.07 ng/sec was obtained for phosphorus [73,74].

An ultrasonic micronebulizer for use with the LC–FPD combination has recently been described by Karnicky *et al.* [75]. It allows effective nebulization of the mobile phase at 2–20 µl/min flow-rate, and the aerosol from the nebulizer is introduced into a modified dual-flame FPD. Compatibility of various mobile phases with the FPD was examined, and acetonitrile and halogenated solvents were found to be unsatisfactory. To minimize flame overloading problems a solvent condenser was employed, but this resulted in loss of up to 50% of the sample. A schematic of the micronebulizer assembly is shown in Fig. 3.11. A dual-wavelength approach [simultaneous monitoring of phosphates (HPO) and potassium emission] was employed to reduce noise resulting from fluctuation in nebulization efficiency. A 5-fold increase in signal to noise (S/N) ratio was obtained. Detection limits down to 0.05 ng/sec for phosphorus were reported.

An application of a chlorine-selective FPD with microcolumn LC has been reported by Folestad *et al.* [76]. In this system the entire column effluent (20–70 µl/min) was introduced into a heated tube filled with silica wool, and the pyrolysis products, including HCl, were made to react with indium, resulting in the formation of InCl. The InCl emission at around 360 nm was monitored through an interference filter. A detection limit of 9 pg/sec ($S/N = 3$) was reported for 1,1,2-trichloroethane in water. Application of the system for determination of therapeutic drugs and chlorinated bases was demonstrated.

The detection limits obtained with an LC–FPD system are generally two orders of magnitude higher than those obtained with GC–FPD. These poorer sensitivities have generally been attributed to inefficiency of the interfacing, and to quenching, both of which are strongly influenced by the flow-rate and composition of the mobile phase.

The use of the FPD with supercritical fluid chromatography has recently been

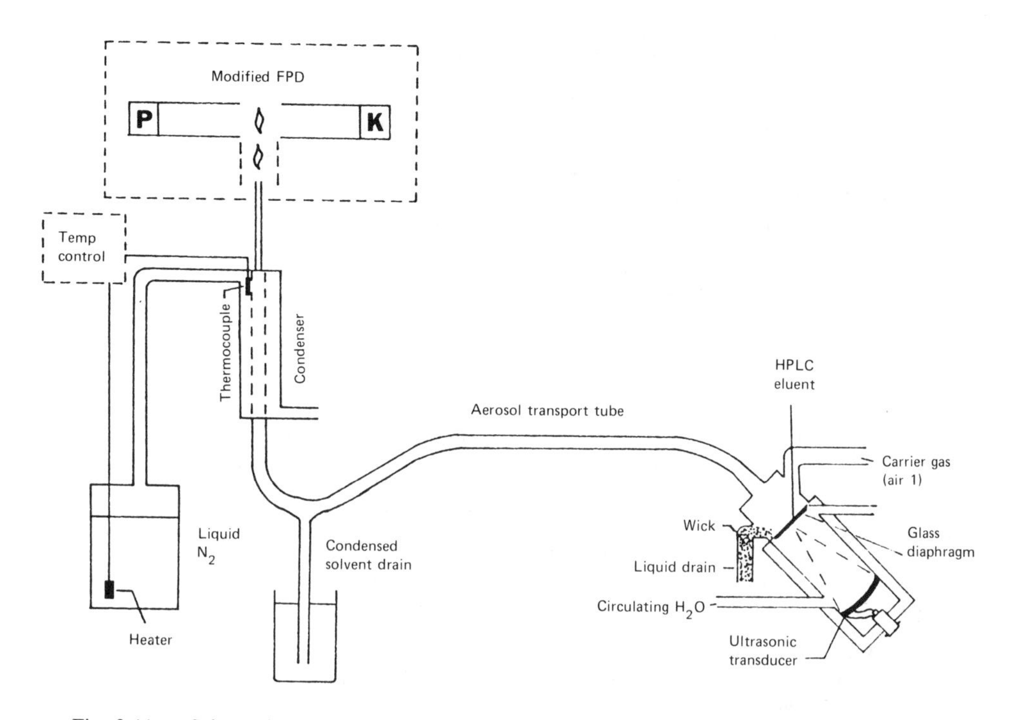

Fig. 3.11 — Schematic of micronebulizer-based LC–FPD interface. (Reproduced with permission, from J. F. Karnicky, L. T. Zitelli and Sj. van der Wal, *Anal. Chem.*, 1987, **59**, 327. Copyright 1987, American Chemical Society.)

demonstrated by Markides *et al.* [77]. A dual-flame FPD was modified by addition of a make-up gas line to improve the transfer of sample to the flame. A detection limit of 25 ng for benzo[*b*]thiophene ($S/N=2$) in the sulphur mode (365 nm) and 0.5 ng for malathione ($S/N=2$) in the phosphorus mode (530 nm) were reported. With the increased interest in supercritical fluid chromatography (SFC), an increase in application of SCF–FPD to the determination of polar and thermally labile S, P and Sn compounds can confidently be expected.

Acknowledgements — The authors wish to thank the University of Missouri, and Dr. Armon Yanders in particular, for allowing the use of Environmental Trace Substances Research Center facilities. A special acknowledgement is made to Dr. Walter Aue for his insightful comments. Thanks are also extended to Ms. Kathy Shaw for competent secretarial assistance.

REFERENCES

[1] H. Draegerwerk (H. and B. Draeger), *FGR Patent* 1133918, 26 July 1962; *Chem. Abstr.*, 1962, **57**, 14443a.
[2] S. S. Brody and J. E. Chaney, *J. Gas Chromatog.*, 1966, **4**, 42.
[3] M. C. Bowman and M. Beroza, *Anal. Chem.*, 1968, **40**, 1448.
[4] W. E. Dole and C. C. Hughes, *J. Gas Chromatog.*, 1968, **6**, 603.
[5] R. Pigliucci, W. Averill, J. E. Purcell and L. S. Ettre, *Chromatographia*, 1975, **8**, 165.
[6] W. P. Cochrane and R. Greenhalgh, *Chromatographia*, 1976, **9**, 255.
[7] R. R. Watts, *J. Assoc. Off. Anal. Chem.*, 1970, **53**, 787.
[8] C. A. Burgett and L. E. Green, *J. Chromatog. Sci.*, 1974, **12**, 356.

[9] S. Hasinski, *Chem. Anal. (Warsaw)*, 1975, **20**, 1135.
[10] V. Joonson and E. Loog, *J. Chromatog.*, 1976, **120**, 285.
[11] S. Hansinski, *J. Chromatog.*, 1976, **119**, 207.
[12] R. B. DeLew, U. S. Patent 3955914, 11 May 1976; *Chem. Abstr.*, 1976, **85**, 86799g.
[13] C. A. Burgett and L. E. Green, *Spectrochim. Acta*, 1975, **30B**, 55.
[14] P. L. Patterson, R. L. Howe and A. Abu-Shumays, *Anal. Chem.*, 1978, **50**, 339.
[15] P. L. Patterson, *Anal. Chem.*, 1978, **50**, 345.
[16] S. O. Farwell and C. J. Barinaga, *J. Chromatog. Sci.*, 1986, **24**, 483.
[17] C. H. Burnett, D. F. Adams and S. O. Farwell, *J. Chromatog. Sci.*, 1977, **15**, 230.
[18] S. O. Farwell, D. R. Gage and R. A. Kagel, *J. Chromatog. Sci.*, 1981, **19**, 358.
[19] C. H. Muller III, K. Schofield, M. Steinberg and H. P. Broida, *Int. Symp. Combust.*, 1979, **17**, 867.
[20] C. H. Muller III, K. Schofield and M. Steinberg, in *Laser Probes for Combustion Chemistry*, D. R. Crosby (ed.), p. 103. American Chemical Society, Washington, D.C. 1980.
[21] T. M. Sugden, E. M. Bulewicz and A. Demerdache, *Chem. Reactions Lower and Upper Atmosphere, Proc. Intern. Symp. San Francisco, Cal.*, 1961, 89; *Chem. Abstr.*, 1962, **56**, 13583.
[22] C. F. Cullis and M. F. R. Mulcahy, *Combust. Flame*, 1972, **18**, 225.
[23] A. I. Mizany, *J. Chromatog. Sci.*, 1977, **8**, 151.
[24] T. Sugiyama, Y. Suzuki and T. Takeucki, *J. Chromatog.*, 1973, **77**, 309.
[25] J. G. Eckhardt, M. B. Denton and J. L. Moyers, *J. Chromatog. Sci.*, 1975, **13**, 133.
[26] C. H. Burnett, D. F. Adams and S. O. Farwell, *J. Chromatog. Sci.*, 1978, **16**, 68.
[27] J. F. McGaughey and S. K. Gangwal, *Anal. Chem.*, 1980, **52**, 2079.
[28] S. K. Gangwal and D. E. Wagoner, *J. Chromatog. Sci.*, 1979, **17**, 196.
[29] C. F. Quincoces and M. G. Gonzalez, *Chromatographia*, 1985, **20**, 371.
[30] M. L. Selucky, *Chromatographia*, 1971, **4**, 425.
[31] H. A. Grieco and W. M. Hans, *Ind. Res.*, 1974, **16**, 39.
[32] J. Ševčík and J. E. Lips, *Chromatographia*, 1979, **12**, 693.
[33] J. M. Raccio and B. Welton, *Instrum. Res.*, 1985, September, 30.
[34] A. Attar, R. Forgey, J. Horn and W. H. Corcoran, *J. Chromatog. Sci.*, 1977, **15**, 222.
[35] M. Maruyama and M. Kakemoto, *J. Chromatog. Sci.*, 1978, **16**, 1.
[35a] P. J. Marriott and T. J. Cardwell, *J. Chromatog.*, 1982, **234**, 157.
[36] J. Ševčík, *Int. J. Environ. Anal. Chem.*, 1983, **13**, 115.
[37] P. L. Patterson, *Standard Practice for using Flame Photometric Detectors in Gas Chromatography*, ASTM Standard E840-81.
[38] G. Salet, *Bull. Soc. Chim. France*, 1869, **11**, 302 (as cited in [41]).
[39] G. Salet, *Ann. Chim. Phys.*, 1873, **28**, 5 (as cited in [41]).
[40] M. Lam Thanh and M. Peyron, *J. Chim. Phys.*, 1963, **60**, 1289.
[41] P. T. Gilbert, *Nonmetals*, in *Analytical Flame Spectroscopy*, R. Mavrodineanu (ed.), Macmillan, London, 1970.
[42] W. A. Aue and C. R. Hastings, *J. Chromatog.*, 1973, **87**, 232.
[43] S. Sass and G. Parker, *J. Chromatog.*, 1980, **189**, 331.
[44] C. R. Vogt and S. Kapila, *J. Chromatog. Sci.*, 1979, **17**, 546.
[45] S. Kapila and C. R. Vogt, *J. Chromatog. Sci.*, 1979, **17**, 327.
[46] R. Belcher, S. L. Bogdanski, O. Osibanjo and A. Townshend, *Anal. Chim. Acta*, 1976, **84**, 1.
[47] A. Karmen and H. Haut, *J. Chromatog.*, 1974, **99**, 349.
[48] D. F. Tomkins and C. W. Frank, *Anal. Chem.*, 1974, **46**, 1187.
[49] P. T. Gilbert, *Anal. Chem.*, 1966, **38**, 1960.
[50] G. Wells, *Anal. Chem.*, 1983, **55**, 2112.
[51] Shimadzu Corp., *Japanese Patent* JP59/121314, 21 March 1984.
[52] E. J. Sowinski and I. H. Suffet, *J. Chromatog. Sci.*, 1971, **9**, 632.
[53] T. Suzuki, Y. Inoue, M. Ura, O. Mitsuru and Y. Sugita, *J. Cryst. Growth*, 1978, **45**, 108.
[54] R. Ross and T. Shafik, *J. Chromatog. Sci.*, 1973, **11**, 46.
[55] W. A. Aue and C. G. Flinn, *J. Chromatog.*, 1977, **142**, 145.
[56] C. G. Flinn and W. A. Aue, *J. Chromatog. Sci.*, 1980, **18**, 136.
[57] C. G. Flinn and W. A. Aue, *Can. J. Spectrosc.*, 1980, **25**, 141.
[58] C. G. Flinn and W. A. Aue, *J. Chromatog.*, 1979, **186**, 299.
[59] C. G. Flinn and W. A. Aue, *J. Chromatog.*, 1978, **153**, 49.
[60] W. A. Aue and C. G. Flinn, *J. Chromatog.*, 1978, **158**, 161.
[61] W. A. Aue and C. G. Flinn, *Anal. Chem.*, 1980, **52**, 1537.
[62] R. M. Dagnall, K. C. Thompson and T. S. West, *Analyst*, 1968, **93**, 518.
[63] R. S. Braman and M. A. Tompkins, *Anal. Chem.*, 1979, **51**, 12.
[64] S. Kapila and C. R. Vogt, *J. Chromatog. Sci.*, 1980, **18**, 144.
[65] B. W. Wright, M. L. Lee and G. M. Booth, *J. High. Resol. Chromatog. Commun.*, 1979, **2**, 189.

[66] H. J. Emeléus and H. L. Riley, *Proc. Roy. Soc.*, 1933, **140A**, 378 (as cited in [41]).
[67] D. O. Duebelbeis, S. Kapila, D. E. Yates and S. E. Manahan, *J. Chromatog.*, 1986, **351**, 465.
[68] D. J. Freed, *Anal. Chem.*, 1975, **47**, 186.
[69] R. P. W. Scott, *Contemporary Liquid Chromatography*, p. 158. Wiley, New York, 1976.
[70] J. B. Dixon and R. C. Hall, *U.S. Patent* US4271022, 2 June 1981; *Chem.* 1981.
[71] Shimadzu Seisakusho Ltd., *Japanese Patent* JP 57/45334, 27 Sept. 1982; *Chem. Abstr.*, 1983, **98,** 13684k.
[72] T. L. Chester, *Anal. Chem.*, 1980, **52**, 1621.
[73] V. L. McGuffin and M. Novotny, *J. Chromatog.*, 1981, **218**, 179.
[74] V. L. McGuffin and M. Novotny, *Anal. Chem.*, 1981, **53**, 946.
[75] J. F. Karnicky, L. T. Zitelli, and Sj. van der Wal, *Anal. Chem.*, 1987, **59**, 327.
[76] S. Folestad, B. Josefsson and P. Marstorp, *Anal. Chem.*, 1987, **59**, 334.
[77] K. E. Markides, E. D. Lee, R. Bolick and M. L. Lee, *Anal. Chem.*, 1986, **58**, 740.

4

Chromatographic detection by atomic plasma emission spectroscopy

Peter C. Uden
Department of Chemistry, Lederle Graduate Research Tower, University of Massachusetts, Amherst, MA 01003, U.S.A.

4.1 ELEMENT-SELECTIVE DETECTION IN CHROMATOGRAPHY

Chromatography must involve a method for qualitative and quantitative detection and characterization of resolved components, a measurement device being placed at the end of the column to respond immediately and predictably to the presence of solute in the mobile phase. The first group of detectors, 'the bulk property detectors', are those which respond to changes produced by solutes in a characteristic mobile phase property. The second group, 'the solute property detectors', can measure a property of the solutes directly.

Detectors may be 'Universal', 'Selective', or 'Specific'. Universal detection of all solutes is not always a desirable goal since it is often necessary to discriminate particular components which are present in a great excess of unresolved matrix species. Such situations are common in environmental analysis where 'trace determinations' are the objective. It may be desirable to analyse for a particular class of analyte ignoring other sample compounds. For true universality, a detector must respond to all solutes but differently to each. Detectors based on techniques such as mass spectrometry and IR spectrophotometry display universality since all solute molecules produce mass or infrared spectra. All solute property detectors are 'selective' to some extent. If the property varies greatly in magnitude for different analytes, selectivity can be high for target compounds but high levels of 'non-selected' co-eluted species will give an interfering response.

Detectors may be 'element-selective', 'structure or functionality selective' or 'property-selective'. *Specific* detectors exhibit a very high degree of selectivity. There are two major objectives of element-selective chromatographic detection. In addition to qualitative and quantitative detection of analytes in interfering background matrices, simultaneous multielement detection can enable the empirical formulae of analytes to be determined. Element-selective detectors in common use in gas chromatography (GC) include the alkali-metal flame ionization detector (AFID), often known as the nitrogen/phosphorus detector (NPD) since it is selective for these elements, the flame photometric detector (FPD) (see Chapter 3) selective for sulphur and phosphorus, and the Hall electrolytic conductivity detector (HECD) which is selective for halogen, nitrogen and sulphur. The wide use of these detectors emphasizes the value of element-selective detection but they are too limited for general empirical formula determinations. In HPLC, no element-specific detectors are in general use.

It is clear that general multielement chromatographic detection is a worthwhile technical objective to complement the molecular and structural specific detection afforded by combination of mass spectrometry (MS) and Fourier transform infrared spectrometry (FTIR). Atomic emission spectroscopy is a natural choice for use in such detection in view of its inherent capacity to monitor all elements. The 'rebirth' of analytical atomic emission techniques during the past decade has re-focused the efforts of chromatographers to employ its capacity in on-line detection.

4.2 ATOMIC EMISSION SPECTROSCOPIC DETECTION IN CHROMATOGRAPHY (AESD)

Spectroscopic monitoring is among the most powerful instrumental analytical techniques used in chromatography. In GC, MS and FTIR detection are widely used and in HPLC, UV absorption and fluorescence are the most common detection modes. Three types of atomic spectrometry have been used in GC and HPLC detection: atomic absorption (AAS) (see Chapter 2), flame emission (FES) (see Chapter 3) and atomic plasma emission (APES). In contrast to AAS, APES has the advantage of multielement capability while maintaining a wide dynamic measurement range. The availability of various plasma sources, particularly in combination with high-resolution monochromators to minimize spectral interferences, has produced a resurgence of analytical application, not least in chromatographic detection.

The three principal plasma sources used in GC detection have been the inductively-coupled argon plasma (ICP), the DC argon plasma (DCP) and the microwave-induced helium plasma, operated at atmospheric or reduced pressure (MIP). The first two of these plasmas have been used effectively as HPLC detectors. The major advantages of combining chromatography with APES include (a) the ability to monitor for many metals and non-metals directly or by derivatization, (b) the ability to tolerate non-ideal elution, the specificity of plasma emission enabling incomplete chromatographic resolution from complex matrixes to be overcome, (c) high elemental sensitivity, (d) multielement detection, and (e) compatability with different chromatographic systems.

4.3 CLASSES OF ATOMIC PLASMA EMISSION DETECTORS

In emission spectroscopy an excitation source transforms a sample from a solid, liquid or gas into a plasma of atoms, ions and molecular radicals which can undergo electron excitation. When these excited states return to the lower energy states they generate light quanta which form an emission spectrum in a suitable spectrometer. Though DC and AC arcs and sparks played a major role in the development of emission spectroscopy, plasma sources now predominate.

4.3.1 The inductively-coupled plasma (ICP) discharge

The ICP was first developed in the 1960s [1] and is now the most widely used spectrochemical source. The ICP discharge is caused by the interaction of a radiofrequency field with a flowing gas, usually argon. The gas flows through a silica tube within a copper coil or solenoid energized by a radiofrequency generator, usually operating at 27 or 41 MHz. This creates a varying magnetic field in the flowing argon, which in turn generates a circulating eddy current in the gas which is initially heated to make it electrically conductive. A very hot, spectrally intense stable plasma discharge is then produced with temperatures up to 10^4 K. The configuration of an ICP torch is shown in Fig. 4.1. Although samples can be introduced as gases, liquids or powdered solids, the most usual arrangement is to employ a spray-chamber nebulizer similar to that used in flame spectrometry to generate an aerosol from a liquid sample stream; this aerosol is carried by the argon flow into the discharge, where the solvent is evaporated and the analyte atomized. Since the ICP discharge is extremely hot and in an inert atmosphere, some problems found in other atomic emission methods are minimized. All compounds should be completely atomized and chemical and molecular interferences should be negligible.

As discussed later, the ICP discharge is a natural complement for liquid chromatographic detection since it is normally designed for use with a liquid inlet stream; HPLC–ICP has in fact attracted considerable attention. As a GC detector it has gained less prominence but may be used successfully, particularly for those elements, primarily metals, for which the sensitivity is high in the argon plasma discharge.

4.3.2 The direct-current plasma jet (DCP) discharge

DC arcs are an established electrical excitation source for atomic emission, consisting typically of a low-voltage (10–50 V), high-current (1–35 A) discharge between a sample electrode and a counter-electrode. The DC arc gives high efficiency of atomization and excitation and is usually used with a solid analyte incorporated into the sample electrode. It generates a line-rich spectrum which makes it valuable for multielement qualitative and quantitative analysis; the spectra are considerably different from those seen in flames because highly populated higher atomic states add complexity to the emission. Application of DC arcs for liquid samples is limited, but a different arrangement, the direct-current plasma jet (DCP) facilitates it [2]. The DCP is a discharge maintained by a DC arc and stabilized by flowing inert gas. In the commercial version which has been favoured for chromatographic use, a cathode jet is placed above two symmetrically placed anode jets in an inverted 'Y' configuration (Fig. 4.2). Flowing argon causes vortexes around the anodes and a 'thermal pinch' gives an arc column of high current density and temperature. Solutions are introduced from nebulizer spray chambers upwards into the junction of the two columns, where analyte spectral emission is observed. The system has been applied in both HPLC and GC, with an elemental range paralleling that of the ICP.

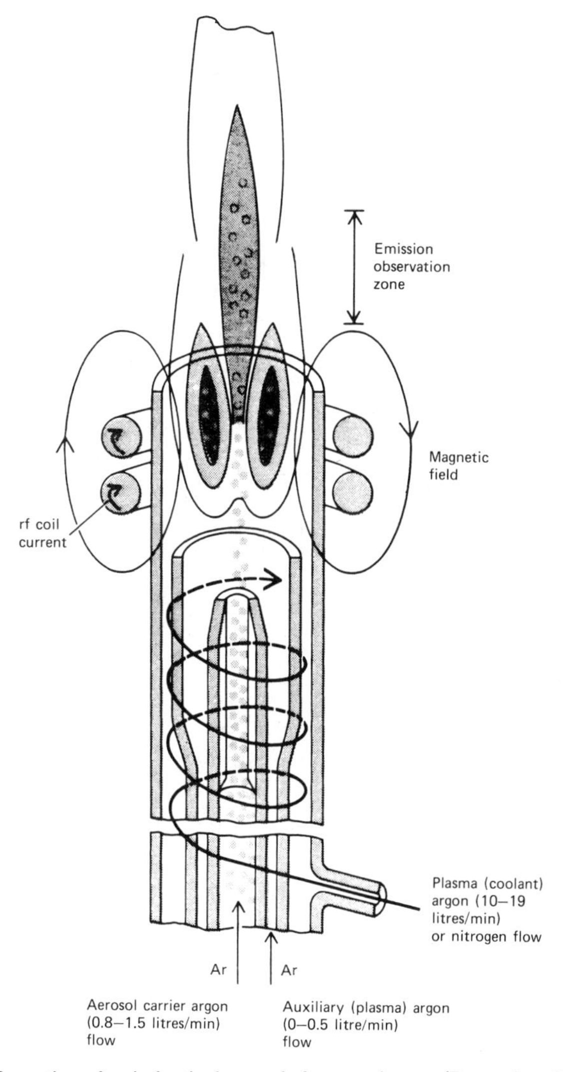

Fig. 4.1 — Configuration of an inductively-coupled argon plasma. (Reproduced by permission, from R. M. Barnes, *CRC Crit. Rev. Anal. Chem.*, 1978, **7**, 203. Copyright 1978, CRC Inc.).

4.3.3 The microwave-induced electrical discharge plasma (MIP) detector

The foremost atom reservoir systems which have been used for gas chromatographic detection have been microwave-induced electrical discharge plasmas (MIP). An argon or helium plasma is maintained within a microwave cavity which serves to focus or couple power from a microwave source, usually operated at 2.45 GHz, into a discharge cell, which is frequently a silica capillary tube. Such microwave plasmas may be operated at atmospheric or reduced pressures, depending on the cavity type [3,4]. Operational power levels for analytical microwave plasmas are considerably

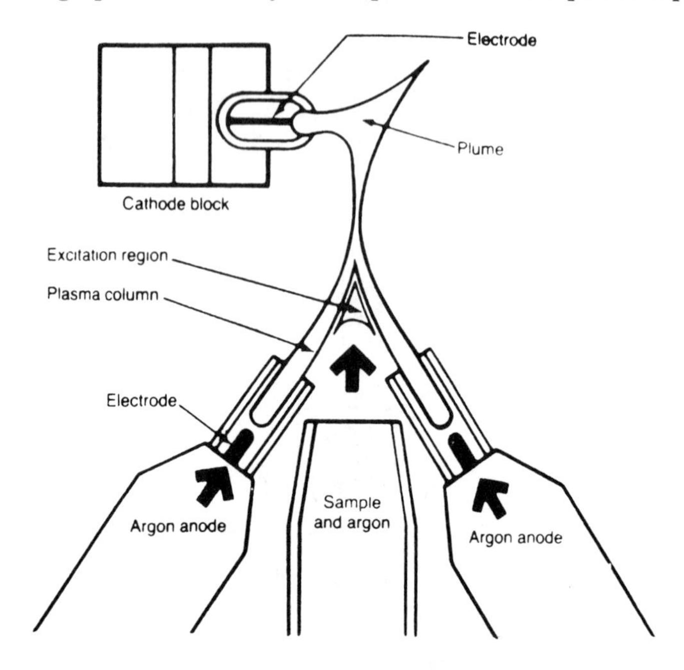

Fig. 4.2 — Configuration of the three-electrode DC argon plasma.

lower (ca. 50–100 W) than for DCP (ca. 500 W) or ICP (1–5 kW) plasmas, making their operation less complex. The power densities are more similar, however, owing to the more compact form of the microwave cavities. Although plasma temperatures are lower in the MIP, high electron temperatures are available, particularly in the helium plasmas, giving high spectral emission intensities for many elements, including many non-metals which respond poorly in the argon ICP or DCP. MIP systems have been found less useful in general for liquid introduction, however, since the plasma usually has insufficient enthalpy to desolvate and vaporize aerosols effectively.

The efficiency of a microwave power source depends on the choice of discharge cavities and waveguides. The latter are metal tubes of various cross-sections which transfer power from a microwave generator to the plasma support gas. If an interruption is made in the waveguide to cause total reflection of energy travelling along it, then standing waves are set up and a resonant cavity is formed. The impedance of the cavity/waveguide structure must be matched to that of the coaxial line of the power supply by use of a coupling device, the efficiency of which is measured in terms of minimization of reflected power. A comparison of various microwave cavities has been made by Risby and Talmi [5], who gave a good overview of MIP characteristics, commenting that the various conditions used for different evaluations have contributed to conflicting conclusions on experimental usefulness. The most widely used cavity for low-pressure operation of helium or argon plasmas has been the 3/4-wave cavity described by Evenson [6] and many of its GC applications are discussed later.

The cavity which has received the greatest study in recent years has been the TM_{010} cylindrical resonance cavity developed by Beenakker [7] and modifications of its original design (Fig. 4.3). This cavity can sustain both argon and helium

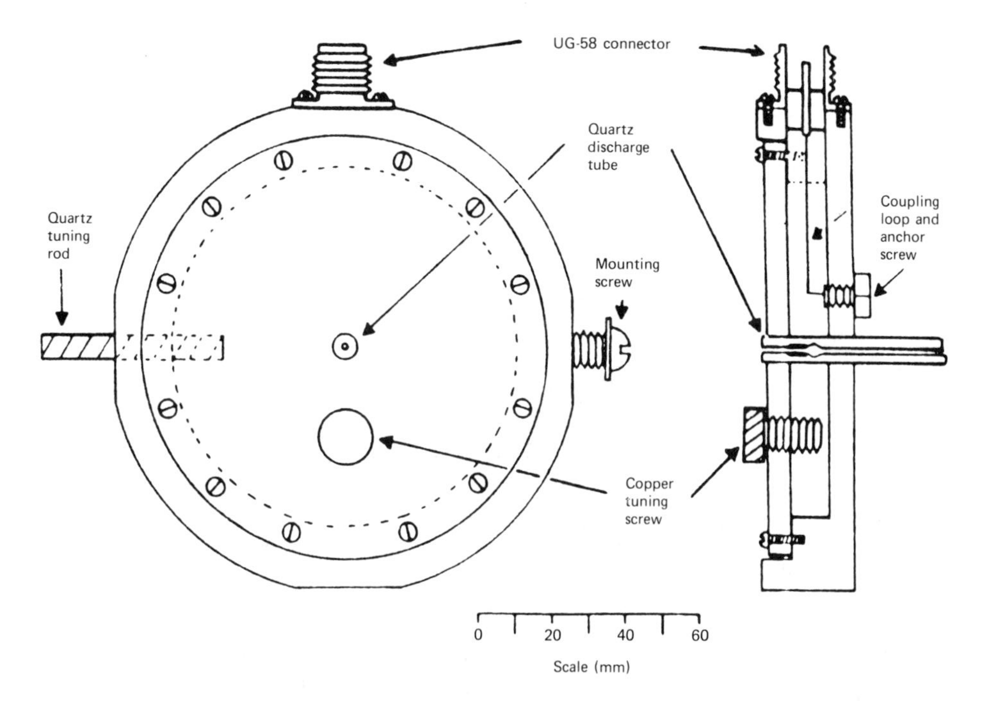

Fig. 4.3 — The Beenakker microwave plasma cavity. (Reproduced by permission, from *J. Chromatog.*, 1982, **239**, 181. Copyright 1982, Elsevier Science Publishers).

atmospheric pressure discharges at low microwave power levels. An advantageous feature of this cavity is that the light emitted can be viewed axially, in contrast to 'transverse viewing', in which spectral monitoring must be made through the cavity walls (the properties of which change with time). With axial viewing, the discharge tube can be made from opaque materials such as alumina or boron nitride, which have been used with advantage in GC–MIP. The advantages inherent in atmospheric-pressure operation greatly simplify GC detection. Another class of atmospheric-pressure microwave plasma cavity which has been used successfully in GC–MIP is the 'Surfatron' which involves surface wave propagation along a plasma column [8].

4.4 PLASMA INTERFACING WITH GAS AND LIQUID CHROMATOGRAPHS

Since the eluate from chromatographic columns is normally at atmospheric pressure, interfacing with reduced-pressure plasmas presents more problems than interfacing with atmospheric-pressure plasmas, for which simplified configurations are possible. A simple direct interface between a packed gas chromatographic column and an atmospheric-pressure DC argon plasma is shown in Fig. 4.4 [9]. Carrier gas and

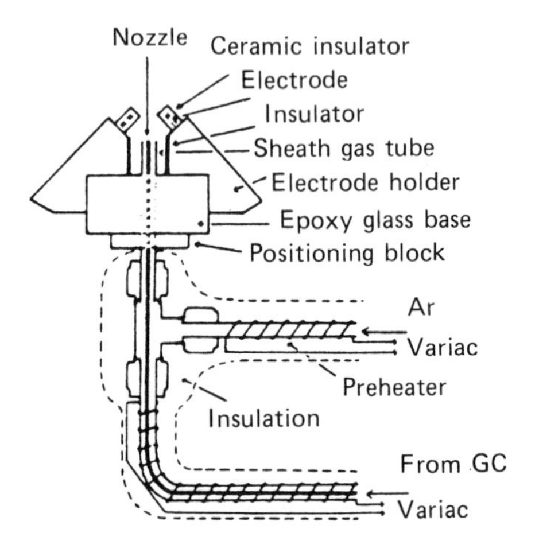

Fig. 4.4 — Diagram of nozzle, sheath-gas preheater, and transfer line from gas chromatograph to DC argon plasma. (Reproduced by permission, from R. J. Lloyd, R. M. Barnes, P. C. Uden and W. G. Elliott, *Anal. Chem.*, 1978, **50**, 2025. Copyright 1978, American Chemical Society).

eluent from a heated transfer line are introduced into the excitation region below the plasma junction, under constraint of an annular argon flow. A dedicated system based on the DC plasma has been described for mercury-specific determination of organomercurials in foodstuffs [10]. An analogous capillary GC configuration uses argon make-up gas at ca. 400 ml/min [11]. As the DC plasma can accommodate liquid inlet sample streams at ml/min flow-rates and higher, it is not adversely affected by vapour pulses from GC solvent peaks, which need not be vented as for microwave-induced plasmas.

The HPLC–DCP interface is simpler in design since a heated transfer line is not required. The main requirement is to reduce post-column peak-broadening by minimizing the interface tube length and volume [12]. The general concept for HPLC interfacing to the ICP is similar to that described by Hausler for size-exclusion chromatography (SEC) [13]. It is recognized that the major limitation to the sensitivity of these HPLC interfaces is the relatively poor (1% or lower) efficiency of transfer of eluent into the plasma excitation region, owing to ineffective nebulization and desolvation. An improved method of HPLC–plasma interfacing uses a 'direct injection nebulizer' (DIN) [14], which can transfer mobile-phase flows of up to 0.5 ml/min into the plasma with an efficiency approaching 100% and minimal peak broadening.

The microwave-induced plasmas (MIP) have found much greater use in GC than in HPLC interfacing, although the compatibility of the direct injection nebulizer with the flow-rates of effluents from microbore columns may expand the potential of these columns. Interfacing of reduced-pressure MIPs with packed column GC typically involves evacuating a silica sample chamber contained within the MIP cavity to a pressure of ca. 1 mmHg [15], but some broadening of the peaks is a drawback. Transverse plasma viewing is through an optical window but optimal sensitivity occurs at different positions for each element. The atmospheric-pressure cavities typified by the Beenakker design are very simple to interface with capillary GC columns since the latter can be terminated within a few millimetres of the plasma. Helium make-up gas or other reactant gases are introduced within the heated transfer line to optimize plasma performance and minimize peak broadening.

The performance of the GC–MIP can be improved if the commonly used capillary torch is replaced by a threaded tangential-flow torch (TFT) [16,17] to give a self-centring plasma, increased emission and better stability. The plasma loses relatively little energy to the walls, thus atom formation and excitation appear to be higher than with the capillary torch. A practical disadvantage, however, is the high consumption of helium as flow-gas in the TFT.

The interfacing of these plasmas with packed GC columns is more complicated since the MIP discharge is extinguished by the vapour burst from the injected solvent and may also fail to tolerate microgram sample peaks. Nevertheless there is a definite need for packed-column applications, particularly if trace determinations are required and the packed-column resolution is adequate. Various methods have been used to overcome this problem. The plasma can be re-ignited after the solvent peak has passed or a valve or alternative venting system can be used to divert the larger peaks away from the plasma. One such system is illustrated in Fig. 4.5, in which a 'fluidic logic' variable flow system is established within an interface oven between the column and the MIP cavity [18]. Such valve systems have also been used to interface an additional parallel or series detector for independent analysis. The incorporation of mass spectral or vapour phase infrared detection in this way would give a very powerful combination of techniques.

4.5 CAPABILITIES OF ATOMIC EMISSION ELEMENT-SPECIFIC DETECTION (AED)

The excellent selectivity and sensitivity of plasma AED makes GC–AED a valuable tool for determination of volatile species in a wide variety of samples. Samples having a complex matrix, e.g. environmental, petrochemical or biological samples, contain many constituents which complicate interpretation of chromatograms. Interferences from unresolved peaks, for components which may be present at much higher levels than the analyte of interest, sometimes make it impossible to determine or even identify the analyte. Element-selective detection can often reduce or even eliminate such interferences.

There are a number of different capabilities shown by plasma AED. Good selectivity for the element to be determined, relative to others present simultaneously in the plasma, is the most important factor. Selectivity is dependent on the emission properties of the element and the spectral resolution and other characteristics of the measurement system. Some regions of the UV–visible emission spectrum are less subject than others to spectral interference due to emission from the helium or argon plasma background or to the line or band spectra of carbon, nitrogen and

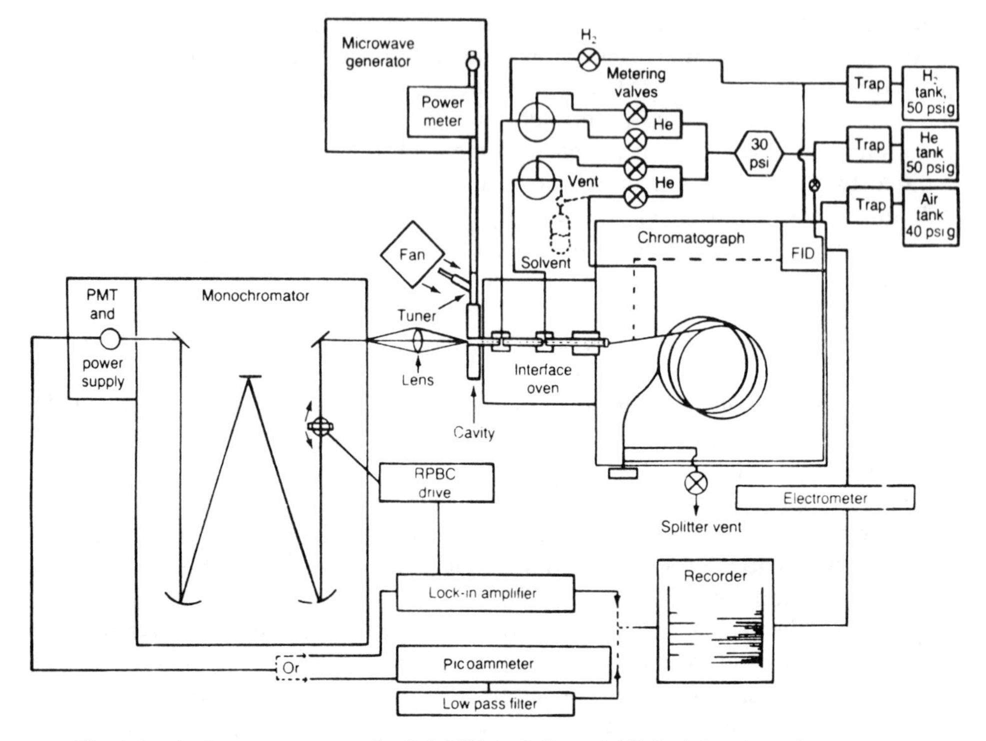

Fig. 4.5 — An instrument system for GC–MIP, including a fluidic logic interface. (Reproduced by permission, from S. A. Estes, P. C. Uden and R. M. Barnes, *Anal. Chem.*, 1981, **53**, 1336. Copyright 1981, American Chemical Society).

oxygen or their molecular combinations. Various definitions of interelement selectivity have been employed but one of the most general defines it as the peak area response per mole of analyte element divided by the peak area response of the 'background' element per mole of that element, at the measured emission wavelength. While high selectivity with respect to carbon is most commonly optimized and reported, other elemental background matrices require establishment of their own selectivity criteria. Selectivities vary greatly from element pair to element pair and their values define analytical usefulness; selectivities also vary between plasmas and with instrumental conditions, so for definitive results calibration measurement is necessary. Care must be taken that both the chromatographic peak characteristics and the plasma emission response for each element are linear; thus reference to the carbon content of fast-eluting organic solvent may give artificially high selectivities of other elements relative to carbon.

Carbon-selective detection may be considered as 'non-selective' detection for organic compounds. The most frequently used carbon emission wavelength is from the carbon ion line at 247.9 nm, but other spectral emission features corresponding to CN, CH etc. may be used [19]. This AED-mode response is analogous to flame ionization but is more completely independent of carbon atom environment and exhibits at least as great sensitivity [4]. Figure 4.6a shows a carbon-selective response by atmospheric-pressure MIP for a leaded gasoline, and Fig. 4.6b shows the lead-specific response obtained under identical chromatographic conditions [20].

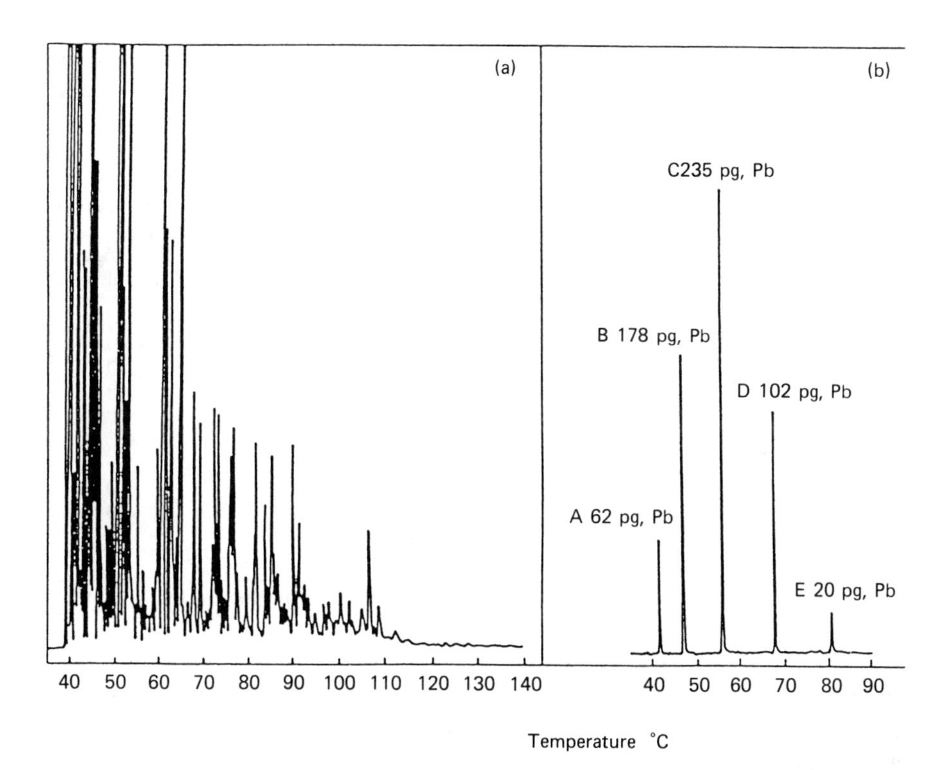

Fig. 4.6 — GC–MIP responses for gasoline samples with a 12.5 m SP-2100 fused silica capillary column, inlet split of 1:100 and a temperature programme from 40 to 100°C at 5°C/min. (a) Carbon-specific response for 0.10 μl at 247.86 nm. (b) Lead-specific response at 283.3 nm. Peak designations; A–tetramethyllead, B–trimethylethyllead, C–dimethyldiethyllead, D–methyltriethyllead, E–tetraethyllead. (Reproduced by permission, from P. C. Uden, *Anal. Proc.*, 1981, **18**, 189. Copyright 1981, Royal Society of Chemistry.

The sensitivity displayed by the AED for an element depends on the spectral intensity at the emission wavelength used. Each element has many possible wavelengths for determination and a scanned spectral survey is needed to choose the optimum line; both sensitivity and selectivity must be considered. The line exhibiting the greatest sensitivity may differ according to whether the ICP, DCP or MIP is used and also on whether the helium or argon plasma is employed. The sensitivity, as defined by the slope of the response curve, is less often used in chromatography than the detection limits expressed as the concentration equivalent to a signal that is three times the standard deviation of the background signal (noise). Limits may be expressed as absolute values of element mass (in a resolved peak) or in mass flow-rate units. The latter is the more useful, since it affords direct comparison with other mass-flow sensitive detectors such as the FID. Detection limits for different elements differ by two or three orders of magnitude, especially for metals, and this range may thus affect interelement selectivity if spectral overlap is present.

The linear dynamic ranges of response for different elements typically extend from the upper load capacity of fused silica capillary columns (ca. 50 ng) down to the elemental detection limits (1–100 pg). Chemical, gas-dopant and wall-interaction effects modify the upper limits of some ranges. Table 4.1 shows elemental detection

Table 4.1 — Elemental detection limits and selectivities for the reduced pressure GC–helium microwave plasma detection system. (Reproduced by permission, from W. R. McLean, D. L. Stanton and G. E. Penketh, *Analyst*, 1973, **98**, 432. Copyright 1973, Royal Society of Chemistry).

Element	Wavelength (nm)	Detection limit (ng/sec)	Selectivity (element/C)
C	247.9	0.06	—
H	486.1	0.02	>500
D	656.2	(0.09)*	880*
O	777.2	4.0	>500
N	746.9	2.8	>500
F	685.6	0.46	>350
Cl	479.4	0.06	485
Br	470.5	0.02	300
I	516.1	(0.05)*	400*
S	545.4	<0.05	50
P	253.6	0.005	—

*Manufacturer's specification for the MPD 850 (Applied Chromatography Systems Ltd.) [15].

limits and selectivities for a reduced pressure microwave-induced helium plasma GC detector (GC–MIP), and Table 4.2 displays selected data for atmospheric-pressure helium microwave-induced plasma detection.

Selective detection can be accomplished in the single-element mode or, if spectroscopic capabilities permit, with multielement monitoring. The latter can be accomplished in a variety of ways. Rapid sequential wavelength-switching is useful provided that the timing is compatible with peak-elution rates [21]. However, the most widely utilized multielement detection has been by means of a diect-readiung polychromator displaying signals from up to 12 monitoring wavelengths simultaneously [4,22]. This has been used with either reduced pressure [22] or atmospheric pressure MIP detection [23,24]. A third approach, utilizing diode-array detection, has shown considerable versatility and sensitivity [25].

Among the most attractive abilities of a multichannel AED detector is the capacity for determination of element ratios to determine empirical formulae of eluted compounds. Such measurements presume that the element response is independent of molecular structure, which is still under debate in some instances and for some plasma conditions. An example of such a determination is shown for the chromatographic separation of methylated chlorination products from fulvic acid in Fig. 4.7 [26]. Simultaneous carbon, hydrogen and chlorine detection is displayed; the peak for methyl trichloroacetate (D) was assigned by retention time standardization and was used as a reference to compute the empirical formulae of other major components, unknowns being identified as chloroform (A), trichloroacetaldehyde

Table 4.2 — Selected detection limits and selectivities for atmospheric-pressure helium microwave plasma GC–detection. (Compiled from [41, 42]).

Element	Wavelength (nm)	Absolute detection limit (pg)	Detection limit (pg/sec)	Selectivity ratio *vs*. C
C	247.9	12	2.7	1
H	656.3	22	7.5	160
D	656.1	20	7.4	194
B(a)	249.8	27	3.6	9.3×10^3
Si	251.6	18	9.3	1.6×10^3
Ge(a)	265.1	3.9	1.3	7.6×10^3
Sn(a)	284.0	6.1	1.6	3.6×10^4
Pb	283.3	0.71	0.71	2.5×10^4
S(b)	545.4	140	52	4.6×10^3
P(a,c)	253.6	56	3.3	1.0×10^4
As(a,c)	228.8	155	6.5	4.7×10^4
F(d)	685.6	—	8.5	3.5×10^3
Cl(d)	481.0	—	16	2.4×10^3
Br(d)	470.5	—	10	1.4×10^3
I	206.2	56	21	5.0×10^3
Se	204.0	62	5.3	1.1×10^4
Mn	257.6	7.7	1.6	1.1×10^5
Fe	259.9	0.9	0.3	2.8×10^5
Hg(c)	253.7	60	0.6	7.7×10^4

(a) For these elements, the plasma was doped with 0.5–1.0 ml/min flow of hydrogen.
(b) These data were obtained by use of a quartz refractor plate background-correction device.
(c) The eluates in these determinations showed band-broadening, making absolute detection limits high.
(d) These data [41] were obtained by use of a valve interface; all other values [42] were obtained by use of a direct capillary interface.

(B) and methyl dichloroacetate (C) from the stoichiometries given in Table 4.3. Such determinations do not attain the accuracy of those obtained from classical mg-level microanalysis but measurements are made directly on GC eluate peaks at sample levels smaller by up to six orders of magnitude! An improved approach to gain greater precision and accuracy uses a multi-referencing method employing pyrolysis products from standard polymers to provide a reproducible reference range; results with relative errors less than 1% are usually obtained [24].

4.6 ELEMENT-SPECIFIC GC–MIP DETECTION OF NON-METALS

Bache and Lisk [3] were among the first to employ such a plasma detector, for determining ng levels of Br, Cl, I, P and S in organic eluates, and the reduced-pressure MIP was demonstrated to be effective for selective detection of P, S, Br, Cl, I, C, H, D, N and O with detection limits in the 0.03–0.09 ng/sec mass flow-rate range [22]. Line emission is observed for all elements in the helium plasma, making it preferable to the less energetic argon MIP in which such elements as Cl, Br, N and O exhibit only weak diatomic molecular emission. Developments by Zeng *et al*. [27], utilizing an analogous reduced pressure system, have focused on improvements in oxygen-specific detection. Highly pure plasma gases and careful exclusion of air improved the limit of detection to 0.03 ng/sec, with a linear dynamic range of three orders of magnitude. Figure 4.8 shows a portion of a chromatogram (with simultaneous detection of C and O) for gasoline with alcohol additives. The importance of

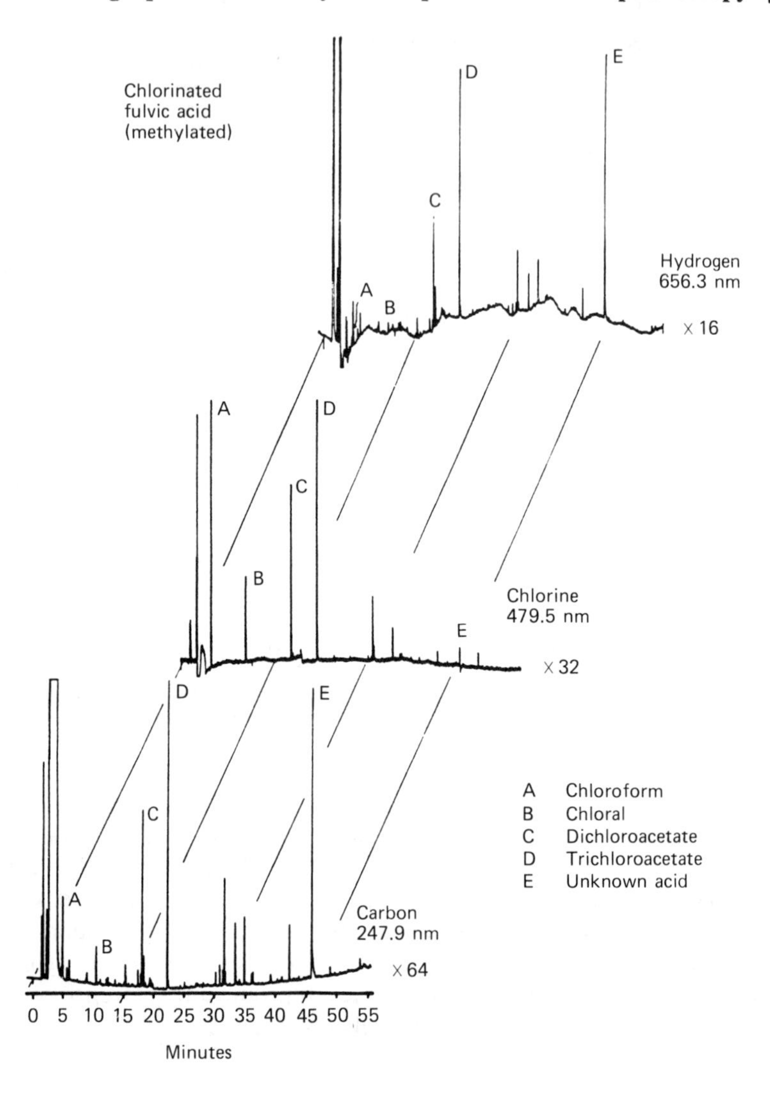

Fig. 4.7 — GC–MIP chromatograms representing simultaneous carbon-, hydrogen- and chlorine-specific detection of fulvic acid chlorination and methylation products: 25 m SE 30 cross-linked fused silica capillary column, inlet split of 1:80 and temperature programme of 25°C for 10 min then 4°C/min to 200°C. Microwave input power 60 W. (Reproduced by permission, from P. C. Uden, K. J. Slatkavitz, R. M. Barnes and R. L. Deming, *Anal. Chim. Acta*, 1986, **180**, 401. Copyright 1986, Elsevier Science Publishers).

this type of element-specific analysis seems certain to grow with the increasing use of oxygen compounds in fuel oils. Oxygen 'fingerprints' of such materials have potential in environmental analyses.

It was noted earlier that when a particular analyte does not possess any elemental content which allows particularly selective or sensitive detection, it may be possible

Table 4.3 — Empirical and molecular-formula determinations of unknown chlorination products of fulvic acid by GC–MIP. (Reproduced by permission, from P. C. Uden, K. J. Slatkavitz, R. M. Barnes and R. L. Deming, *Anal. Chim. Acta*, 1986, **180**, 401. Copyright 1986, Elsevier Science Publishers).

Compound	Formula[a]	R.s.d. (%)[b]	
		H	Cl
$C_3H_3Cl_3O_2$ (trichloroacetate)	Reference		
$C_3H_4Cl_2O_2$ (dichloroacetate)	$C_3H_{3.9}Cl_{2.0}O_x$	1.8	3.6
C_2HCl_3O (chloral)	$C_2H_{0.9}Cl_{2.9}O_y$	8.7	1.1
$CHCl_3$ (chloroform)	$C_1H_{1.1}Cl_{2.9}$	11.1	0.4

[a]Mean of 3 results.
[b]Relative standard deviation.

to use chemical derivatization of molecular functional groups to introduce such elements. Elegant methods have been developed for introduction of labels to identify compounds otherwise difficult to determine by GC–MIP, notably those having a simple C,H,O,N composition; an example is dual derivatization to introduce both chlorine and fluorine to enhance selectivity in the analytical determination of derivatized amino-acids [28].

An extensive investigation of the atmospheric-pressure Beenakker microwave cavity for packed column GC detection of trihalomethanes and other purgeable organics in drinking water showed the utility of the technique [29]. With 10-ml water samples, detection limits were below 1 ng/ml and the technique showed constant relative responses for chlorine, bromine and iodine in the various volatile molecules studied. It would be predicted that since the MIP can be considered as a classical mass-flow dependent detector, such detection limits would also apply for capillary columns; this was borne out by subsequent work. The advantage of the MIP detector over the other primary GC detectors used for halo-organics, the electron-capture detector (ECD) and the Hall electrolytic conductivity detector (HECD), are clear. Although the MIP lacks the sensitivity of the ECD for polyhalogenated compounds, it has the advantage of uniform response to element content for each halogen, irrespective of analyte molecular structure, in most cases investigated. The key advantage over the HECD is the element specificity for individual halogens, that is not seen in the halogen-detection mode of that detector. A further halogen-specific application of the MIP system, operated with glass capillary columns, was in the examination of chlorinated products of natural humic materials in potable waters [30]. Humic substances, which are amorphous, hydrophilic, acidic, polydisperse substances with molecular weights ranging from several hundreds to tens of thousands, are derived from soil, vegetation and algae and are found in most natural freshwater sources. Their structures vary with location and formation conditions but they consist mainly of aromatic polyhydroxy, polymethoxy, polycarboxylic acids with smaller amounts of carbohydrates, nitrogen bases and nucleotide residues. The range of potential products formed by chlorination during water treatment procedures continues to challenge environmental chemists and Fig. 4.9 illustrates an example of the analytical utility of GC–MIP. The most distinctive feature of the upper chromatogram, of a non-methylated chlorinated humic acid extract, is the broad chlorine-containing band labelled 2. Methylation with diazomethane gives the

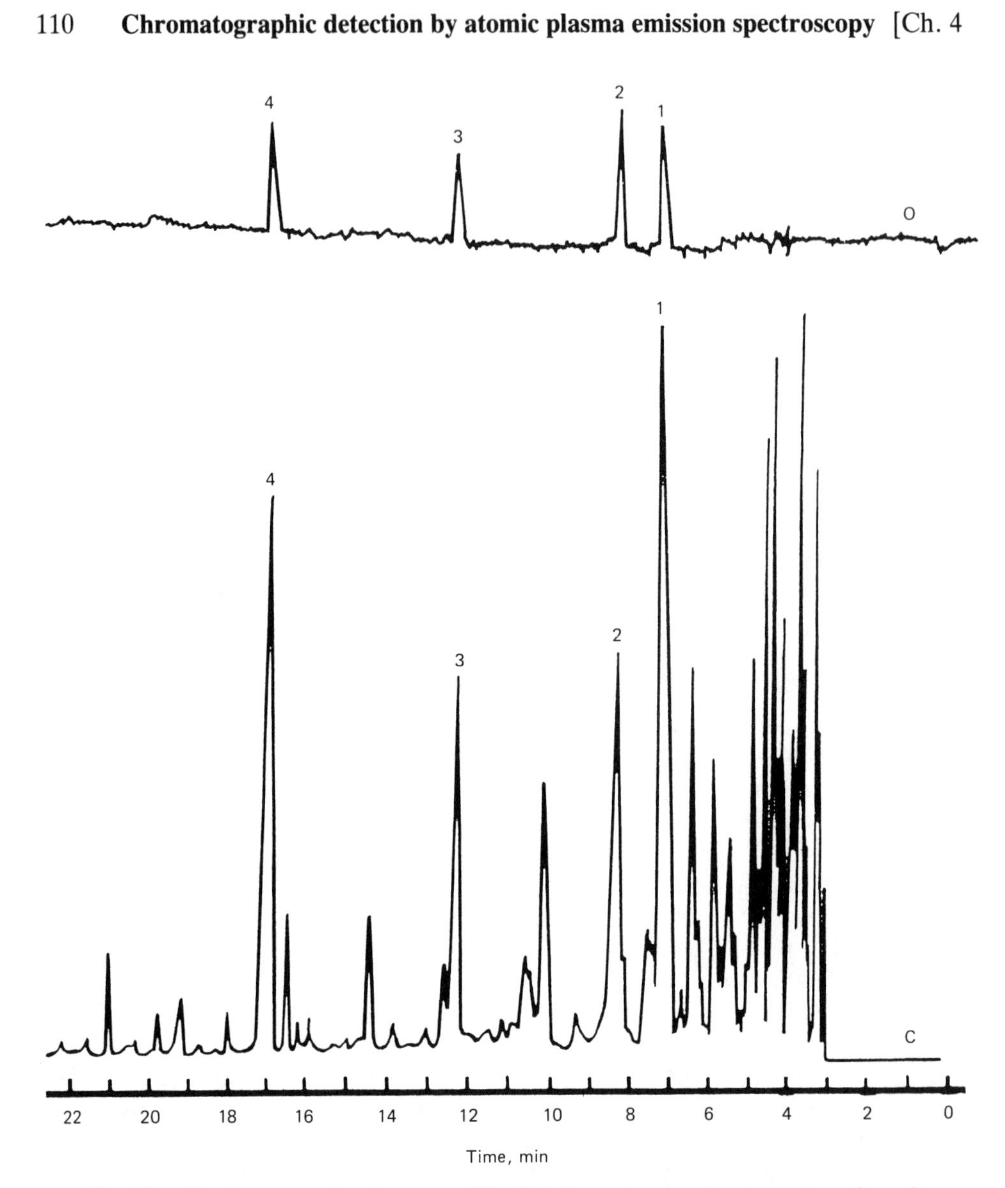

Fig. 4.8 — Simultaneous reduced-pressure GC–MIP chromatograms showing carbon- (lower) and oxygen-specific (upper) detection of gasoline with added alcohols: 32.5 m polyethylene glycol 20M SCOT column, 53°C for 2 min, then to 80°C at 5°C/min, then to 150°C at 12°C/min; 1 — *tert*-butanol, 2 — ethanol, 3 — *sec*-butanol, 4 — n-butanol. (Reproduced by permission from Zeng Kewei, Gu Qingyu, Wang Guochen and Yu Weile, *Spectrochim. Acta*, 1985, **40B**, 349. Copyright 1985, Pergamon Press.)

product producing peak 2 in the lower chromatogram, which is shown to be methyltrichloroacetate, derived from trichloracetic acid. The amount of chlorine present in the compounds in band 2 of the upper chromatogram emphasizes the relative importance of this chlorination product.

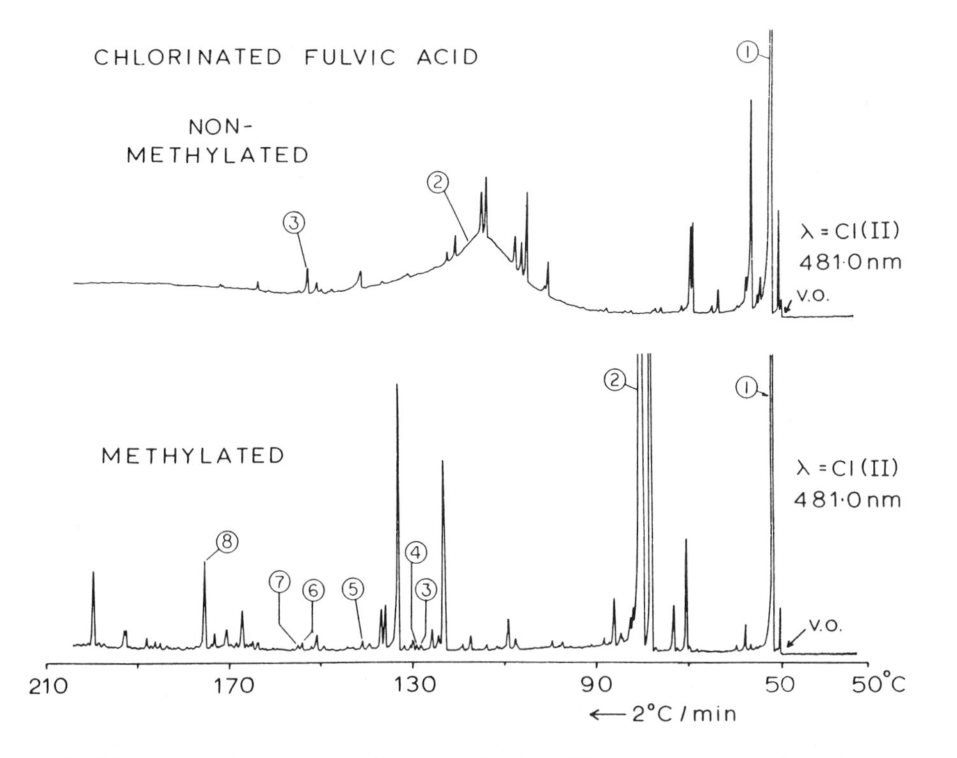

Fig. 4.9 — Atmospheric-pressure chlorine-specific GC–MIP of non-methylated (upper) and methylated (lower) chlorinated fulvic acid extracts. 100 m OV 225 glass SCOT column. Peak identities; above, 1 — chloroform, 2 — trichloroacetic acid, 3— 1-chlorophenol; below, 1 — chloroform, 2 — methyl trichloroacetate, 3 — 2,4,6-trichlorophenol methyl ether, 4 — methyl-2-chlorobenzoate, 5 — methyl-3,5-dichlorobenzoate, 6 — 1-chlorophenol, 7 — 2-chlorophenol, 8 — pentachlorophenol methyl ether. (Reproduced by permission, from B. D. Quimby, M. F. Delaney, P. C. Uden and R. M. Barnes, *Anal. Chem.*, 1980, **52**, 259. Copyright 1980, American Chemical Society).

For some non-metals, notably S, F, Cl and Br, oscillating refractor plate background-correction gave an increase in selectivity relative to carbon, though with some sacrifice in detection limit. A rapid-scanning spectrometer has been applied to scan a preselected spectral window to provide information in the multielement mode [21]. Elemental response per mole of C, Cl and Br for a range of halogenated hydrocarbons and pesticides was found to be independent of molecular structure. Another development is the application of near infrared (NIR) atomic emission in GC–MIP. A cooled Beenakker cavity and TFT were used at high power (370 W) with a Fourier-transform NIR spectrometer [31]. Spectral and chromatographic results were obtained from a series of time-resolved interferograms. Computer-generated element-specific chromatographic reconstructions for eight non-metals (C, H, N, O, F, Cl, Br and S) were obtained from a single injection. Atomic emission intensity was plotted against chromatographic retention time, for each of eight or more optical frequencies chosen to coincide with an element-specific spectral NIR region. Both the FT approach and the spectral-region method for non-metallic elements suggest a worthwhile extension of GC–MIP application, which could be particularly valuable for environmental chromatography.

The Beenakker cavity has been the subject of a number of detailed investigations for element-specific GC detection, including boron-specific detection for diol boronate esters and other compounds [32], and pyrolysis–GC detection of Si, P, Cl and Br in pyrolysates of silicone and polyphosphazene polymers [33].

After a series of studies on low-frequency, high-voltage electrodeless discharge afterglows sustained in nitrogen and argon, as potential GC detectors, sufficiently encouraging data from a similar helium afterglow detector were obtained for it to be commercially developed [34]. Absolute detection limits in the low pg range were obtained for F, Cl, Br, I, C, P, S, Si, Hg and As, and undesirable contamination and solvent extinction of the discharge were eliminated.

An argon MIP has been used to analyse halogen-containing compounds by a combination of spectral approaches monitoring the iodine atom line (206.2 nm), a bromine band (298.5 nm), a CCl molecular band (278.8 nm) and fluorine indirectly by the Si atomic emission at 251.6 nm resulting from SiF formation [35]. Halogenated, phosphorus- and sulphur-containing pesticides at the sub-ppm level have been added as spikes to various agricultural products and determined after extraction; the recoveries were above about 75% [3]. Trace quantities of polybrominated biphenyls have been determined by reduced-pressure helium GC–MIP [36]. Various studies have evaluated the applicability of GC–MIP for analysing phosphorus-containing pesticide residues. In early work using an argon MIP [37], untreated agricultural extracts were directly injected to give clean element-specific chromatograms despite the large impurity concentrations present, and reasonable recoveries of down to 0.03 ppm of pesticide were obtained. Quantitative analysis of sulphur-based pesticides by GC–AED was also performed, with use of the 216.7 nm line; three pesticides were determined with a single working curve and despite the need to correct the background for carbon signals, detection limits of 0.02 μg of sulphur were obtained [38].

Arsenic and antimony have been determined in various environmental samples [39]. Tervalent arsenic and antimony were derivatized and reacted to form stable triphenylarsine and stibine, which were extracted and determined by reduced-pressure GC–MIP at 228.8 and 259.8 nm respectively, with detection limits of 20 and 50 pg. Elemental selectivities were also very high. A parallel study [40] determined alkylarsenic acids in commercial pesticides and environmental samples by borohydride reduction.

As a general conclusion it may be noted that with the exception of such elements as N and O, the reduced-pressure MIP plasmas are likely to receive less attention in future development than the atmospheric-pressure plasmas. The necessity to maintain an efficient vacuum system interfaced to the GC is a distinct disadvantage which is not in general offset by the analytical performance.

4.7 ELEMENT-SPECIFIC GC–MIP DETECTION OF METALS

As seen in Table 4.2 [41,42], detection limits and selectivities for metal-specific detection in the atmospheric-pressure MIP system are usually better than for non-metals, by virtue both of emission intensity and absence of background in the spectral region monitored. GC–MIP data have been obtained for many transition and main group metals [41,42], but environmental interest focuses on a relatively small number of elements, with particular emphasis on lead, tin and mercury. Each of these elements is determinable by GC–MIP with Beenakker-type cavities, with sub-pg/sec detection limits. The facility for the determination of alkyllead compounds in gasolines was noted earlier [20] and another example of an analysis of environmental concern is shown in Fig. 4.10, where lead- and carbon-specific

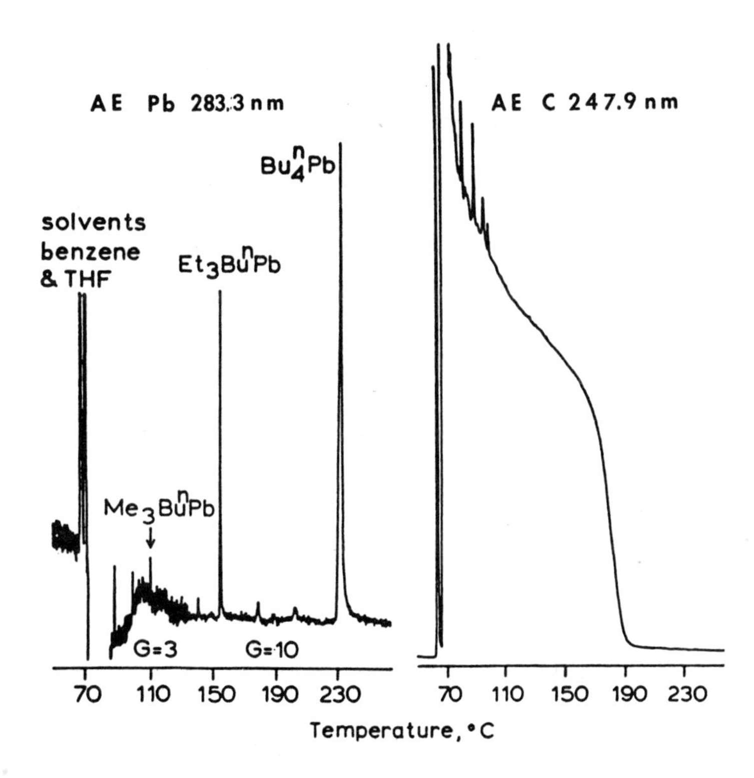

Fig. 4.10 — Simultaneous atmospheric-pressure GC–MIP chromatograms showing lead- (left) and carbon- (right) detection of industrial plant water effluents derivatized as tributylalkylleads. (Reproduced by permission, from S. A. Estes, P. C. Uden and R. M. Barnes, *Anal. Chem.*, 1982, **54**, 2402. Copyright, 1982, American Chemical Society.

detection is compared for trialkyllead chlorides extracted from an industrial plant effluent and derivatized with a butyl Grignard reagent to form their analogous trialkylbutyllead compounds [43]. The extent of chromatographic interference from the high level of carbon compounds in this sample would prevent any qualitative or quantitative detection of the trialkyllead compounds by GC–ECD or GC–MS without extensive clean-up and probable loss of analyte. Even GC–FID analysis of this sample proved impossible.

An important comparison of the performance of reduced- and atmospheric-pressure MIP systems for the detection of organomercury compounds was made by

Olsen *et al.* [44]. They found the latter system to be the better, giving a 1 pg detection limit for mercury, with a selectivity factor of 10^4 relative to carbon. Figure 4.11 shows

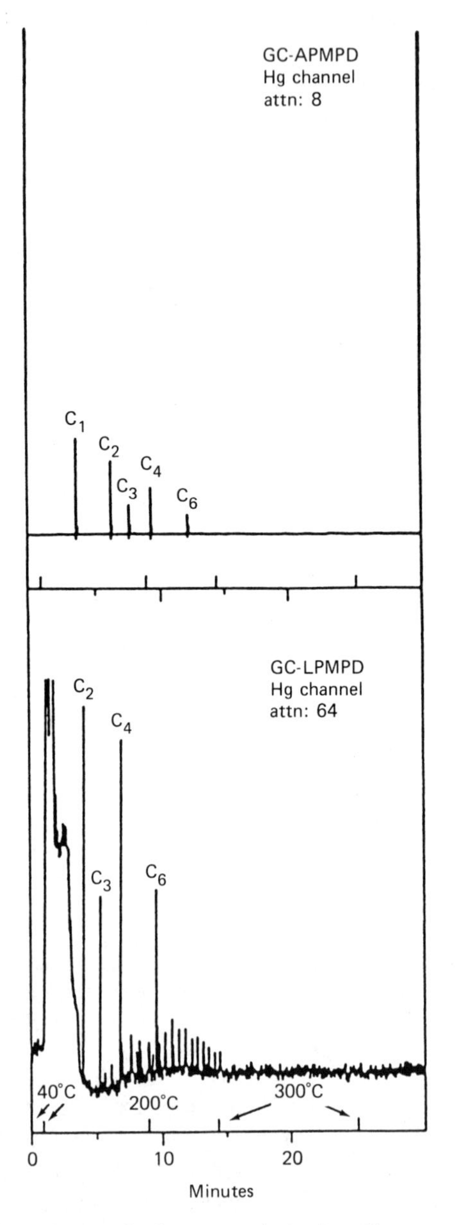

Fig. 4.11 — Comparison of atmospheric-pressure (upper) and low-pressure (lower) GC–MIP for dialkylmercury compounds in a shale oil matrix. Mercury-specific detection at 253.6 nm. C(1), C(2), ... refer to dimethylmercury, diethylmercury, etc. (Reproduced by permission, from K. B. Olsen, D. S. Sklarew and J. C. Evans, *Spectrochim. Acta*, 1985, **40B**, 357. Copyright 1985, Pergamon Press).

a comparison chromatogram of dialkylmercury compounds in a shale oil matrix. GC–MIP of volatile elemental hydrides of germanium, selenium and tin has been found to give sub-ng detection [45] and there is considerable potential for the determination of these elements in environmental matrices.

The study of metal chelates with sufficient volatility and thermal stability for them to be separated by GC has received much attention over the past 20 years, with most emphasis on complexation with 2,4-pentanedione (acetylacetone) and its analogues; this area has been comprehensively reviewed recently [46] but has had limited impact as yet on environmental analysis. Among examples of application has been GC–MIP analysis of blood plasma for chromium (converted into its trifluoracetylacetonate) with excellent accuracy and precision [47]. Trace determinations of beryllium, copper and aluminium have also been reported.

A very comprehensive compilation of information on coupled gas chromatography–atomic spectrometry has been published, which covers all modes of atomic absorption, atomic emission and atomic fluorescence [48].

4.8 PROSPECTS FOR HPLC–MIP

As will be discussed in subsequent sections of this chapter, HPLC has been most extensively interfaced with the high-power DCP and ICP argon plasmas, which are able to tolerate mobile phase flow-rates and the solvents used in the various liquid chromatographic procedures. The low-power helium MIP, however, cannot be directly interfaced to the HPLC column, since continuous introduction of ml/min liquid flow-streams into the discharge will quench it. Some solutions to this problem have been considered. One approach employed high-power (kW) discharges, operating in the radiofrequency or microwave range, and accommodated continuous solvent-flow provided that there was efficient nebulization. The only direct introduction of liquid into the MIP was accomplished by using a heated wire over which the HPLC effluent flowed [49], and the vapour was carried into the discharge by a cross-stream of helium. This system shows some potential for reversed-phase separations.

There remains one considerable incentive for the development of a viable HPLC–helium MIP interface, namely the potential for monitoring effluent for non-metallic elements, which is difficult or impossible with the argon DCP and ICP. The direct-injection nebulizer (DIN) noted earlier [14] may provide one answer. Another involves the removal of the solvent, and it is possible that the 'thermospray' approach now used widely in HPLC– MS could be adapted; in fact the problems of interfacing the lower-powered MIPs are largely parallel to those in HPLC–MS and HPLC–FTIR; cryo-focusing, as used in the latter, is another possibility. Investigations are also under way on alternative plasma cavities which may be able to sustain the helium MIP under conventional HPLC flow conditions. Capillary HPLC columns with mobile phase flow-rates of a few μl/min are another interesting possibility for helium MIP interfacing, but sample capacity may limit any application for trace determinations.

4.9 DC ARGON PLASMA DETECTOR (DCP) FOR GC AND HPLC

As pointed out in the introduction to this chapter, the DC argon plasma has received substantial attention as a chromatographic detector, particularly in its commercially available forms of the earlier two-electrode DC plasma jet and the later more efficient three-electrode version. Though the DCP has not been as widely adopted as the ICP in recent years for general plasma spectrochemical analysis, it nevertheless has considerable advantages in ruggedness, simplicity of operation, ease of interfacing and relatively low cost of operation. Since chromatographic interfacing was first reported [9,12], there has been considerably more interfacing described for GC than for the ICP and almost as much for HPLC. While the energetics of the argon plasmas are particularly suited for metal determination there have also been some important applications to non-metal determinations.

4.10 GAS CHROMATOGRAPHIC APPLICATIONS OF THE DCP

The configuration of the three-electrode DCP is shown in Fig. 4.2 (p. 100), and for GC detection a transfer line exit-orifice is situated directly beneath the plasma excitation region. Early versions [9] for packed column work used heated stainless-steel transfer lines but fused silica transfer lines connected directly to capillary columns through zero-volume connectors have proved best for capillary GC work [11]. Since the DCP jet must usually be located one or two feet distant from the GC column oven, particular attention must be given to minimizing the contribution of the transfer line to extra-column peak broadening. The GC–DCP system has proved suitable for detecting metallic elements that can be efficiently excited, and also elements, such as boron and silicon, present in GC derivatizing groups. For silicon, the absence of interfering spectral response from the silica discharge tube often used in the MIP is an added advantage; a selectivity of 2×10^6 for silicon with respect to carbon and a detection limit of 25 pg/sec, have been reported [11]. Other results obtained in the capillary GC mode with the 2-electrode and 3-electrode plasma jets are shown in Table 4.4. In comparison with GC–MIP, the absolute detection limits

Table 4.4 — Results obtained for metals by capillary GC–DCP. (Reproduced from [11] by permission of the author).

Element	Detection wavelength (nm)	Detection limit (pg/sec)	Selectivity
Three-electrode plasma jet			
Cr	267.7	4	4×10^8
Sn	286.3	60	2.5×10^6
Pb	368.3	100	5×10^5
B	249.8	3	3×10^5
Two-electrode plasma jet			
Cu	324.7	5.6	$>10^6$
Ni	341.4	320	$>10^6$
Hg	253.6	65	6×10^5

for some elements such as lead and mercury are one or two orders of magnitude poorer but those for others such as boron are similar. For individual analyses the GC–DCP and the GC–MIP systems have contrasting meritorious features. The atmospheric-pressure MIP is well suited to high-resolution capillary GC and its low detection limits allow trace determinations to be made with very small injected samples. It is limited, however, in ability to handle larger samples, and packed column applications require solvent venting. The DCP system is somewhat less convenient to interface, but for metals and most metalloid elements the sensitivities of the DCP system are close to those of the MIP. The extremely high selectivities (with respect to carbon) typically found give the method wide appeal for samples with high unresolved chromatographic backgrounds.

An application of high metal selectivity in GC–DCP was the determination of the organomanganese gasoline additive MMT (methylcyclopentadienylmanganese-tricarbonyl) [50]. A dual-detection chromatogram with manganese-specific DCP detection at 279.83 nm and FID reference is shown in Fig. 4.12. Direct injection of

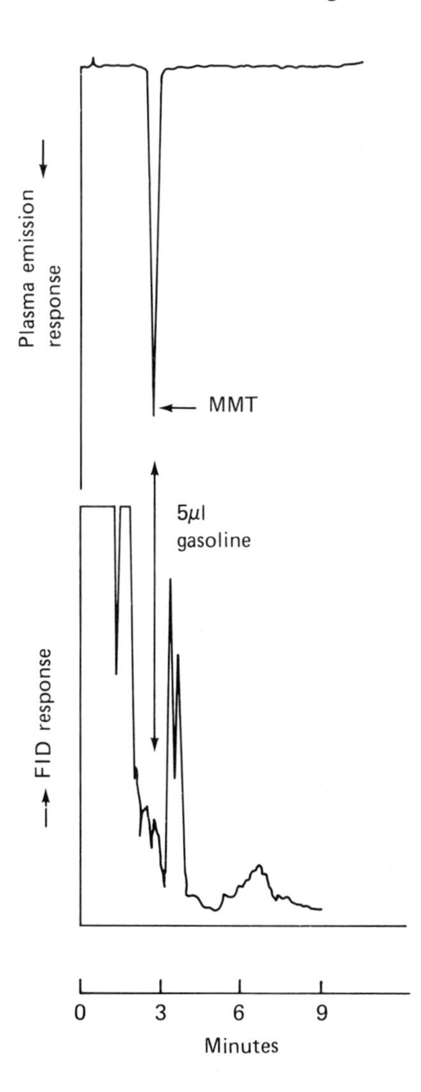

Fig. 4.12 — Dual-detector gas chromatogram of unleaded gasoline containing MMT: 1.5 μl of gasoline. Effluent split 1:1 between flame ionization and DC argon plasma detectors. Mn 279.83 nm line monitored. (Reproduced by permission, from P. C. Uden, R. M. Barnes and F. P. D. Sanzo, *Anal. Chem.*, 1978, **50**, 852. Copyright 1978, American Chemical Society.)

5-μl samples of gasoline with a related organomanganese compound added as an internal reference enabled MMT to be determined at levels of a few ppm with a relative standard deviation of 0.8–3.4%. The environmentally significant metals and metalloids silicon, germanium, tin and lead have been examined in a study of redistribution reactions of their alkyl and halo compounds [51]. The versatility of GC–DCP has been shown in the development of a simple low-cost GC system dedicated to the DCP for specific determination of methylmercury compounds in fish [10]. Comparisons of data were made with those obtained by GC with electron-capture detection and by total mercury cold-vapour atomic absorption, and artifact formation upon injection was investigated. A chromatogram comparing the GC–DCP and GC–ECD data is shown in Fig. 4.13.

4.11 LIQUID CHROMATOGRAPHIC APPLICATIONS OF THE DCP

The first study of HPLC interfaced with the DC argon plasma [12] used standard nebulization for reversed-phase elution, but an impact device for normal phase elution of hydrocarbon and halocarbon solvents. Metal complexes of Ni, Cu, Hg and Cr were determined with mass-flow detection limits of 0.3 ng/sec for copper and 1.25 ng/sec for chromium. The applicability of ion-pair HPLC procedures has been demonstrated for the determination of cationic surfactants in fuel oils by use of an ion-pairing reagent which incorporates a specific tagging element, such as boron, to achieve good sensitivity [52]. Some results have been reported for HPLC–DCP applications to determination of sulphate, nitrate and acetate as their cadmium salts, but as the minimum detectable levels are in the 100 ppm range, any environmental applicability is limited [53]. Determination and speciation of Cr(III) and Cr(VI) (as chromate) by HPLC–DCP has been based on reversed-phase ion-pairing separations, with detection in the 5–15 ng/ml range [54]. Applications included biological samples from ocean floor drillings, chemical dump sites, surface well-water and waste water samples. An interesting application of the DCP detection technique involved the determination of tin levels down to 10 ng/ml by a combination of HPLC with continuous on-line hydride generation, followed by DCP measurement. The method was found suited to determination of alkyltin chlorides as well as tin(II) and tin(IV) ions [55].

4.12 THE INDUCTIVELY-COUPLED ARGON (ICP) PLASMA DETECTOR FOR GLC AND HPLC

4.12.1 Gas chromatographic applications of the ICP

Despite its wide adoption as a spectroanalytical emission source, the inductively-coupled plasma (ICP) has been little used as a GC detector; as is stressed later, this is in contrast to its wide adoption in HPLC monitoring. However, it offers the advantage of withstanding organic solvents more readily than does the MIP, because of its higher gas temperature. In the first effective effort to evaluate the potential of GC–ICP, a packed column was interfaced to a demountable ICP torch through a T-junction which enabled make-up argon to be added [56]. The optical system incorporated a 0.35 m scanning monochromator and a 1.5 m, 0.02 nm resolution multichannel direct-reading spectrometer. This study focused on the non-metallic

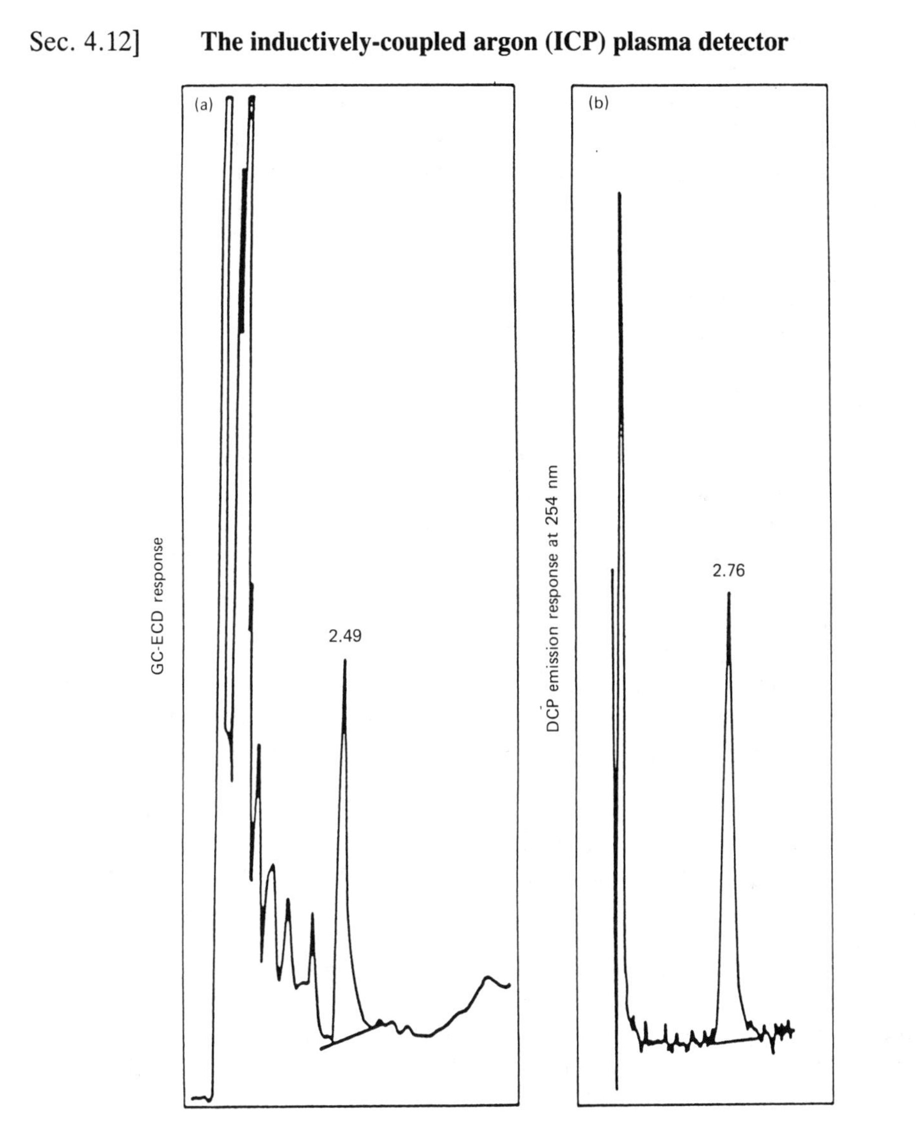

Fig. 4.13 — GC–ECD (standard work-up) (left) and GC–DCP (standard=rapid work-up) chromatograms of NBS–RM50 tuna-fish for methylmercury. (Reproduced by permission, from K. W. Panaro, D. Erikson and I. S. Krull, *Analyst*, 1987, **112**, 1097. Copyright 1987, Royal Society of Chemistry).

elements Br, Cl, F, I, H, Si, and C, together with the metals Sn, Pb, and Fe. Although near-IR lines were observed for the halogens, they were found to be too weak for useful data acquisition; thus the UV–visible spectral region used for MIP detection was employed. Predictably, results for the metallic elements were the more encouraging. Detection limits for Si, Fe and Sn were at the ng level, as were those for C and H. Iodine was detectable at 24 ng, but limits for F, Cl and Br were at or above the microgram level. Linear dynamic ranges reached four orders of magnitude for the metals. Selectivity with respect to carbon was also variable, being poor for Cl, but reaching 100 or more for other elements. Selectivity ratios were found to be limited by background changes induced by introduction of large quantities of organic material. In contrast to the reduced-pressure MIP, oxygen or nitrogen did not have to be added to the plasma to reduce deposits. While deposits sometimes formed on the inside of an extended ICP coolant tube, they caused no problem since they formed well above the observation zone. The disappointing observation, however, was that performance levels for non-metals were inferior to those with the MIP and this has certainly reduced the subsequent investigation of the argon ICP.

Both all-argon and nitrogen-cooled ICPs have been used for determination of tetra-alkylleads in gasolines, and studies of zinc and nickel as diethyldithiocarbamates have been made [57]. GC–ICP for oxygen-specific detection has been investigated by use of near-IR emission, with oxygen entrainment being limited by an extended torch design [58], and hydrides of Ge, As and Sb have been measured at the low ng/ml level by using a slew scanning monochromator; EPA control reference materials for these elements were analysed by a hydride generation procedure [59].

4.12.2 Liquid chromatographic applications of the ICP

In contrast to the dominance of the microwave-induced plasmas as element-selective GC detectors, in HPLC detection by far the most attention has been paid to the ICP. A survey of the literature of successful HPLC–plasma detection also emphasizes that metal-specific detection is predominant and will probably remain so until interface systems can be devised which will successfully remove HPLC mobile phases but quantitatively transfer eluate peaks to a plasma favouring non-metals (such systems are currently being investigated). Such an interface may incorporate an improved version of the moving-band eluate transport devices which have been used in other HPLC detection modes, notably HPLC–MS, or it may be based on thermospray or electrospray technology.

The major difficulty experienced in HPLC–plasma interfacing is incompatibility of the system with mobile phases at typical analytical flow-rates. All element-specific atomic spectroscopic detectors, except batch-mode graphite-furnace atomic absorption (GFAAS) are based on on-line nebulization and excitation of small volumes (5–200 μl) of liquid, which are converted into an aerosol and then introduced into an atomization–excitation cell. The major reason why relatively poor detection limits have been reported even for HPLC–ICP is the relatively ineffective conversion of effluent flow into aerosol and its transport to the plasma; typically only 1–5% of the sample reaches the plasma torch. A frequent observation has been poor tolerance of the ICP for common solvents used in HPLC, notably in ion-pairing and size-exclusion chromatography. Solutions to these problems lie in more quantitative nebulization, atomization and excitation of HPLC samples, as well as in improved transport systems.

Many publications on HPLC–ICP have appeared since 1979 [60] but detection limits obtained for many elements have been of marginal use for elemental speciation in real samples at levels of environmental significance. Current developments, however, suggest that a substantial enhancement in working sensitivities is occurring.

Much of the earliest HPLC–ICP study [61] was on simulated peaks and not column elution. The major portion of the study focused on detection of copper chelates, and typical ICP conditions involved sample introduction at 2 ml/min into a 1.2-kW argon plasma. Detection limits in the 'chromatographic mode' (in μg/l.) were Cu (6.8), Ni (43), Co (21), Zn (19), Cd (89), Cr (20) and Se (280). Subsequent detection for real chromatographic systems has not generally bettered these values; the minimum detection limits are typically two or more orders of magnitude poorer than for continuous flow ICP–AES. Most workers have used aqueous mobile phases with performance characteristics that are familiar from standard sampling procedures. However, normal phase HPLC in which organic solvents such as hexane or methyl isobutyl ketone (MIBK) are used is a greater challenge, since the ICP behaviour is then less well defined and spectral background interference is greater.

A successful approach to this problem [62] was to use microbore HPLC at lower flow-rates and a simple T-junction interface; for test samples of copper and zinc diketonates and dithiocarbamates, peak broadening was minimized by optimal design of the interface, connecting tubing, nebulizer, spray chamber and plasma torch to give virtually constant peak-width ratios for ICP and UV detection at different HPLC flow-rates. It was also noted that sensitivity was independent of flow-rate above 15 μl/min, showing the ICP to operate as a mass-flow sensitive detector; below that flow-rate, however, it behaved as a concentration-sensitive detector.

HPLC–ICP has been effective for determination of metalloids; a detection limit for arsenic in organoarsenic acids was found to be 130 ng/ml in 100-μl injected samples, the arsenic signal being independent of molecular form [63]. The spectrometer was a simultaneous unit with 48 channels operating at 1.2 kW, allowing chromatograms simultaneously sampled for As, Se and P to be displayed on-line. Figure 4.14 shows a reversed-phase chromatogram of arsenic species on a resin-based column. Integration times for each point displayed were 5 sec. A single-wavelength study [64] measured arsenic and cadmium compounds at the As emission wavelength of 228.812 nm and the Cd emission line at 228.802 nm. A 1.2-kW ICP was used and mutual interferences at each wavelength were determined. Detection limits for 50-μl injections were 2.6 ng/sec (3.1 ng/μl) for As as arsenite and 0.059 ng/sec (0.12 ng/μl) for Cd as the nitrilotriacetate. Reversed-phase determination of organo-anions of phosphorus and sulphur has also been accomplished [65].

Interfaced ICP has proved valuable for elemental speciation in size-exclusion chromatographic profiling of fuel liquids and oils [13]. Simultaneous analysis of the molecular size distribution of sulphur-, vanadium- and nickel-containing compounds in petroleum crudes and residues has been accomplished and the method appears of potential value in fingerprint identification of oils in environmental circumstances. In overcoming the inherent inefficiencies in obtaining quantitative transfer of HPLC eluates to the ICP, a promising development has been the direct injection nebulizer (DIN) [14], a total-injection microconcentric nebulizer which gives close to 100% nebulization and transport efficiency. Detection limits ranging from 164 ng/ml for sulphur to 4 ng/ml for zinc have been reported. Figure 4.15 shows an ion-pair reversed-phase chromatogram of a range of sulphur anions at a sulphur concentration of 10 μg/l. The ICP measurement was at 182.0 nm with an argon purge. Many approaches to enhance the efficiency of HPLC–ICP are under consideration; these include miniaturized plasma torches, water-cooled spray chambers, low-power and low-flow torches, aerosol cooling and oxygen feeding. There is no doubt that such enhancements will be needed to offset some of the inherent difficulties with practical

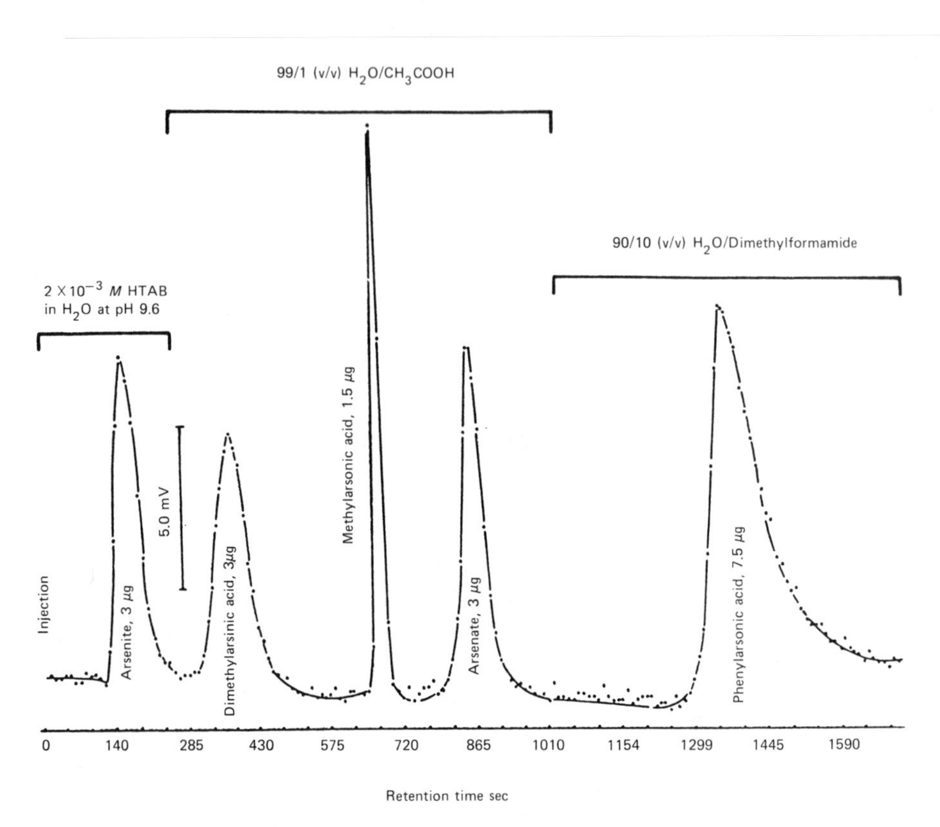

Fig. 4.14 — HPLC–ICP separation of arsenite, arsenate, methylarsonic acid, dimethylarsinic acid and phenylarsonic acid. Reversed-phase, resin-based column; injection volume 100 µl. (Reproduced by permission, from K. J. Irgolic, R. A. Stockton, D. Chakraborti and W. Beyer, *Spectrochim. Acta*, 1983, **38B**, 437. Copyright, 1983 Pergamon Press).

HPLC–ICP conditions, such as the limitation of aspiration rates of solvents such as methanol and acetonitrile to 0.1–0.2 ml/min.

Perhaps the most sophisticated application of HPLC–ICP yet described has been that of interfaced HPLC–mass spectroscopy [66]. With this technique detection limits as low as 100 pg have been found for arsenic and 700 pg for lead. Figure 4.16 shows a dual-detection ion-pair chromatogram for arsenic and selenium anions, obtained by using an anion-switching mass-monitoring mode.

4.13 PLASMA EMISSION DETECTION FOR SUPERCRITICAL FLUID CHROMATOGRAPHY (SFC)

Although the first SFC separations were reported in the early 1960s, it has only been in recent years, with the advent of viable capillary SFC systems and commercialization of instrumentation, that it has attracted renewed attention. High-resolution

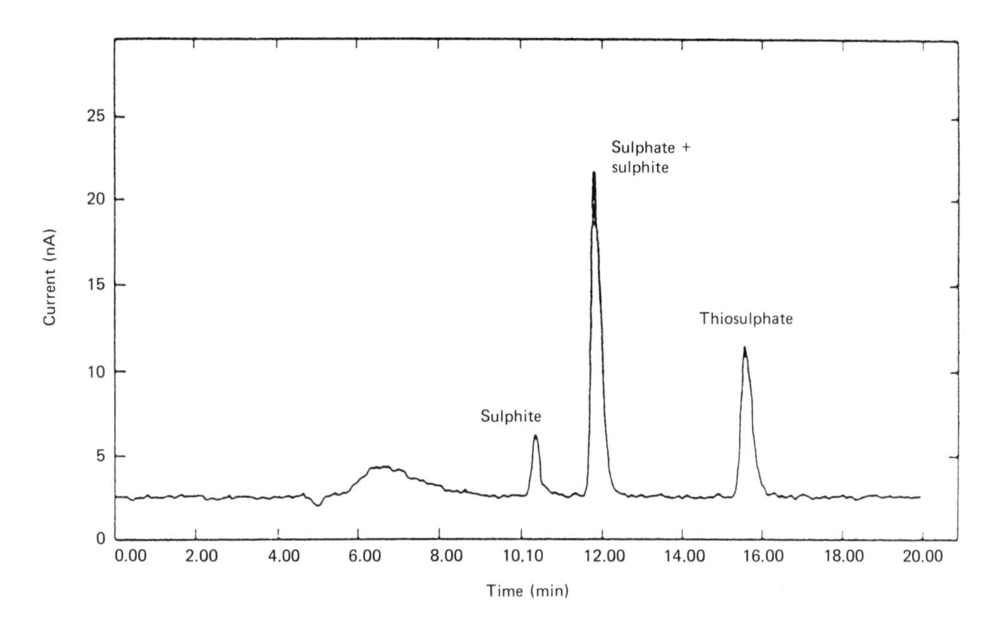

Fig. 4.15 — Ion-pair reversed-phase, HPLC–DIN–ICP chromatogram of several S-containing anions at 10 μg/ml each (as S): mobile phase, 5m*M* tetrabutylammonium phosphate in (95/5) water/methanol; column flow-rate 0.7 ml/min (ca. 15% to plasma); wavelength 182.0 nm (Ar purge). (Reproduced by permission, from K. E. LaFreniere, V. A. Fassel and D. E. Eckels, *Anal. Chem.*, 1987, **59**, 879. Copyright 1987, American Chemical Society).

capillary and packed microcolumn SFC, supercritical extraction methods, and their combination for analysis of complex samples promise to extend separation capabilities in areas where neither GC or HPLC may be suitable. Adoption of detectors for SFC has followed various directions; where the methodology and instrumentation have been derived from GC, as in capillary SFC, the flame-ionization detector has been favoured. For development related more to HPLC, the UV–visible spectrophotometric detector has been adopted. Plasma emission, in particular ICP, seems a natural development by virtue of its use in both GC and HPLC. An initial report has appeared describing such an interface [67]. Supercritical fluid transfer to the ICP allowed sample to be introduced in an easily atomized form in a gas (formed from the supercritical carbon dioxide mobile phase) with close to 100% efficiency. Modification of plasma excitation by SFC solvents appears to be less troublesome than that by typical organic HPLC solvents. It seems likely that as SFC becomes more widely adopted, element-specific detection by atomic plasma emission will become a useful option.

4.14 FUTURE DIRECTIONS OF PLASMA SPECTROMETRY FOR CHROMATOGRAPHIC DETECTION

In practical terms, for environmental as well as other areas of analysis, the broader adoption of plasma spectral detection will depend on the introduction of standard-

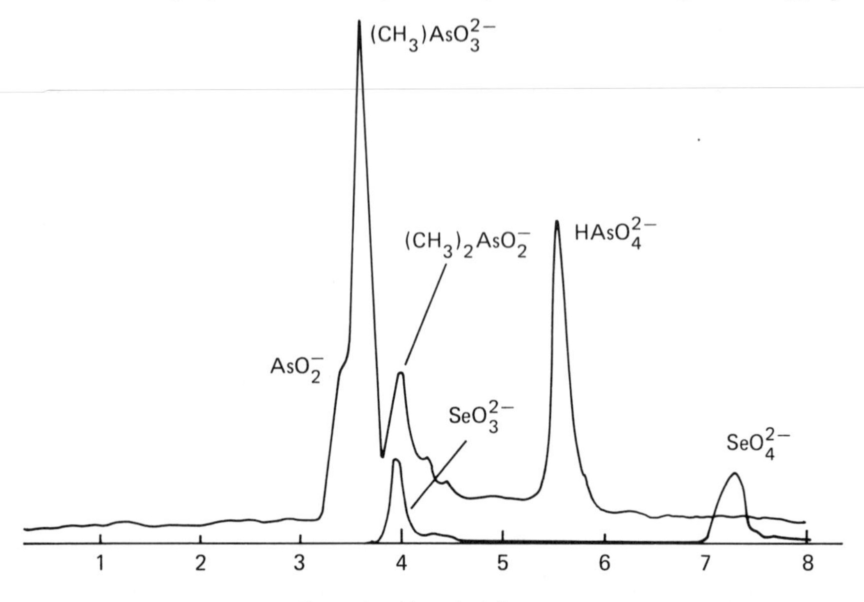

Fig. 4.16 — Detection of As and Se species by peak-switching HPLC–ICP–MS. ^{76}Se was monitored. Amount injected was 10 μl of 0.5 mg/l. solution (5 ng as element.) Mobile phase 5% methanol, 0.005*M* ion-pair reagent P1C–A, flow-rate 1.0 ml/min. (Reproduced by permission, from J. J. Thompson and R. S. Houk, *Anal. Chem.*, 1986, **58,** 2541. Copyright 1986, American Chemical Society).

ized commercial instrumentation which will enable interlaboratory study of repeatability of data and development of 'recommended' methods of analysis which can be widely accessible. Environmental analysis in particular is subject to restrictions designed to ensure high levels of accuracy and precision in analysis. Plasma chromatographic detection has already demonstrated its wide utility, as shown by research publications from academic, governmental and industrial laboratories. It can be stated with some confidence that GC–microwave-plasma emission systems in particular are close to wide adoption and prospects for economic commercialization appear good, despite earlier set-backs. Fully integrated units which circumvent the necessity for analysts to interface their own chromatograph, emission device and spectrometer, may become as familiar in the future as GC–MS and GC–FTIR systems are now. Integrated HPLC and SFC systems will be delayed somewhat longer, but their eventual adoption seems very probable.

REFERENCES

[1] R. M. Barnes, *CRC Crit. Rev. Anal. Chem.*, 1978, **7**, 203.
[2] R. J. Decker, *Spectrochim. Acta*, 1980, **35B**, 19.
[3] C. A. Bache and D. J. Lisk, *Anal. Chem.*, 1967, **39**, 786.
[4] W. R. McLean, D. L. Stanton and G. E. Penketh, *Analyst*, 1973, **98**, 432.
[5] T. H. Risby and Y. Talmi, *CRC Crit. Rev. Anal. Chem.*, 1983, **14**, 231.
[6] F. C. Fehsenfeld, K. M. Evenson and H. P. Broida, *Rev. Sci. Instr.*, 1965, **36**, 294.
[7] C. I. M. Beenakker, *Spectrochim. Acta*, 1976, **31B**, 483.
[8] M. H. Abdallah, S. Coulombe and J. M. Mermet, *Spectrochim. Acta*, 1982, **37B**, 583.

[9] R. J. Lloyd, R. M. Barnes, P. C. Uden and W. G. Elliott, *Anal. Chem.*, 1978, **50**, 2025.
[10] K. W. Panaro, D. Erikson and I. S. Krull, *Analyst*, 1987, **112**, 1097.
[11] J. O. Beyer, *Ph.D. Dissertation*, University of Massachusetts, 1984.
[12] P. C. Uden, B. D. Quimby, R. M. Barnes and W. G. Elliott, *Anal. Chim. Acta*, 1978, **101**, 99.
[13] D. W. Hausler, *Spectrochim. Acta*, 1985, **40B**, 389.
[14] K. E. LeFreniere, V. A. Fassel and D. E. Eckels, *Anal. Chem.*, 1987, **59**, 879.
[15] *MPD 850 Manual*, Applied Chromatography Systems Ltd., Luton, U.K.
[16] A. Bollo-Kamara and E. G. Codding, *Spectrochim. Acta*, 1981, **36B**, 973.
[17] S. R. Goode, B. Chambers and N. P. Buddin, *Spectrochim. Acta*, 1985, **40B**, 329.
[18] S. A. Estes, P. C. Uden and R. M. Barnes, *Anal. Chem.*, 1981, **53**, 1336.
[19] Reference 5, p. 250.
[20] P. C. Uden, *Anal. Proc.*, 1981, **18**, 189.
[21] M. Zerezghi, K. J. Mulligan and J. A. Caruso, *J. Chromatog. Sci.*, 1984, **22**, 348.
[22] K. S. Brenner, *J. Chromatog.*, 1978, **167**, 365.
[23] K. J. Slatkavitz, L. D. Hoey, P. C. Uden and R. M. Barnes, *Anal. Chem.*, 1985, **57**, 1846.
[24] H. J. Perpall, P. C. Uden and R. L. Deming, *Spectrochim. Acta*, 1987, **42B**, 243.
[25] Y. Takigawa, T. Hanai and J. Hubert, *J. High Resol. Chromatog. Chromatog. Commun.*, 1986, **9**, 698.
[26] P. C. Uden, K. J. Slatkavitz, R. M. Barnes and R. L. Deming, *Anal. Chim. Acta*, 1986, **180**, 401.
[27] Zeng Kewei, Gu Qingyu, Wang Guochuen and Yu Weile, *Spectrochim. Acta*, 1985, **40B**, 349.
[28] D. F. Hagen, J. S. Marhevka and L. C. Haddad, *Spectrochim. Acta*, 1985, **40B**, 335.
[29] B. D. Quimby, M. F. Delaney, P. C. Uden and R. M. Barnes, *Anal. Chem.*, 1979, **51**, 875.
[30] B. D. Quimby, M. F. Delaney, P. C. Uden and R. M. Barnes, *Anal. Chem.*, 1980, **52**, 259.
[31] D. E. Pivonka, W. G. Fateley and R. C. Fry, *Appl. Spectrosc.*, 1986, **40**, 291.
[32] S. W. Jordan, I. S. Krull and S. B. Smith, Jr., *Anal. Lett.*, 1982, **15**, 1131.
[33] G. D. Riska, S. A. Estes, J. O. Beyer and P. C. Uden, *Spectrochim. Acta*, 1983, **38B**, 407.
[34] G. W. Rice, A. P. D'Silva and V. A. Fassel, *Spectrochim. Acta*, 1985, **40B**, 1573.
[35] A. J. McCormack, S. C. Tong and W. D. Cooke, *Anal. Chem.*, 1965, **37**, 1470.
[36] K. J. Mulligan, J. A. Caruso and F. L. Fricke, *Analyst*, 1980, **105**, 1060.
[37] C. A. Bache and D. J. Lisk, *Anal. Chem.*, 1965, **37**, 1477.
[38] H. E. Taylor, J. H. Gibson and R. K. Skogerboe, *Anal. Chem.*, 1970, **42**, 1569.
[39] Y. Talmi and V. E. Norvall, *Anal. Chem.*, 1975, **47**, 1510.
[40] Y. Talmi and D. T. Bostick, *Anal. Chem.*, 1975, **47**, 2145.
[41] B. D. Quimby, P. C. Uden and R. M. Barnes, *Anal. Chem.*, 1978, **50**, 2112.
[42] S. A. Estes, P. C. Uden and R. M. Barnes, *Anal. Chem.*, 1981, **53**, 1829.
[43] S. A. Estes, P. C. Uden and R. M. Barnes, *Anal. Chem.*, 1982, **54**, 2402.
[44] K. B. Olsen, D. S. Sklarew and J. C. Evans, *Spectrochim. Acta*, 1985, **40B**, 357.
[45] R. B. Robbins and J. A. Caruso, *J. Chromatog. Sci.*, 1979, **17**, 360.
[46] P. C. Uden, *J. Chromatog.*, 1984, **313**, 3.
[47] M. S. Black and R. E. Sievers, *Anal. Chem.*, 1976, **48**, 1872.
[48] L. Ebdon, S. Hill and R. W. Ward, *Analyst*, 1986, **111**, 1113.
[49] H. A. H. Billiet, J. P. J. van Dalen, P. J. Schoemakers and L. deGalen, *Anal. Chem.*, 1983, **55**, 847.
[50] P. C. Uden, R. M. Barnes and F. P. DiSanzo, *Anal. Chem.*, 1978, **50**, 852.
[51] S. A. Estes, C. A. Poirier, P. C. Uden and R. M. Barnes, *J. Chromatog.*, 1980, **196**, 265.
[52] J. J. Kosman, Jr., *Ph.D. Dissertation*, University of Massachusetts, 1982.
[53] I. S. Krull in *Liquid Chromatography in Environmental Analysis*, J. F. Lawrence (ed.), Chapter 5, Humana Press, Clifton, N.J., 1983.
[54] I. S. Krull, K. W. Panaro and L. L. Gershman, *J. Chromatog. Sci.*, 1983, **21**, 460.
[55] I. S. Krull and K. W. Panaro, *Appl. Spectrosc.*, 1985, **39**, 960.
[56] D. L. Windsor and M. B. Denton, *J. Chromatog. Sci.*, 1979, **17**, 492.
[57] K. Ohls and D. Sommer, in *Developments in Atomic Plasma Spectrochemical Analysis*, R. M. Barnes (ed.), pp. 321–326, Heyden, London, 1981.
[58] R. M. Brown, Jr. and R. C. Fry, *Anal. Chem.*, 1981, **53**, 532.
[59] M. A. Eckhoff, J. P. McCarthy and J. A. Caruso, *Anal. Chem.*, 1982, **54**, 165.
[60] L. Ebdon, S. Hill and R. W. Ward, *Analyst*, 1987, **112**, 1.
[61] D. M. Fraley, D. Yates and S. E. Manahan, *Anal. Chem.*, 1979, **51**, 2225.
[62] K. Jinno, H. Tsuchida, S. Nakanishi, Y. Hirata and C. Fujimoto, *Appl. Spectrosc.*, 1983, **37**, 258.
[63] K. J. Irgolic, R. A. Stockton, D. Chakraborti and W. Beyer, *Spectrochim. Acta*, 1983, **38B**, 437.
[64] W. Nisamaneepong, M. Ibrahim, T. W. Gilbert and J. A. Caruso, *J. Chromatog. Sci.*, 1984, **22**, 473.
[65] D. R. Heine, M. B. Denton and T. D. Schlabach, *J. Chromatog. Sci.*, 1985, **23**, 454.
[66] J. J. Thompson and R. S. Houk, *Anal. Chem.*, 1986, **58**, 2541.
[67] J. W. Olesik and S. V. Olesik, *Anal. Chem.*, 1987, **59**, 796.

5

Atomic fluorescence detectors

Alessandro D'Ulivo
Istituto di Chimica Analitica Strumentale del C.N.R., Via Risorgimento, 35, 56100 Pisa, Italy

Atomic-fluorescence spectrometry (AFS) was first used as an analytical technique in 1964 [1,2], and many papers about analytical AFS have been published since then. The current image of analytical AFS is that of a very flexible technique usable for single or multielement determinations with a wide variety of light sources, atomizers, optical designs and electronics. Most AF spectrometers are simple and inexpensive to assemble and their performance combines good detection limits with wide dynamic ranges, and good precision and accuracy.

However, AF instruments do not appear to be successful from the commercial point of view, compared with other atomic-spectrometric techniques, and currently only one AF spectrometer is marketed. Atomic-absorption spectrometry (AAS) has assumed unmatched popularity in laboratory practice for single element determination, as has inductively coupled plasma atomic-emission spectroscopy (ICP–AES) for multielement determination. It is beyond the scope of this chapter to analyse the causes of this, but it is clear that the lack of commercially available spectrometers has caused the application of AFS to be restricted in the field of element-specific detection for chromatography, because most detectors are assembled by interfacing commercially available AA or ICP–AE spectrometers with the chromatographic apparatus. So far, only a few works report the use of AFS as an element-specific detection system for chromatography, and in all cases home-made spectrometers were employed. There is no doubt that the potential of this analytical technique has not yet been fully exploited for chromatographic detection.

This chapter aims to give some insight into the capabilities of AFS for chromatographic detection, and give indications for the assembling of experimental apparatus. The first section gives a brief summary of the theory of atomic fluorescence: interested readers should refer to the excellent paper of Omenetto and Winefordner [3] for further information on the subject. The second section, without being exhaustive, deals with basic AF instrumentation, with emphasis on particular features that appear most promising for chromatographic applications. Finally, the third section reviews applications of AFS as an element-specific detector for gas or liquid chromatography.

5.1 THEORY OF ATOMIC FLUORESCENCE

5.1.1 Atomic-fluorescence transitions

The term atomic fluorescence refers to the radiational de-excitation of free atoms that have been electronically excited by radiation of suitable energy, or in other words it is an atomic emission stimulated by light. Atomic-fluorescence radiation is emitted isotropically by the atomic vapour and can be observed from all directions around the atom cell.

There are many types of atomic fluorescence, each characterized by the pathways followed by atoms in the excitation and de-excitation processes [3,4]. Some examples of atomic-fluorescence transitions are shown in Fig. 5.1. In *resonance fluorescence*, the radiational excitation and de-excitation of the atomic system are between the same upper and lower energy levels and the wavelengths of the absorbed radiation λ_A and of the fluorescence radiation λ_F are the same. The atomic excitation and de-excitation can be a combination of radiational and collisional processes. When the upper level can be reached by means of collisional processes and is followed by the radiational process, the transition is said to be *excited*. Alternatively, when the upper level is reached by radiational processes followed by the collisional ones, the transition is said to be *thermally assisted*. A transition is termed *Stokes* if $\lambda_F > \lambda_A$ and *anti-Stokes* if $\lambda_F < \lambda_A$. If only the upper level is common to the radiational excitation and de-excitation processes the transition is termed *direct-line*. If different upper levels are involved in radiational excitation and de-excitation processes the transition is termed *stepwise-line*. A particular type of excitation is the two-photon process (Fig. 5.1) but it is produced only with high-intensity sources, and in practice it is observed only with laser excitation. Another type of fluorescence transition, *sensitized fluorescence*, is observed when an atom of one species is excited radiationally and then transfers its excitation energy to an atom of the same or another species, that is de-excited radiatively according to the following scheme, where the asterisk denotes an electronically excited atom.

$$A + h\nu_A \rightarrow A^*$$
$$A^* + B \rightarrow A + B^*$$
$$B^* \rightarrow B + h\nu_F$$

where h is Planck's constant.

Since not all these processes are active simultaneously an atomic-fluorescence spectrum consists typically of only a few lines, the relative intensity of which depends on the type of atomizer and light source. Indeed, the simplicity of atomic-fluorescence spectra is one of the most important features of this technique.

5.1.2 Atomic-fluorescence radiance

The intensity of fluorescence radiation produced in a transition depends on several factors such as the intensity of the excitation source, the concentration of atoms in

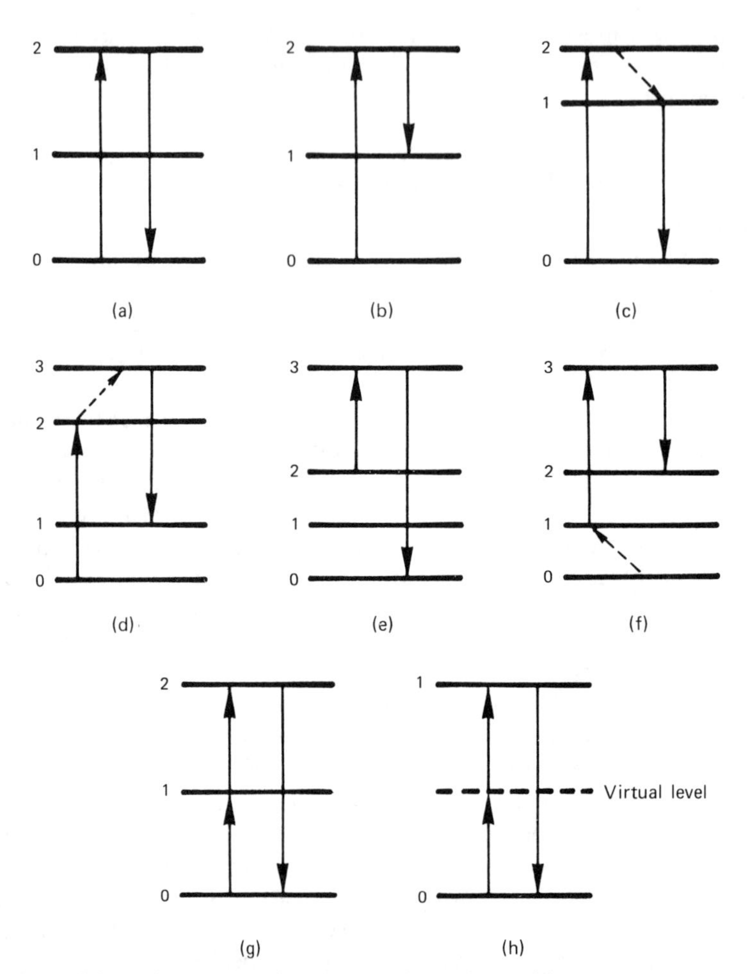

Fig. 5.1 — Schematic representation of some AF transitions. (a) resonance fluorescence; (b) direct-line fluorescence; (c) stepwise-line fluorescence; (d) thermally assisted stepwise-line fluorescence; (e) direct-line anti-Stokes fluorescence; (f) excited-state direct-line fluorescence; (g) two-photon transition involving real states; (h) two-photon transition involving virtual level. Full lines, radiational processes. Dashed lines, collisional processes.

the atomizer, the quantum efficiency of the process (quantum efficiency being the ratio of the energy emitted in fluorescence to the energy absorbed, per unit of time) and the extent of self-absorption occurring in the atomizer. Let us consider an idealized case of a simple two-level system where (1) is the lower level and (2) the upper level. If the atomic vapour is assumed to be dilute and thin [3,4], the fluorescence radiance will be given by:

$$B_{\mathrm{F}} = \frac{l}{4\pi} Y_{21} E_{\nu_0}[\int k(\nu)\mathrm{d}\nu] \tag{5.1}$$

where B_F is the absolute radiance (erg.sec^{-1}.cm^{-2}.sr^{-1}), l is the thickness of the fluorescing layer (cm), Y_{21} is the quantum efficiency for the transition 2→1, E_{ν_0} is the spectral irradience of the source (erg.sec^{-1} cm^{-2} Hz^{-1} and $k(\nu)$ the atomic-absorption coefficient (cm^{-1}). The term in square brackets in Eq. (5.1) is the absorption coefficient integrated over the width of the absorption line, which is a function of the Einstein B_{12} absorption coefficient, of the statistical weights of states (1) and (2) and their population, and of the energy $h\nu_0$ of the exciting radiation.

Equation (5.1) gives some important information. The fluorescence radiance is linearly dependent on the source irradiance and the quantum efficiency of the transition: thus to obtain large fluorescence signals it is necessary to have high-irradiance light-sources and atomizers with high quantum efficiency in which collisional de-excitation processes are kept to a minimum. Collisional de-excitation is commonly called *quenching* and is due to collisions of the excited atoms with molecules, radicals and elements present in the atomizer.

Equation (5.1) holds for low-irradiance sources, such as common continuum or spectral-line sources, in which the population of the upper state n_2 is negligible compared with that of the lower state n_1. If the irradiance of the source is enhanced, the population of the states changes until the limiting condition of saturation is reached, in which $n_1 = n_2$. To take account of this, a more general expression must be considered for the fluorescence radiance [3]:

$$B_F = \frac{l}{4\pi} Y_{21}[\int K(\nu)d\nu]\, E_{\nu_0}[E^s_{\nu_0}/(E_{\nu_0} + E^s_{\nu_0})] \quad (5.2)$$

where $E^s_{\nu_0}$ is called the saturation parameter (erg.sec^{-1}.cm^{-2}.Hz^{-1}). It is simple to demonstrate that Eq. (5.2) reduces to Eq. (5.1) in the case of a low-irradiance source, i.e. when $E_{\nu_0} \ll E^s_{\nu_0}$. Equation (5.2) must be used whenever E_{ν_0} is not negligible compared to $E^s_{\nu_0}$ so that the population n_2 is not negligible compared to n_1. Clearly, the dependence of B_F on E_{ν_0} is non-linear. In the limiting case of saturation, $E_{\nu_0} > E^s_{\nu_0}$, and we have for the maximum value of fluorescence radiance the expression:

$$|B_F|_{max} = \frac{l}{4\pi} h_{\nu_0} A_{21} \left(\frac{n_T}{2}\right) \quad (5.3)$$

where A_{21} is the Einstein coefficient for spontaneous emission (sec^{-1}) and n_T is the total atomic population ($n_1 + n_2$). In the saturation condition, B_F is independent both of the source radiance and of the quantum efficiency, but remains linearly dependent on the total atomic population n_T as long as the optical density is low.

5.1.3 Fluorescence growth curves

The fluorescence radiance B_F is linearly dependent on the atomic population in the atomizer while the concentration of atoms remains low. Under this condition linear graphs are obtained on plotting B_F *vs.* atom concentration. When the atom concentration is increased further, growth curves bend toward the concentration axis. The shape of the growth curves depends on many factors [3] such as the geometry of the illuminated volume and of the observed volume, the type of excitation source (line or continuum), and its irradiance.

The optical pathway geometry is important because complete and uniform illumination and observation of the entire atomizer cell gives the longest linear range for the growth curves, and this is especially important in analytical applications [3].

The type of source influences the shape of the curves only at high atom concentration: with both line and continuum sources, linear behaviour is observed at low concentrations (log–log plot). At high atom concentrations, line sources produce a maximum followed by a slope of -0.5, whereas continuum sources give a maximal constant response, with B_F independent of the concentration (see Fig. 5.2).

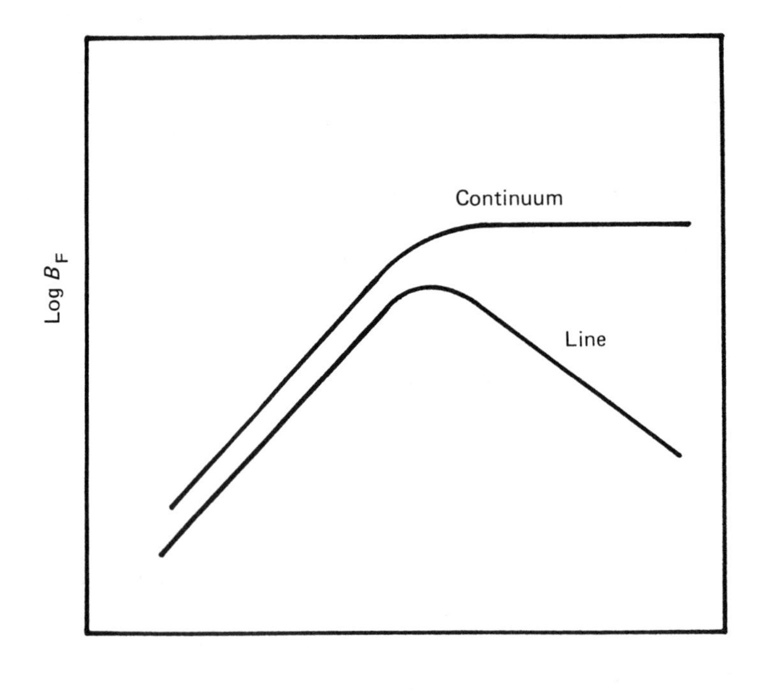

Fig. 5.2 — Shape of AF growth curves with line or continuum light source.

Increasing the source intensity usually results in wider linear ranges, because the lower part of the curve is extended but the atomic concentration at which self-absorption causes curvature remains the same. If the intensity of the source allows saturation conditions to be established (as with laser excitation) the absorption coefficient goes to zero [Eq. (5.3)] and self-absorption does not occur. This should result in an improvement by extending both ends of the linear range, but in practice these phenomena are not always observed because the laser beam geometry hinders complete illumination of the atom cell [5].

5.2 ATOMIC-FLUORESCENCE INSTRUMENTATION

5.2.1 The atomic-fluorescence spectrometer

Figure 5.3 shows a diagram of the most common apparatus used for single-element AF determination. Although this apparatus could be improved, it is sufficient to allow the working principles of an AF spectrometer to be described.

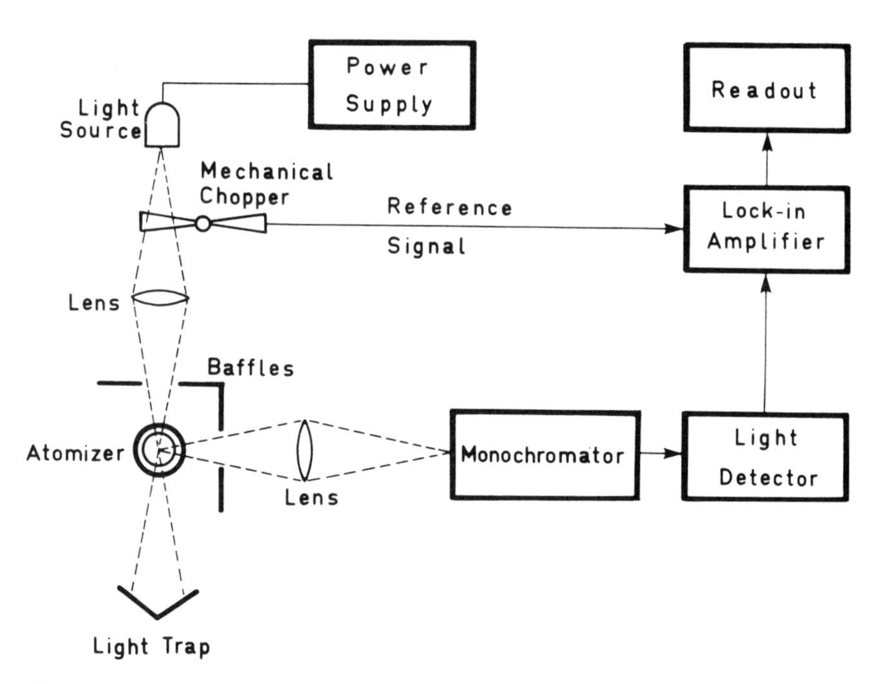

Fig. 5.3 — Schematic diagram of a single-element dispersive AF spectrometer using mechanical modulation of the light source.

The radiation from the light-source is focused on the atom cell by a lens, to enhance the spectral irradiance E_{ν_0}. A mechanical chopper provides amplitude modulation of the light source at a fixed frequency f_0. The AF signal will be modulated at the same frequency; this provides a method of distinguishing the AF signal from any atomic emission and other noise originating in the atomizer. A fraction of the total AF radiation is collected by another lens and focused onto the dispersive system, where a suitable AF line is isolated. The intensity of the line is measured by a light detector, typically a photomultiplier tube (PMT). At this stage, the output signal of PMT contains both the AF signal and the noise. The selective and discriminant amplification needed to improve the signal-to-noise ratio is usually provided by a lock-in amplifier. Baffles, diaphragms and light traps are often used to minimize reflection or back-diffusion of light coming directly from the source toward the AF collecting lens and then toward the light detector.

The most common modifications to the spectrometer depicted in Fig. 5.3 are: (a) replacement of mechanical modulation of the light-source by electronic modulation (through power supply modulation); (b) removal of the monochromator, or its replacement with a filter to give a non-dispersive spectrometer, and (c) the use of gated electronics, boxcar integrators or a photon-counter in place of the lock-in amplifier. These modifications, combined with several types of light-sources, atomizers and optics, offer to the user a wide variety of configurations, the merits of which will now be discussed.

5.2.2 Atomizers

To be useful for AFS, an atomizer must have several properties, some of which are common to the other atomic spectrometric techniques and some of which are specific [3,4]. The sample utilization efficiency, atomization efficiency, quantum efficiency, and residence time of atoms in the optical path should all be maximized in order to obtain large signals. At the same time, the background emission in the relevant spectral region should be minimized to obtain a low noise level. The atomizer system should also provide a signal of suitable precision, free from interference effects. Furthermore, the atomizer geometry and the atom cell illumination are important, as already illustrated, for obtaining large signals and wide dynamic ranges. Since no single atomizer has all these features, some sort of compromise must be reached for each analytical problem.

When gas or liquid chromatography is interfaced with an AF spectrometer (used as an element-specific detector), the atomizer represents the coupling point between the two pieces of apparatus. Additional characteristics may be required of such an atomizer. The most relevant one is that the atomizer must be compatible with a continuous flow of sample in order for it to function as an 'on-line' detector. Practically all types of atomizers are suitable for interfacing with gas-chromatographic apparatus, whereas only premixed flames, plasmas and continuous-flow furnaces possess the requirements for functioning with a liquid chromatograph.

5.2.2.1 Premixed laminar flames

Turbulent flames, combined with total consumption burners, were used in early studies [6]. They are prone to interference and scattering problems [7,8] and were replaced by premixed flames supported on laminar flow burners.

Premixed laminar flames suffer from poor sample utilization in the nebulization step: typically not more than about 5% of the aspirated solution is converted into a fine mist and only a fraction of it is then desolvated and atomized in the flame. At the same time, high gas flow-rates (10–15 l./min) produce undesirable dilution of the atomic vapour and a low residence time for the atoms in the optical path (of the order of msec), the overall effect being reduction of sensitivity. In addition, most premixed laminar flames, in particular the acetylene-based flames, exhibit high background emission due to radicals of the secondary reaction zone (OH, C_2, CH, CN) and give poor detection limits. Separation of the secondary reaction zone by argon or nitrogen sheathing (Fig. 5.4) results in a noticeable improvement of sensitivity [9,10]. Since nitrogen is an efficient quencher, argon should be preferred, but in most cases nitrogen offers quite acceptable performance, at lower cost. A general advantage of the flame system is its high precision, expecially when steady signals are obtained with continuous-flow nebulization. Other characteristics such as temperature, ato-

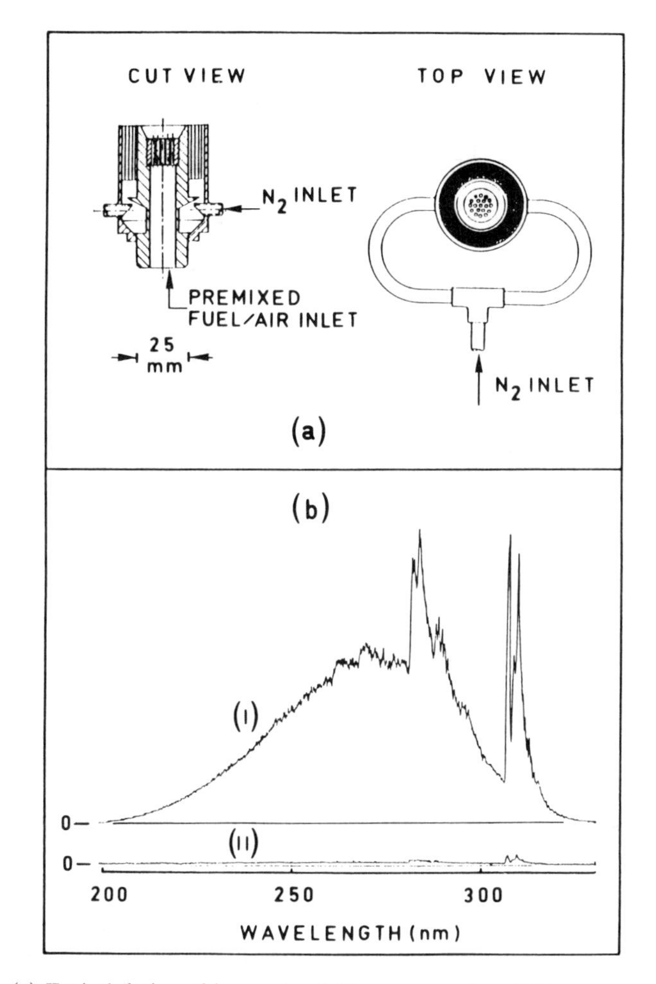

Fig. 5.4 — (a) Typical design of burner head (titanium) used in AFS to support sheathed flames; (b) effect of gas sheathing on the flame background of the air–acetylene flame. (I) unsheathed flame, (II) sheathed flame. (Reprinted from P. L. Larkins, *Spectrochim. Acta*, 1971, **36B**, 477, by permission. Copyright 1971, Pergamon Press.)

mization efficiency, quantum efficiency, interference effects and background emission depend very much on flame type and are summarized in Table 5.1.

Burner geometry is also important; the slot-type burner heads used in AAS are not well suited for the geometrical requirements of AFS. Round burners (Fig. 5.4), provided with gas sheathing [10,11], are quite satisfactory, simple to make and easily fitted onto a commercially available AAS premix chamber already provided with nebulizer, gas-handling system and safe-operating devices.

5.2.2.2 Miniature diffusion flames

These flame systems were specially developed to be used in AFS combined with particular sampling techniques. Their use results in improved sensitivity since higher

Table 5.1 — Summary of characteristics of premixed laminar flames for AFS

Flame	Max. temp. (°C)	Advantages	Disadvantages	Notes
Air–acetylene (argon-sheathed)	≃2300	Good atomization efficiency and acceptable sensitivity for most elements. Low interference effects.	Reduced atomization efficiency for refractory elements.	The usual flame for AFS.
N_2O–acetylene (argon-sheathed)	≃2700	Good atomization efficiency also for refractory elements. Very low interference and scattering problems.	Detection limits worse than for air–acetylene, owing to lower quantum efficiency and higher background emission. Ionization of some elements.	Practical use limited to refractory elements or very complex matrices
Ar–O_2–acetylene (argon-sheathed)	2200–2600	Flame temperature varies with Ar/O_2 ratio. Detection limits 2–5 times better than for air–acetylene flame [10a].	More hazardous and expensive than air–acetylene flame. Careful operation required [10b].	Useful flame system for all elements.
Air–H_2 (argon-sheathed)	≃2000	Low background emission and good quantum efficiency. Good detection limits for volatile elements.	Atomization efficiency not suitable for many elements. Interference and scattering problems.	Practical use limited to As, Se, Te, Hg, Cd and other volatile elements in fairly pure solutions.
Ar–H_2–entrained-air	⩽1400	High quantum efficiency and very low background emission. Good transparency to UV radiation even for λ<200 nm.	Poor atomization efficiency for most elements. High interference and scattering problems with solution nebulization.	Practical use limited mainly to hydride generation technique with good detection limits.

residence times, smaller atom cell volumes, lower background emission and higher quantum efficiencies are achieved, compared with similar flames operated at conventional gas flow-rates.

Tsujii and Kuga [12] developed a special burner (Fig. 5.5), supporting a miniature argon–hydrogen–entrained air flame, to be used in combination with the hydride generation method and non-dispersive apparatus. In this system, an argon

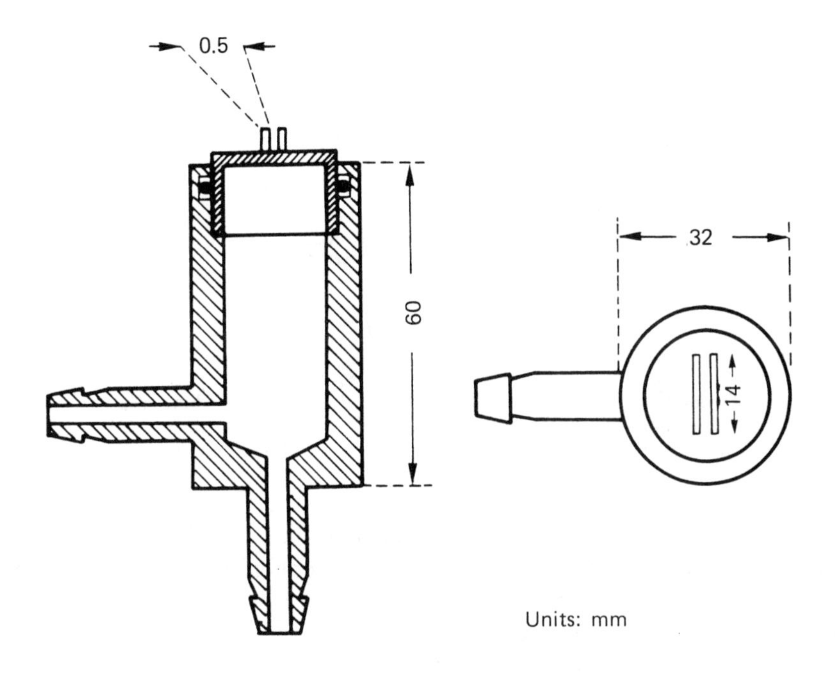

Fig. 5.5 — Burner head used to support a laminar argon–hydrogen miniature flame. (Reprinted from K. Tsujii and K. Kuga, *Anal. Chim. Acta*, 1978, **97**, 51, by permission. Copyright, 1978, Elsevier Science Publishers.)

flow of about 1 l./min acts both as support gas for the flame and as carrier gas for the hydrides. The decrease of the hydrogen flow-rate affords a remarkable improvement in the AF signal (Fig. 5.6) and a noticeable decrease in the baseline background noise. Compared with the conventional Ar/H_2 flame using 1–2 l./min flow of hydrogen, an improvement by a factor of 30–40 in the absolute detection limits is obtained for As, Sb and Sn [12,13] with a hydrogen flow-rate of 0.25 l./min. A further fivefold improvement is obtained by a slight modification of the burner design to allow a hydrogen flow-rate of 0.15 l./min; the detection limit for As is 10 pg [14] and 20 pg for Sb [15]. The argon–hydrogen miniature flame can also be supported on narrow glass or silica tubing (i.d. 4–8 mm) to obtain similar improved performances [16,17]. The major disadvantage of miniature flames is that only a small sample volume (0.5–5 ml) can be used with the direct sweeping of hydrides, because larger samples produce a pressure increase in the hydride generation and the flame can be extinguished as a result. Trapping the volatile hydrides at liquid-nitrogen tempera-

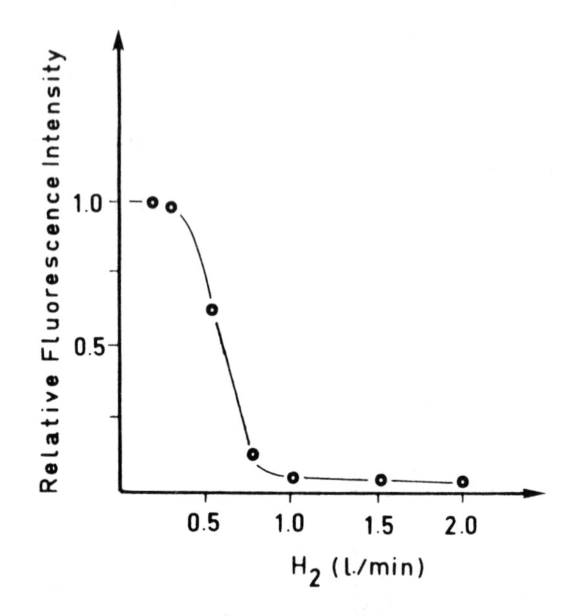

Fig. 5.6 — Dependence of AF signal on the hydrogen flow-rate for the burner head shown in Fig. 5.4 in combination with hydride generation. (Reprinted from K. Tsujii and K. Kuga, *Anal. Chim. Acta*, 1978, **97**, 51, by permission. Copyright, 1978, Elsevier Science Publishers.)

ture, followed by thermal stripping, circumvents this problem, and also larger volumes of sample solution can be used with a consequent improvement in concentration detection limits. When a glass tube is used as burner, the risk of extinguishing the flame can be reduced in part by using a glass tube with a larger internal diameter. Thus, at constant gas flow-rate, the residence time is also increased and the sensitivity improved [18,19].

A high temperature (2400°C), nitrous oxide–hydrogen miniature laminar diffusion flame (H_2, 0.16 l./min, N_2O, 1.2 l./min) has been used in combination with a graphite-cup atomizer to prevent light-scattering by the volatile matrix components (e.g. NaCl) or from the small carbon particles derived from graphite itself [19a]. Very good detection limits (0.2–0.4 ng) and low amounts of scatter were obtained for Cd, Cu and Mn with a continuum light source, and there is no doubt that noticeable improvement could be obtained with a more intense spectral line source.

Miniature flames fit very well with the purpose of gas chromatographic detection and their use gives sensitivities comparable to or better than those obtained with the gas chromoatography–electrothermal AAS detector (see Section 5.3.2).

5.2.2.3 Inductively coupled argon plasma (ICP)

The ICP was first used as an atomizer for AFS by Montaser and Fassell [20] to overcome several disadvantages related to the use of flames. The radiofrequency powered ICP can provide an effective temperature of 6000–10000°C, and therefore it gives lower interference, fewer scattering problems and higher volatilization–atomization efficiency than do flames. Since the plume of hot gases mostly consists of monatomic species, it also has a good quantum efficiency. The only disadvantage associated with this kind of atomizer is the high background emission, which increases with the rf power dissipated in the plasma. The noise–power spectrum of an ICP atomizer has a strong component that is inversely dependent on the measurement frequency (f) (flicker noise) in the range below 200 Hz, and at above 200 Hz the predominating noise component is almost independent of f (white noise). The signal-to-noise ratio (SNR) can therefore be improved by source modulation above 200 Hz with gated electronics [21]. Other characteristics such as stability of signal, precision and residence time of atoms are comparable to those in flame systems. The ICP is an excellent general purpose atomizer because both refractory elements and easily atomizable and ionizable elements can be determined. The optimum conditions for the determination of each are readily found by regulating the radiofrequency power applied and the height, above the load coil, at which the observation is made [20,22]. In many cases the element ionization produced in the ICP can yield better detection limits. The population of ions can be excited by means of a dye laser tuned to the correct ionic absorption wavelength, and detection of a suitable ionic fluorescence line [23,24]. As an alternative to a dye laser, another ICP, continuously supplied with a solution of the element (10–20 g/l.), can be used as the atomic or ionic line source [25,26], but with this latter technique a worsening of the detection limits was observed.

The combination of an ICP atomizer/ionizer with a tunable dye laser to excite the ionic or atomic fluorescence represents one of the most powerful atomic spectrometric techniques for single-element determination [23]. It provides detection limits and dynamic ranges comparable with or better than for ICP–AES but has higher selectivity, lower interference effects, and the possibility of background signal correction [24].

5.2.2.4 Electrothermal atomizers

Carbon rods [27–31] and graphite furnaces [32] can be considered the most useful atom reservoirs for AFS. High atomization efficiency, resulting from the combined effect of high temperature (up to 3000°C) and the reducing properties of carbon or graphite, high quantum efficiency, high residence time of atoms in the optical path, high efficiency of sample utilization and small volume of the atomic cell all combine to yield large AF signals. Although the typical flame background is absent, a high black-body emission from the atomizer surface can significantly degrade the SNR. A careful choice of the observation zone is then needed, in addition to modulation and synchronous detection of the AF signal, in order to minimize the background-induced shot-noise. The main disadvantage of these atomizers is their extreme sensitivity to the matrix composition: sometimes large errors occur from chemical and physical effects (negative errors) and light-scattering (positive errors). The need for small discrete samples (~1–10 μl) and the stepwise operating conditions (drying, ashing, atomizing) make these atomizers incompatible with 'on-line' connection to a liquid chromatograph. Direct connection with a gas chromatograph does not present the same problem: in this case the atomizer can be operated at constant temperature and 'on line' detection is applicable (see Section 5.3.2).

A graphite tube atomizer [32] developed to be specifically used in non-dispersive AFS is shown in Fig. 5.7. It consists of a graphite tube 50 mm long with 4.5 mm outer

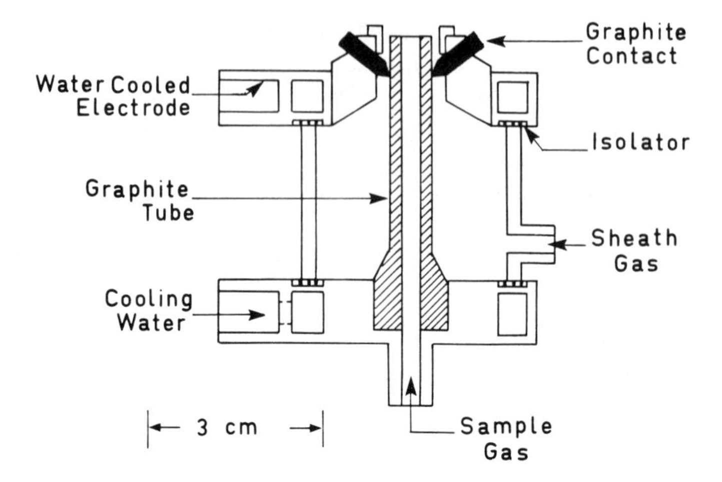

Fig. 5.7 — Cut view of graphite-furnace atomizer used in AFS. (Reprinted from K. Kuga and K. Tsujii, *Anal. Lett.*, 1982, **15**, 47, with permission from the coryright holders, Marcel Dekker Inc.)

and 2.5 mm internal diameter. The tube is positioned vertically in an outer mantle that protrudes sightly beyond the graphite tube. To protect the graphite from combusion and to prevent air entrainment in the observation zone, situated just above the top of furnace, a sheathing flow of argon is used; this also acts as a carrier for the atoms. A 10-μl sample is placed in a graphite dish positioned in the middle of the tube. A maximum temperature of 2500°C can be reached and detection limits of 20 pg of lead, 0.2 pg of zinc and 0.1 pg of cadmium are obtained. The same atomizer, but with a pyrolitic graphite coating, and operated at a constant temperature of 1200°C can be used in conjunction with the hydride generation technique to obtain a 10-pg detection limit for arsenic [33]. It is likely that it could also provide excellent performance in conjunction with a gas chromatograph. Because the favourable geometry ensures a long contact time with the analyte, it is possible to use a low carrier gas flow-rate (0.1 l./min), and the reduced dead volume minimizes dilution of the effluent from the gas chromatograph.

5.2.2.5 Continuous-flow furnaces

Continuous-flow furnaces may offer a suitable alternative to flame devices for use in on-line AF element-specific detectors for liquid chromatography. Such an atomizer combines most of the favourable features of flame and discrete flameless atomizers. Molnar and Winefordner [34] used a pneumatic nebulizer with a conical design to force-feed the sample solution through the sample transport needle with the aid of a syringe pump. The sample aerosol was then passed thorough a stainless-steel desolvation chamber heated at 393°C (gas temperature 246–278°C) before introduction into the vitreous carbon furnace (about 5 mm bore and 40 mm long) electrically heated to 1500–1800°C. With a sampling flow-rate of 0.12 ml/min the efficiency of nebulization and desolvation is 93%. The absolute detection limits can reach the sub-nanogram level (see Table 5.3, p. 149) and reproducibility is greatiy improved in comparison with discrete flameless atomizers. Good performances have also been obtained [35] with use of a pyrolytically coated graphite tube capable of being heated to 3000°C, but in this case an argon–methane carrier gas mixture had to be used to prolong the life of the pyrolytic coating. The system, capable of operating with sampling flow-rates as high as 2.7 ml/min and with no background emission at wavelengths shorter than 300 nm, gives detection limits at ng/ml level with good precision, by using a low-intensity xenon-arc lamp (150 W) and a photon-counting detection system. Improved detection limits could be obtained by using a more intense light-source than the xenon arc. It is important to underline that the sampling flow-rate for these devices, covering the range 0.1–2.7 ml/min, would be compatible with the carrier flow-rates typically employed in HPLC (0.2–2 ml/min).

5.2.3 Light-sources

Intense light-sources are needed in AFS in order to achieve low detection limits and wide dynamic ranges, but in a general evaluation of a given light-source, other characteristics must also be taken into account, such as light-output stability, lifetime, ease of operation and relative cost. A comparison of most of the suitable light-sources is given in Table 5.2. Pulsed tunable dye lasers are also included, although generally the high costs and difficulty of operation make them unsuitable for routine utilization in most laboratories. However, the potential for application of this 'non-traditional' light-source will give rise to substantial improvements in AFS once these inconveniences are overcome [36]. The development and working principles of light-sources are not discussed here; the reader may refer to specialized works on non-laser sources [37] and laser sources [38].

5.2.4 Interferences

In AFS, as in AES, the major interferences are caused by matrix components able to reduce the free atom population by means of chemical and physical interactions. In addition, in AFS the 'quenching' effect may also give rise to interference. Spectral interferences are seldom caused by the line-source, as is the case in AAS, but may be a problem with non-dispersive optics or in dispersive apparatus with continuum sources [3,4].

Light-scattering in AFS causes non-specific enhancement of the signal and is one of the most serious problems because correction is difficult. The extent of light-scattering depends on the type of atomizer employed and, as discussed in Section 5.2.2, flameless atomizers are more subject to light-scatter than flames, whereas the ICP can be considered to be an ideal solution to the problem. Many strategies have been proposed for correcting the scattering error; these have been reviewed recently [39]. The simplest way to correct for the scattering, when a spectral line-source is used for excitation, is to filter out from the source all the lines except for a resonance one and then to make the detection only with a non-resonance line, which will obviously be at a different wavelength. However, in this way the light intensity of the source is greatly reduced and a substantial worsening in the detection limit is to be expected. With continuum sources the use of a scanning monochromator allows the scattering to be measured in a portion of the spectrum adjacent to the AF line, thus providing a correction procedure. With a tunable dye laser as source, many correction strategies are possible. If a non-resonance AF line is available for the

Table 5.2 — Suitable light-sources for AFS

Light source	Advantages	Disadvantages	Notes
High-pressure focused xenon-arc lamp, Eimac type (150–500 W)	Low cost, long lifetime, very stable, simple to operate. Fairly intense at >250 nm. One source can excite many elements.	Poor intensity at <250 nm. Pulsed operation gives worsening in detection limits. Spectral interferences.	Continuum source. Requires the use of monochromator to avoid spectral interferences
Hollow-cathode lamp, continuous-wave operated (cw-HCL)	Very stable, simple to operate, moderate price. Available for most elements.	Poor intensity for most AF applications. Limited lifetime and shelf-life. Scattering correction troublesome, as for all spectral line sources.	Spectral line source. Usable with both dispersive and non-dispersive apparatus.
Hollow-cathode lamp, pulse operated (p-HCL)	As for cw-HCL. More intense than corresponding lamp cw-operated.	Limited lifetime and shelf-life. Self-reversal of spectral lines may cause unsatisfactory performance.	As for cw-HCL.
Boosted-output HCL	As for cw-HCL, but more intense by about one order of magnitude, with little or no self-absorption, and sharper lines.	Limited lifetime (500 hr) and shelf-life (1 year). Lamps for volatile elements (As, Se) may have less favourable performances.	As for cw-HCL. Now commercially available.
Demountable cathode lamp (DCL)	Improved version of boosted-output HCL. Very intense, sharp line for both volatile and non-volatile elements. One or two orders of magnitude more intense than cw-HCL.	More expensive and complicated to operate than cw-HCL or boosted-output HCL.	As for cw-HCL. Now commercially available together with interchangeable single- and multielement cathode.
Microwave-excited electrodeless discharge lamp (MW-EDL)	Highly intense and sharp line source. Long lifetime and long shelf-life. For some volatile elements up to 2000 times more intense than corresponding cw-HCL.	Need for expensive microwave generator. Tedious to prepare and operate. Limited to volatile elements and elements forming volatile halides. Temperature control necessary to obtain acceptable stability of output intensity and shorter warm-up time. Pulsed operation not reported so far.	As for cw-HCL. The most studied source for AFS. As, Sb, Bi, Se, Te, Ge, Sn, Pb, Hg, Cd, Zn most suitable elements.
Radiofrequency-excited electrodeless discharge lamp (RF-EDL)	Intense and sharp line source. Long shelf-life. For some volatile elements up to 100 times more intense than corresponding cw-HCL. Pulsed operation results in improved performances also for commercially available models.	Limited to volatile elements and elements forming volatile halides. Commercially available RF-EDL more expensive than HCL. Stability and warm-up time less favourable than HCL.	As for cw-HCL. Most suitable elements as for MW-EDL.
Thermal gradient lamp (TGL)	Very intense and stable. Sharp line. Absence of self-absorption.	Available only for more volatile elements such as As, Se, Te, Cd and Zn	As for cw-HCL. Now commercially available.
Pulsed tunable dye lasers. (e.g. dye laser pumped with excimer-laser or Nd–YAG laser.)	Several orders of magnitude more intense than MW-EDL. Possibility of working at or near saturation condition. Excellent selectivity. Good strategies for scattering correction.	Elements having absorption lines at $\lambda < 220$ nm (e.g. As, Se, Te and Zn) cannot be directly excited. Relatively low output stability. Expensive and requiring expensive signal-processing electronics. Require expert operator. Limited life of dyes. Danger from laser radiation.	Annotation refers to the present state of the art.

detection no problems of scattering will be observed, the source being tuned at a different wavelength from the detector. If a resonance line is measured, the source can be slightly detuned from the excitation wavelength so that the scattering effect can be measured alone and a correction made.

Interfacing AFS, or other atomic spectrometric techniques, with chromatography greatly reduces the chances of interference effects. In an ideal chromatographic process, all the components of the mixture injected will be separately eluted from the column, so a time-resolved atomization will be achieved. In practice some overlap may occur between the analyte of interest and the interfering components, but still the selectivity is greatly enhanced. Any interference effects associated with the carrier medium should give a constant background value, and although this will worsen the detection limit by degrading the SNR, it should be easily differentiated from the specific signal associated with a peak.

5.2.5 Atomic-fluorescence detection

The simplicity of AF spectra makes the isolation and detection of the selected AF spectral lines easier then in other atomic spectrometric techniques such as AES and AAS. A simple, low-cost monochromator is typically employed, the most important feature of which is a high light-throughput rather than a high resolution power. As the line profile of the light-source is not detected, a wide spectral bandpass (1–5 nm or more) can be adopted without deleterious effect on the linear range of calibration graphs. The major limitation to the use of wide spectral bandpass in AFS comes from atomizers with a high background emission in the wavelength region of interest. In this case a worsening in the SNR is obtained if a wide spectral bandpass is used [4,40]. A further limitation could be the simultaneous presence of an AF line emitted by another element in the same spectral bandpass. This line would not be separated, so a pure spectral interference would be originated. Such inconveniences are typical of continuum sources, but are seldom observed with spectral line sources and are effectively eliminated or corrected with tunable dye laser sources [3].

The monochromator may be replaced by an interference filter; this gives a simple non-dispersive apparatus with a large optical aperture. However, filters do not transmit ultraviolet radiation well (e.g. a maximum of 10% at 250 nm with about 10 nm bandwidth). The detector noise may then become the factor that limits the sensitivity: a more intense light-source and/or more efficient collection of the fluorescence emission than is possible with a simple lens [4] must then be considered. Interference filters are also quite expensive, so that, unless an instrument is dedicated to a single-element determination, it is cheaper to use a simple monochromator than a complete set of filters.

The radiation is usually detected with a photomultiplier. This device possesses many favourable properties: high sensitivity, low dark current, large dynamic range, wide range of spectral sensitivity. Solid state detectors (e.g. the silicon-intensified photodiode), in spite of recent improvements, have not yet reached the performance of photomultiplier tubes with respect to SNR and dynamic range [41,42].

Non-dispersive instruments that employ only a solar-blind photomultiplier (Hamamatsu type R166 or R759 with a spectral range of 160–320 nm) and collect the AF radiation directly on the photocathode through a simple quartz lens are quite popular. Wide optical aperture, high energy throughput, high optical transmission even in the far ultraviolot region, and the simultaneous detection of many fluorescence lines for the elements that have complex AF spectra (e.g. As, Sb, Se) can result in an improved sensitivity compared with dispersive systems, but this is true only if very low background atomizers can be used [9,10]. The ability to detect several AF lines simultaneously can be an advantage also in multielement instruments, as will be illustrated in Section 5.2.7, but it must be emphasized that both in single- and multielement systems very pure spectral line sources are needed for reasonable selectivity to be obtained. Non-dispersive instruments, however, are more prone than dispersive ones to interference and scattering problems, and they are better suited for use in conjunction with special sampling techniques such as hydride generation and chromatography, where the risk of interference is greatly reduced and the use of a low-noise atomizer is possible.

In most cases, the spectral bandpass of a solar-blind or an ordinary response photomultiplier can conveniently be narrowed by using a wide-bandpass filter [40,43] to minimize some specific source of noise. A typical example of such noise arises from the OH radicals that are present in all flame systems. These emit radiation in the spectral ranges 280–295 and 305–323 nm and are therefore also detected by the solar-blind photomultiplier. Vickers *et al.* [40] found that the use of a chlorine filter narrows the useful spectral bandpass to about 160–280 nm; in this region the transmission characteristics remain excellent. The filter consists simply of a silica tube (25 mm in diameter, 52 mm long) provided with a silica window, filled with chlorine at about 1 atm pressure, and interposed between the atomizer and the photocathode. Excellent results are obtained by using such a filter with non-sheathed H_2–air or H_2–O_2–Ar flames for As and Hg determination. Less benefit is observed for the C_2H_2–air and H_2–N_2O flames because their continuum background emission extends into the 200–300 nm region (see, for example, Fig. 5.4b). In this case the only alternative way to reduce flame noise is to use argon or nitrogen gas sheathing [10].

5.2.6 Signal-processing electronics

Lock-in amplification, photon counting and gated synchronous detection have proved to be the most suitable techniques for signal processing. The choice among them depends on the light-source and atomizer characteristics [4].

Lock-in amplification is used with both mechanically and electronically modulated sources having fairly intense light output. The choice of the correct modulation frequency may be decisive for obtaining the optimum signal-to-noise ratio (SNR). For example, the noise power spectrum of many flame atomizers shows a strong '1/*f*' component (flicker noise) in the region below 1 kHz. In this case, and if the limiting noise is due to the flame background, a modulation frequency greater than 1 kHz will result in an improved SNR. An interesting study on the characteristic noise spectra of common analytical spectrometric sources was made by Talmi *et al.* [44]. High-frequency shot-noise can conveniently be reduced by using large integration time constants if steady signals or transient signals with suitably long lifetimes are processed.

Photon counting has been found to be superior to the lock-in system only when low-intensity light-sources such as conventional continuous-wave hollow-cathode lamps or low-power xenon-arc lamps are employed in combination with low-background atomizers [45].

With pulsed light-sources operated at a duty cycle much lower than 0.5 (duty cycle = pulse-on time/pulse period), lock-in amplification gives unsatisfactory responses because for most of the time period between pulses only the noise component is detected. In this case, the use of electronics gated to the light-source pulse gives an improved SNR. The fluorescence signal pulse background noise is integrated only during the source pulse; background noise is integrated for the same length of time between pulses and subtracted. For most common pulsed light-sources such as hollow-cathode lamps (HCLs) or rf electrodeless discharge lamps (rf EDLs), where pulse lengths are of the order of msec, the light-source power supply and the integrator can easily be synchronized by using general purpose digital electronics, and mini- or micro-computers. This approach is particularly suitable for multielement instruments in which many light-sources and integrators have to be synchronized sequentially [45]. With pulsed tunable dye lasers as sources, the approach must be different. Such devices have pulse lengths of the order of nsec, and a repetition rate of 10–100 Hz and the duty cycle can consequently drop to values of 10^{-8}–10^{-6}. The only solution appears to be the boxcar integrator [5]. This instrument is a versatile gated integrator for the measurement of repetitive signals, but its cost is much greater than that of a lock-in system.

The pulsed-source gated detector systems give an improvement in SNR relative to continuous sources operated at the same average power, provided that the background emission of the atomizer is high. The extent of enhancement, G, is estimated to be [4]

$$\frac{(\mathrm{SNR})_{\mathrm{PULSED}}}{(\mathrm{SNR})_{\mathrm{CW}}} = \frac{\overline{i_p}}{\overline{i_c}}\sqrt{\frac{1}{ft_p}} \tag{5.4}$$

where $\overline{i_p}$ is the average photodetector current (A) due to AF for the pulsed source, $\overline{i_c}$ is the average photodetector current (A) due to AF for the continuous-wave source, f is the repetition rate (Hz) of the pulsed-source and gated detector, and t_p is the pulse-width (sec) of the fluorescence pulses due to the pulsed source. The reader should refer to the excellent book of Malmstadt *et al.* [46] for detailed information on the electronics employed in SNR enhancement.

5.2.7 Multielement instruments

The isotropic nature of the AF radiation is very convenient for simultaneous multielement detection. Many light sources and many light detectors can be positioned symmetrically around the atomizer. In this way, simple and inexpensive instruments are easily assembled for the simultaneous determination of up to 3 or 4 elements. Several multielement AF spectrometers have been described; dispersive and non-dispersive optics are used in combination with line or continuum sources, flame and non-flame atomizers. Non-dispersive instruments, owing to their characteristics of simultaneous detection for many AF lines and high energy throughput, combined with inherent design simplicity, are attractive. It is not accidental that the only two commercial AF instruments (Technicon AFS-6 and Baird ICP-AFS) are multielement and non-dispersive. An excellent review of multielement atomic-fluorescence spectrometry is available [47].

Multielement non-dispersive instruments can utilize two different approaches, namely frequency multiplexing and time multiplexing.

In the frequency multiplexing approach, each source (HCL, EDL) is modulated at a different frequency f_i, and the atomic fluorescence of all elements is simultaneously excited by the light-sources and detected by a solar-blind photomultiplier or other broad-response photomultiplier fitted with suitable filters. Individual lock-in amplifiers, each tuned and phase-locked to the relevant light-source modulation frequency f_i, separate the mixed analytical signals and, at the same time, reject the noise. This approach is expensive owing to the total cost of several lock-in amplifiers. A more convenient, home-made, apparatus for simultaneous determination of up to four elements by the use of commercially available rf EDLs is depicted in Fig. 5.8 [16]. The control unit (Fig. 5.9) consists of a Colpitts oscillator supplying the 450-

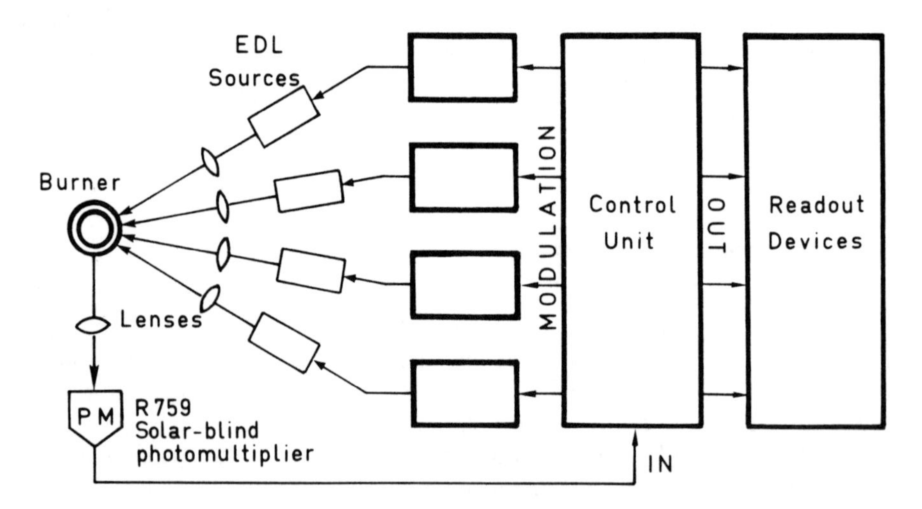

Fig. 5.8 — Schematic diagram of a four-channel non-dispersive AF spectrometer. (Reprinted from A. D'Ulivo, P. Papoff and C. Festa, *Talanta*, 1983, **30**, 907, by permission. Copyright, 1983, Pergamon Press.)

kHz driving frequency to the frequency-dividers for the four different channels. In this mode, four square-waves were generated at frequencies of 7031, 7500, 8036 and 8544 Hz. The square-wave signals controlled both the modulation of the radio-frequency power supplying each lamp and the switching of the phase-sensitive detector in the corresponding amplifier. The composite signal output of the photo-multiplier was pre-amplified and fed to a battery of four bandpass filters each tuned

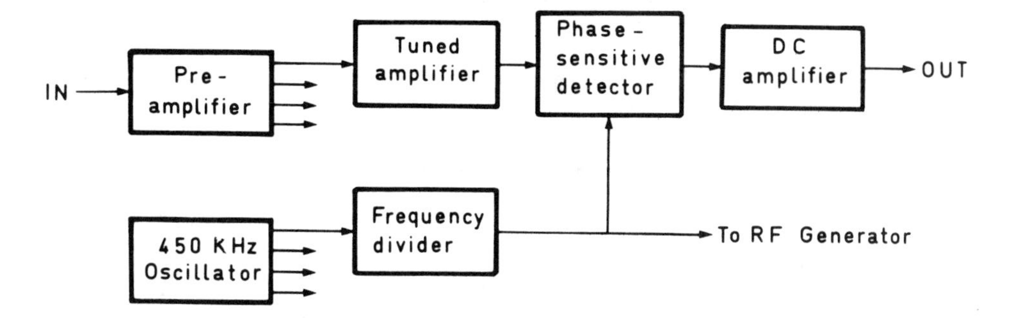

Fig. 5.9 — Block diagram of the control unit of the four-channel non-dispersive AF spectrometer shown in Fig. 5.8.

to the modulation frequency of one of the lamps. The d.c. signals from the tuned filters were then further amplifed and filtered. In order to obtain increased sensitivities, the system has been used with hydride generation, coupled to a low-background argon–hydrogen/entrained air miniflame as the atomizing system. Arsenic, selenium, tin and mercury could be determined simultaneously in 1-ml samples of natural water; detection limits were 0.04–0.1 ng/ml. Dynamic ranges of up to three decades from the detection limits [17] were readily obtained, and interferences were low. Lead and ionic alkyllead compounds were also determined with the same apparatus, but in the single-element mode. An aqueous 5-ml sample gave detection limits of 0.06 ng/ml Pb^{2+} [18] and 0.003–0.005 ng/ml of triethyl, trimethyl-, diethyl and dimethyllead [17] with the aid of hydride generation. The use of this multichannel apparatus in multielement gas-chromatographic detection is described in Section 5.3.2.

In time-multiplexing apparatus the light-sources (HCL or EDL) are pulsed sequentially and the resulting pulsed fluorescence signals are detected, again by a solar-blind photomultiplier. Each signal is measured by a gated integrator synchronized with the firing of the relevant lamp. In this mode of operation, the atomic fluorescence of just a single element is excited and detected at a given time. Thus, the technique is not a true simultaneous multielement detection technique. The first example of a non-dispersive time-multiplexed AF spectrometer [45] is shown as a block diagram in Fig. 5.10. The system uses pulsed HCLs, and a microcomputer both for control of operations and for calculation. The pulsing scheme used is depicted in Fig. 5.11, along with an overview of the time-multiplexing concept. By using pulses of 2–10 msec duration, followed by a delay time of 6–30 msec before the next lamp was turned on, the system was rapid enough to allow multielement determinations on a transient atom population, such as that produced from a flameless atomizer [45]. Therefore this system could be used for chromatographic detection where, typically, the lifetimes of transient signals range from a few sec (for capillary gas chromatography) to several tens of sec or more (for HPLC).

Demers *et al.* [21] have described a multielement instrument capable of simultaneous determinations of up to twelve elements, based on time-multiplexing and non-

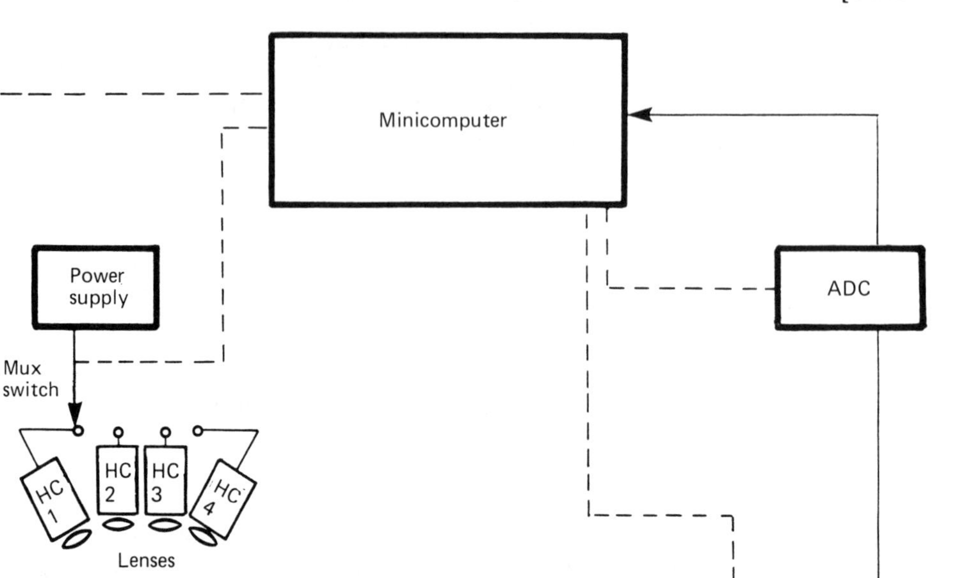

Fig. 5.10 — Block diagram of a computer-controlled time-multiplexed multichannel non-dispersive AF spectrometer. (Reproduced with permission from E. F. Palermo, A. Montaser and S. R. Crouch, *Anal. Chem.*, 1974, **46**, 2154, Copyright 1974, American Chemical Society.)

dispersive optics. This instrument uses as atomizer an ICP provided with a pneumatic cross-flow nebulizer. Encircling the torch are up to twelve element modules, one of which is shown in Fig. 5.12. Each module can be hand-held and comprises an HCL, a photomultiplier tube, an optional interference filter, and lenses. It is independent of the other element modules and can be adjusted to observe any desired height in the plasma. An optical interference filter serves to improve the SNR by excluding the IPC background noise, which exhibits a strong '$1/f$' component at frequencies below about 200 Hz. For this reason, to achieve further improvement in the SNR, the HCLs are pulsed at a frequency of about 500 Hz. Detection limits are similar to those of flame atomic-absorption for all elements. Compared with ICP/AES, its detection limits were inferior for many refractory metals, but were better for the alkali metals. An instrument of this type is one of the only two AF instruments commercially available at present (Baird, MA, USA). Recently, the performance of this apparatus has been greatly improved. About one order of magnitude improvement in detection limits has been achieved by using ultrasonic nebulization followed by desolvation [48] and a further 3–5 times improvement can be obtained by replacing the HCL with a commercially available boosted-output HCL [49]. Doping the argon carrier gas with propane at a flow-rate of 10–40 ml/min has proved to be effective in improving the detection limits for the refractory elements [50].

An eight-channel non-dispersive atomic-fluorescence spectrometer capable of

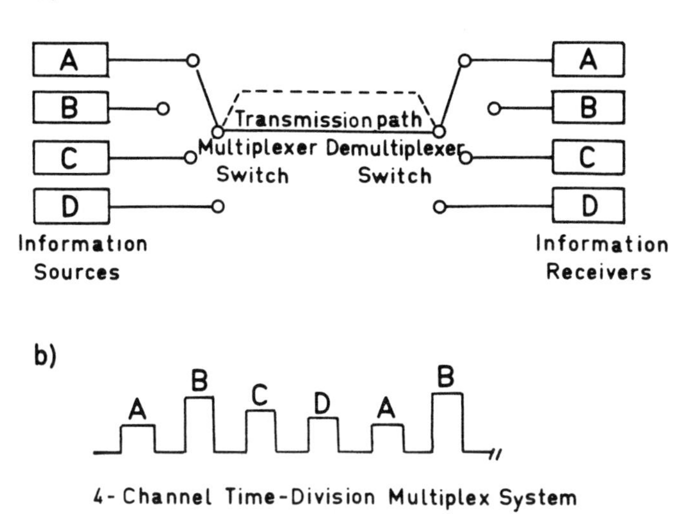

Fig. 5.11 — (a) General diagram of the 4-channel time-division multiplex system. (b) Waveforms in the transmission path for pulse amplitude modulation. (Reproduced with permission from E. F. Palermo, A. Montaser and S. R. Crouch, *Anal. Chem.*, 1974, **46**, 2154. Copyright 1974, American Chemical Society.)

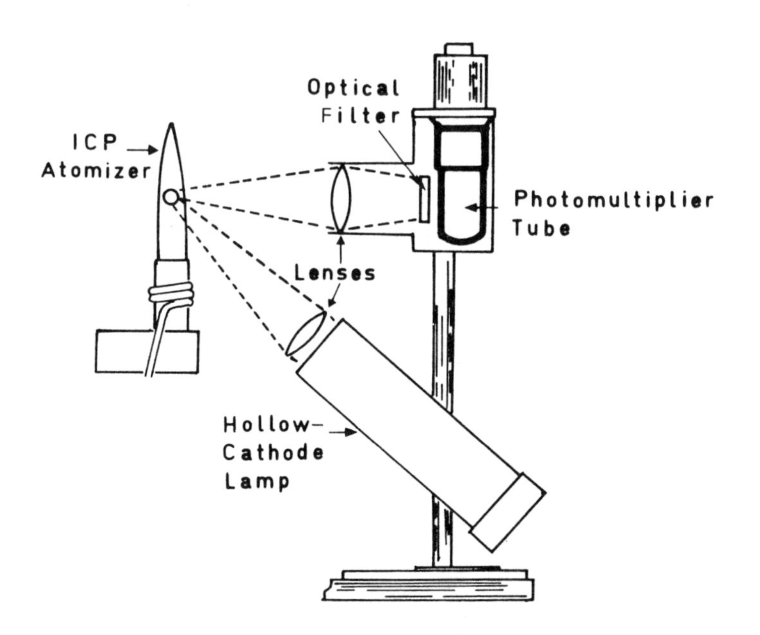

Fig. 5.12 — Schematic representation of one of the twelve elements of the Baird multielement AF spectrometer. (Reprinted from D. R. Demers, D. A. Busch and C. D. Allemand, *Int. Lab.*, 1982, **12**, May, 41, by permission. Copyright 1982, International Scientific Communications Inc.)

using both pulsed rf EDLs and HCLs has been described [51]. The instrument is based on the time-multiplexing principle and uses a Commodore PET 2001 microcomputer to control the lamp firing and data manipulation. A single solar-blind photomultiplier is employed to collect the fluorescence radiation. An accurate study has been made of the behaviour of ten pulsed EDLs (for As, Sb, Bi, Se, Te, Sn, Pb, Hg, Cd and Zn) as a function of the average rf-power, mode of pulsing, and step frequency. With an argon-sheathed air/C_2H_2 flame, detection limits comparable to those of ICP–AES, ICP–AFS, and flame AAS are obtained.

5.2.8 Detection limits

Table 5.3 reports detection limits for some elements commonly studied in the natural environment. For each element, the detection limits for different types of instruments are given, as a guide to suitability for application to specific analytical problems, Dynamic ranges are not reported; typically, under correct conditions of atomic cell illumination, these cover 3–5 decades from the detection limits, and are sometimes wider for some laser-excited AF instruments [5].

Table 5.3 requires some comments. The existence of remarkable differences between detection limits is peculiar to AFS, because the atomizer and the light-source characteristics are both important for obtaining improved sensitivities. For example, the use of the cold vapour technique or hydride generation, allowing a low-background atomizer to be employed, leads to very low detection limits. In addition, the fact that intense light-sources such as microwave-excited EDLs or rf EDLs are available for As, Sb, Bi, Se, Te, Sn, Pb and Hg is also important.

A detailed comparison between detection limits for AFS and AAS or ICP–AES is beyond the scope of this contribution, but some general remarks can be made about flame and flameless systems. For flame systems, AFS can be considered to be superior to AAS, especially with such intense light-sources as boosted-output HCLs or demountable cathode lamps [37], and for many elements it is also comparable with ICP–AES. For flameless systems, AAS is usually superior, because use of the graphite furnace yields detection limits at picogram or sub-picogram level for most elements. Only for a few elements, such as Cd and Zn, are the detection limits for AFS superior to those of AAS.

This discussion also applies to use of non-laser sources. In contrast, with laser-excited AFS it is now possible to obtain exceptionally low detection limits with flame, plasma and graphite atomization systems [23,52,63], and the detection of as little as a few atoms is expected in the near future [36].

5.3 ATOMIC FLUORESCENCE IN CHROMATOGRAPHIC DETECTION

5.3.1 Liquid chromatography

Van Loon *et al.* [64] first proposed using an AF spectrometer as an element-specific detector for chromatography, with the aim of achieving speciation of several elements simultaneously. An inexpensive three-channel non-dispersive AF spectrometer, working in frequency-multiplexing mode, was assembled from individual lock-in amplifiers, HCL as excitation sources, and a nitrogen-sheathed air–acetylene flame as atomizer. It was interfaced to an HPLC by direct connection of the column outflow to the nebulizer capillary of the burner; the column flow-rate was

Table 5.3 — Detection limits achievable in AFS by using different experimental instruments

Element	Detection limit† (ng/ml)	(pg)	Light source‡	Atomizer*	Apparatus§	Notes	References
Arsenic	320	—	p-RF-EDL	Air/C_2H_2	ND,M		[51]
	400	—	p-HCL	ICP	ND,M		[21]
	15(16)	—	RF-EDL(TGL)	Air/C_2H_2	ND,S		[53]
	0.02	8	p-RF-EDL	m-Ar/H_2	ND,M	Hydride generation	[16,17]
	—	10	cw-RF-EDL	m-Ar/H_2	ND,S	Hydride generation	[14]
	0.1	100	MW-EDL	Ar/H_2	D,S	Hydride generation	[54]
Antimony	50(150)	—	B-HCL	H_2/O_2/Ar(Air/C_2H_2)	D,S		[55]
	140	—	p-RF-EDL	Air/C_2H_2	ND,M		[51]
	10	—	B-HCL	Air/C_2H_2	ND,S		[37]
	250	—	p-HCL	ICP	ND,M		[21]
	—	20	cw-RF-EDL	m-Ar/H_2	ND,S	Hydride generation	[15]
	0.1	100	MW-EDL	Ar/H_2	D,S	Hydride generation	[54]
Bismuth	3	6.6×10^3	MW-EDL	C-GFA	D,S	Photon counting	[45]
	170	—	p-RF-EDL	Air/C_2H_2	ND,M		[51]
	2	—	MW-EDL	C-GFA	D,S		[34]
	80	—	p-HCL	ICP	ND,M		[21]
	0.005	100	MW-EDL	Ar/H_2	ND,S	Hydride generation	[56]
	—	10	MW-EDL	CRA	D,S		[57]
Selenium	16(30)	—	TGL(RF-EDL)	Air/C_2H_2	ND,S		[53]
	25	—	p-HCL	ICP	ND,M	Ultrasonic nebulizer	[48]
	40	—	p-RF-EDL	Air/C_2H_2	ND,M		[51]
	0.015	300	MW-EDL	Ar/H_2	ND,S	Hydride generation	[58]
	0.06	60	MW-EDL	Ar/H_2	D,S	Hydride generation	[54]
	0.01	20	p-RF-EDL	m-Ar/H_2	ND,M	Hydride generation	[16,17]
Tellurium	10	10^3	MW-EDL	C-GFA	D,S		[34]
	0.1	2×10^3	MW-EDL	Ar/H_2	ND,S	Hydride generation	[59]
	200	—	p-HCL	ICP	ND,M		[21]
	23	—	p-RF-EDL	Air/C_2H_2	ND,M		[51]
	0.08	80	MW-EDL	Ar/H_2	D,S	Hydride generation	[54]
Germanium	200	—	p-HCL	ICP	ND,M		[21]
	100	—	MW-EDL	N_2O/C_2H_2	D,S		[57]
Tin	35	—	p-HCL	ICP	ND,M	Ultrasonic nebulizer	[48]
	50	—	p-RF-EDL	Air/C_2H_2	ND,M		[51]
	0.1	100	p-RF-EDL	m-Ar/H_2	ND,M	Hydride generation	[16,17]
	10	10^3	MW-EDL	C-GFA	D,S		[34]
	20	—	Xe	C-GFA	D,S	Photon counting	[35]
	3	—	PTDL	ICP	D,S	Boxcar integrator	[24]

Element	Detection limit† (ng/ml)	(pg)	Light source‡	Atomizer*	Apparatus§	Notes	References
Lead	5	500	MW-EDL	C-GFA	D,S		[34]
	10	—	Xe	C-GFA	D,S	Photon counting	[35]
	900	—	p-RF-EDL	Air/C_2H_2	ND,M		[51]
	20	—	p-HCL	ICP	ND,M		[48]
	0.02	—	PTDL	Air/C_2H_2	D,S	Boxcar integrator	[24]
	0.0006	0.006	PTDL	CRA	D,S	Boxcar integrator	[24]
Pb^{2+}	0.06	—	p-RF-EDL	m-Ar/H_2	ND,S	Hydride generation	[18]
R_3Pb^+,R_2Pb^{2+}	0.005	—	p-RF-EDL	m-Ar/H_2	ND,S	Hydride generation	[19]
	—	20	MW-EDL	GFA	ND,S		[32]
Mercury	600	—	p-RF-EDL	Air/C_2H_2	ND,M		[51]
	400	—	p-HCL	ICP	ND,M		[21]
	0.2	—	p-HCL	ICP	ND,M	Cold vapour	[59]
	0.001	10	cw-RF-EDL	Ar	D,S	Cold vapour, gold trap	[60]
	—	15	p-RF-EDL	m-Ar/H_2	ND,M	Cold vapour	[17]
	—	7	MW-EDL	CRA	D,S		[31]
Cadmium	10	—	Xe	Air/C_2H_2	D,S		[61]
	—	0.0015	MW-EDL	CRA	D,S		[57]
	—	0.1	MW-EDL	GFA	ND,S		[32]
	1	—	Xe	C-GFA	D,S	Photon counting	[35]
	0.1	—	p-HCL	ICP	ND,M	Ultrasonic nebulizer	[48]
	0.6	—	p-RF-EDL	Air/C_2H_2	ND,M		[51]
	0.001	—	MW-EDL	Air/H_2	D,S		[62]
Zinc	6	—	Xe	Air/C_2H_2	D,S		[61]
	—	0.02	MW-ERDL	CRA	D,S		[57]
	—	0.2	MW-EDL	GFA	ND,S		[32]
	5	—	Xe	C-GFA	D,S	Photon counting	[35]
	0.1	—	p-HCL	ICP	ND,M	Ultrasonic nebulizer	[48]
	0.4	—	p-RF-EDL	Air/H_2	D,S		[51]
	0.04	—	MW-EDL	Air/H_2	D,S		[62]
Copper	5	—	MW-EDL	Ar/H_2	D,S		[62]
	2	—	Xe	C-GFA	D,S	Photon counting	[35]
	0.4	—	p-HCL	ICP	ND,M	Ultrasonic nebulizer	[48]
	0.06	0.3	MW-EDL	CRA	D,S		[31]
	0.002	0.15	PTDL	CRA	D,S	Boxcar integrator	[63]
	0.2	—	B-HCL	Air/C_2H_2	ND,S		[37]

† Detection limits are those reported in the original literature and therefore they may be differently defined.
‡ p-RF-EDL, pulsed radiofrequency excited electrodeless-discharge lamp; p-HCL, pulsed hollow-cathode lamp; TGL, thermal gradient lamp; cw-RF-EDL, continuous wave RF-EDL; B-HCL, boosted output HCL; MW-EDL, microwave excited EDL; Xe, xenon-arc continuum source; PTDL, pulsed tunable dye laser.
* ICP, inductively coupled plasma; m-Ar/H_2 miniature argon–hydrogen–entrained air flame; GFA, electrothermal graphite furnace atomizer; C-GFA, continuous flow GFA; CRA,

kept compatible with the nebulizer flow-rate. Interfacing with a gas chromatograph was much easier; the effluent could be introduced into the flame through a port at the base of the burner. A specific application of this detector, in which all its capabilities are demonstrated, is the simultaneous detection of several metals as their EDTA (Cu, Zn, Ni), glycine (Cu, Zn, Ni) and trien complexes (see Fig. 5.13). This type of

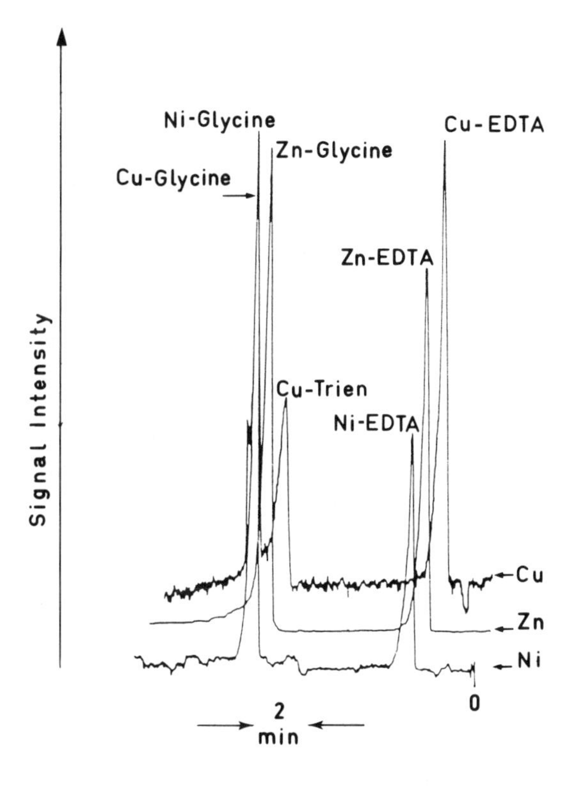

Fig. 5.13 — Typical chromatograms for Cu, Zn and Ni complexes. (Reprinted from J. C. Van Loon, J. Lichwa and B. Radziuk, *J. Chromatogr.* 1977, **136**, 301, by permission. Copyright 1977, Elsevier Science Publishers.)

chemical system could arise in clinical samples obtained during the diagnosis and treatment of metal poisoning. The sample (25 μl) was injected into a Partisil-10 SCX cation-exchange column at 55°C, with the mobile phase flowing at 4 ml/min. It consisted of pure water for the first minute, then a convex gradient up to 100% 1*M* NH_4NO_3 over 5 min. The HCLs were modulated at 325, 285 and 80 Hz and the AF radiation of Zn and Ni was detected by a solar-blind photomultiplier (R166); an ordinary-response photomultiplier tube (R106) was used for Cu. The difficult separation of these metal chelates results in very similar retention times and makes the conventional UV–visible detector unsuitable. An AA element-specific detector *was* used successfully but three successive chromatographic runs, each with different conditions (change of lamp, wavelength and slit-width), had to be made to obtain the same information as that given by a single run with the multielement AF detector.

A single-element dispersive AF spectrometer consisting of a pulsed Eimac xenon continuum arc lamp (300 W), lock-in amplifier and air–acetylene flame has been used [65] in combination with HPLC as a specific detector for iron to study the acetylation of ferrocene with acetic anhydride. About 5 μl of reaction mixture were injected into a Chromasep S column (50 cm × 2 mm) packed with 10 μm pellicular silica gel and operated at a flow-rate of 1.5 ml/min 40:1 v/v diethyl ether/methanol eluent. Ferrocene, 1-acetylferrocene and 1,1′-diacetylferrocene were separated and detected, first by an ultraviolet detector operated at 254 nm and then by the AF detector, the outlet of the ultraviolet detector being connected to the capillary of the nebulizer system. The flame was run slightly rich, with 150 ml/min acetylene, 3 l./min air and 1.5 ml/min diethyl ether. The iron AF resonance line at 248.3 was used, but a trap was needed to remove phosphine from the acetylene, because of interference by the molecular fluorescence signal from the PO species (220–275, 323–330 nm), The AF detector was found to work better than the ultraviolet detector, as the former was insensitive to organics co-eluted with the iron compounds and its integrated response depended only on the iron content, irrespective of the molecular environment containing it.

Following the general suggestions made by Van Loon *et al.* [64], Mackey assembled a four-channel non-dispersive AF spectrometer [66] for use as a simultaneous element-specific detector (Cu, Zn, Fe and Mg) for liquid chromatography. The spectrometer (see Fig. 5.14) used conventional HCLs, each modulated at a different frequency, and the light was focused onto an air–acetylene flame, nitrogen-sheathed, by 40-mm diameter, 50-mm focal length quartz lenses. The light-intensity was approximately doubled by using mirrors to reflect the light back through the flame. The mirrors were 35 mm in diameter with a radius of curvature of 103 mm and were coated with aluminium and a protective layer of either UV-transmitting glass or magnesium fluoride. The mirrors could be replaced by lenses and additional lamps to obtain an eight-channel detector, but then the system had reduced sensitivity. The AF of copper was detected by a wide spectral response photomultiplier (R106) shielded by a Schott UG11 filter and a 6-mm iris. For zinc, iron and magnesium the AF radiation was detected by a solar-blind photomultiplier (R166) shielded by a 9 mm iris. The electronics consisted of three individual lock-in detectors (IM-1, Varian AA4 amplifier) and power supplies (MLS1-A, Varian AA4) for Cu, Fe and Zn. The modulation frequency for Zn was 285 Hz and for Cu and Fe 263 and 313 Hz respectively, and was obtained by altering the standard reference frequency of 285 Hz provided by the IM-1 indicator modules. For Mg detection, a three-channel indicator unit, a lamp supply and a photomultiplier high-voltage source (GBC Scientific) were used, but only one of the three channels available was used. In this case the frequency could easily be varied from 180 to 33 Hz and was set at 217 Hz. The nitrogen-sheathed burner was made of titanium and was identical to that described by Larkins [10] (Fig. 5.4). It was fitted onto a commercial premix chamber (Varian AA4), with fuel and oxidant flow-rates controlled by a GCU-2 unit. The output of the chromatographic column was connected directly to the adjustable tantalum nebulizer and its uptake rate was made identical to the chromatographic

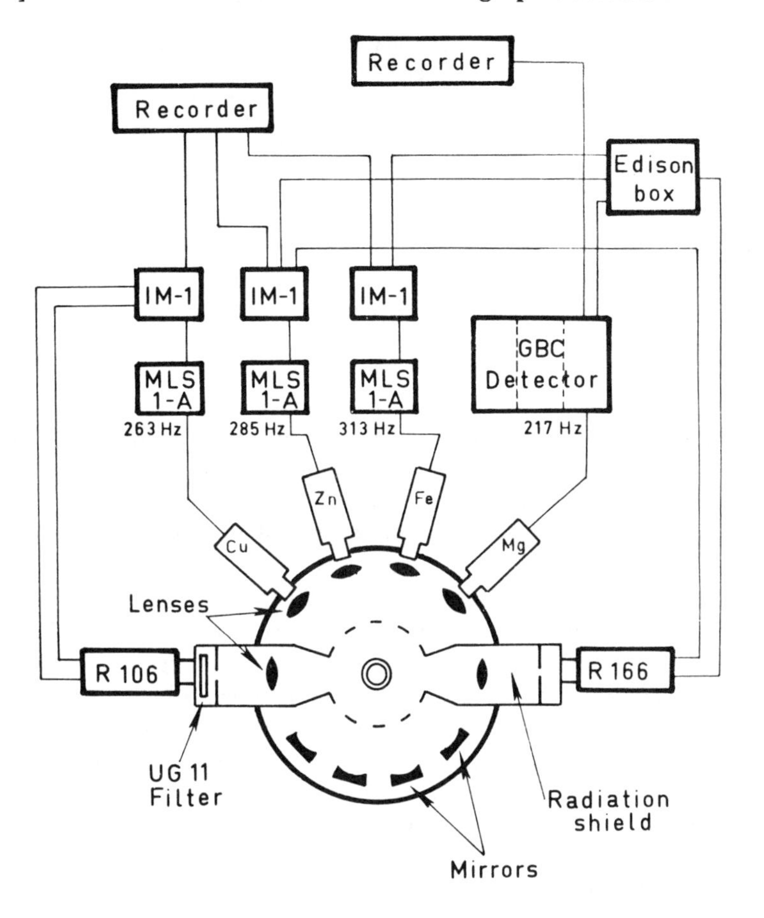

Fig. 5.14 — Schematic diagram of the four-channel non-dispersive AF spectrometer employed by Mackey as a liquid chromatographic detector. (Reprinted from D. Mackey, *J. Chromatog.*, 1982, **236**, 81, by permission. Copyright 1982, Elsevier Science Publishers.)

column flow-rate (*ca.* 1.7 ml/min). The detection limits ranged from 2 to 25 ng/ml, but recently Mackey has greatly improved the sensitivity of this detector by using more intense light-sources. Boosted output HCLs (Photron) and demountable cathode lamps (SGE) were tested and the results are reported in Table 5.4.

Mackey [68] applied the multichannel AF detector (employing an HCL) coupled with HPLC to the study of the natural occurrence of metal complexes with organic ligands present in sea-water, e.g. humic and fulvic acids, proteins, tetrapyrroles (derived from chlorophyll) and specific ion-complexing agents (e.g. siderophores). The low natural concentration of such metal–organic complexes necessitated a preconcentration step for them to be detectable by HPLC–AFS. Several litres of sea-water were passed through a cartridge containing a suitable hydrophobic resin. The sea salts were then removed with 50 ml of water and the adsorbed material was eluted with 5 ml of methanol. The methanol was evaporated by a stream of filtered nitrogen,

Table 5.4 — Detection limits for a multielement non-dispersive AF liquid-chromatographic detector (ng/ml) [67]

Element	HCL	Boosted output HCL	Demountable cathode lamp
Cu	20	2	1.5
Zn	5	0.25	0.05
Fe	25	7	6
Ni	25	5	1
Mg	2	—	—

and the residue was taken up in about 1.5 ml of water by ultrasonic treatment; the solution was filtered and injected into the chromatograph. The suitability of several adsorbents for preconcentration of metal–organic complexes has been studied by liquid chromatography with an AF detector [66,69,70]. Octadecyl-bonded silica was found to be the most suitable adsorbent and was used both in the preconcentration step (Sep-Pak C_{18} cartridges) and as support in the chromatographic column (Rad-Pak 10-μm particles). However these 'reversed-phases' showed several limitations. Not all organic complexes were retained and so only a restricted determination was possible [68]. In addition there were large systematic variations in the chromatograms as the column aged. This was probably due to an increased cation-exchange capacity of the C_{18} phase caused by free silanol groups competing with organic ligands for trace metals [71]. However, Mackey was able to demonstrate the existence in sea-water of organic complexes of Cu, Zn, Fe, Mg, Ni and Mn, but no complexes were found of Cd and Cr [68,71]. A typical chromatogram is shown in Fig. 5.15. In these investigations the AF detector was placed in series with ultraviolet and molecular-fluorescence detectors, to allow guesses to be made about the nature of the organic ligands.

5.3.2 Gas chromatography

Coupling a gas chromatograph with any atomic spectrometer greatly simplifies the task of the atomizer since the effluents supplied to it are already in the vapour phase. In principle, any kind of atomizer can be coupled with a gas chromatograph to give an on-line element-specific detector, but in practice, careful choice of the atomizer is necessary to obtain good sensitivities and selectivities for specific analytical problems.

The suitability of various atomization systems for connection to a gas chromatograph was tested by Radziuk *et al.* [72] for the determination of alkyllead compounds by AFS and AAS. Air–acetylene flames supported on a 10-cm single-slot burner, for AAS, and on a round burner provided with nitrogen sheathing, for AFS, were tested as flame systems. The non-flame atomizers used for both AAS and AFS were a carbon rod with graphite cup (CRA 63 cup, Varian) and a silica furnace, heated by a resistance wire and covered with asbestos (Fig. 5.16). In addition, a carbon-rod tube (CRA 63 tube, Varian), a graphite-tube furnace (HGA 2100, Perkin-Elmer), and a silica tube heated in an air–acetylene flame were tested for AAS. The timing circuit

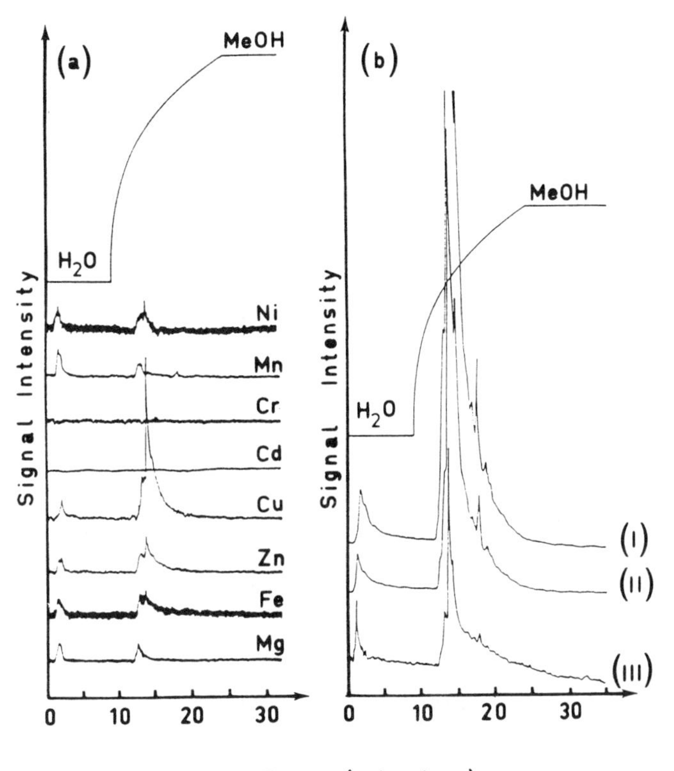

Fig. 5.15 — Chromatograms of solutions of material extracted from sea-water by Sep–Pak cartridges. (a) Atomic-fluorescence detector responses from two separate 1000-μl loadings; (b) response of molecular fluorescence detector at different loadings and sensitivities, (I) 100 μl, 1 μA, (II) 400 μl, 1 μA, (III) 25 μl and 0.1/0.2 μA with a change of sensitivity after the first peak. Excitation at 360 nm with detection above 470 nm. An 8-mm i.d. Rad-Pak C18 column was used. (Reprinted from D. Mackey, *Mar. Chem.*, 1983, **13**, 169, by permission. Copyright 1983, Elsevier Science Publishers.)

was modified to allow the non-flame atomizers to be heated continuously at the charring temperature for preset periods. The glass chromatographic column (150 cm long, 6 mm in diameter) was packed with 3% OV-101 on Chromosorb W (80–100 mesh). The chromatographic effluents were transferred to the atomizer system through a heated stainless-steel tube (1.6 mm bore) and led into the space immediately below the burner slot at the centre of the burner head, for the AA flame. The tube was bent at a right angle inside the burner to point upwards towards the burner slot so that the effluent entered the gas stream and mixed with it. A similar arrangement was used for the AF flame. For the carbon-rod atomizer (type CRA 63), the stainless-steel tubing was inserted into a hole bored through one of the electrodes and the side of the cup or the tube. The electrodes of the standard CRA 63 working head were relocated at the top of the terminals in order to make the AF detection possible at a convenient angle (Fig. 5.17). The head was aligned horizon-

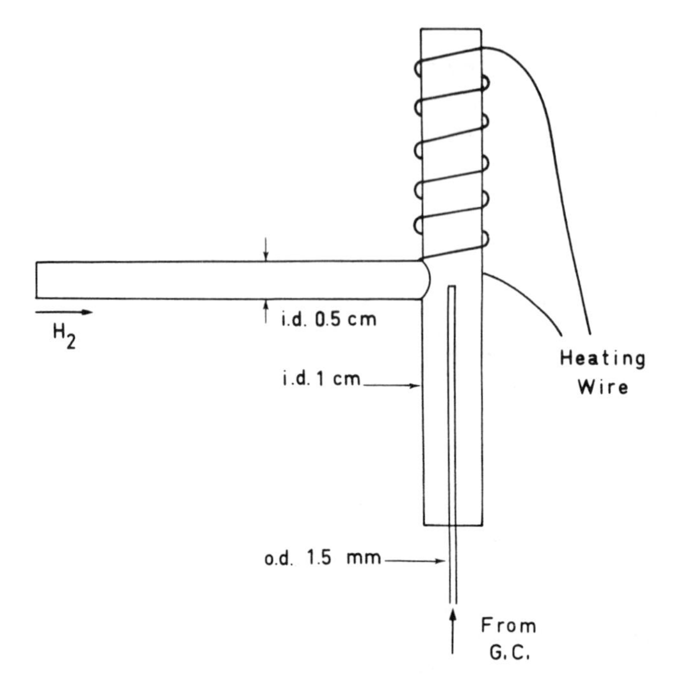

Fig. 5.16 — Schematic diagram of silica furnace used for AF gas-chromatographic detection. (Reprinted from B. Radziuk, Y. Thomassen, L. R. P. Butler, J. C. Van Loon and Y. K. Chau, *Anal. Chim. Acta*, 1979, **108**, 31, by permission. Copyright 1979, Elsevier Science Publishers.)

tally in the light-beam by replacing the foot adaptor with a piece of aluminium rod (25 mm diameter) and the masking plates were redesigned to minimize the stray light. The sensitivity of the atomizer system was tested by injecting 1–5 μl of trimethylethyllead standard solution into the gas-chromatographic column, with programmed temperature (from 50 to 200°C at 40°C/min), and with nitrogen carrier-gas flow-rates of 140 ml/min. As reported in Table 5.5, and in agreement with the general trend, AFS is superior to AAS for flame systems but the reverse is true for non-flame atomizers. The alkyllead was found to decompose in the heated stainless-steel transfer line, and this caused uncorrectable solvent peaks due to remobilization of lead by solvent containing methyl or ethyl groups, such as methanol, ethanol and acetone. However the phenomenon was serious only with high concentrations of alkyllead, and it decreased with the type of material used for the transfer line, in the following order: silica > alumina > stainless-steel > carbon > tantalum.

Rigin [73] used a multielement non-dispersive AF spectrometer for determination of As, Sb, Bi, Ge, Sn, Pb, Se and Ta after gas-chromatographic separation of the volatile hydrides. The AF spectrometer consisted of several modules, each containing an EDL, an interference filter and a photomultiplier [76]. The hydrides were atomized by a hot helium-gas atomizer at 1800 K [75]. The atomizer was interfaced to the gas chromatograph by a silica transfer line. For the generation of volatile hydrides, 20 ml of aqueous sample were placed in a glass reaction vessel, purged with argon at 40 ml/min and pretreated with 3 ml of 20% potassium iodide

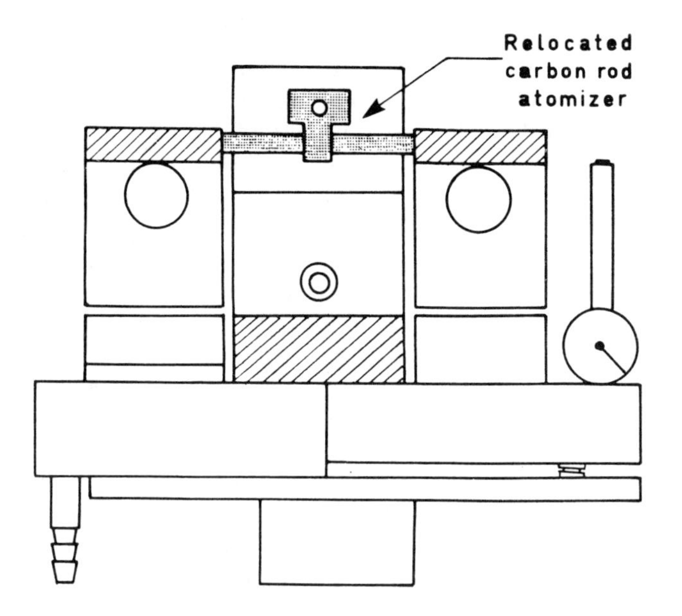

Fig. 5.17 — Modified CRA 63 atomizer employed for AF gas-chromatographic detection. (Reprinted from B. Radziuk, Y. Thomassen, L. R. P. Butler, J. C. Van Loon and Y. K. Chau, *Anal. Chim. Acta*, 1979, **108**, 31, by permission. Copyright 1979, Elsevier Science Publishers.)

Table 5.5 — Detection limits for (SNR=2) trimethylethyllead by AAS and AFS. Data taken from B. Radziuk, Y., Thomassen, L. R. P. Butler, J. C. Van Loon and Y. K. Chau, *Anal. Chim. Acta*, 1979, **108**, 31, by permission. Copyright 1979, Elsevier Science Publishers

Atomizer	Gas flow-rates (l./min)	Atomization temperature (°C)	Detection limit (ng)	
			AAS	AFS
Air/C_2H_2	11–12/2	2400	1.5	0.3
Air/C_2H_2 (for silica tube)	11/5.5	970	0.09	—
HGA 2100	—	1500	0.03	—
CRA 63 tube	—	1500	0.07	—
CRA 63 cup	—	1500‡ (1000)†	0.15	0.2
Silica furnace	0.6(H_2)† 15(nitrogen sheathing)†	1000	0.1	0.1

RF-EDL was used as light-source.
†Condition used in AFS measurement.
‡Condition used in AAS measurement.

solution, 8 ml of concentrated hydrochloric acid and 1 ml of 5% thiourea solution to prevent interference effects from Cu, Ag, Au and platinum group metals. The analytes were then converted into volatile hydrides by the addition of 5 ml of 8% sodium borohydride in 2% sodium hydroxide solution, 5 ml of 40% sodium hydroxide solution after 2 min, 2 ml of 5% tartaric acid solution after a further 2 min and 5 ml of 50% orthophosphoric acid followed by 2 ml of 8% sodium borohydride in 2% sodium hydroxide solution after a further 3 min. After the last addition of reagent, the argon was allowed to flow for 5 min more. The evolved hydrides were dried by flushing the argon stream through a cartridge packed with zeolite of NaA type, then trapped in a gas-chromatograph column packed with Chromosorb 102 and kept at 253 K. The column was inserted into the gas-chromatograph oven and the separation performed by increasing the temperature at a rate of 20°C/min, with helium at 30 ml/min as carrier gas. The detection limits were 30 pg for Se, 50 pg for As, 100 pg for Bi, Sn and Te, 150 pg for Ge and 200 pg for Pb and Sb. The elements Cu, Ag, Au, Fe, Ni, Co, Pt, Pd and Os did not interfere even in 10^6-fold mass ratio to the analyte. In a more recent work Rigin [76] used another non-dispersive AF instrument for determination of several elements after separation of their volatile fluorides. The light-source was a xenon arc lamp (3000 W) and the atomizer was a miniaturized ICP silica torch, provided with air and water cooling, powered at 450 W with a radiofrequency of 56 MHz and operated at 0.85 l./min flow of argon. The gas-chromatograph column was made of nickel (180 cm long, 3.8 mm external diameter, 0.4 mm wall thickness) and packed with caesium fluoride. For the determination of Ge, Re, V, U, Mo and W the sample was decomposed by heating with xenon tetrafluoride at 773 K for 30 min in a miniature nickel autoclave. The system was then cooled to room temperature and the excess of gas discarded. The autoclave was connected to the gas-chromatograph column and the volatile fluorides were stripped by passing a stream of F_2/He (1:5 v/v) at 100 ml/min into the autoclave heated at 600 K. The fluorides trapped by the chromatographic column were separated with a flow of 120 ml/min of a mixture of F_2 and Ar (1:5 v/v) with an inverse spatial thermal gradient of 26°C/100 mm applied along the column. With 50–100 mg of sample, the detection limits ranged from 2 ng/g for W to 80 ng/g for Ge.

Simultaneous speciation of several volatile hetero-organic compounds containing various metallic or non-metallic elements (e.g. alkylleads, alkyltins, alkylselenides, alkymercury etc.) can readily be achieved by interfacing a gas chromatograph with a multichannel AF spectrometer. The major problem here is the choice of a suitable material for a general purpose transfer line compatible with all compounds to be determined.

Alkyllead, alkyltin and alkylselenide compounds were detected simultaneously by gas chromatography with the multichannel non-dispersive AF spectrometer, described in Fig. 5.8 [77], as a simultaneous multielement detector. A wide-bore capillary column (Hewlett Packard HP-1, 0.53 mm i.d., 10 m long, film thickness 2.65 μm of methylsilicone oil) was used instead of a packed column. The transfer line from the chromatograph to the atomizer was the capillary column itself: about 1 m of the capillary column led out of the chromatograph oven through an empty detector port, and was inserted directly into the burner through a chromatographic rubber septum (Fig. 5.18a). This external portion of the capillary column was inserted into a stainless-steel tube of the same length (1.5 mm i.d.) to protect it mechanically. The

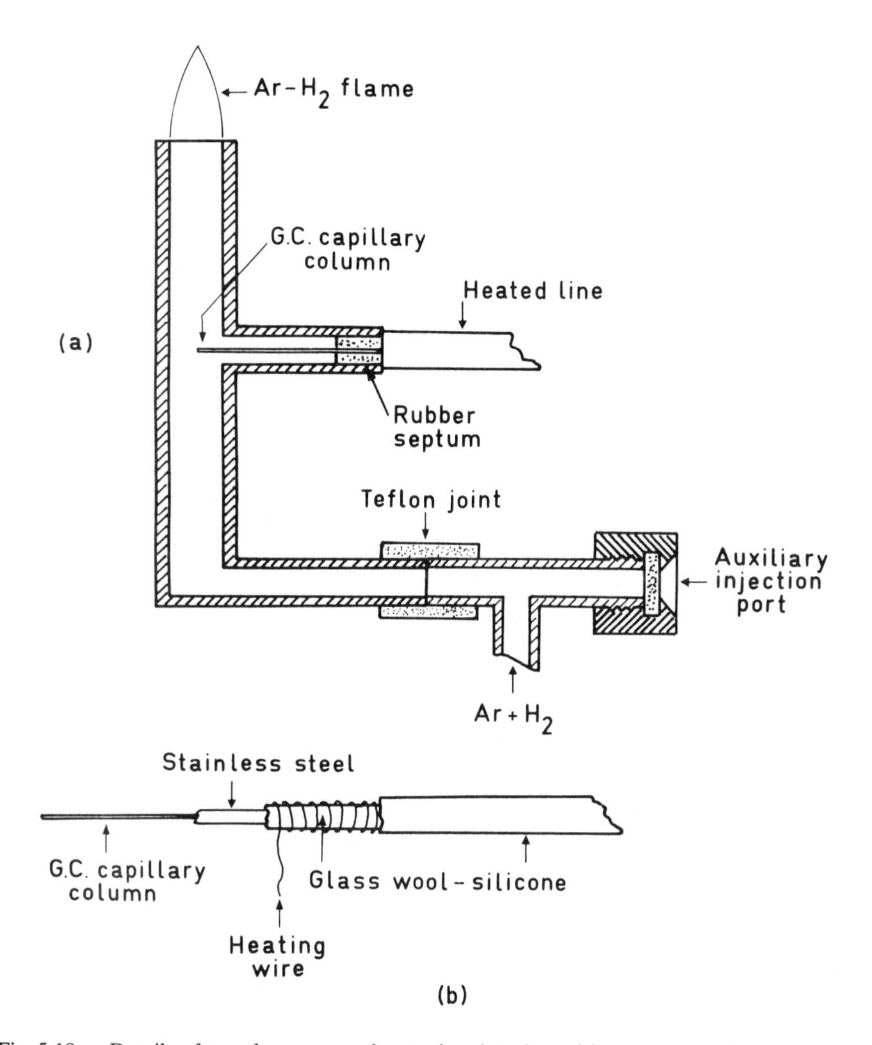

Fig. 5.18 — Details of gas chromatograph–atomizer interface: (a) connection of wide-bore capillary column to the glass burner; the auxiliary injection port was employed only in some interference studies; (b) detail of the heated transfer line. (Reproduced by permission from A. D'Ulivo and P. Papoff, *J. Anal. At. Spectrom.*, 1986, **1**, 479. Copyright 1986, Royal Society of Chemistry.)

stainless-steel tube was forced into a flexible glass-wool–silicone tube (2 mm i.d.) wrapped with nichrome wire, and the whole was covered with another flexible glass wool–silicone tube (4 mm i.d.) (Fig. 5.18b). The nichrome wire was connected to a Variac autotransformer to allow temperature control of the external portion of the capillary column. This method avoided problems related to the transfer line material, since the wide-bore capillary column was more inert than a packed column. In addition, the wide-bore capillary columns gave better performance than the packed ones, in efficiency, resolution and analysis time, and the load capacity was satisfactory. The atomizer, made of borosilicate glass (4 mm i.d.), was the same as the one used for hydride atomization [16,17], and supported a miniature argon–hydrogen–entrained air flame obtained with flows of 0.4 l./min argon and 0.2 l./min hydrogen. These flow-rates gave a small flame volume and low dilution of the gas-chromatographic effluent (argon, 6 ml/min). The apparatus was tested by injecting, directly into the column, 0.5–2 μl of a standard solution containing $PbMe_4$, $PbEt_4$, $SnMe_4$, $SnEt_4$, Me_2Se and Me_2Se_2 (Me = methyl, Et = ethyl). The column temperature was increased from 60 to 150°C at 15°C/min and then maintained at 150°C for 5 min. The external portion of the capillary gas-chromatographic column was kept at 120°C. The atomizer was wrapped with nichrome wire, insulated with asbestos tape and kept at 200°C to avoid any peak broadening of less volatile compounds such as $PbEt_4$, $SnEt_4$ and Me_2Se_2, which would probably be caused by their partial condensation on the cold surfaces. A typical chromatogram is shown in Fig. 5.19. A phototransistor detector (Philips BPX25) was used to monitor the molecular emission of the solvent in the flame in the visible-red spectral region. In spite of its poor sensitivity, this simple device was found to be effective for monitoring the appearance of the solvent in the flame, and proved very useful in studies of interferences. Because of the low atomizer temperature particular attention was paid to interference effects. Selectivity was measured for many organic compounds (see Table 5.6), and was $>10^6$ in all cases. These selectivity values are remarkable for a non-dispersive apparatus without background correction. No interelement effect between analytes or instrumental cross-talk between the channels of the AF spectrometer was observed. The most serious interference effect was observed when the peak for the solvent (0.5–2 μl) and the analyte peak overlapped. In this case, the fluorescence signal was greatly reduced, probably owing to quenching effects, but molecular absorption and chemical effects may also have contributed. However, this effect could be eliminated by using another solvent with a different retention time. It seems unlikely that problems would be encountered with any organic compound at trace levels. Detection limits, expressed as three times the standard deviation of the baseline noise, were 30 pg of Pb, 50 pg of Sn and 10 pg of Se. They compared very favourably with the literature values for electrothermal GC–AAS detection for Pb and Sn, and were 15 times better for Se, confirming that the miniature argon–hydrogen flame can be used successfully for atomization of some organometallic compounds as well as for volatile hydrides.

5.4 CONCLUDING REMARKS

Although many AFS instruments have been described in the literature, few of these have found application as element-specific detectors for chromatography. This may be because of the general lack of commercial equipment with the necessary sensitivity, specificity and convenience.

In practice, only air–acetylene flames as atomizers and low intensity light-sources such as HCls or continuum xenon-arc lamps have been used in combination with HPLC. Only recently have more intense light-sources such as boosted output HCLs or demountable cathode lamps been applied to HPLC–AFS (with excellent results). In GC–AFS, where rf EDLs have been tested in combination with several flame and

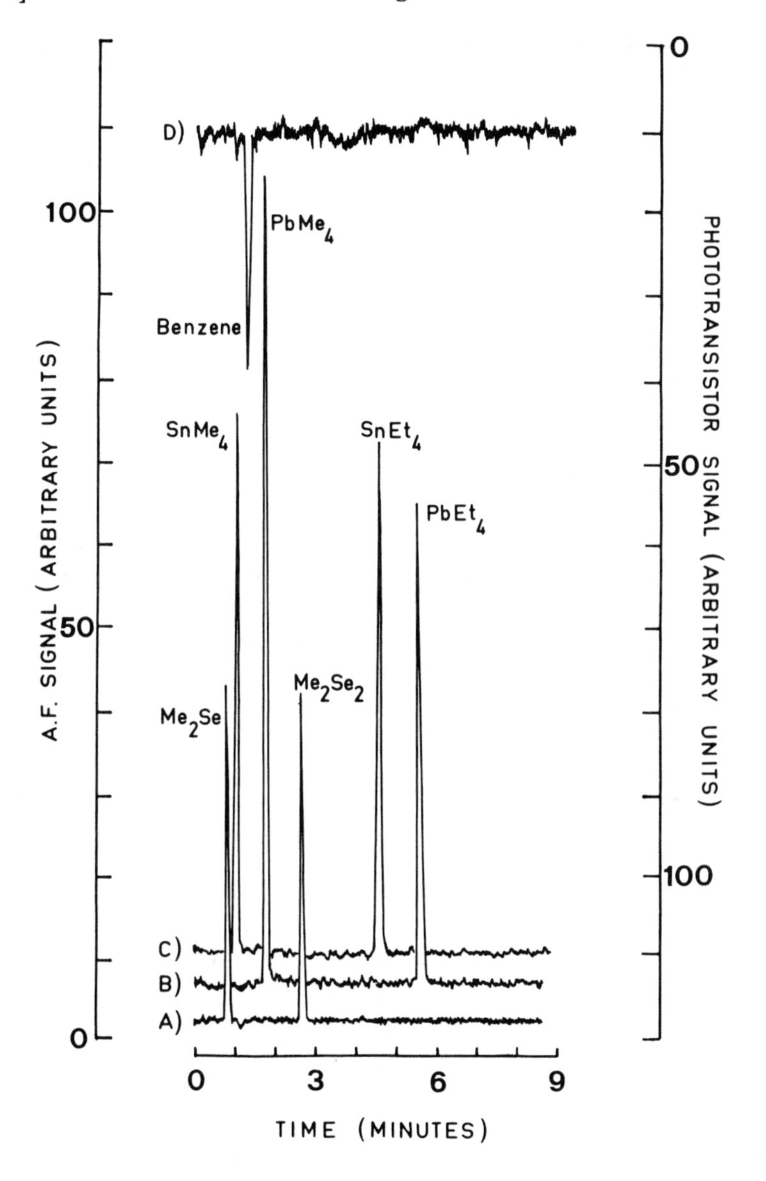

Fig. 5.19 — Typical chromatograms obtained with the AF gas-chromatographic detector: (A), selenium channel, 0.3 ng Se for each peak; (B) lead channel, 1 ng Pb for each peak; (C) tin channel, 1.5 ng Sn for each peak; (D) phototransistor response for solvent, 1 µl of benzene. (Reproduced by permission from A. D'Ulivo and P. Papoff, *J. Anal. At. Spectrom.*, 1986, **1**, 479. Copyright 1986, Royal Society of Chemistry.)

non-flame atomizers, interesting results have been obtained in terms of sensitivity and selectivity. However, these few applications do not begin to illustrate the potential for use of AFS in chromatographic detection.

For liquid-chromatographic detection, AFS can be considered to be competitive

Table 5.6 — Selectivity of multielement AF gas-chromatographic detector for some organic compounds

Organic compound	Amount injected (μl)	Calculated selectivities†		
		Pb	Sn	Se
n-Hexane	0.5–2	10^8	10^8	10^7
Benzene	0.5–2	10^8	4×10^7	9×10^7
CS_2	0.2–1	2×10^6	10^6	10^7
CCl_4	0.5–1	2×10^7	3×10^7	10^8
CH_3CCl_3	0.5–2	4×10^7	10^8	10^8
Ethanol	0.5–2	10^8	10^8	10^8
Methyl acetate	0.5–2	10^8	10^8	10^8
Butan-2-one	0.5–2	10^8	10^8	10^8

† Selectivity is calculated as the amount of the organic compound (ng) that gives the same signal as 1 ng of the organometallic compound (as the element).

with AAS, since the graphite-furnace atomizer cannot be used in a convenient 'on-line' detector. Thus, partly owing to the current commercial availability of intense light-sources, the sensitivity of flame AFS is superior to that of flame AAS. In addition, flame AFS gives wider dynamic range and better multielement capability. Other atomizers seem promising for AFS detection in liquid-chromatography; these are the continuous-flow furnaces and the ICP, both coupled with intense light-sources. The multielement ICP–AFS instrument developed by the Baird Co. could provide a good AF detector for liquid chromatography, if ultrasonic nebulization and boosted output HCLs are also used. The performance is comparable with that of ICP–AES for detection limits and dynamic ranges, and it suffers less from interference problems.

For gas-chromatographic detection, in practice any kind of atomizer can be employed. Thus, non-flame AAS with graphite or silica electrothermally heated atomizers can be used, at least for single-element detection systems. An interesting atomizer for GC–AFS is the argon–hydrogen miniature flame, when volatile, readily atomizable compounds are to be determined. This simple atomizer gives sensitivities the same as or better than those obtained with non-flame AAS, when it is used for the atomization of hydrides and volatile organometallic compounds such as alkylleads, alkyltins and alkylselenides.

A non-dispersive optical system is preferred to a dispersive one in most of the chromatographic detectors, particularly because its inherent simplicity is a help if multielement apparatus has be be assembled. It also gives better detection limits when used in combination with low background atomizers as employed in most chromatographic detectors.

Application of laser-excited AFS to liquid- or gas-chromatographic detection could lead to exceptionally low detection limits, probably several orders of magnitude better than those of AFS with conventional light-sources. However, at present laser-excited AFS has practical limitations, such as the high cost of laser instrumentation and the difficulties involved in the use of pulsed tunable dye lasers. In addition, laser-excited AFS does not allow simultaneous multielement determinations. However, it is a field in rapid progress, and it will probably be the one in which we will find the next significant improvements.

REFERENCES

[1] J. D. Winefordner and T. J. Vickers, *Anal. Chem.*, 1964, **36**, 161.
[2] J. D. Winefordner and R. A. Staab, *Anal. Chem.*, 1964, **36**, 165.
[3] N. Omenetto and J. D. Winefordner, *Prog. Anal. Atom. Spectrosc.*, 1979, **2**, 183.
[4] R. F. Browner, *Analyst*, 1974, **99**, 617.
[5] N. Omenetto and J. D. Winefordner in *Analytical Laser Spectroscopy*, N. Omenetto (ed.), p. 167. Wiley, New York, 1979.
[6] C. Veillon, J. M. Mansfield, M. L. Parsons and J. D. Winefordner, *Anal. Chem.*, 1966, **38**, 204.
[7] N. Omenetto, L. P. Hart and J. D. Winefordner, *Appl. Spectrosc.*, 1972, **26**, 612.
[8] D. R. Demers and D. W. Ellis, *Anal. Chem.*, 1968, **40**, 860.
[9] P. L. Larkins and J. B. Willis, *Spectrochim. Acta*, 1971, **26B**, 491.
[10] P. L. Larkins, *Spectrochim. Acta*, 1971, **26B**, 477.
[10a] D. J. Johnson and J. D. Winefordner, *Anal. Chem.*, 1976, **48**, 341.
[10b] K. A. Saturday and G. M. Hieftje, *Anal. Chem.*, 1977, **49**, 2013.
[11] D. W. Brinkman and R. D. Sacks, *Anal. Chem.*, 1975, **47**, 1279.
[12] K. Tsujii and K. Kuga, *Anal. Chim. Acta*, 1978, **97**, 51.
[13] K. Tsujii and K. Kuga, *Anal. Chim. Acta*, 1978, **101**, 199.
[14] K. Tsujii and E. Kitazume, *Anal. Chim. Acta*, 1981, **125**, 101.
[15] K. Tsujii, E. Kitazume and K. Yagi, *Anal. Chim. Acta*, 1981, **128**, 229.
[16] A. D'Ulivo, P. Papoff and C. Festa, *Talanta*, 1983, **30**, 907.
[17] A. D'Ulivo, R, Fuoco and P. Papoff, *Talanta*, 1985, **32**, 103.
[18] A. D'Ulivo, P. Papoff and C. Festa, *Talanta*, 1985, **32**, 383.
[19] A. D'Ulivo, R. Fuoco and P. Papoff, *Talanta*, 1986, **33**, 401.
[19a] S. K. Hughes and R. C. Fry, *Appl. Spectrosc.*, 1981, **35**, 26.
[20] A. Montaser and V. A. Fassel, *Anal. Chem.*, 1976, **48**, 1490.
[21] D. R. Demers, D. A. Busch and C. D. Allemand, *Int. Lab.*, 1982, **12**, May, 41.
[22] D. R. Demers and C. D. Allemand, *Anal. Chem.*, 1981, **53**, 1915.
[23] N. Omenetto, H. G. C. Human, P. Cavalli and G. Rossi, *Spectrochim. Acta*, 1984, **39B**, 115.
[24] H. G. C. Human, N. Omenetto, P. Cavalli and G. Rossi, *Spectrochim. Acta*, 1984, **39B**, 1345.
[25] M. A. Kosinski, H. Uchida and J. D. Winefordner, *Anal. Chem.*, 1983, **55**, 688.
[26] S. Greenfield and M. Thomsen, *Spectrochim. Acta*, 1985, **40B**, 1369.
[27] T. A. West and X. K. Williams, *Anal. Chim. Acta*, 1969, **45**, 27.
[28] R. G. Anderson, I. S. Maines and T. S. West, *Anal. Chim. Acta*, 1970, **51**, 355.
[29] J. F. Alder and T. S. West, *Anal. Chim. Acta*, 1970, **51**, 365.
[30] C. J. Molnar, R. D. Reeves, J. D. Winefordner, M. T. Glenn, J. R. Ahlstrom and J. Savory, *Appl. Spectrosc.*, 1972, **26**, 606.
[31] B. M. Patel, R. D. Reeves, R. F. Browner, C. J. Molnar and J. D. Winefordner, *Appl. Spectrosc.*, 1973, **27**, 171.
[32] K. Kuga and K. Tsujii, *Anal. Chim. Acta*, 1976, **81**, 305.
[33] K. Kuga and K. Tsujii, *Anal. Lett.*, 1982, **15**, 47.
[34] C. J. Molnar and J. D. Winefordner, *Anal. Chem.*, 1974, **46**, 1419.
[35] S. A. Clyburn, B. R. Bartschmid and C. Veillon, *Anal. Chem.*, 1974, **46**, 2201.
[36] S. Greenfield, G. M. Hieftje, N. Omenetto, A. Scheline and W. Slavin, *Anal. Chim. Acta*, 1986, **180**, 69.
[37] J. V. Sullivan, *Prog. Anal. Atom. Spectrosc.*, 1981, **4**, 311.
[38] C. A. Sacchi and O. Svelto, in *Analytical Laser Spectroscopy*, N. Omenetto (ed.), p. 2. Wiley, New York, 1979.
[39] N. Omenetto and J. D. Winefordner, *Prog. Anal. Atom. Spectrosc.*, 1985, **8**, 371.
[40] T. J. Vickers, P. J. Slevin, V. I. Muscat and L. T. Farias, *Anal. Chem.*, 1972, **44**, 930.
[41] Y. Talmi and R. W. Simpson, *Appl. Opt.*, 1980, **19**, 1401.
[42] Y. Talmi, *Appl. Spectrosc.*, 1982, **36**, 1.
[43] P. D. Warr, *Talanta*, 1970, **17**, 543.
[44] Y. Talmi, R. Crosmum and N. M. Larson, *Anal. Chem.*, 1976, **48**, 326.
[45] M. K. Murphy, S. A. Clyburn and C. Veillon, *Anal. Chem.*, 1973, **45**, 1468.
[45a] E. F. Palermo, A. Montaser and S. R. Crouch, *Anal. Chem.*, 1974, **46**, 2154.

[46] H. V. Malmstadt, C. G. Enke and S. R. Crouch, *Electronic Measurements for Scientists*, p. 816. Benjamin, Menlo Park, California, 1974.
[47] A. H. Ullman, *Prog. Anal. Atom. Spectrosc.*, 1980, **3**, 87.
[48] E. B. Jansen and D. R. Demers, *Analyst*, 1985, **110**, 541.
[49] R. L. Lancione and D. M. Drew, *Advances in the application of atomic fluorescence to the analysis of precious metals*, presented at IPMI's 'Analytical Seminar '84', Newport, 20 September 1984.
[50] D. R. Demers, *Spectrochim. Acta*, 1985, **40B**, 93.
[51] D. A. Naranjit, B. H. Radziuk, J. C. Rylaarsdam, P. L. Larkins and J. C. Van Loon, *Appl. Spectrosc.*, 1985, **39**, 128.
[52] N. Omenetto and H. G. C. Human, *Spectrochim. Acta*, 1984, **39B**, 1333.
[53] T. Norris and J. V. Sullivan, *Am. Lab.*, 1982, **14**, No. 12, 67; *Int. Lab.*, 1983, **13**, Jan., 76.
[54] K. C. Thompson, *Analyst*, 1975, **100**, 307.
[55] D. Kolihová and V. Sychra, *Anal. Chim. Acta*, 1972, **59**, 477.
[56] S. Kobayashi, T. Nakahara and S. Musha, *Talanta*, 1979, **26**, 951.
[57] G. F. Kirkbright and T. S. West, *Chem. Brit.*, 1972, **8**, 428.
[58] T. Nakahara, S. Kobayashi, T. Wakisawa and S. Musha, *Spectrochim. Acta*, 1981, **36B**, 661.
[59] T. Nakahara, T. Wakisawa and S. Musha, *Spectrochim. Acta*, 1981, **36B**, 661.
[59a] R. L. Lancione and D. M. Drew, *Spectrochim. Acta*, 1985, **40B**, 107.
[60] R. Ferrara, A. Seritti, C. Barghigiani and A. Petrosino, *Anal. Chim. Acta*, 1980, **117**, 391.
[61] D. J. Johnson, F. W. Flankey and J. D. Winefordner, *Anal. Chem.*, 1974, **46**, 1898.
[62] K. E. Zacha, M. P. Bratzel, Jr., J. D. Winefordner and J. M. Mansfield, Jr., *Anal. Chem.*, 1968, **40**, 1733.
[63] M. A. Bolshov, A. V. Zybin and I. I. Smirenkina, *Spectrochim. Acta*, 1981, **36B**, 1143.
[64] J. C. Van Loon, J. Lichwa and B. Radziuk, *J. Chromatog.*, 1977, **136**, 301.
[65] D. D. Siemer, P. Koteel, D. T. Haworth, W. J. Taraszewski and S. R. Lawson, *Anal. Chem.*, 1979, **51**, 575.
[66] D. Mackey, *J. Chromatog.*, 1982, **236**, 81.
[67] D. Mackey, *Personal communication*, 1986.
[68] D. Mackey, *Mar. Chem.*, 1983, **13**, 169.
[69] D. Mackey, *J. Chromatog.*, 1982, **237**, 79.
[70] D. Mackey, *J. Chromatog.*, 1982, **242**, 275.
[71] D. Mackey, *Mar. Chem.*, 1985, **16**, 105.
[72] B. Radziuk, Y. Thomassen, L. R. P. Butler, J. C. Van Loon and Y. K. Chau, *Anal. Chim. Acta*, 1979, **108**, 31.
[73] V. I. Rigin, *Zh. Analit. Khim.*, 1986, **41**, 46.
[74] V. I. Rigin, *Zh. Analit. Khim.*, 1980, **35**, 64.
[75] V. I. Rigin, *Zh. Analit. Khim.*, 1980, **35**, 863.
[76] V. I. Rigin, *Zh. Analit. Khim.*, 1986, **41**, 788.
[77] A. D'Ulivo and P. Papoff, *J. Anal. At. Spectrom.*, 1986, **1**, 479.

6

Interfaces between liquid chromatography and atomic absorption

Les Ebdon and Steve Hill
Department of Environmental Sciences, Plymouth Polytechnic, Drake Circus, Plymouth PL4 8AA, England

6.1 INTRODUCTION

Although gas chromatography (GC) has frequently been coupled with analytical atomic spectrometry for determination of trace metal speciation [1], it is limited to volatile and thermally stable organometallic species or metal chelates. The use of liquid chromatography (LC), in particular high-performance liquid chromatography (HPLC), considerably increases the number of types of chemical and physical species which may be studied. The separation of ions and non-volatile organometallic species with high molecular weight, in addition to volatile species, is possible by use of one or other of the popular LC configurations. Adsorption, ion-exchange, gel-permeation, normal and reverse-phase chromatography have all been used in conjunction with atomic spectrometry.

A number of criteria may be used for the characterization of a good HPLC detector. First, the noise level governs the detection limit available with a particular detector. A chromatographic peak can only be recognized as such if its height is at least twice that of the highest noise peak. In addition to purely electrical sources, air-bubbles and impurities may also cause noise. A drift in the baseline resulting from slow changes in the flow-rate or ambient temperature is also undesirable. In consideration of sensitivity, distinction must be made between the absolute and the relative sensitivity of a detector. The absolute sensitivity is a function of the instrument design, measuring technique and noise level, whilst the relative sensitivity depends on the amount of substance that is just detectable under a defined set of conditions. However, there are other factors to consider, such as band-spreading in the detector or coupling, dependence of the response on external parameters, and convenience of servicing. Recent studies [2] have suggested that the important dispersion characteristics of a coupled system using high-efficiency columns are not those that occur in the column, but those associated with the interface and detector. Clearly any such dispersion is important since it not only reduces element sensitivity but can also destroy the separation originally attained in the column.

The coupling of liquid chromatography with atomic spectrometry has been reviewed in a number of publications [3–8]. Generally its aim is to combine the capabilities of chromatography for species separation with the selectivity and sensitivity of atomic spectrometry for detection and determination. The ideal interface system may be identified as one which offers high sensitivity with on-line, real-time analysis, and thus produces continuous chromatograms at practical levels. The detector cell must be capable of efficiently accepting and atomizing large volumes of a liquid mobile phase containing traces of analytes. Thus the atom cells used (flame or furnace) must be capable of handling flows of solvent, typically in the range 0.1–3.0 ml/min, which may be aqueous or organic in nature. It has often been stated that the use of flame atomic-absorption spectrometry (FAAS) is limited by lack of sensitivity, and that for many environmental studies the higher sensitivity offered by electrothermal furnace devices must be utilized. However, recent advances in which particular attention has been paid to the atom cell have produced a range of interfaces for HPLC–FAAS which are sensitive (detection limits in the pg range), yet simple to construct and readily demountable. This chapter looks at the techniques available for directly coupled HPLC–atomic-absorption spectrometry and the utilization of these techniques for various applications.

6.2 HIGH-PERFORMANCE LIQUID CHROMATOGRAPHY DIRECTLY COUPLED WITH FLAME ATOMIC-ABSORPTION SPECTROMETRY

In general, HPLC offers acceptable recoveries of nanogram amounts of analytes, provided that an appropriately sensitive detector is available. Many types of detector have been coupled to HPLC for determining trace metals. These include devices that use UV absorption, fluorescence, electrochemical techniques, chemiluminescence, mass spectrometry, flame ionization, electron capture, and electron-spin resonance as the detection method. However, none of these adequately satisfies the criteria cited above for the ideal detector, and notably, all fail to satisfy the need for unambiguous element selectivity. The intrinsic selectivity of flame atomic absorption readily overcomes this problem, and the technique has the additional advantage of readily accepting a continuous flow of liquid samples. Thus coupled HPLC–FAAS systems have much potential for metal speciation studies.

The simplest interface of this type may be constructed by connecting a transfer line directly between the nebulizer and the HPLC column. Such an approach may lead to a number of problems, however. Often the nebulizer is 'starved' of liquid, i.e. the column flow-rate is less than the nebulizer uptake rate. This is not uncommon, because optimal flow-rates in HPLC are set at values determined by the criteria of chromatographic separation but the nebulizer uptake rate is usually higher and has been designed to maximize sensitivity for the aspiration of standard solutions, as is normal in AAS. If the burner is starved in this way, a reduced-pressure region may result at the end of the column, and this may lead [9] to gas-bubble formation, which would be particularly deleterious if a post-column flow-cell detector were operated in series with the AAS detector. This type of interface was used by Jones *et al.* [10] in a study of the effect of column flow-rate on nebulization efficiency for a number of transition metals. It was found that decreasing the column flow-rates caused the nebulization efficiency to increase, but that lower analytical signals were observed because of the lower sample volumes introduced. Various mobile phases were investigated in this study. For methanol, ethanol, chloroform and benzene, 100% nebulization efficiencies could be achieved at flow-rates of 1 ml/min, whereas for water at the same flow-rate only 32% was nebulized into the flame. These figures illustrate the higher efficiencies often obtained with the starved mode (cf. 10–20% at 5 ml/min, common for many conventional nebulizers). In a further study, Yoza and Ohashi [11] preferred to operate their nebulizer at the same flow-rate as the chromatographic pump, but found balancing of the two flow-rates to be difficult. Their solution was to incorporate an additional solvent reservoir at the end of the column, from which any additional solvent required by the nebulizer could be drawn (Fig. 6.1). This, however, leads to further dilution, which is obviously undesirable for trace analysis.

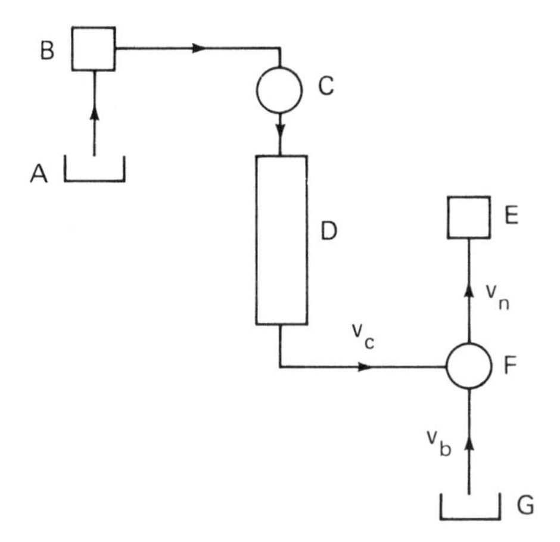

Fig. 6.1 — A schematic representation of flow system. A; Eluent reservoir, B; pump, C; sample injector, D; separation column, E; nebulizer, F; three-way glass tubing, G; water reservoir. (Reprinted from N. Yoza and S. Ohashi, *Anal. Lett.*, 1973, **6**, 595, by permission of the copyright holder, Marcel Dekker Inc.)

An alternative to these approaches would be to operate the nebulizer at a flow-rate lower than the column flow-rate, i.e. to flood the nebulizer. Since LC flow-rates are generally below 3 ml/min, a very low aspiration rate would be necessary to encompass the entire range while still maintaining aspiration. Koropchak and Coleman [9] reported that operating a nebulizer at slight back-pressure not only eliminated the use of a post-column diluter to match LC flow-rates with nebulizer uptake rate, but also gave improved signal-to-noise ratios with a standard nebulizer arrangement. However, such improvements are dependent on the design of the nebulizer, since changes in the shape of the capillary tip influence both droplet characteristics and transport efficiencies.

Despite the disadvantages of such simple couplings, many of the early studies in this area were based on such interfaces. Several workers utilized FAAS in conjunction with simple ion-chromatography. Manahan and Jones [12] noted the potential of FAAS as a metal-specific detector for the determination of ethylenediaminetetra-acetic acid (EDTA) and nitrilotriacetic acid (NTA) separated as their copper complexes by ion-exchange chromatography. The column was constructed from a 1.00 ml syringe, the resin being held in place by a plug of glass wool (Fig. 6.2). A

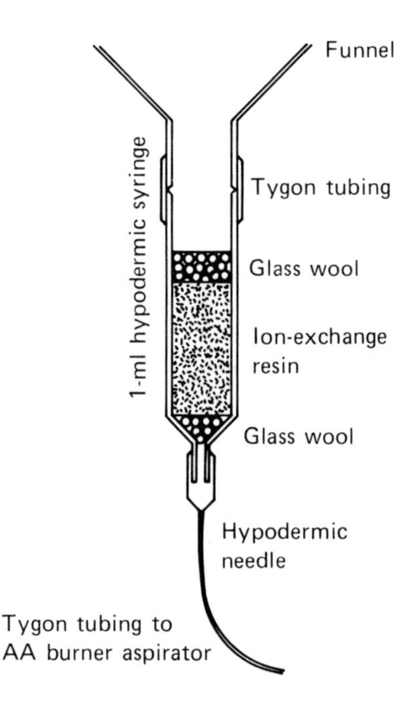

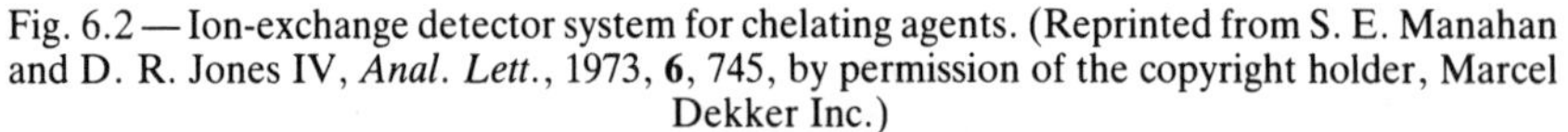

Fig. 6.2 — Ion-exchange detector system for chelating agents. (Reprinted from S. E. Manahan and D. R. Jones IV, *Anal. Lett.*, 1973, **6**, 745, by permission of the copyright holder, Marcel Dekker Inc.)

hypodermic needle was used to attach the column to the tubing from the nebulizer of the atomic-absorption instrument. Later, the same workers [13] used HPLC, incorporating an anion-exchange column, to examine various copper-chelated and spiked sewage effluents (Fig. 6.3). This work also extended the range of aminocarboxylic acid–copper chelates studied, to include ethylene glycol-bis(2-aminoethyl ether)tetra-acetic acid (EGTA) and 1,2-diaminocyclohexanetetra-acetic acid (DCTA), and demonstrated that organic mobile phases, toluene/pyridine, could be used for the separation of various chromium chelates [14]. Pankow and Janauer [15] also employed an ion-exchange system to separate and preconcentrate chromate in natural waters and reported detection limits of 0.1 mg/ml.

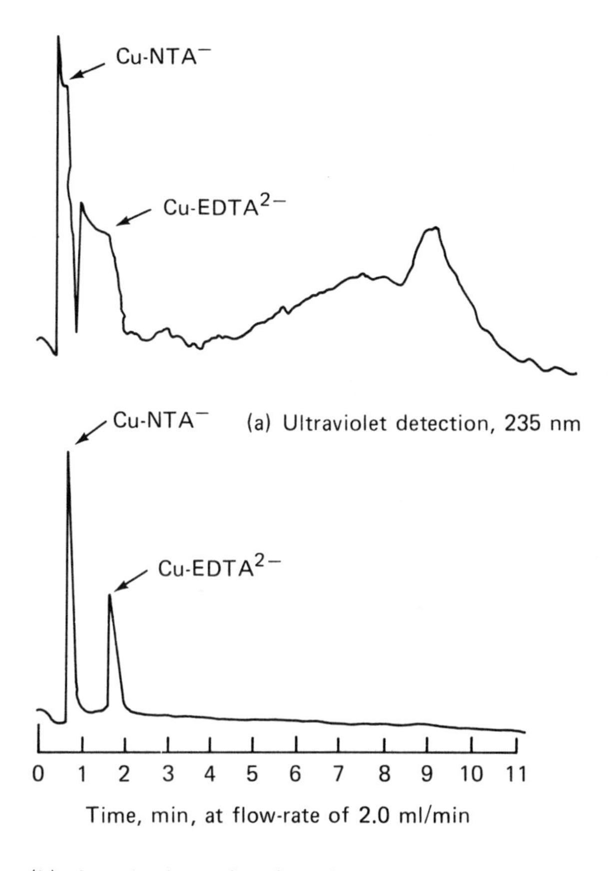

Fig. 6.3 — Chromatograms of sewage spiked with copper chelates of NTA and EDTA. Ultraviolet detector (A) and copper AAS detector (B): Aminex A-14 anion exchange resin. $0.05M(NH_4)_2SO_4$ column, 2.0 ml/min flow-rate (Reprinted by permission from D. R. Jones IV and S. E. Manahan, *Anal. Chem.*, 1976, **48**, 1897. Copyright 1976, American Chemical Society.)

The potential of FAAS as a detector for gel chromatography was first noted by Yoza and Ohashi in 1973 [11] in a study of magnesium and potassium in chloride solution. The outlet of the separation column was attached directly to the nebulizer, and the flows were balanced with an additional solvent reservoir as described above. Later, a similar system was used for the determination of condensed phosphate anions (diphosphate, triphosphate etc.) measured as magnesium complexes [16]. Yoshimura and Tarutani [17] have also used gel chromatography, in a study of the separation of isothiocyanate complexes of chromium(III), although here the effluent was collected in an automatic fraction collector for subsequent analysis by AAS.

The separation of tetra-alkyllead compounds in petrol has been used to demonstrate a direct coupling of LC to FAAS [18]. Once again, few modifications were made, except that in this case an internal flow-spoiler was removed from the burner to allow a higher fraction of the effluents to reach the flame. Cassidy *et al.* [19] used molecular-sieve and reversed-phase chromatography for the separation of various organosilicon compounds. The organosilicons were first preconcentrated on porous polymer columns before HPLC separation. After elution of the sorbed organosilicons (with 4-methylpentan-2-one), the HPLC effluent was fed directly to the nebulizer and into a nitrous oxide–acetylene flame. Detection limits reported for the various organosilicons ranged from 0.5 to 5 μg. Van Loon *et al.* [20] used direct coupling of the column eluent to the nebulizer to monitor copper–amino-acid complexes, used in the treatment of metal poisoning, and also to study zinc aryl and alkyl compounds in lubricating oils. A similar coupling was utilized in the preconcentration and speciation of gold and platinum complexes from aqueous solutions [21].

An important step forward in direct coupling was reported by Slavin and Schmidt in 1979 [22]. They appreciated both the problems discussed above and the advantages of operating the nebulizer in a starved mode. During a study of metal labelling of amino-acids, the simple direct coupling was found to be troublesome owing to the mismatch of flow rates — in this case the natural uptake rate of the nebulizer was about 8 ml/min, whereas the chromatography was optimum at a flow-rate of 1 ml/min. Their solution was to use the discrete-injection method of sample introduction, a technique they had recently published for liquid chromatography, where it was used to measure silica in the eluate from the column [23]. The system used is shown in Fig. 6.4. The effluent from the column is collected as a drop at the end of the column.

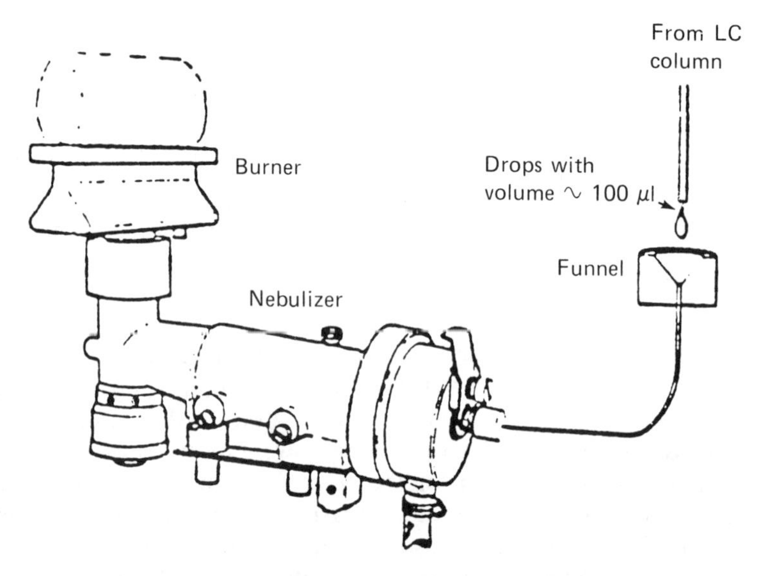

Fig. 6.4 — The system used for the atomic-absorption injection method. The eluent from the column is collected in approximately 100 μl drops by the drop former. When the drop falls into the funnel it is aspirated into the flame at the top of the atomic-absorption burner. (Reprinted from W. Slavin and G. J. Schmidt, *J. Chromatog. Sci.*, 1979, **17**, 610, by permission. Copyright 1979, Preston Publications Inc.)

When a drop of the optimum size (about 100 μl) has been formed, it falls into the conical PTFE cup and is sucked into the atomic-absorption flame, producing a transient signal. The concept of metal labelling of species to enable their determination by atomic spectrometry has great potential, but the relatively low sensitivity of flame AAS may well be a problem in the determination of amino-acids in body fluids at natural levels.

We have recently reported a modified coupling also based on discrete-volume, starved nebulization, which prevents blocking of the funnel by particulates such as dust, and facilitates greater flexibility in the positioning of the instruments. Here the funnel is replaced by a vented capillary tube between the HPLC column and the nebulizer (Fig. 6.5). In addition, we also incorporate a slotted-tube atom trap

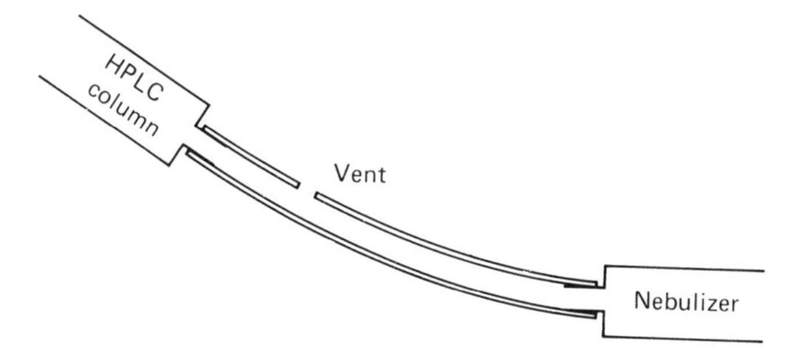

Fig. 6.5 — Schematic diagram of HPLC–nebulizer interface.

(STAT) above the flame, to achieve an increase in sensitivity for those elements readily dissociated to their ground-state atoms. A second advantage of using such tubes may be identified. Since the flame is a dynamic entity and the samples are passed through it in a plug fashion, a series of transient signals is produced. Thus the recorder response must be faster than, and consistent with, the amplifier time-constant, as in systems using electrothermal atomization. By increasing the residence time of the atoms in the optical path, the demands made on the electronics are greatly reduced and peak distortion is avoided. Although simple in design, this interface lends itself to many applications, and has successfully been used in our laboratories for the determination of organotin species at below ng/ml levels [24]. This was achieved by using a Whatman Partisil-10 SCX column (25 cm×4.6 mm i.d.) and a mobile phase of methanol:water (80:20 v/v) with ammonium acetate (0.1*M*) as buffer. No sample pretreatment was required, other than preconcentration during the extraction step. An initial extraction was made into chloroform, which was then evaporated to allow the sample to be redissolved in methanol before the injection. This method has been applied to an extensive survey of organotin levels in estuarine waters and in shellfish; routine measurements down to 20 ng/l. were made.

A number of other methods of direct coupling through the nebulizer have been published. A flow-injection sample manipulator (FISM) was used by Renoe *et al.* [25] as an interface between the chromatograph and spectrometer. This FISM allowed the addition of matrix modifier, in this case acidified lanthanum chloride, to the HPLC eluate before its introduction into the nebulizer of the spectrometer.

Several of the more recent publications on coupled LC–FAAS have stressed relatively simple interface systems, and have reported increased sensitivity by attention to the atom cell. The most common approach has been the use of hydride generation; this has been used for the determination of a number of metals and metalloids. The technique offers several advantages over conventional solution nebulization, including the capability for preconcentrations of the analyte, the elimination of chemical and spectral interferences and the presentation of the analyte in desolvated form to the atomization source [26]. These advantages have led to improvement of detection limits by as much as three orders of magnitude [27].

6.3 HIGH-PERFORMANCE LIQUID CHROMATOGRAPHY WITH DIRECTLY COUPLED HYDRIDE-GENERATION ATOMIC-ABSORPTION SPECTROMETRY

Hydride generation involves the formation of the gaseous hydride by chemical reduction of the sample, and this hydride is then entrained in a current of inert gas and led into the observation zone. Here it is decomposed by heat to form the atomic vapour. Several different methods have been based on this principle, although they differ in the ways in which both the reduction and the atomization are accomplished. The major advantage of hydride generation is the increased efficiency in sample transport, compared with nebulization. The review of hydride generation by Godden and Thomerson [28] covers the design of reaction vessels, methods of atomization, interferences and applications in some detail. Various collection devices for storing the hydrides before transfer to the atom cell have been used, although more recent systems [29,30] have excluded such devices by sweeping the hydride directly into the atom cell. This has been made more popular by the now universal use of acidified sodium tetrahydroborate as the reductant. Generation of hydrides with sodium tetrahydroborate is more rapid and efficient than with other reductants, so simpler systems without collection devices can be used. This reagent is also more suitable for continuous hydride generation systems, and these have been shown to give speedier analysis, improved precision and greater freedom from interferences [31] than manual injection.

The use of hydride generation after separation by HPLC has been reported by several groups of workers. In a comprehensive study of organotin compounds, Burns *et al.* found that coupled hydride–electrothermal atomization (ETA)–AAS readily permitted the determination of tin in organotin compounds after mineralization. The furnace consisted of a fused-silica T-tube, heated by a single-filament winding and insulated by a soft firebrick housing. This technique was found to be superior to GC–AAS for the determination of the methyltin series, since redistribution reactions did not take place on the chromatographic column [33]. For tetramethyl- and tetraethyltin, the response was similar to, but not identical with, that of inorganic tin, but for other compounds the response, although linear, was a function of the thermal stability and volatility of the alkyltin hydrides produced.

Probably the most popular use of hydride generation after separation by liquid chromatography has been for the determination of reducible arsenic species. Ricci *et al.* [34] used ion-chromatography in conjunction with hydride generation and an electrothermally heated silica tube for atomic-absorption detection (Fig. 6.6). The

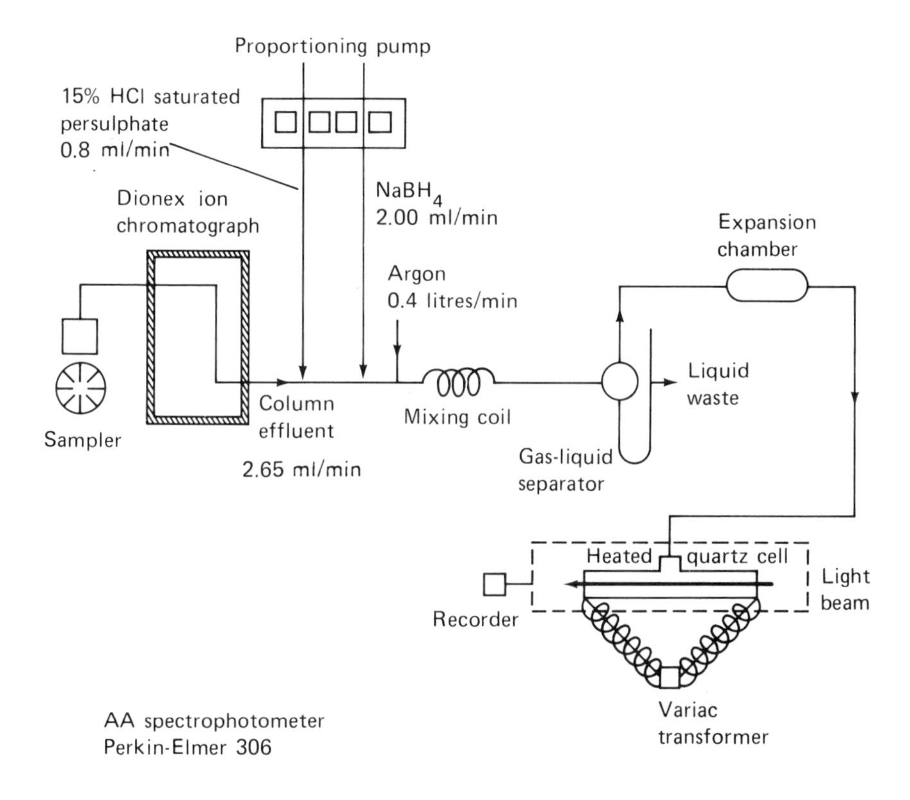

Fig. 6.6 — Dionex ion-chromatograph coupled with automated arsine generation and AAS detector. (Reprinted by permission from G. R. Ricci, L. S. Shepard, G. Colovos and N. E. Hester, *Anal. Chem.*, 1981, **53**, 610. Copyright 1981, American Chemical Society.)

separation of reducible arsenic species required either a gradient elution (necessitating column restabilization for one hour between determinations) or two separate isocratic separations. With the isocratic approach the column required re-equilibration for one hour after every 10–15 samples. Thus, although satisfactory sensitivity had been achieved, the sample throughput time was low, and the need for repeated column equilibration makes this technique somewhat unattractive for routine work.

A simple and sensitive continuous-flow hydride system based on two peristaltic pumps has been developed in our laboratories [35]. This consists of a modified form of the system developed by Thompson *et al.* [36] for use with an inductively coupled plasma. A small nitrogen/hydrogen/entrained-air flame was used as the atom cell and sodium tetrahydroborate was the reductant. The continuous generation of the hydride improves precision by avoiding the need for discrete injections, and the accuracy is also improved since the zero level is unambiguously defined. Such a system is thus well suited for use with coupled HPLC systems and has been successfully used in a number of studies. The construction of the HPLC–hydride–FAAS interface as used for the determination of arsenic species is shown in Fig. 6.7.

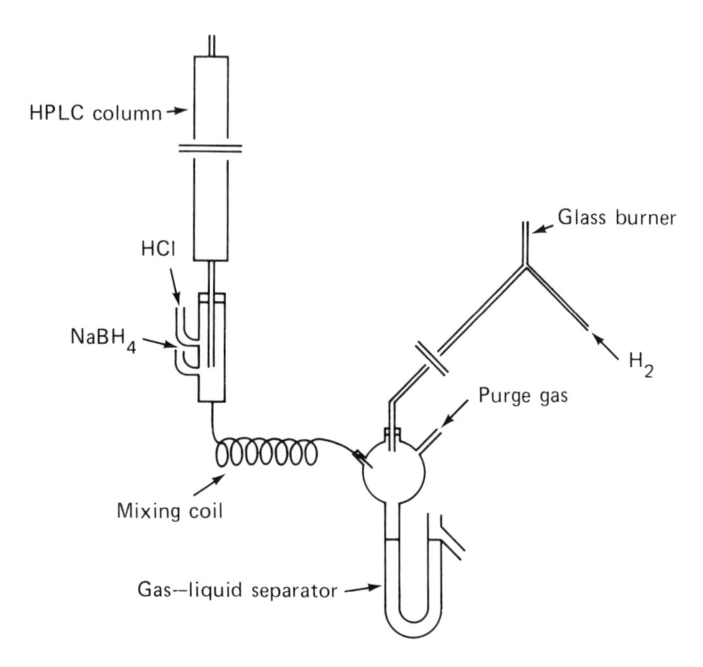

Fig. 6.7 — Schematic diagram of HPLC–hydride–FAAS/FAFS coupling. Reprinted from S. Hill, L. Ebdon and P. Jones, *Anal. Proc.*, 1986, **23**, 6, by permission. Copyright 1986, Royal Society of Chemistry.)

The basic system consists of two peristaltic pumps which deliver the acidified sample (5*M* in hydrochloric acid) at 7.0 ml/min, and sodium tetrahydroborate solution, at 2.5 ml/min, to a mixing coil, and then into a conventional gas–liquid separator. Argon purge gas (120 ml/min) then carries the volatile hydrides into a small hydrogen diffusion flame burning on an inverted 'Y' glass burner (8.5 mm i.d., 100 mm high). The side-arms of the burner act as the gas inlets, a fuel-gas flow of 180 ml/min being used. The burner is then located vertically in the spectrometer at the focal point of the entrance-slit lens, replacing the conventional burner/nebulizer assembly. The eluate from the high-performance liquid chromatograph is introduced into the reaction manifold just before the point at which the tetrahydroborate solution is introduced, as shown in Fig. 6.7.

Various separation systems have been reported in the literature for arsenic species. The technique used here at Plymouth [37–39] employs elution with 0.1*M* ammonium carbonate from a resin-based strong anion-exchange (SAX) BAX 10 column. This is a polystyrene-based material (mean particle diameter 3 μm) with quaternary ammonium substituents. Although the system gives acceptable separations, the use of a precolumn packed with Zipax, a silica-based anion-exchange material (40 μm) and a step elution system of sulphuric acid (10^{-4}%)/ammonium carbonate (0.1*M*), results in a preconcentration step on the Zipax, which in addition gives protection to the general column. The conditions used for both hydride generation and separation of the arsenic species are summarized in Table 6.1. The

Table 6.1 — Conditions for HPLC–AAS

GC–AAS	Hydride generation
Column packing: 10% OV101 on Chromosorb W(80–100)	Reagents: (flow-rate 2.5 ml/min)
	HCl (1.0*M*)
Temperatures:	$NaBH_4$ (4.0%)
Column 30°C isothermal	Buffer, pH 5.0
Injector 50°C	(sodium acetate/acetic acid)
Interface 60°C	
Gas flow-rates:	Drying agent: NaOH pellets/water bath
N_2 30 ml/min	
H_2 100 ml/min	
Air 4.0 1./min	Trapping system:
C_2H_2 0.6 1./min	Packing: glass beads (40 mesh)
Light source: As (HCL) 8 mA	
Wavelength: 193.7 nm	Trapping period: 5 min
Bandpass: 0.5 nm	Purge flow (N_2): 300 ml/min
Deuterium arc background correction applied	

results obtained from the system for the separation of a mixture of arsenite, arsenate, dimethylarsinic acid (DMA) and monomethylarsonic acid (MMA) are shown in Fig. 6.8. The complete separation is achieved in less than 4 min, but because of the arbitrary nature of switching the eluents, the retention times for the last three peaks are best measured with respect to the final peak. With this system arsenic has been speciated in a range of samples, including pore waters, estuarine waters, urine, fruit and vegetables.

For the separation of arsenic species, the coupled HPLC–hydride–AAS system avoids the problems of low nebulization efficiency normally encountered with FAAS, allowing sensitive detection along with real-time detection. However, such a technique is only suited to the determination of reducible species which form *volatile* hydrides. In many instances, the hydrides are not volatile, even though the associated metal may be readily reduced. Examples of such compounds are the tributyl compounds of tin. Although Sn(IV) will readily form a volatile hydride, the hydrides formed from tributyltin species are liquids which are not very volatile, so are not detected with this system with good sensitivity. Thus it is necessary to modify the techniques so as to heat the gas–liquid separator or to incorporate some means of degrading the tributyltin before the hydride-generation stage.

A technique has been developed in Plymouth [38] that uses ultraviolet irradiation to degrade compounds such as tributyltin to species which may then be detected by using conventional hydride generation. The organotin species are separated by

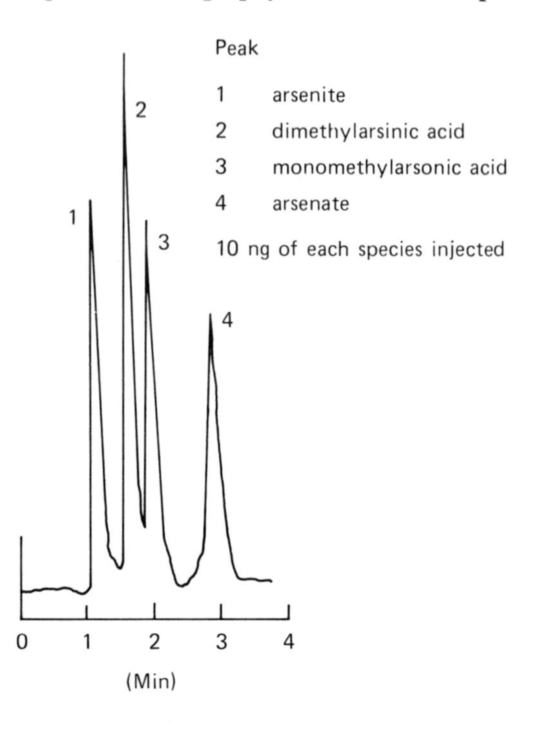

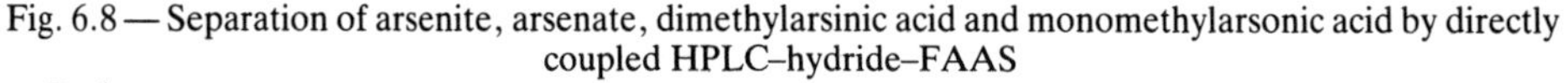

Fig. 6.8 — Separation of arsenite, arsenate, dimethylarsinic acid and monomethylarsonic acid by directly coupled HPLC–hydride–FAAS

Peak
1 arsenite
2 dimethylarsinic acid
3 monomethylarsonic acid
4 arsenate

10 ng of each species injected

HPLC, then passed through a fused-silica coil wound onto a glass former. The diameter of the coil is large enough to allow an ultraviolet lamp to be inserted down its centre. To avoid diffusion and thus retain sample integrity in the coil, the flow is segregated with small air-bubbles. After passage through the photolysis unit, the HPLC eluate passes into the continuous-flow hydride generator and the hydrides are then detected in the conventional way in a small hydrogen-diffusion flame. A schematic diagram of the complete interface is shown in Fig. 6.9. Detection limits of 2 ng for tributyltin have been obtained with this system.

6.4 TRANSPORT DETECTOR INTERFACE SYSTEMS FOR FLAMES

The most widely reported type of transport detector used in chromatography is one in which the sample is conveyed, e.g. by a moving wire or belt, to the detector [40–46]. The substantial difference between this and more conventional detectors is that the eluate does not enter the sensing element of the detector directly from the

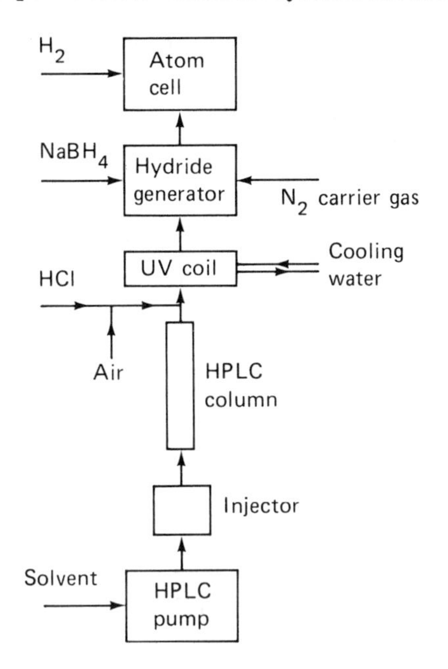

Fig. 6.9 — Schematic diagram of complete interface incorporating the ultraviolet irradiation coil.

column, but is transported through a thermal zone which removes the mobile phase, and through a zone where either vaporization or pyrolysis of the sample takes place. Although there has been renewed interest in the use of belts as transport devices for LC–MS [47,48], wire transporters [41–43] were used in most early designs, (and also in the only commercially available LC–flame ionization detector of this nature — now discontinued). A range of other devices such as chains [44], discs [46] and a wire helix [60] have also been used.

Today, many of the problems associated with moving-belt interfaces for LC–MS have been eliminated; for example, the thermospray approach sprays the sample onto the belt, and most of the solvent is vaporized and removed without being deposited as a liquid on the moving surface. However, many of the early moving-wire detectors for chromatography proved unsuccessful. Problems with ineffective pyrolysis of the solutes and difficulties in coating the moving carrier with effluent were studied by a number of workers [50,51] and although some improvements were made, the difficulties were never fully eliminated. Two other major problems were encountered. First, the amount of effluent coated on the wire remains the same, irrespective of effluent flow-rate; in other words, the split ratio (fraction of the solute deposited on the wire) will vary with the flow-rate. Secondly, there is a tendency for the solute to creep during the period of drying, causing extremely high concentrations to occur at irregular intervals along the moving carrier. This effect was seen in the chromatogram as sharp 'spikes' superimposed on the solute peaks. A number of methods were devised to overcome these problems, such as spraying the column effluent onto the wire [52], coating the wire with a layer of porous sorbent [53], and feeding the wire directly into a combustion chamber, the total combustion products being swept by a stream of air into the FID [41]. However, the technique remained unreliable and is now little used, the last reports of the technique having appeared in the literature during the mid-1970s.

The basic principle of the moving-wire techniques is to collect and transport a *continuous* stream of effluent. However, if the effluent is regarded as a series of discrete aliquots, the nature of the interface may be modified to take advantage of the successful procedures used in atomic absorption for analysis of microsamples. Such techniques have the advantage that sample introduction avoids the use of a nebulizer, with its inherently low efficiency, so the detection limits are 1–2 orders of magnitude better than those of conventional flame AAS [54]. In recent years, various direct sample-insertion devices have been reported, employing platinum loops [55], nickel cups [56] and tantalum boats [57]. The best known of these is probably the use of a nickel microcrucible, the so called 'Delves cup' [56], for the determination of lead in biological and environmental solutions. Other devices have also been reported for electrothermal operation in graphite furnaces, such as the carbon rod, carbon filament and tantalum strip atomizers [58].

An interface we have developed at Plymouth [38,59,60] utilizes the benefits of microsample insertion devices by collecting the HPLC eluate as a series of discrete amounts which are then transported by a series of rotating spirals into the flame. Although this interface is based on a similar principle, most of the disadvantages described above for the moving-wire technique are overcome, although certain desirable features have been retained, such as the desolvation stage before atomization.

The system consists of eight platinum wire spirals onto which the eluate is collected as it drops from the end of the HPLC column. Each of the spirals is mounted at 45° to the next on a rotating disc. The disc is then mounted on a small stepper motor and located above the spray chamber so that each spiral may rotate in turn to occupy a position in the flame directly below a small orifice in a silica tube, which acts as an atom trap (Fig. 6.10). Movement of the stepper motor is controlled by a simple microprocessor which governs both the speed of rotation and the time the spirals spend in each location. Since each spiral has a maximum loading to avoid loss of sample, a minibore (2 mm i.d.) HPLC column is used to avoid constraints on the column flow-rate and hence the chromatography. Once loaded onto the spirals, samples may be desolvated by heating in a gentle flame from a microburner before moving into the flame, although radiation heating from the flame is often sufficient. This interface has been used for determination of organolead compounds, including di- and trialkyllead compounds that are normally only determined as derivatives (Fig. 6.11 [59]) and for the determination of metals associated with various protein fractions in blood plasma [60].

6.5 HIGH-PERFORMANCE LIQUID CHROMATOGRAPHY COUPLED WITH ELECTROTHERMAL-ATOMIZATION ATOMIC-ABSORPTION SPECTROMETRY TECHNIQUES

Electrothermal atomization, mainly with graphite furnaces, offers the advantage of high sensitivity for a small portion of sample, although the conventional drying and

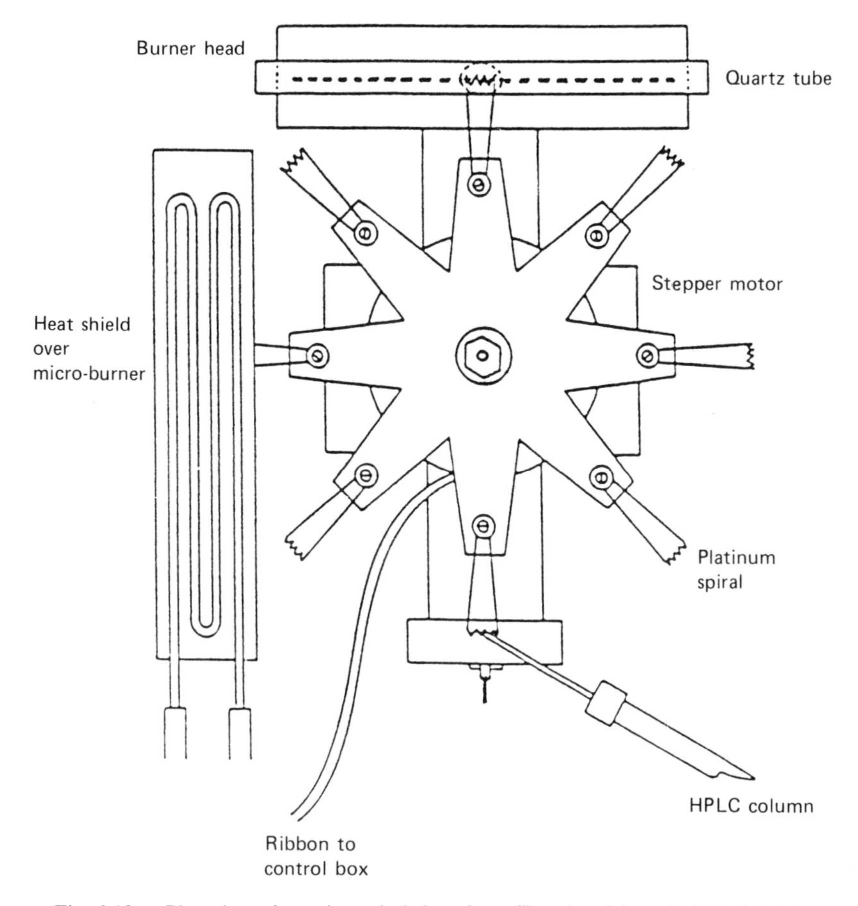

Fig. 6.10 — Plan view of rotating spirals interface. (Reprinted from S. Hill, L. Ebdon and P. Jones, *Anal. Proc.*, 1986, **23**, 6, by permission. Copyright 1986, Royal Society of Chemistry.)

ashing stages prior to atomization make it difficult to couple the chromatograph directly to a furnace. Various indirect couplings have been designed to overcome this problem, often by installing some form of sampling device between the LC and furnace. Usually the sample to be injected into the furnace is extracted in some way from the continuous flow of HPLC eluate. One of the first examples was reported by Cantillo and Segar [61], who used a multiport sampling and injection valve controlled by a sequencer. This device took an aliquot of eluate and injected it into the furnace, while the main LC eluate stream was stopped or directed through a by-pass during the heating cycles of the atomizer.

Various HPLC–ETA–AAS couplings have been developed for the speciation of arsenic, based on the two couplings developed by Brinckman *et al.* [62]. The first of these utilized a PTFE flow-through cell, the eluate from which was periodically sampled and injected into a graphite furnace, in a so-called pulsed-mode operation (Fig. 6.12a). In the second, termed survey-mode, the eluate was collected by an auto-

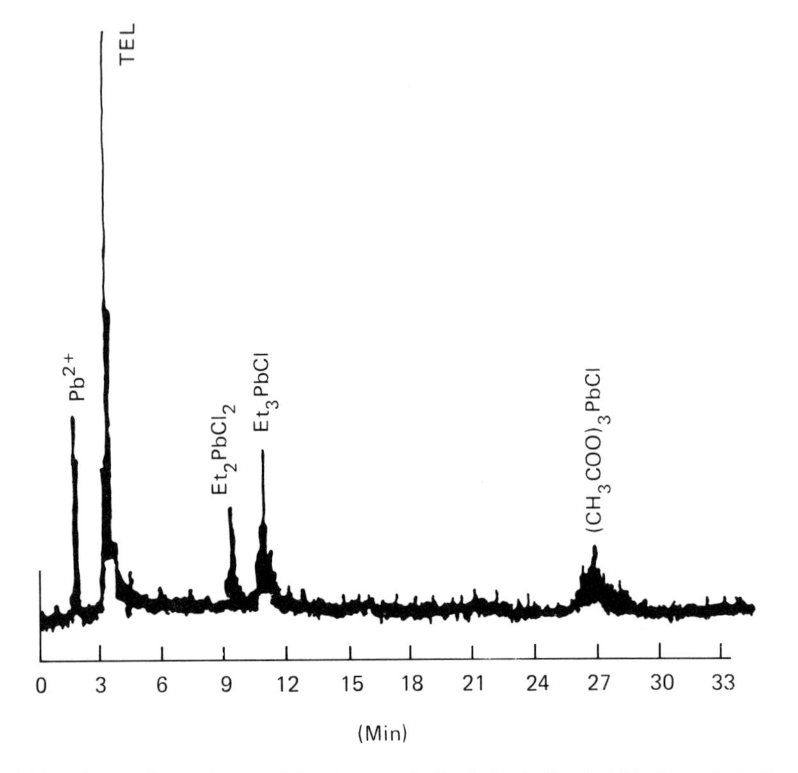

Fig. 6.11 — Separation of inorganic lead, tetraethyllead, diethyllead dichloride, triethyllead chloride and triacetyllead chloride by directly coupled HPLC–FAAS with a computer-controlled interface. (Reprinted from L. Ebdon, S. Hill and P. Jones, *J. Anal. At. Spectrom.*, 1987, **2**, 205, by permission. Copyright 1987, Royal Society of Chemistry.)

sampler and each fraction was analysed by ETA–AAS (Fig. 6.12b). Initially these two sampling modes were demonstrated for the speciation of various arsenic, tin, mercury, and lead compounds, but the survey-mode of operation was also used for the speciation of organometallic polymers, organotin and silicates, in later work [63]. A flow-through PTFE sampling cell arrangement has also been used as an interface between a low capacity anion-exchange column and a graphite furnace by a number of other groups when speciating organic and inorganic reducible forms of arsenic in pesticide residues [64,65], and soil and water samples [66]. Full details of the clean-up procedure for use in this analysis have been published, and the flow-through PTFE sampling cup has been produced commercially [67]. It is worth noting, however, that the extremely high background molecular absorption levels encountered with the ion-pair reagents, such as tetraheptylammonium nitrate (THAN), used in some of this work, required Zeeman-effect background correction, since conventional continuum-source correction proved insufficient.

Koizumi *et al.* [68] also used HPLC coupled with Zeeman-effect ETA–AAS for the speciation of tetra-alkyllead compounds in gasoline. A 10-μl sample was taken from each 250-μl fraction and vaporized in a high-temperature furnace. Any

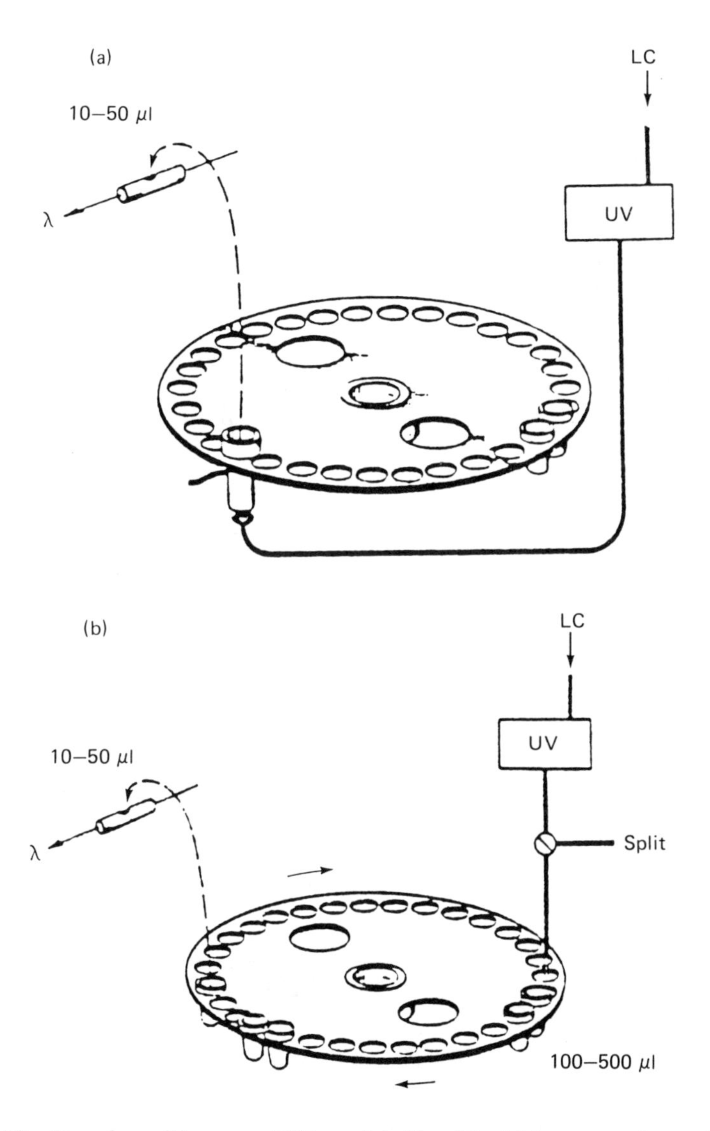

Fig. 6.12 — Two views of the carousel GF sample holder of the AS-1 auto-sampler are depicted. A, in the *pulsed* (periodic stream sampling) *mode*, the AS-1 sampling pipette traverses the arc (-----) between the GF tube orifice (at arrowhead) and the conical aperture to the well sampler. Thus, 10–50 μl effluent samples are reproducibly and periodically introduced for automatic, programmed GFAA analysis at the chosen wavelength (λ). Here, effluent from the HPLC assembly continuously passes through the well sampler with the AS-1 carousel in a fixed position. B, in the *survey* (segmental stream analytical) *mode*, the AS-1 carousel is permitted to revolve normally at a rate mainly dependent on column flow, extent of effluent stream splitting, needed fraction size (or chromatogram resolution), analyte concentration, and GFAA sample size chosen. The period between presentation of a new cup to the effluent is set by the combination of the GF atomization timing program and the number of replicate GF samples pipetted from each cup. Conventional 1 ml manufacturer's polycarbonate carousel cups are employed if these are compatible with mobile phase liquids. (Reprinted from F. E. Brinckman, W. R. Blair, K. L. Jewett and W. P. Iverson, *J. Chromatog. Sci.*, 1977, **15**, 493, by permission. Copyright 1977, Preston Publications Inc.)

interference caused by background absorption was effectively handled by the background correction system. Vickrey's group used Zeeman-effect background correction for a number of applications [69–71]. They described an interface device with a sampling valve, timing circuit and a facility to dispense automatically an additional solution [69]. This might be a matrix modifier, which they termed a co-analyte (such as a nickel solution in the case of selenium speciation [69]). Later, microprocessor control of this interface was introduced [71], 37-μl samples being injected into the furnace from each 100 or 220 μl of eluate. Stream splitting of chromatographic peaks was also used [47] before atomization for the speciation of tetraphenyllead, and Brinckman's so-called pulsed-mode operation was utilized for speciation of Cr(III) and Cr(VI) [70], with sampling of the eluate every 30 or 120 sec. The nature of the concentration profile obtained by taking samples from the effluent stream by this technique is shown in Fig. 6.13. Several users have reported ways of

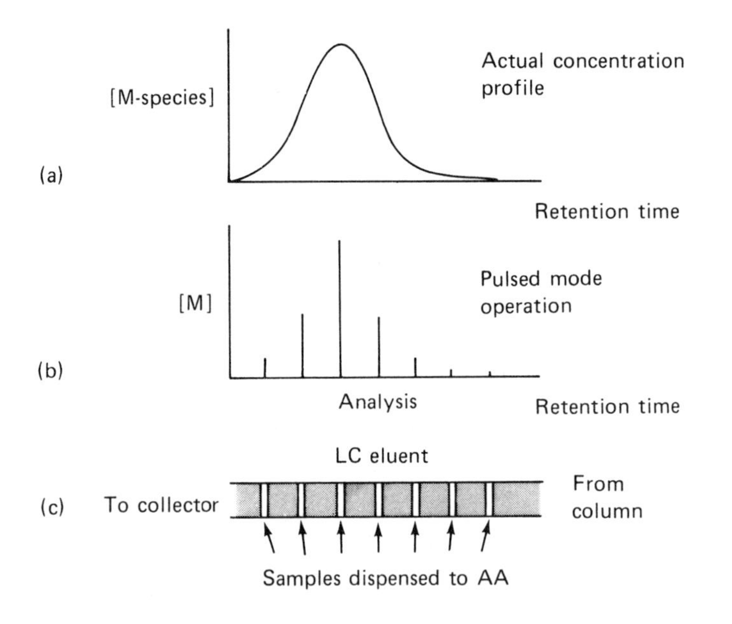

Fig. 6.13 — Description of the "pulsed" sampling mode. A, actual concentration profile of metallospecies leaving the column. B, the measured concentration profile from the intermittent removal of aliquots of the eluent stream. C, a diagram of the eluent stream showing the aliquots which are removed and analysed by the graphite-furnace atomic-absorption spectrometer. (Reprinted from T. M. Vickrey and W. Eve, *J. Autom. Chem.*, 1979, **1**, 198, by permission. Copyright 1979, Taylor & Francis.)

enhancing the performance of coupled HPLC–ETA–AAS systems. The addition of iodine before atomization has been found to enhance both the signal and its precision in determination of tetra-alkyllead species, and a similar effect is found by using zirconium-coated cuvettes in the speciation of organotin compounds [72]. Although the exact reasons for these effects were not demonstrated unequivocally, it is reasonable to speculate that the iodine converts the volatile organolead forms into inorganic species, thus preventing premature loss of analyte before atomization, and producing lead in a form amenable to reduction in the furnace to atomic lead. A zirconium coating on graphite tubes reduces the activity of the carbon surface and hence the possibility of forming further tin–carbon bonds which might reduce the tin signal by delaying or preventing atomization. Zirconium-coating of tubes is commonly advocated for enhancing sensitivity for those elements which form refractory carbides, such as the rare-earth elements, but its use in tin determinations is not so common.

Recent work on coupled LC–ETA–AAS systems has either moved towards dual-detector systems or has emphasized the use of microprocessors. Fish and co-workers [73,74] have used ETA–AAS in conjunction with a rapid-scan UV–visible detector to investigate vanadyl and nickel compounds in heavy crude petroleum and asphaltenes, although here the sample was loaded into the furnace with the auto-sampler. A coupled HPLC–ETA–AAS system constructed in our laboratories [75] was based on an earlier design [76] although the injector sequence, valve operation, and activation of the pneumatic injector were all microprocessor-controlled. This allowed automatic sampling of the eluate stream without the need for conventional fraction collection. The furnace used for this work [77] had to be modified slightly so that an injector could be fixed to the faceplate and aligned with the cuvette sample-injector opening (Fig. 6.14). In addition, the vertical access port was replaced by a borosilicate glass tube which allowed nitrogen to be blown into the chamber through a stainless-steel lance to accelerate cooling. The increased gas-flow reduced the cooling time to about 20 sec. The interface itself consisted of two Altex (4-way) slide injection valves with pneumatic actuators. The sample (76.6 µl) and co-analyte [5 µl of 0.5% $Ni(NO_3)_2$ solution] loops were of 0.8-mm bore PTFE cut to appropriate lengths. Delivery of the co-analyte and sample was by nitrogen pressure through a 1/16 in. o.d. 316 stainless-steel tube activated by a solenoid. The co-analyte followed the sample through the system into the cuvette, thus reducing the possibility of intersample contamination. The analysis sequence of the entire system was controlled by a microprocessor control system. The use of A/D convertors allowed the determination of peak area from the atomic-absorption signal, and the determination of furnace temperature. Status lines were also used to inform the computer of (a) whether the furnace door was open and ready for the next injection cycle, and (b) when the atomization of the sample was about to occur, so that data acquisition could begin. The results obtained were recorded on a standard chart recorder, or integrator.

By blowing the sample into a hot cuvette it is possible to gain a number of advantages. First, a larger volume of sample may be accommodated by the cuvette as vaporization of solvent occurs almost immediately, thus increasing the effective sensitivity. Additionally, the analysis sequence time for any one determination is reduced both by the shortening of the drying time and the decrease in the cooling range. These effects plus the increased rate of cooling achieved by introduction of extra nitrogen coolant gas reduced the total cycle time from over 3 min to approximately 50 sec. A schematic representation of the complete system is shown in Fig. 6.15, and typical results obtained from the basic interface for the determination of arsenic are shown in Fig. 6.16.

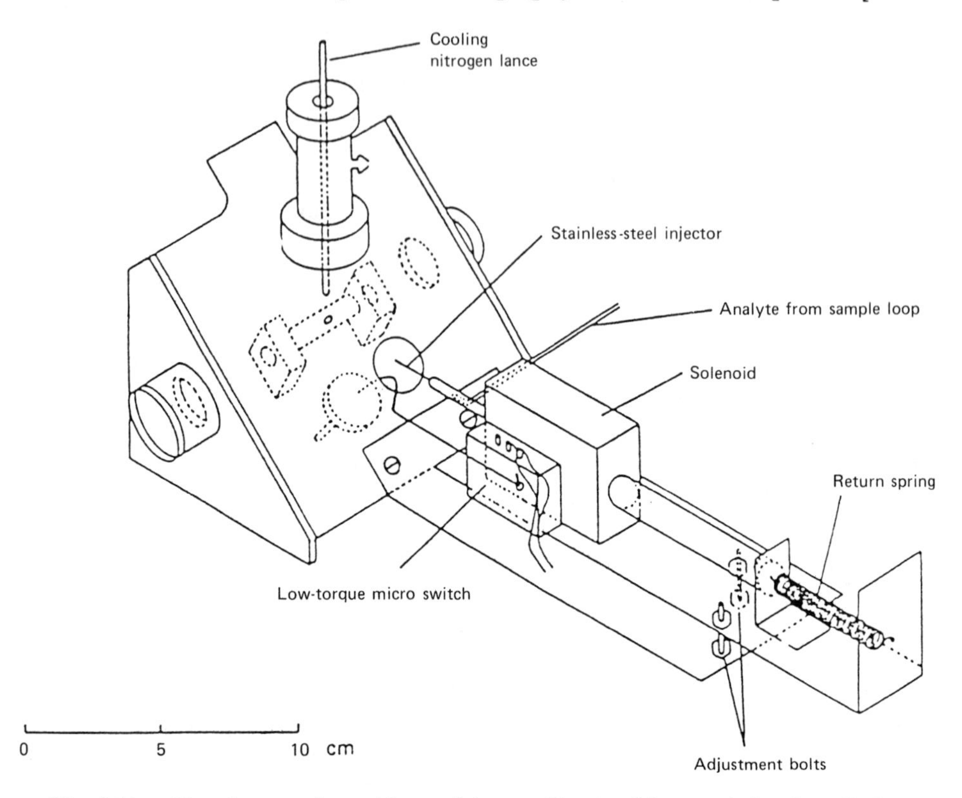

Fig. 6.14 — Flameless atomizer with auto-injector. (Reprinted by permission from S. J. Haswell, R. A. Stockton, K. C. C. Bancroft, P. O'Neill, A. Rahman and K. J. Irgolic, *J. Autom. Chem.*, 1987, **9**. Copyright 1987, Taylor & Francis.)

This interface emphasizes the complexity of coupling a liquid chromatograph to a graphite furnace which cannot continually monitor directly the eluate from the column. Although collection of the HPLC effluent in some form of auto-sampler, followed by subsequent discrete injection of the collected fractions into an electrothermal atomizer has provided an interim solution to the need to improve on early HPLC–FAAS couplings, there are several practical problems associated with this approach. In addition, the results are not obtained in real-time and the non-continuous nature of the detector makes the work tedious and is likely to lead to peak-broadening. The use of minibore HPLC to achieve the separation at low eluent flow-rates may prove of interest in re-evaluating coupled HPLC–ETA–AAS systems, which are at present limited by the inability to handle a continuous flow of sample. Most laboratory HPLC pumps have been developed for operation at flow-rates in the range 0.5–4 ml/min, but many of these will operate with good precision at lower flows, down to 0.25 ml/min, although many require recalibration for this range. In contrast to microbore HPLC, where flows in the μl/min range are used, minibore HPLC utilizes the 250–1000-μl/min range, which is within the capability of many conventional pumps provided they are recalibrated by simple volumetric measurement. The injected volume may be reduced in minibore HPLC by a factor of

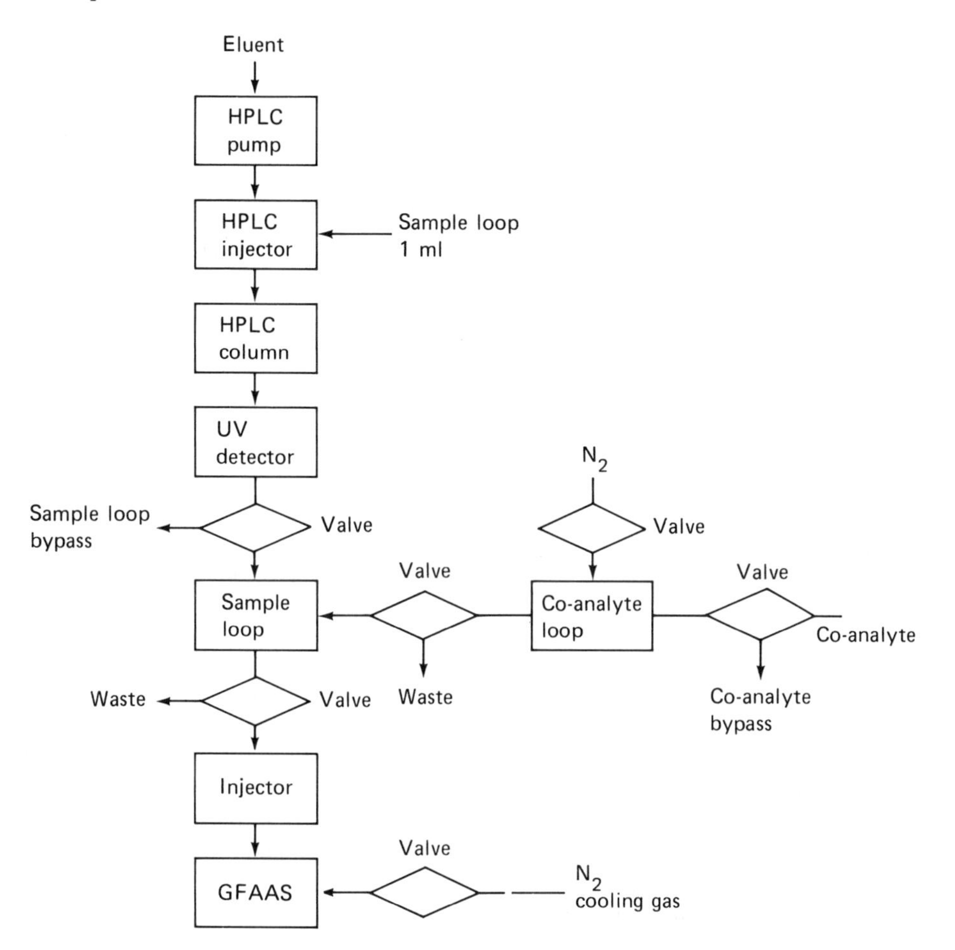

Fig. 6.15 — Schematic diagram of HPLC–ETA–AAS interface for the determination of arsenic [75].

10 compared with conventional HPLC (e.g. 100 μl compared with 1 ml for trace metal speciation), and this too is not incompatible with ETA–AAS, where 10–50 μl samples are optimal. In addition, the development of suitable electrothermal atom cells which can be maintained at the atomization temperature, without destruction of the analytical cuvette, would also be advantageous. Although the practical problems at present are formidable the advantages of high sensitivity and closely controllable chemical environment emphasize the attractions of this approach.

6.6 CONCLUSION

The interface arrangements described in this chapter demonstrate the possibility of metal-specific detection for trace-metal speciation by liquid-chromatography/atomic-

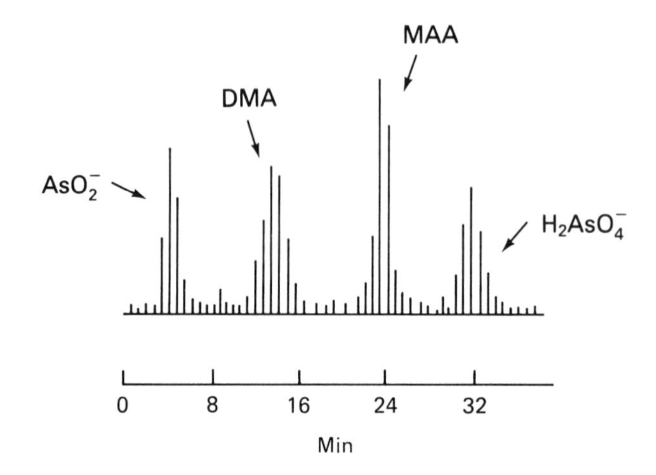

Fig. 6.16 — Separation by HPLC-ETA-AAS of arsenite, dimethylarsinic acid, methanearsonic acid and arsenate. (Reprinted by permission from R. Iadevaia, N. Aharonson and E. A. Woolson, *J. Assoc. Offic. Anal. Chem.*, 1980, **63**, 742. Copyright 1980, Association of Official Analytical Chemists.)

absorption interface systems. The advantages of such an approach include unequivocal chromatographic interpretation and the ability to accept less than optimal chromatographic resolution, since if two species are co-eluted and only one contains the metal of interest, the use of metal-specific detection ensures that only the metal-containing species is detected. The coupling of gas chromatography with atomic spectrometry is well suited to the separation of organometallic species with thermal stability and favourable gas–liquid partition coefficients such as tetra-alkyllead compounds. In most real-life situations, however, high-performance liquid chromatography is the separation technique of choice, and direct couplings have now been developed for flame instruments which can not only be interfaced through the nebulizer, but also incorporate hydride generation stages, and in some cases novel transport systems to carry the eluate directly into the flame, thus avoiding the problems associated with nebulization. Couplings for electrothermal atomizers have also been developed, many of them microprocessor-controlled, although these tend to suffer from inability to handle the continuous flow of cluatc from an IIPLC and are often complex in construction. Careful atom-cell design to increase atom residence times has now made it possible to devise systems which actually give better detection limits by use of flame atomizers (see Table 6.2). New approaches to electrothermal atomizers designed specifically as HPLC detectors may yet reverse this state of affairs.

The systems described above, although all different in approach, demonstrate the potential and versatility of directly coupled liquid-chromatography/atomic-spectrometry. Many of the interfaces described are simple and reliable, permit real-time analysis, and produce continuous chromatograms. Most are readily assembled in any laboratory equipped with HPLC and AAS, and are just as readily demounted when the instruments are again required for conventional use. Although only a

Table 6.2 — Linear working range and detection limits (as Pb) for four different atom cells. (Reproduced with permission from L. Ebdon, R. W. Ward and D. A. Leathard, *Analyst*, 1982, **107**, 129. Copyright 1982, Royal Society of Chemistry).

Atom cell	Linear range (ng)	Detection limit (pg)	
		Tetramethylead	Tetraethyllead
I	10–300	1000	2000
II	1.0–50	58	75
III	0.8–20	48	71
IV	0.1–15	17*	17*

*Equivalent to 82 fmole of compound.

limited number of examples is given in the text, there are numerous applications in many fields, including clinical, industrial, forensic and environmental analysis.

REFERENCES

[1] L. Ebdon, S. Hill and R. Ward, *Analyst,* 1986, **111**, 1113.
[2] E. D. Katz and P. W. Scott, *Analyst,* 1985, **110**, 253.
[3] L. Ebdon, S. Hill and R. Ward, *Analyst,* 1987. **112**, 1.
[4] I. S. Krull and S. Jorden, *Int. Lab.*, 1980, Nov./Dec., 13.
[5] F. J. Fernandez, *Chromatog. Newsl.*, 1977, **5**, 17.
[6] F. J. Fernandez, *At. Abs. Newsl.*, 1977, **16**, 33.
[7] J. C. Van Loon, *Anal. Chem.*, 1979, **51**, 1139A.
[8] S. J. Hill and L. Ebdon, *European Spectrosc. News,* 1985, **58**, 20.
[9] J. A. Koropchak and G. N. Coleman, *Anal. Chem.*, 1980, **52**, 1252.
[10] D. R. Jones IV, H. C. Tung and S. E. Manahan, *Anal. Chem.*, 1976, **48**, 7.
[11] N. Yoza and S. Ohashi, *Anal. Lett.*, 1973, **6**, 595.
[12] S. E. Manahan and D. R. Jones IV, *Anal. Lett.*, 1973, **6**, 745.
[13] D. R. Jones IV and S. E. Manahan, *Anal. Chem.*, 1976, **48**, 1897.
[14] D. R. Jones IV and S. E. Manahan, *Anal. Lett.*, 1975, **8**, 569.
[15] J. R. Pankow and G. E. Janauer, *Anal. Chim. Acta,* 1974, **69**, 97.
[16] N. Yoza, K. Kouchiyama, T. Miyajima and S. Ohashi, *Anal. Lett.*, 1975, **8**, 641.
[17] K. Yashimura and T. Tarutani, *J. Chromatog.*, 1982, **237**, 89.
[18] C. Botre, F. Cacace and R. Cozzani, *Anal. Lett.*, 1976, **9**, 825.
[19] R. M. Cassidy, M. T. Hurteau, J. P. Mislan and R. W. Ashley, *Chromatog. Sci.*, 1976, **14**, 444.
[20] J. C. Van Loon, B. Radziuk, N. Kahn, J. Lichwa, F. J. Fernandez and J. D. Kerber, *At. Abs. Newsl.*, 1977, **16**, 79.
[21] N. Kahn and J. C. Van Loon, *Anal. Lett.*, 1978, **11**, 991.
[22] W. Slavin and G. J. Schmidt, *Chromatog. Sci.*, 1979, **17**, 610.
[23] J. G. Atwood, G. J. Schmidt and W. Slavin, *J. Chromatog.*, 1979, **171**, 109.
[24] L. Ebdon, S. J. Hill and P. Jones, *Analyst,* 1985, **110**, 515.
[25] B. W. Renoe, C. E. Shideler and J. Savory, *Clin. Chem.*, 1981, **27**, 1546.
[26] M. H. Hahn, K. J. Mulligan, M. E. Jackson and J. A. Caruso, *Anal. Chim. Acta,* 1980, **118**, 115.
[27] W. B. Robbins and J. A. Caruso, *Anal. Chem.*, 1979, **51**, 889A.
[28] R. G. Godden and D. R. Thomerson, *Analyst,* 1980, **105**, 1137.
[29] K. C. Thompson and D. R. Thomerson, *Analyst,* 1974, **99**, 595.
[30] D. E. Fleming and G. A. Taylor, *Analyst,* 1978, **103**, 101.
[31] F. D. Pierce and H. R. Brown, *Anal. Chem.*, 1977, **49**, 1417.
[32] D. T. Burns, F. Glockling and M. Harriott, *Analyst,* 1981, **106**, 921.
[33] D. T. Burns, F. Glockling and M. Harriott, *J. Chromatog.*, 1980, **200**, 305.
[34] G. R. Ricci, L. S. Shepart, G. Colovor and N. E. Hester, *Anal. Chem.*, 1981, **53**, 610.
[35] L. Ebdon, J. R. Wilkinson and K. W. Jackson, *Anal. Chim. Acta,* 1982, **136**, 191.
[36] M. Thompson, B. Pahlavanapour, S. J. Walton and G. F. Kirkbright, *Analyst,* 1978, **103**, 705.
[37] R. W. Ward, *Ph.D. Thesis,* Plymouth Polytechnic, 1982.
[38] S. Hill, L. Ebdon and P. Jones, *Anal. Proc.*, 1986, **23**, 6.
[39] C. T. Tye, S. J. Haswell, P. O'Neill and K. C. C. Bancroft, *Anal. Chim. Acta,* 1985, **169**, 195.

[40] A. Stolyhwo, O. S. Privett and W. L. Erdahl, *J. Chromatog. Sci.*, 1973, **11**, 263.
[41] K. Slais and M. Krejei, *J. Chromatog.*, 1974, **91**, 181.
[42] A. T. James, J. R. Ravenhill and R. P. W. Scott, *Chem. Ind. London,* 1964, **18**, 746.
[43] T. E. Young and R. T. Maggs, *Anal. Chim. Acta,* 1967, **38**, 105.
[44] E. O. A. Haahti, T. Nikkari, J. Kärkkäinen, in *Gas Chromatography, 1964,* A. Goldup, (ed.) p. 190. Institute of Petroleum, London, 1965.
[45] H. Coll, H. W. Johnson, A. G. Polgar, E. E. Seibert and K. H. Stross, *J. Chromatog. Sci.*, 1979, **7**, 30.
[46] H. Dubsky, *J. Chromatog.*, 1972, **71**, 395.
[47] M. J. Hayes, E. P. Lankmayer, P. Vouros, B. L. Karger and I. M. McGuire, *Anal. Chem.*, 1983, **55**, 1745.
[48] L. Yang, G. J. Fergusson and M. L. Vestal, *Anal. Chem.*, 1984, **56**, 2632.
[49] B. M. Lapidus and A. Karmen, *J. Chromatog.*, 1972, **10**, 103.
[50] R. P. W. Scott and J. G. Lawrence, *J. Chromatog. Sci.*, 1970, **8**, 65.
[51] A. T. James, J. R. Ravenhill and R. P. W. Scott, in *Gas Chromatography 1964,* A. Goldup (ed.), p. 205. Institute of Petroleum, London, 1965.
[52] J. H. Van Dijk, *J. Chromatog. Sci.*, 1972, **10**, 31.
[53] V. Pretorius and J. F. J. Van Rensburg, *J. Chromatog. Sci.*, 1973, **11**, 355.
[54] H. Berndt and J. Messerschmidt, *Spectrochim. Acta,* 1979, **34B**, 241.
[55] S. Gücer and H. Berndt, *Talanta,* 1981, **28**, 334.
[56] H. T. Delves, *Analyst,* 1970, **95**, 431.
[57] H. L. Kahn, G. E. Peterson and J. E. Schallis, *At. Abs. Newsl.*, 1968, **7**, 35.
[58] M. Kahl, D. G. Mitchell, G. I. Kaufman and K. M. Aldous, *Anal. Chim. Acta,* 1976, **87**, 215.
[59] L. Ebdon, S. Hill and P. Jones, *J. Anal. At. Spectrom.*, 1987, **2**, 205.
[60] L. Ebdon, S. Hill and P. Jones, *Analyst,* 1987, **112**, 437.
[61] A. Y. Cantillo and D. A. Segar, *Proc. Int. Conf. Heavy Metals in the Environment,* Toronto Canada, 1975, I–183–204.
[62] F. E. Brinckman, W. R. Blair, K. L. Jewett and W. P. Inverson, *J. Chromatog. Sci.*, 1977, **15**, 493.
[63] E. J. Parks, F. E. Brinckman and W. R. Blair, *J. Chromatog.*, 1979, **185**, 563.
[64] E. A. Woolson and N. Ahronson, *J. Assoc. Off. Anal. Chem.*, 1980, **63**, 523.
[65] R. Iadevaia, N. Aharonson and E. A. Woolson, *J. Assoc. Off. Anal. Chem.*, 1980, **63**, 742.
[66] F. E. Brinckman, K. L. Jewett, W. P. Inverson, K. J. Irgolic, K. C. Ehrhardt and R. A. Stockton, *J. Chromatog.*, 1980, **191**, 31.
[67] *Waters Technical Bulletin,* 82291/J91/Jan. 1982.
[68] H. Koizumi, R. D. McLaughlin and T. Hadeishi, *Anal. Chem.*, 1979, **51**, 387.
[69] T. M. Vickrey, M. S. Buren and H. E. Howell, *Anal. Lett.*, 1978, **11**, 1075.
[70] T. M. Vickrey and W. Eue, *J. Autom. Chem.*, 1979, **1**, 198.
[71] T. M. Vickrey, H. E. Howell and M. T. Paradise, *Anal. Chem.*, 1979, **51**, 1880.
[72] T. M. Vickrey, H. E. Howell, G. V. Harrison and G. R. Ramelow, *Anal. Chem.*, 1980, **52**, 1743.
[73] R. H. Fish and J. T. Komlenic, *Anal. Chem.*, 1984, **56**, 510.
[74] R. H. Fish, J. J. Komlenic and B. K. Wines, *Anal. Chem.*, 1984, **56**, 2452.
[75] S. J. Haswell, *Ph.D. Thesis, Plymouth Polytechnic,* 1983.
[76] R. A. Stockton and K. J. Irgolic, *Intern. J. Environ. Anal. Chem.*, 1979, **6**, 313.
[77] S. J. Haswell, R. A. Stockton, K. C. C. Bancroft, P. O. O'Neill, A. Rahman and K. J. Irgolic, *J. Autom. Chem.*, 1987, **9**, 6.

7

Tin and germanium

O. F. X. Donard
Groupe d'Océanographie Physico-Chimique, Laboratoire de Chimie Physique A, Université de Bordeaux I, 33405 Talence, France
R. Pinel
Laboratoire de Chimie Analytique, Faculté des Sciences et Techniques, 64000 Pau, France

7.1 INTRODUCTION

The last decade has seen a considerable increase in the development of applications of organotin in the chemical industry [1]. The production of these compounds has increased from 27000 tons in 1976 and was expected to reach 63000 tons in 1986 [2]. Organic compounds of metals are usually more toxic than their inorganic counterparts and their input into the environment is of considerable interest.

As a response to these major environmental concerns, recent developments in analytical chemistry have put a strong emphasis on determination of the chemical species in a wide variety of matrices. These developments have also contributed to a better understanding of the natural biogeochemical cycles of trace metals in aquatic systems. Methylation of mercury received considerable attention from 1950 to 1960 after the contamination of Minamata by methylated forms of mercury. Similarly, there is a growing interest in methyltins in the environment, either from natural tin occurring in sediments, or from degraded man-made organotin products [3,4], especially in view of the toxicity of these compounds. In general, the toxicity is a function of the number of alkyl groups linked to the tin atom [5]:

$$R_4Sn = R_3SnX > R_2SnX_2 > RSnX_3 \gg SnX_4$$

and to the type of alkyl group:

$$\text{ethyl} > \text{methyl} > \text{propyl} = \text{butyl} > \text{phenyl} \gg \text{octyl}$$

Germanium also has a large number of known organic derivatives but in comparison to organotin compounds, none has achieved similar significant industrial importance [6]. Most analytical efforts in determination of germanium and organogermanium compounds deal with global aspects of oceanic fluxes [7, 8].

7.1.1 Industrial applications of tin and germanium

Analytical chemistry has so far been much more concerned with use of tin and organotins, rather than germanium and organogermanium compounds, in the chemical industry. The use of germanium for computer microchips is well known, but the variety of use of its organic forms is much narrower than the wide range of applications of organotin compounds. The properties and applications of organotins differ according to the length of the alkyl chain [9]. Mono-organotins ($RSnX_3$) exhibit low toxicity, and received little attention until recently, but they are now used in addition to diorganotin compounds for increased efficiency in PVC [poly(vinyl chloride)] stabilization [10]. Diorganotin compounds (R_2SnX_2) were the first to find commercially significant industrial application, because they are used mainly as dioctyl- or dibutyldithioglycollates to stabilize PVC polymers [1]. Growing current concern is focused on the triorganotin compounds which, because of their highly toxic properties towards molluscs and fungi, have been applied as biocides. Triethyltin hydroxide and acetate are used as wood preservatives. Similarly, triphenyltin hydroxide, and tricyclohexyltin exhibit excellent agricultural fungicide properties [11]. However, there is some controversy over their widespread use, because many of the compounds utilized as acaricides or insecticides are considered too toxic for these types of application [12]. Antifouling paints also represent a very effective application of the biological properties of triorganotins. Bis(tributyltin) oxide (TBTO) and tributyltin fluoride are incorporated in numerous antifouling coatings. These compounds are of recent special concern since significant correlations have been observed in France between failing oyster fisheries and exposure to antifouling paints [13–15]. Tetraorganotins (R_4Sn) are employed as intermediate products and have little commercial use to date. Other applications use organotins as catalysts, fire retardants and reducing agents, as well as in the pharmaceutical and ceramic and glass industries. The industrial uses have recently been reviewed [5,11].

7.1.2 Environmental aspects

In 1976, approximately 4300 tons of organotins escaped into the environment. An important fraction of the total of organotin compounds used as biocides and algicides in antifouling paints directly entered the aquatic environment [16]. As a result, there is an international controversy about their fate and impact. After the 1972 Oslo convention dealing with aspects of the human environment, organotin compounds were on the grey list of dangerous substances, but are now assigned to the black list after the 1980 EEC Conference. However, some of the decisions have been based only on production figures, because of the lack of general understanding of the aquatic chemistry of these compounds [17]. Opinions differ about the degradation of these compounds. The degradation processes occur through removal of an organic group from the tin atom and ultimately lead to non-toxic SnO_2 [18]. Biological cleavage and ultraviolet radiation significantly contribute to their degradation, simultaneously lowering the biological toxicity of the compounds [4,19].

In addition to man-made inputs, alkyltins may arise naturally from chemical or enzymatic methylation reactions [20–22]. The size of the alkyl group not only influences the toxicity of the chemical species to aquatic life [12] but also regulates their dissolved/particulate equilibrium in waters and hence their biogeochemical fate [23,24]. The development of appropriate specific and sensitive methods contributes to a better understanding of the natural biogeochemical cycles of elements. Thus, Brinckman *et al.* [25] have detected volatile species such as Me_2SnH_2 and Me_3SnH in the waters of Baltimore harbour, suggesting new environmental pathways for tin. Man-made products may also be hydrogenated to give volatile species, e.g. n-$BuSnH_3$ [25]. Methylation reactions can also affect butyltins; Maguire [26] has detected $(n\text{-}Bu)_3MeSn$ and $(n\text{-}Bu)_2Me_2Sn$, which are very unlikely to be of direct man-made origin.

It is important to note that the most sensitive techniques have been developed by oceanographic groups in order to comprehend detailed reactive aspects of global oceanic chemistry. Byrd and Andreae [27] first estimated general tin fluxes from industrial sources to the ocean. Andreae and his group have also studied the biogeochemistry of germanium in coastal [28,29] and oceanic [8] waters. Studying the chemical speciation of this element will bring some clues to understanding its general fluxes in the global oceanic cycles, mainly by differentiating between continental weathering and inputs through hydrothermal activity. By improving their analytical methods, Hambrick *et al.* [30] were able to report the novel occurrence of methylated forms of germanium in natural waters. The behaviour of methylgermanium contrasts with that of other methylated species, and the comparison may bring some new insights into the general methylation problem.

7.1.3 Methodology

As a response to increased environmental pressure, a wide variety of techniques has been developed, based on recent progress in gas chromatography (GC), liquid chromatography (LC) and high-pressure liquid chromatography (HPLC), to separate organotin compounds [31–33]. Electron-capture, mass spectrometric, flame photometric, or atomic-absorption detectors are used. Fernandez [34], Van Loon [35], Ebdon *et al.* [36] and more recently Harrison *et al.* [37] and Thompson *et al.* [1] have reviewed the different analytical methods developed for the determination of organotin compounds. There are several trends towards use of different combinations of separation and detection methods. Two are of particular interest and can be classified on the basis of the separation technique used.

First organotins can be speciated by gas chromatography. Parris *et al.* [38] used a graphite furnace as GC detector for trace organometallic compounds. Different methods include a combination of cryogenic trapping of volatile derivatives of organometallic species, and separation on the basis of their boiling points, by warming the trap and/or by direct chromatography. Detection is generally achieved with a flame photometric detector (FPD), a silica or graphite furnace aligned in an atomic spectrophotometer, or by a hydrogen-rich flame emission detector. Braman and Tompkins [39] used on-line derivatization of organotins with sodium borohydride, cryogenic trapping, separation on chromatographic packing material, and detection by means of the Sn–H band emission in the hydrogen-rich emission detector. Hodge *et al.* [40] adapted a silica furnace/AAS system to detect nanograms of alkyltin compounds in environmental samples. Soon after, Andreae and Froelich [41] determined inorganic germanium with a similar set-up, by using a graphite furnace as detector. Improvements of this apparatus allowed the determination of methylated tin compounds [42] in oceanic waters, with a graphite furnace, a silica furnace, or a flame emission spectrometry detector. Optimization of the technique allowed the first identification of methylated germanium species in environmental waters [30]. Recent improvements have allowed simultaneous determination of methyl- and n-butyltin compounds with increased sensitivity [43]. Further gain in sensitivity can be obtained with an all-Teflon compact system for the identification of n-butyltin compounds [44]. After hydride generation, trapping can be achieved on Tenax at room temperature [45].

A slightly different approach is based on the derivatization of organotin compounds by reaction with a Grignard reagent in a solvent medium. Thus, after extraction from the aqueous phase, organotins are converted into their tetra-alkylated forms before injection into a GC or a GC/AAS apparatus. Müller [46] used a similar derivatization technique to determine butyltins by GC/FPD. To avoid potential losses of derivatized species, butyltin ions can be converted into high molecular-weight mixed tetraorganostannanes of the type (n-pentyl)$_x$SnBu$_{4-x}$ [47]. Recently, Matthias *et al.* [48] used a simultaneous hydride-formation/extraction procedure for the determination of butyl- and methylbutyltins from environmental waters. Also, methylated tins can be butylated before GC/AAS detection, as described by Chau *et al.* [49]. Finally, Müller [50] has also recently extended the detection of organotins (butyltins, cyclohexyl- and phenyltins) from environmental samples by ethylation and determination by use of high-resolution gas chromatography and flame photometric detector (HRCG/FPD). These methods, using preliminary solvent extractions, are suitable for the determination of organotin compounds in environmental samples, with preconcentration of the analyte solution.

A second important trend combines the separation facilities of liquid chromatography (and more recently HPLC) with the specificity of AAS detection. The success of the method relies on the capability to separate high molecular-weight, non-volatile organotin compounds in complex matrices. The major drawback of this technique is the interface problem generated by the connection between the 'continuous' aspect of the eluent flow of the HPLC column and the 'sequential' aspect of the atomization cycles in the detection with the graphite furnace. Several solutions have been found to this problem. On-line detection can be achieved by nebulizing the eluent directly into the flame of an AAS. This technique has been applied, for example, by Lattard and Rocca [51] to the determination of tributyl- and triphenyltin compounds. An alternative on-line determination of tin and tributyltin compounds injects the eluent into a cup for vaporization into a silica atom trap heated by an air–hydrogen flame [52]. These methods, however, exhibit low sensitivity, and an elegant alternative method using on-line derivatization of the eluent by hydride generation and subsequent detection in a heated silica furnace achieves higher sensitivity [53]. At least three orders of magnitude improvement in sensitivity can be gained by graphite-furnace atomization of organotin compounds, instead of flame detection.

Such techniques raise other questions to be solved, such as the type of extraction to be performed on the sample, the choice of the solvents and type of chromatography columns, the mode of sampling of the eluent, the interfacing device and the need for matrix modifiers to obtain complete atomization of the alkyltin compounds. The various responses to these problems differ mainly in the type of interface developed. Cantillo and Segar [54], after the separation, use an injection valve to transfer aliquots of the eluent into the graphite furnace. A different device has been developed by Vickrey and co-workers [55, 56]. Improvements include a 2-position, 8-port valve allowing the effluent to be collected in a fraction collector or to bypass the detector. Brinckman *et al.* [57, 58], Parks *et al.* [59] and Jewett and Brinckman [60] have also extensively improved this type of instrumentation. Their interfacing system relies on an autosampler injector. A recent modification of this interfacing technique includes a fraction microcollector controlled by a photocell [61].

7.2 ANALYTICAL TRENDS

Independent of the type of set-up developed, based on either gas or liquid chromatography, all methods have four basic and critical analytical steps: sample preparation, separation of analytes, detection, and electronic processing of the signal (Fig. 7.1).

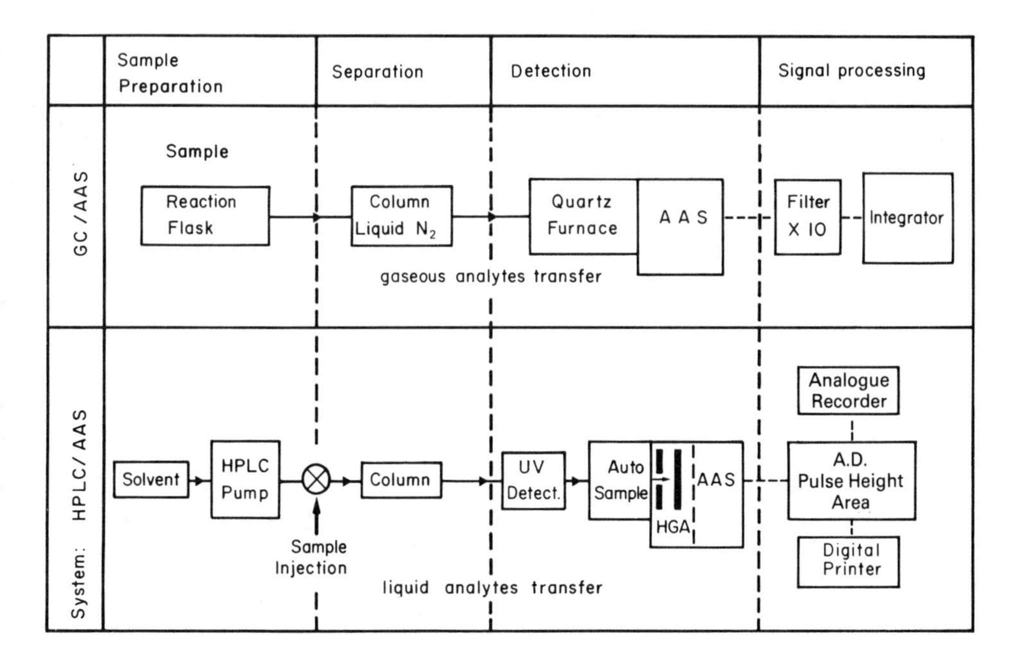

Fig. 7.1 — Schematic analytical systems for the determination of organotin compounds.

This section will describe in detail the two emerging trends using either gas or liquid chromatographic separation or organotin of organogermanium compounds, and the different solutions used for their detection. 'On-line' systems will refer hereafter to apparatus where the derivatization of samples is directly connected to the separation and detection device, as opposed to an 'off-line' set-up where analytes can be alkylated independently of the analytical system.

7.2.1 The gas chromatography/atomic-absorption system (GC/AAS)

7.2.1.1 Apparatus

Several set-ups using *on-line* hydride generation have been successfully used to speciate and quantify tin and alkyltin, germanium and methylgermanium compounds in water samples. The apparatus (Fig. 7.2) generally uses a reaction vessel for

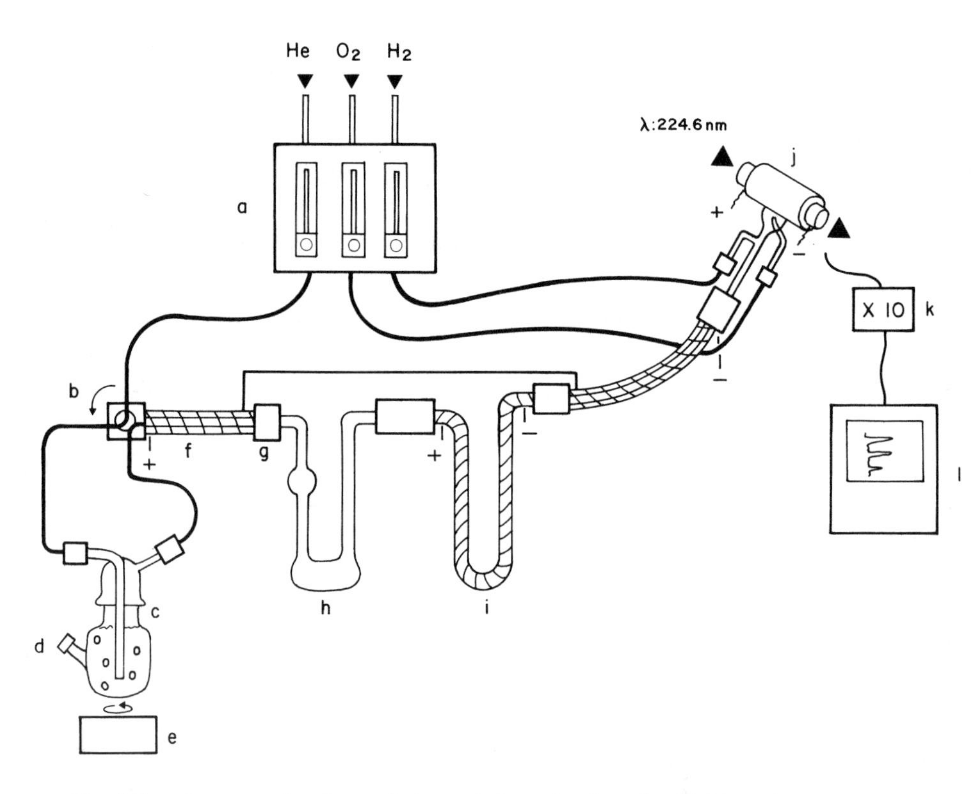

Fig. 7.2 — Apparatus for the speciation and determination of methyltin and butyltin compounds: a, flowmeters; b, four-way valve; c, hydride-generation reaction flask; d, $NaBH_4$ injection port; e, magnetic stirrer; f, heated Teflon transfer lines; g, Teflon connectors; h, optional water trap in dry-ice/acetone bath; i, speciation trap in liquid nitrogen; j, silica furnace (QFAAS); k, low-pass filter and amplifier ($\times$ 10); l, integrator.

the hydride generation. The reaction flask for volatilization of the organometallic compounds is constructed of borosilicate glass. Different sizes of reaction vessel (50–250 ml) can be attached to the apparatus but 100 ml is generally suitable. The reaction flask should be shaped so that any possible entrapment of generated volatile species is avoided [62]. The bubbler can either enter the sample solution or be terminated slightly above it. The latter is preferred when dealing with excessively foaming samples. Helium is used as a carrier gas to avoid potential condensation in the cold-trap. The reaction vial has an injection port on the side, sealed by a septum, to allow direct injection of an aqueous solution of sodium borohydride. Higher efficiency in the reduction of organometallic compounds is obtained with continuous stirring of the solution.

The carrier gas sweeps volatile organometallic hydrides from the reaction vessel through an optional U-shaped (45 cm, 6 mm i.d.) Pyrex water trap immersed in a dry-ice/acetone bath (−32°C). This trap protects the coated stationary phase from deterioration by contact with water. However, it should be used with caution, since only low-boiling compounds such as methyltin or methylgermanium hydrides will not be at least partly trapped in it. Removal of the trap is necessary for higher boiling point compounds such as butyl- or ethyltin hydrides [43,44]. Deterioration of the packing phase then occurs rapidly and it needs to be replaced every 15 runs (instead of 60 when the water trap is used).

Organotin or organogermanium hydrides are then trapped in a U-shaped borosilicate glass (45 cm, 6 mm i.d.) cold-trap packed with chromatographic stationary phase held in place by plugs of silaned glass wool. Details of the mesh size and coating phase will be given later (see Section 7.2.1.4). After removal from the liquid nitrogen bath (−192°C) the cold-trap is heated with a coil of nichrome or chromel wire (28 gauge, 5–10 Ω total resistance) powered by a variable voltage transformer. The length and size of the transfer lines should be reduced to a minimum in order to reduce the dead volume and gain sensitivity [63]. Teflon transfer lines (2.5 mm i.d.) can be connected to borosilicate glass or Teflon devices by Omnifit Teflon variable bore connectors. Extreme care should be taken when assembling the apparatus, in order to avoid potential leaks due to the high pressure created during generation of the hydrides or by the back-pressure occurring during the expansion of the stationary phase when the cold-trap is heated. The transfer lines can be heated, when analysing for high-boiling compounds, by inserting the 2.5 mm i.d. tubing into larger bore Tygon or Teflon tubing wrapped with 26 gauge nichrome wire [43,44,64]. A four-way valve allows the reaction flask to be by-passed and thus prevents accumulation of water in the system during purging. Except for the bubbler and the reaction vessel, deactivation of the glass surfaces is required for the determination of germanium compounds [30,41]. Silylation of the glassware with a 5% solution of dimethyldichlorosilane in toluene prevents peak tailing. Passivation of the cold-trap was found unnecessary for methyltin compounds [43] but is critical for higher boiling-point compounds. Activation of wet and heated glass surfaces by alkaline sodium borohydride and subsequent loss of n-butyltin compounds by decomposition on the glass surface is a possible reason for the lack of reproducibility. Strong bases (B^-) catalyse the decomposition of tin hydrides as follows [65]:

$$2Bu_3SnH \xrightarrow{B^-} Bu_6Sn_2 + H_2$$

$$Bu_3SnH + ROH \xrightarrow{B^-} Bu_3SnOR + H_2$$

Decomposition of butyltin hydrides is avoided by using an 'all-Teflon' set-up [44,66]. Separation of volatile species occurs after removal and warming of the chromatographic trap. The compounds are eluted according to their boiling points and are detected by various types of detector described in a later section.

A slightly modified approach uses an 'off-line' sample pretreatment, separation in a commercial GC and detection with a silica furnace aligned with an AAS or a flame photometric detector. These techniques involve solvent extraction of organotin compounds from the aqueous matrix, preconcentration by gentle evaporation if needed, and derivatization with a Grignard reagent. Gaseous samples can be directly concentrated in a cold-trap installed before the GC apparatus [49,67]. Another purge/trap method collects tin hydrides at room temperature in a commercial automatic purge-and-trap sampler packed with Tenax GC [45,68]. The analytes are desorbed by rapid warming of the sample trap and directly swept into the GC column. Transfer lines connect the sample trap to the GC column injection port, or the column outlet to the silica furnace. Several types of small-bore Teflon, stainless-steel or glass tubing are adequate but care should be taken in the choice of the material used, to avoid potential decomposition of the analyte on the transfer line. The transfer line can be heated with regular heating tape. Separation of liquid samples occurs after a straightforward injection of the analytes through the injection port of the GC. Detection is most often achieved with a modified FPD [46,47,50, 69–71].

7.2.1.2 Derivatization

Derivatization techniques allow the quantitative transformation of polar organotin or organogermanium compounds to either increase or decrease their volatilities and render them suitable for GC separation.

Hydride generation

The highest sensitivity to date is achieved after hydride generation. On-line reduction of alkyltin compounds with an aqueous solution of sodium borohydride yields volatile organotin hydrides.

$$R_xSn^{(4-x)+} \xrightarrow[H^+]{NaBH_4} R_xSnH_{4-x} + H_2$$

$$x = 1, 2, 3 \qquad R = Me, n\text{-}Bu, Ph$$

Hydride generation has recently received considerable attention for several important reasons: (i) ease of operation, (ii) good preconcentration and separation of the analytes of interest from potential matrix interferences, (iii) high efficiency of introduction of elements after cold trapping in the detector [72]. The hydride-generation method has been successfully applied to the determination of inorganic tin, n-butyl-, ethyl-, and phenyl-tin compounds in aqueous samples [40,43,44,66,73–75]. Elution and detection of compounds is continuous and the signal is easily integrated for quantification (Fig. 7.3). Similarly, hydride generation

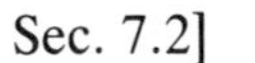

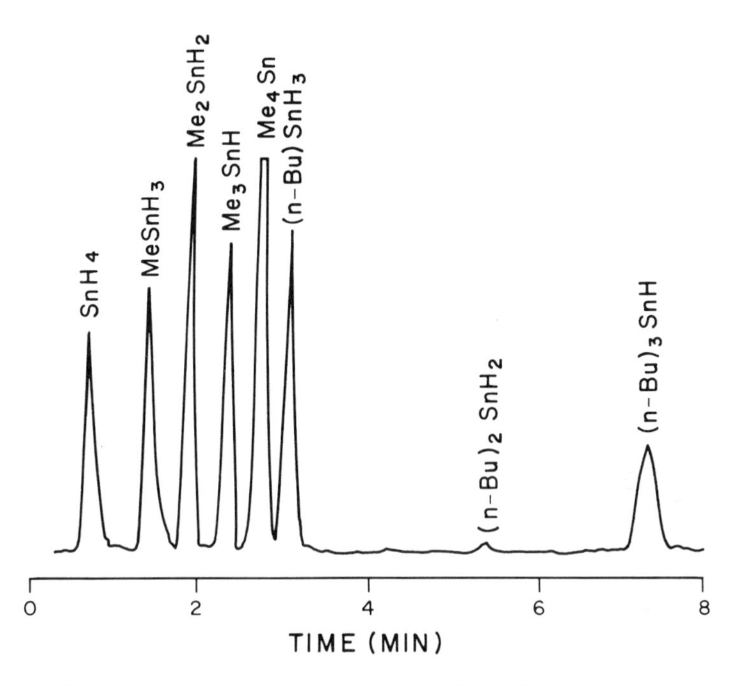

Fig. 7.3 — Speciation of total recoverable inorganic tin (TRISn), methyltin and butyltin compounds with a gas chromatograph/silica furnace atomic-absorption (224.6 nm) system (GC/QFAAS). The amount of the compounds present (as Sn) was: TRISn, 3.5 ng; $MeSn^{3+}$, 3.5 ng; Me_2Sn^{2+}, 4 ng; Me_3Sn^+, 3.5 ng; Me_4Sn, 4 ng; $(n\text{-}Bu)Sn^{3+}$, 4 ng; $(n\text{-}Bu)_2Sn^{2+}$, 0.2 ng; $(n\text{-}Bu)_3Sn^+$, 7.5 ng. (Reproduced by permission from O. F. X. Donard, S. Rapsomanikis and J. H. Weber, *Anal. Chem.*, 1986, **58**, 772. Copyright 1986, American Chemical Society.)

is efficient for the speciation of inorganic germanium and methylgermanium compounds in ocean water samples [30,41,74,76].

Typically, carrier gas is bubbled through the water sample for a few minutes to strip oxygen or any other volatile gas from the solution. Then an excess of aqueous $NaBH_4$ solution is directly injected on-line in the reaction flask, either by syringe through the injection port or with an automated injection device. The hydrides generated are then swept to the cold-trap. Some experimental conditions are presented in Table 7.1.

The sodium borohydride concentration and the water sample acidity may be varied according to the element determined (Table 7.1). Comparison between the pH and the pK_a of a species indicates that the reduction is usually performed at a pH that is a few units below the pK_a of the species of interest [76]. Sodium borohydride solution is added in excess, under acidic conditions, to yield volatile organotin or organogermanium hydrides. A 4% aqueous solution is adequate for most elements determined (Table 7.1). Acidic conditions contribute to neutralizing the NaOH (1%) added to stabilize the $NaBH_4$ solution, and generate hydrogen by decomposition of the $NaBH_4$. The nature and concentration of the acid may also affect the efficiency of hydrogenation of the organometallic species. Pinel *et al.* [77] have

Table 7.1 — On-line hydride generation conditions for water samples in GC/AAS systems.

Species	Sample size (ml)	Sample pretreatment	Hydride generation reaction	References
TRISn $Me_xSn^{(4-x)+}$ $Et_xSn^{(4-x)+}$ n–$Bu_xSn^{(4-x)+}$	100	1 ml of 2*M* acetic acid	1 ml of 4% aqueous $NaBH_4$	[40]
TRISn $Me_xSn^{(4-x)+}$	100	pH 6.5 with 4 ml of 2*M* Tris-HCl	2 × 1 ml of 1% aqueous $NaBH_4$	[39]
TRISn $Me_xSn^{(4-x)+}$	100	pH 2 with 0.2 ml of 5*M* HNO_3	1 ml of 4% $NaBH_4$ in 0.02*M* NaOH	[42]
TRISn $Me_xSn^{(4-x)+}$	10	none	100–500 μl of 4% aqueous $NaBH_4$	[45]
TRISn $Me_xSn^{(4-x)+}$ n–$Bu_xSn^{(4-x)+}$	100	pH 2 with 0.2 ml of 5*M* HNO_3	2 × 1.5 ml of 4% aqueous $NaBH_4$	[43]
n–$Bu_xSn^{(4-x)+}$ Et_3Sn^+	100	pH 1.6 with 2 ml of 5*M* HNO_3	2 × 2.5 ml of 6% aqueous $NaBH_4$	[44]
Ge $Me_xGe^{(4-x)+}$	25–250	5 ml of 1.9*M* Tris-HCl + 10 ml of 300 g/l. NaCl + 1 ml of 0.2*M* EDTA per 100 ml of sample	6 ml of 20% $NaBH_4$ in 0.06*M* NaOH per 100 ml of sample	[30]

$x = 1,2,3$; TRISn is total recoverable inorganic tin.

shown that acetic acid (up to 0.2*M*) is preferable to hydrochloric acid or nitric acid, giving a higher yield of SnH_4 from an initial solution containing 100 ng of inorganic tin. Similar results may also be obtained by using a buffer solution such as Tris-HCl [76] but any additional reagent is likely to increase the blank signal. It is possible to avoid the use of a buffer by starting with a water sample acidified to pH 2. The sodium borohydride solution should be gently introduced into the reaction flask. The method of introduction may vary, according to different authors. In any case, the sodium borohydride solution should be delivered below the liquid level of the sample, for higher efficiency. Continuous stirring during the injection and the total stripping time ensures good reproducibility. Sterile disposable plastic syringes do not present contamination problems if changed often enough.

It must be noted that butyltins or methylgermanium compounds require drastic reducing conditions, probably for steric reasons in the first case, and the strength of the Ge–CH_3 bond in the second. An initial low pH (1.6) and higher concentration of $NaBH_4$ (6%) is necessary to ensure complete volatilization of n-butyltin hydrides from the water sample. Under these conditions high molecular-weight tributyltin compounds can be fully recovered as Bu_3SnH in the gas phase after hydrogenation [43,44]. Even more drastic conditions are needed for the hydrogenation of methylgermanium derivatives from water samples. A 100-ml sample is first stabilized with 5 ml of Tris-HCl, 10 ml of a saturated solution of NaCl, and 1 ml of 0.2*M* EDTA. The reaction requires a slow injection of 6 ml of 20% $NaBH_4$ solution [30]. The stripping time for removal of hydrides from the solution depends on several factors, such as sample size, stripping flow-rate, and initial pH. In general, removal of volatile species from simple water matrices is rapid, but most authors allow an 8 min stripping time to ensure complete recovery of the organometallic compounds. Methylgermanium compounds are an exception, for they require 25 min to be completely removed from the solution [30]. A limiting factor in this case is the potential accumulation of water in the trap. A low starting pH is efficient in achieving rapid removal of hydrides from the solution, but the large excess of hydrogen generated requires a gas-tight system to avoid potential leaks that will alter the reproducibility. Further, if stannane is very rapidly evolved, it might not be completely trapped in the cold-trap. A possible alternative to hydride generation for determination of organotin compounds could be a direct derivatization with an aqueous solution of tetraethylborate [24], as developed for methyllead ions in water.

'On-line' hydride generation is a straightforward sensitive method, but is not free from interferences. Various authors [72,78,79] have reported possible inhibition of the hydride generation by metals. Though interference from dissolved trace metals is unlikely to occur owing to their low concentration in natural water samples, it can take place significantly with metal-enriched leachates from solid matrices such as sediments [80]. Organic or sulphide-rich matrices may also partially react [81] and compete non-quantitatively during hydride generation, and thus decrease reproducibility.

For simple solution matrices, some questions also remain to be solved. In contrast to arsenic, reduction with sodium borohydride cannot differentiate between Sn(II) and Sn(IV) [25, 43, 76]. Results for inorganic tin should actually be regarded as 'Total Recoverable Inorganic Tin' (TRISn) without any specific attribution of the oxidation state of the inorganic tin present in the water sample. Another important problem likely to interfere with the correct determination of alkyltin hydrides is the possible redistribution of alkyl groups during the hydride-formation reaction, or later in the cold-trap [1]. Soderquist and Crosby [73] mention that polyhydride derivatives are not stable when prepared in a batch mode. However, these authors operated with higher concentration of phenyltin hydrides than those usually found in analysis of environmental samples. We have not observed any evidence of redistribution reactions occurring during analysis of aerobic waters but have noticed redistribution reactions when determining organotin compounds in few strongly anaerobic samples.

Alkylation with a Grignard reagent

An alternative derivatization method uses off-line reaction of organotins with a Grignard reagent in a solvent medium to convert alkyltins, R_xSnX_{4-x}, into mixed tetra-alkyltins ($R_xSnR'_{4-x}$) which are more suitable for GC separation. The procedure includes solvent extraction, complete alkylation with a Grignard reagent, reduction of interferents and concentration of analytes by gentle evaporation of the solvent prior to injection into the gas chromatograph. These methods are generally focused on the determination of high boiling-point compounds such as butyltins, and have been applied to water and sediment samples.

The first important step is complete recovery of the analytes by liquid–liquid partition. The samples are acidified with hydrobromic acid and the analytes are extracted into an organic solvent. Water sample sizes may vary from 25 to 500 ml. Tropolone solution in benzene (0.05–1%) yields high recoveries for methyltin or butyltin compounds [47,49,82]. The tropolone–benzene solution should be prepared less than 5 min before the extraction is done [47]. The extraction efficiency for butyltins is enhanced by the acidification with hydrobromic acid [82]. However, these conditions are not desirable for methylated tins, which require a combination of hydrobromic, hydrochloric, acetic and sulphuric acids. Improvements can be made by using the 'salting-out' method to obtain high extraction efficiencies for highly polar solvated methylated tin species from waters [83]. Tributyltin species can also be extracted from the aqueous phase by concentration on a macroreticular resin, and subsequently eluted with acidified diethyl ether [46].

Quantitative derivatization is performed on the organic solvent extract by addition of Grignard reagent:

$$R_xMe^{(4-x)+} \xrightarrow[\text{solvent, } H^+]{R'MgX} R_x'MeR'_{(4-x)}$$

$$x = 1,2,3 \qquad R = \text{Me, n-Bu, Ph} \qquad R' = \text{n-Bu, Me, n-Pe, Et}$$

The reaction is rapid and requires only gentle stirring. Butyltins can be methylated with 2.5*N* methylmagnesium bromide in diethyl ether [82] or with 20% methylmagnesium chloride solution in tetrahydrofuran (THF) [46]. Methylmagnesium chloride also efficiently derivatizes triphenyltin hydroxide [71]. The permethylated products can be fairly volatile compared with the solvent and extreme care is then required in the procedures. Maguire and Huneault [47] rendered the butylin species less volatile by using an n-pentyl Grignard reagent, thus facilitating the solvent evaporation step. A recent development in 'off-line' volatilization methods combines simultaneous hydride formation and extraction for environmental butyltin and methylbutyltin species. The organotins are extracted and derivatized with a mixed solution of dichloromethane and 4% aqueous sodium borohydride solution. This combined step gives higher analyte recoveries than the two-step extraction/derivatization sequence does [48]. A later improvemernt by Müller [50] uses ethylmagnesium bromide (2*M* in THF). Ethylation was chosen as a derivatizing step because of the possible natural methylation of tin and butyltin compounds in the environment. Furthermore, ethylation will facilitate the later GC identification of the tetrasubstituted compounds.

After derivatization, the organic phase is usually separated and dried with a drying reagent such as anhydrous calcium chloride or sodium sulphate, and reduced to small volume by evaporation of the solvent. The analytes are thus ready for injection into the gas chromatograph. The factors for concentration from the initial water samples reach $5 \times 10^2 - 5 \times 10^3$, allowing the determination of alkyltins at below the ng/l. level in environmental waters.

Authors variously use one or two initial extractions of the natural water sample and recoveries after completion of all sample preparation steps range from 85 to 100% for alkyltins. In some cases, however, compounds such as Me_3Sn^+ and inorganic tin may exhibit poor recovery. Low pH conditions are not desirable for the recovery of methyltins, according to Chau *et al.* [49]. On the other hand, Maguire *et al.* [84] state that for inorganic tin, recoveries are low and variable at pH 5–7, probably because of the formation of unextractable SnO_2. Here also it is not possible to differentiate between Sn(II) and Sn(IV) and inorganic tin is considered as 'Total Recoverable Inorganic Tin'.

Both derivatization methods (with either sodium borohydride or a Grignard reagent) are successfully applied to environmental water samples. The hydride generation method is rapid and more sensitive because the entire sample is analysed. Also, increased sensitivity can be obtained by collecting the hydrides from successive aliquots of the sample. The collected hydrides are then determined in one measurement on warming of the cryogenic trap. Peralkylation is slightly inferior in sensitivity because here only a fraction of the final solution is effectively used for analysis. Derivatization reactions involve many sample preparation steps, leading to potential loss of analytes by adsorption on the glassware or by evaporation during reduction of the solvent volume, and require considerable time and effort. Also determination of lower boiling-point compounds may be obscured by co-chromatography of the solvent. However, the tetra-alkyltin species obtained are reported to be quite stable in the organic solvent [1]. They are also suitable for direct determination by GC/MS analyses and have been successfully applied in the analysis of complex matrices such as sediments [26, 85, 86].

7.2.1.3 Internal standards

Calibration is achieved in various ways. For off-line and on-line systems, standardization is often performed by use of a solution of one or more adequately derivatized compounds. Some controversy exists, however, as to the effect of the sample matrix on the calibration slope. Brinckman *et al.* [25] noticed that the calibration slopes for

Sn(IV), Sn(II) and Me_2Sn^+ increase when the solvent medium is changed from distilled water to 0.2*M* sodium chloride salinity solution. This effect was not observed by Donard and co-workers [23,43] when using a different set-up.

Besides potential matrix sample effects, all of the systems above use several pretreatment steps, leading to possible errors during quantification. Since many parameters can alter the sensitivity, the addition of an internal standard can confirm the proper operation of the system and allow direct calibration for each individual compound. Meinema *et al.* [82] used $HexBu_2SnMe$ as internal standard for the determination of $Bu_xSnMe_{(4-x)}$ ($x = 1,2,3$). To improve the precision of the method, $HexBu_2SnCl$ was added prior to the methylation reaction [46,50]. Similarly, Me_4Sn and Et_3SnBr respectively were used as internal standards for the quantification of methyltin and n-butyltin species [43,44]. The standards were prepared in methanol in 10-ml 'Hypovials' sealed with crimp-on Teflon-lined septa, to give a final concentration of 100 μg/ml as Sn. When stored in the dark at 4°C all solutions were stable for several months. Standards were injected with a μl syringe through the septum of the reaction flask before hydride generation.

7.2.1.4 Separation conditions

A clean separation of the analytes is dependent on the choice of the stationary phase. In general, good resolution is achieved with non-polar phases for both organotin or organogermanium compounds.

Off-line systems use long columns (2 m) packed with chromatographic support coated with silicone-based stationary phases such as OV 1, OV 3, OV 101, SE 50, SE 54, or SP 2100. The loading and mesh-size used vary from author to author and the type of compound, but a 100–120 mesh size and a light loading (3–5%) should be an adequate initial combination for chromatographic separation of organotin compounds.

Difficulties may arise in attempting to separate alkyltins [87, 88]. Co-chromatography of the sample and the solvent is another potential restriction on the method, as well as possible on-column redistributions in the presence of tetramethyltin [53]. Tam *et al.* [89] resolved the chromatographic separation of methylated tin compounds by coupling two 2-m columns. Additional success was due to use of lower oven and inlet temperatures and to extensive silanation of the packing support. Light loadings are generally preferred for well defined peaks but insufficient stationary phase may cause increased tailing [90]. Tailing of the analyte signal may also be associated with the detector memory rather than chromatographic retention [38, 46]. It is worthy of notice that capillary columns have improved the off-line separation of derivatized alkyltins [46, 50, 66, 71]. In the last case, two columns were utilized. A 30 m, 0.3 mm i.d. column was coated with either a 0.15 μm film of Pluronic L 64 or a 0.5 μm film of PS 255, the first being used to obtain a compromise between resolution of critical pairs of organotin compounds, and the second to obtain sufficient retention time for low boiling-point compounds together with elution of higher boiling-point compounds [50]. The on-line set-up uses a slightly different approach for speciation of analytes. Here, compound separation relies more on the differences in volatility of the analytes rather than on the chromatographic properties on the column itself. The U-shaped traps do not contain more than 3 g of packing material and component resolution is related to the rate of warming (1–1.8°C/sec) applied through a Variac after removal of the trap from the liquid nitrogen. Under these conditions, a 30 cm fully-packed U-trap column has an efficiency of approximately 1300 theoretical plates [39]. Hodge *et al.* [40] initially used a simple hydride trap packed with glass wool. They were able to separate methyl-, ethyl-, and butyltin hydrides but only 5 samples could be run before replacement of trap was needed. Rapid deterioration of the glass wool resulted in poor reproducibility for higher boiling-point compounds. A short glass wool plug (5–8 cm) is enough to trap inorganic germanium as germane at liquid-nitrogen temperature if no separation is needed [41].

One of the main problems associated with cold-trapping is the possibility of rapid decomposition of the hydrides on the packing phase and on the walls of the trap (see Section 7.2.1.2) during warming of the trap. The stationary phase is not only important for its chromatographic properties but also for preventing catalytic decomposition of the hydrides (as mentioned earlier) by reaction with the siliceous solid support. Here again, authors agree on using non-polar stationary phases similar to those employed with the off-line system packing material (see above). Usually 10–20% loading provides well resolved peaks for lower boiling-point compounds such as methyltin [23] or methylgermanium compounds [30]. However, lower loadings (3–5%) are required for the speciation of higher boiling-point compounds such as butyltins [43, 44, 80]. To elute these species, the speciation trap must undergo a 500°C temperature change within 4 min, leading to rapid expansion of the packing material, poor reproducibility and important tailing of higher boiling-point compounds. These drawbacks may be overcome by using a high initial flow-rate of the carrier gas, a high pressure in the system, and larger particle size (40–60 mesh) for the packing support [44].

7.2.1.5 Determination

Several types of detectors have been coupled to both on-line and off-line chromatographic systems. The ideal detector must be both sensitive and specific with regard to the organometallic species determined. Three detectors fit these criteria: the graphite furnace, the electrothermal silica furnace, and the flame photometric detector (FPD). Their performances depend on several factors, such as the design of the detector, and the composition and flow-rate of the incoming gas.

The graphite furnace detector

A chromatograph is connected to a graphite furnace with a simple heated transfer line from the column to the graphite furnace injection port. Parris *et al.* [38] connected a small bore (0.027 in. i.d.) heated stainless-steel tube from the column outlet of the flame ionization detector collector to the internal gas purge of the graphite furnace. Similarly, Andreae and Froelich [41] adapted a graphite furnace to a cold-trapping system for the determination of inorganic and methylated germanium species. The outlet from the hydride generation apparatus is directly connected to the left-hand internal purge opening of the graphite tube by thick-wall glass tubing (1 mm i.d.). An argon inlet is provided on the transfer tube before the injection port to purge the furnace after atomization. The graphite furnace program cycle is lengthened to match the long peak elution time required, and the inorganic and methylgermanium species are atomized at 2700°C.

In general, atomization of the analytes is regarded as a single process:

$$C_xH_yM_z \xrightarrow{2700°C} xC + yH + zM$$

However, Parris *et al.* [38] noted that these compounds decompose at well below the volatilization temperature of the metal (<1900°C). They suggest that the addition of hydrogen as reactive gas to the carrier-gas stream may assist the decomposition reaction by scavenging reactive free alkyl radicals (C_nH_m).

The electrothermal silica furnace detector
An efficient alternative to the graphite furnace is the silica furnace (Fig. 7.4). It is

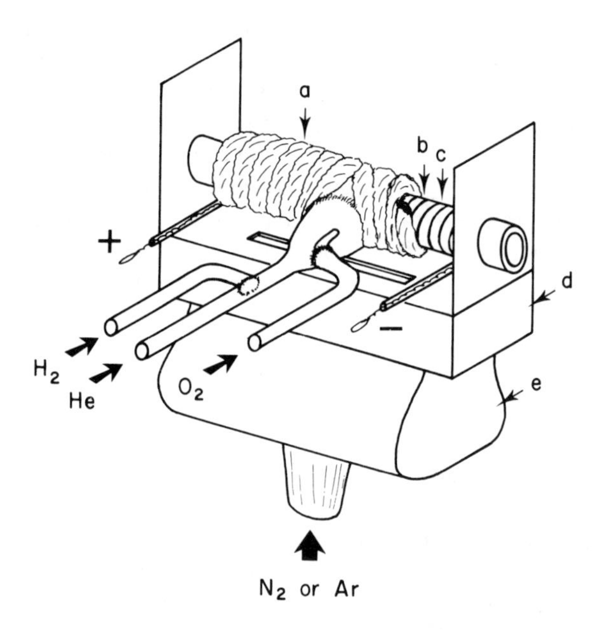

Fig. 7.4 — Detail of silica furnace mounted on top of a flame AAS burner. a, insulating cord; b, insulating tape; c, heating wires; d, stainless-steel cradle; e, burner head.

mounted on a stainless-steel cradle on top of the AAS burner head [40,43,53,67] and can be directly connected to the cold-trap. Careful choice of the silica tube size to match closely the contours of the AAS beam will help to increase the sensitivity [63, 91]. Furthermore, an important gain in atomization efficiency is achieved by addition of reactive gases such as oxygen and hydrogen to the furnace. The mixing of the gas is realized in a mixing chamber before detection in the AAS beam and improves the signal-to-noise ratio. The detection of methyltin and butyltin compounds requires a large flow of hydrogen (1200 ml/min) but only a low flow of oxygen (90 ml/min). The presence of oxygen is critical and is the most important factor regulating the overall sensitivity of the system [43]. The H_2/O_2 ratio is not fixed, and has to be slightly readjusted with every furnace replacement. Sheathing the furnace by a stream of nitrogen round the burner head stabilizes the flame at both ends of the silica tube, decreases the noise level, and helps to extend its life. Power is supplied to the silica tube with a Variac and the operating temperature is 950°C.

Here also, tin atoms can be detected at well below the temperature of vaporization of the metal. The significant and decisive effect of the addition of hydrogen and oxygen, on the response, suggests that atomization could occur through collision with H· free radicals as previously described for arsine [92] and selenium [93]. For example, a dialkyltin dihydride could react according to the equation

$$R_2SnH_2 + 4H\cdot \xrightarrow{O_2,\ H_2} 2RH + 2H_2 + Sn$$

The use of the silica furnace detector provides excellent sensitivity combined with the versatility and specificity of atomic-absorption spectrometry. The background correction mode can be used and the identity of a peak can be confirmed by repeating the analysis at a different wavelength. Also calibration ranges covering several order of magnitudes can be obtained by changing the wavelength from that of the most sensitive resonance line (224.6 nm for tin) to that of a less sensitive line (235.48 > 284.00 nm for the tin EDL) [43,94].

The flame photometric detector

Tin is difficult to excite thermally in conventional flames but several authors such as Dagnall *et al.* [95], Aue and Flinn [70], Kapila and Vogt [96], and Braman and Tompkins [39], have adapted and modified a conventional FPD for the determination of organotin compounds. Quantification of tin is realized by monitoring the red fluorescence molecular-band emission of Sn–H species at 609.5 nm in a nitrogen-hydrogen diffusion flame. An additional scan in the broad and less sensitive 360–490 nm blue region (corresponding to the SnO band) may be used to confirm the presence of tin compounds [96]. The tin signal is markedly affected by the geometry of the silica enclosure of the FPD [39, 70] and the highest sensitivities are obtained with hydrogen-rich flames [50]. According to Aue and Flinn [97] the response to tin compounds is related to luminescence occurring at the surface of the silica. Improvements in sensitivity can be obtained by using the surface luminescence effect, by inserting a loose plug of silica wool immediately above the shielded flame of the FPD [97]. In general, conventional FPD detectors are used with a 600 nm cut-in interference filter (band-pass 600–2000 nm) [45, 48].

Andreae [76] compared the detection limits obtained for tin compounds with the three types of detector. The FPD yields the lowest value. However, this type of detector presents some drawbacks. It is not as specific as atomic-absorption detectors and suffers possible interferences from carbon dioxide, hydrocarbons, and gaseous sulphur [42]. Further, the spectral emission band of GeH (615 nm) slightly overlaps that of SnH. As a result, the germanium signal has to be monitored simultaneously at 650 nm with an 11 nm band-pass filter and be electronically subtracted from the tin signal [42].

In terms of routine applications, the graphite furnace system, although less sensitive, provides wider linear calibration ranges than those obtained with an FPD but requires frequent costly replacements. In comparison, the silica furnace system is cheaper and its sensitivity compares with that of the FPD detector. Both systems are, however, subject to decreases in sensitivity, as a result of contamination by solvents or organotin deposits, but the cost and ease of replacement of the silica furnace should make it the detector of choice for the determination of organotin compounds [98].

7.2.1.6 Signal processing

Other improvements for further gain in sensitivity can finally be obtained by careful processing of the electronic signal. Often, the detector output is directly connected to an integrator and the organotin peaks quantified by measurement of the peak area. By insertion of a low-pass filter to remove electronic noise, and an amplification stage prior to integration, the detection limit can be lowered and the sensitivity increased by a factor of up to 5 [80,96].

7.2.2 The high-performance liquid chromatography/graphite furnace atomic-absorption system (HPLC/GFAAS)

Liquid chromatography also seems to be an attractive alternative for the separation of organotin compounds. High-performance liquid chromatography (HPLC) is commonly used for the fractionation of high molecular-weight and large organic molecules. In general, when organotin halides are adsorbed from the gas phase, the high Lewis acidity of the tin will be likely to generate a large tailing of the signal. Also, some authors [87, 99] consider that potential adsorption and decomposition problems associated with gas chromatographic separation of alkyltin halides should partly be avoided by using liquid chromatography. However, even if HPLC is adequate for the speciation of organotin compounds, the lack of an effective detection interface has restrained the development of this technique. Nevertheless, several systems have been designed and have proved to be useful for the determination of high molecular-weight organotin compounds, but the low sensitivity of this technique compared to GC/AAS makes it unsuitable for the direct determination of tin and germanium in environmental waters, with omission of derivatization and preconcentration steps.

7.2.2.1 Apparatus

The equipment used by different authors is based on standard commercial HPLC chromatographs coupled with an atomic-absorption spectrophotometer equipped with either a flame or graphite furnace detector.

When flame atomic-absorption is used for the detection, a simple and direct connection is made between the outlet of the column and the nebulizer capillary. The flow-rate of the chromatographic effluent is regulated to an appropriate value. Unfortunately, the flame detection limits (D.L.) are often poor for determination of total metals in environmental samples unless preconcentation is used (e.g. D.L. = 200 ng as Sn [52]). Also, important peak spreading may be observed, resulting from dilution of the sample in the solvent when passing through the column [100].

As a consequence, several authors have turned to graphite furnace detection because its detection limits are potentially three orders of magnitude lower.

However, in this case, a continuous system (HPLC) is associated with a sequential detection system (GFAAS) and results in a fractionated organotin signal (Fig. 7.5).

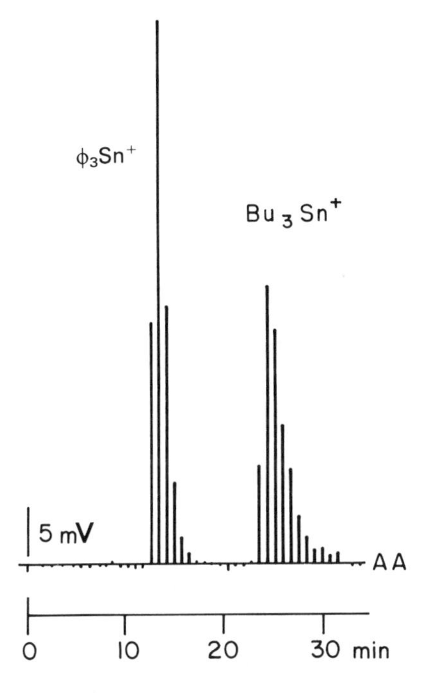

Fig. 7.5 — Speciation of a 1:1 solution of triphenyltin (ϕ_3Sn^+) and tri-n-butyltin (n-Bu_3Sn^+) cations (500 ng as Sn) with a high-performance liquid chromatography/graphite furnace atomic-absorption system (HPLC/GFAAS). (Adapted from Jewett and Brinckman [60]).

Different interfaces developed for organotin determination will be described later in this chapter. To avoid problems associated with the sequential aspect of GFAAS detection, Burns *et al.* [53] included on-line generation of organotin hydrides and a gas–liquid separation before silica furnace detection. They achieved excellent sensitivity (2–20 pg as Sn) but this technique is not applicable to larger organotin molecules.

7.2.2.2 Sample pretreatment

Total tin determination does not take into account the chemical species occurring in natural environmental matrices, and samples can be treated with drastic leaching procedures such as oxidative acid attack, without any problems. On the other hand, the speciation of organotin compounds from environmental samples requires an extraction method that will both yield high recoveries of analytes and avoid possible displacements of the equilibria between the different species extracted.

In general, the concentration of organotin compounds occurring in natural water is much too low for direct chromatographic separation and detection. Extraction allows a 500–1000-fold preconcentration of the organotin species (see also Section

7.2.1.2) and avoids possible interference from various ions present in the sample (especially sulphate). The most commonly used method was proposed by M and T Chemical Ltd. in their standard test method [101]. The sample is acidified with hydrobromic acid and the organotin species are concentrated by extraction into toluene; total tin is determined in a second sample by extraction with a toluene solution of tropolone. The extracts are transferred to glass vials and can be stored in the laboratory freezer until they are analysed [102]. Other solvents may also be used, such as n-pentane, chloroform, hexane, benzene, or methyl isobutyl ketone [103]. However, toluene should be considered as the solvent of choice, because of its high extraction efficiency and the rapid separation of the two phases. Other methods of sample preparation include use of an ion-exchange resin to trap and concentrate organotin compounds from dilute aqueous solutions [104], or an apolar resin (Bio Beads SX-2) as proposed [46] for the extraction of trialkyl-, dialkyl- and monoalkyltin compounds, with tropolone as a suitable complexing agent.

7.2.2.3 Separation conditions

To date, there is no HPLC technique that offers complete speciation of all organotin compounds likely to be found in the environment, though several approaches have been proposed in the literature. Burns *et al.* [53] separated methyltin and ethyltin compounds (R_xSnX_{4-x}) on non-polar ODS Spherisorb S5W by reversed-phase chromatography, using acetone–pentane mixtures (60:40 and 70:30 v/v) as eluent, without redistribution of the ethyl and methyl groups, but their attempt to extend the method to butyltin derivatives failed. Separation of organotins has also been achieved on alkyl- or cyanoalkyl-bonded silica phases, but the speciation usually concerns homologous series of compounds, e.g. tetraorganotins [105], triorganotins [51,61,106], diorganotins [107] and mono-organotins [108]. For example, Ph_3SnCl, Bu_3SnCl and Et_3SnCl are well separated on a Micropack MCH-5 Cap column (methanol as eluent). Similarly, Me_3SnCl, Et_3SnCl and Ph_3SnCl have been separated from Bu_3SnX on a Nucleosil C_{18} 5 μm column with methanol–water as eluent.

Other separation techniques have also been applied. Ion-exchange chromatography on Partisil 10 SCX with methanol–water (70:30 v/v) as eluent was employed [60] to speciate traces of di- and triorganotin compounds in water. Interesting separations were also obtained [109] by gel permeation chromatography on micro Styragel (a copolymer of styrene and divinylbenzene suitable for use with organic mobile phases) with tetrahydrofuran as solvent for organotin compounds, but some problems were encountered with chlorinated derivatives. Although separation techniques based on molecular size appear to be promising for the speciation of Bu_3, Bu_2, and Bu organotin chlorides with toluene as eluent, some difficulties arise in choosing the elution conditions, because the separation is not simply achieved by direct size-exclusion chromatography but also by important irreversible adsorption of the most acidic species on the micro Styragel support. A possible solution to these difficulties is complexation of the organotin compounds with morin [107] or tropolone [61].

7.2.2.4 Atomization efficiency

Atomic-absorption determination of tin with the graphite furnace is well known to be troublesome, for both simple and complex matrices. The major problem seems to be due to the tendency of tin to form volatile compounds which are easily lost during the drying and ashing steps, and to its interactions with the wall of the furnace. These two processes were studied in detail by Lundberg *et al.* [110], who used radioactive ^{113}Sn to monitor the losses of tin during the temperature cycle. Several approaches to solution of these problems have been described. They include: (i) the separation of tin from the matrix by liquid–liquid extraction, (ii) pretreatment of the graphite furnace by coating its inner wall with salts of refractory metals (W, Ta, Zr, Mo, La, Nb) that will form refractory carbides and thus prevent interaction with tin and reduce sensitivity effects between the different organotin compounds [56, 111–121], (iii) the possible use of a stabilized-temperature platform furnace (STPF) [113], (iv) the addition of a matrix modifier to the sample. This last method is the simplest and requires minimal sample treatment before injection into the furnace. It prevents volatilization of tin during the atomization programme and decreases interferences. Many organic and inorganic compounds have been proposed as matrix modifiers for this purpose. The most important are solutions of ascorbic acid (10%) [122], ammonia (10% v/v s.g. 0.88 solution) [110], citric acid (8%) in hydrochloric acid (10% v/v) [123], ammonium nitrate [114], phosphoric acid [124], ammonium dihydrogen phosphate [125], ammonium dihydrogen phosphate/magnesium nitrate in 1% v/v nitric acid [126] and recently 0.04% potassium dichromate in 2% v/v nitric acid [127].

The matrix modifier is introduced into the sample in the cup of the fraction collector. The movable support holding the aspiration capillary has an additional capillary tube (Fig. 7.6) carrying a gentle flow of compressed air to homogenize the

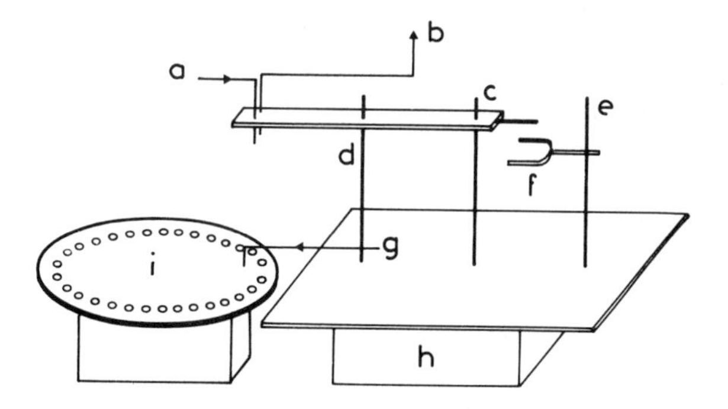

Fig. 7.6 — Schematic of an HPLC/GFAAS interface assembly for the determination of organotin compounds: a, compressed air inlet; b, to GFAAS; c, movable rod; d, fixed guiding rod; e, fixed support rod; f, photoelectric cell; g, HPLC effluent; h, auto sampler; i, fraction collector. (After Pinel *et al.* [61].)

sample before aspiration. The addition of the matrix modifier slightly dilutes the collected fraction but enhances the signal and improves its reproducibility.

In complex matrices, sulphate ions are responsible for suppression of the tin atomic-absorption signal [114] and all tin measurements fail in the presence of a sulphate concentration higher than 10 mg/l. In non-aqueous extracts, no modifier was used until the recent work of Parks *et al.* [128] who proposed the addition of ammonium dichromate to a toluene–propanol extract, resulting in an interesting enhancement of the tin signal. Also, the addition of a toluene solution of picric acid to complex sample matrices was proved to be a very efficient means of obtaining a sensitive tin signal and excellent reproducibility [127].

7.2.2.5 Data processing

The sequential aspect of the signal can be treated in a similar fashion to that used for continuous gas chromatograms. HPLC 'peak area' can be measured by planimetry of the 'peak' joining the tops of the GFAAS 'multiplet' or by summation of the heights of the individual peaks [57, 58]. Later developments include presentation of the results as a histogram. An integrator printer can be used to convert the histogram into chromatographic peak area measurements [60].

7.3 DETECTION LIMITS, SENSITIVITY AND PRECISION

The detection limits obtained with the various systems developed vary considerably. Figure 7.7 presents a compilation of detection limits for inorganic tin and organotin compounds (expressed as Sn) given by the authors using various techniques. The range of absolute detection limits is also given in Tables 7.3 and 7.4 (p. 217). The wide scatter observed within one technique is relevant to discrepancies between the different sensitivities of similar types of systems and to the species of organotin compound determined. Most of these detection limits are based on a theoretical value determined as 2 or 3 times the standard deviation of the blank, and therefore yield low values for organotin compounds. In general, detection limits for inorganic tin are somewhat higher, owing to the presence of traces of inorganic tin in the reagents. The lowest detection limits for tin are achieved with the GC/FPD and GC/QFAAS systems and are in the low pg range. It is, however, difficult to establish a true comparison, for many of these techniques depend on preliminary extraction and/or concentration steps. The wide difference in sensitivity observed between the GC and LC separation methods is related to the fact that in the first the analytes are concentrated, and in the second, they are diluted in the solvent. Instead of being competitive these techniques are complementary, for they do not deal with the same molecular weight range of organotin compounds and the same type of matrices. The detection limits obtained for inorganic and methylgermanium compounds vary from 120 to 75 pg of Ge respectively [30, 41], and are within the same range as those for tin compounds.

Limitation of the tin blank is critical for good sensitivity in inorganic tin determination. Reagents such as buffers, acids, sodium borohydride, and the sodium hydroxide often used to stabilize sodium borohydride solutions, contain traces of inorganic tin at levels that are unacceptable in view of the sensitivity of the system. Tin levels can be reduced in Tris-HCl buffer by co-precipitation with lanthanum hydroxide [42]. High-quality nitric acid is preferred for acidification of the samples, to ensure low tin blanks. Reduction of the tin blank in the sodium borohydride solution may be achieved in several ways. After dilution, the borohydride solution can be filtered through a 0.2-μm membrane and nitrogen passed through it to remove

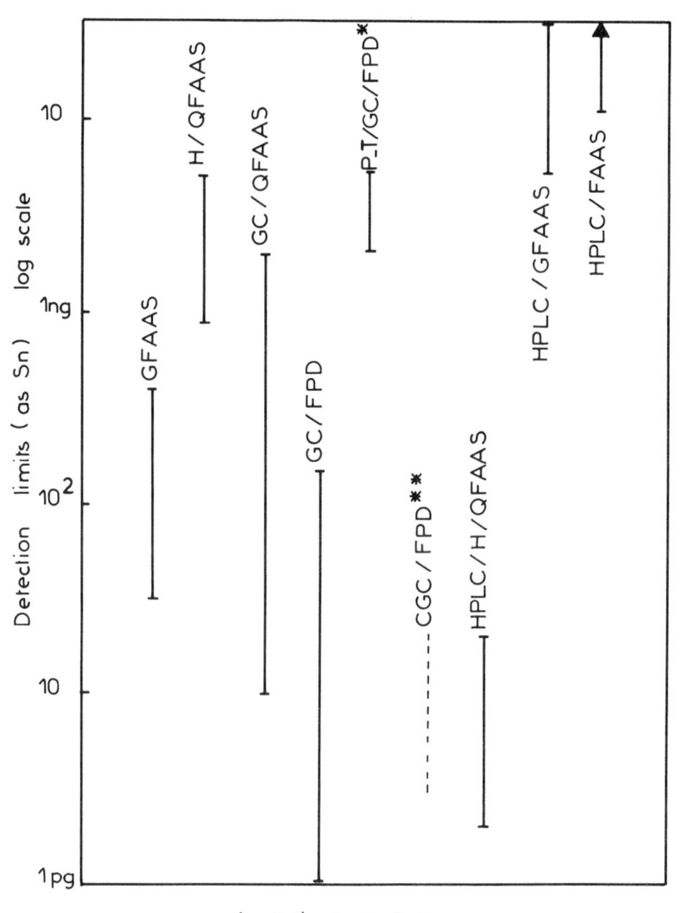

Fig. 7.7 — Detection limit ranges obtained with various types of apparatus for tin and organotin compounds (as Sn). GFAAS is graphite furnace atomic-absorption spectrophotometry; H is hydride generation; GC is gas chromatography; QFAAS is silica furnace atomic-absorption spectrophotometry; FPD is flame photometric detection; P–T is purge trap system; CGC is capillary gas chromatography; HPLC is high-performance liquid chromatography; FAAS is flame atomic-absorption spectrometry. *Detection limits based on a 10-ml sample size; **estimated from authors' data.

any tin hydride. Allowing the solution to stand overnight in a glass flask will also significantly reduce the tin level of adsorption. Solutions may also be purified by electrolysis for an hour (at 3 V) with continuous passage of helium [77]. Sodium borohydride solutions are not very stable and should be prepared daily. Some authors (see Table 7.1, p. 198) stabilize the borohydride solution by addition of sodium hydroxide [129], but this reagent is most often contaminated with a high level of tin.

The atomization efficiency may decrease with increasing size of the alkyl group in organotin compounds [43,130]. Although the most sensitive linear calibration ranges are fairly similar for both GC/QFAAS and GC/FPD techniques [98], the use of atomic-absorption spectrometry allows extension of the dynamic linear range by changing the wavelength to a less sensitive one [43].

The precision of the methods presents wide variations (between 4 and 30%), depending on the element and on the technique used. Most of these variations are due more to the preliminary sample treatment steps and the numerous operations to be performed during the analysis than to the intrinsic variability of the apparatus. Automation of the systems should result in increased reproducibility of the analysis.

7.4 ENVIRONMENTAL ANALYSIS

The GC/AAS system allows the handling of gaseous, liquid, and solid samples. Considerably fewer applications have been developed for the determination of organotin compounds in solid matrices by use of the LC/AAS system.

7.4.1 Air samples

Inorganic germanium can be collected from air by adsorption at room temperature on 60–80 mesh gold-coated glass beads packed in 12 cm (10 mm o.d.) borosilicate glass tubes, connected at one end to a continuous sampling pump. The germanium is quantitatively removed from the gold-coated beads by gentle warming with 10 ml of 0.1*M* sodium hydroxide, followed by a 40-ml water rinse. This solution is then analysed by the usual procedure after acidification of the sample with hydrochloric and oxalic acids [6]. Airborne particulates are collected by addition of a filter holder ahead of the sampling tube. Germanium is then leached from the filter and determined as described above.

The volatile organotin compounds in air can be cryogenically trapped after chromatographic separation and directly desorbed into the detection system for determination. Aerosols can be collected on Teflon PTFE filters (1 μm nominal pore size); the filter are leached with 6*M* hydrochloric acid and sonication for 2 hr, and the resulting solutions are analysed [131].

7.4.2 Water samples

The low concentrations of organotin and organogermanium compounds in environmental waters require extreme care to be taken during sample collection.

Germanium

Contamination problems are not serious for determination of inorganic and methylgermanium compounds [30]. Water samples are collected with an acid-cleaned stainless-steel/polycarbonate sampler, stored in acid-cleaned polyethylene bottles and acidified to pH 2 with hydrochloric acid [132].

Tin

Contamination problems are more likely to occur with tin compound determinations. Poly(vinyl chloride) water samplers and containers must be avoided. Water can be collected with polycarbonate, polyethylene, or Teflon samplers. Samples should immediately be filtered and acidified to pH 2 with nitric acid. Storage prior to analysis should be reduced to a minimum. Maguire and Tkacz [85] do not filter their samples before determination of butyltins by derivatization in an 'off-line' system. The samples are collected in amber bottles and acidified to pH 1, and can be stored for up to three months in the dark at 4°C before extraction. Hydrochloric acid can be washed with 1% tropolone solution in benzene to remove the inorganic tin blank [86]. Both analytical techniques have been widely used for the determination of organotin compounds in environmental waters. Detailed analytical procedures have been given in Section 7.2.1.2.

7.4.3 Sediment samples

In comparison to water samples, very little work has yet been published on the extraction methods and sample pretreatment required for the determination of organotin compounds in sediment matrices. Hydride generation has mostly been used for the determination of methyltin in sediments. Organotins are usually extracted with an acidic leaching solution. Tugrul *et al.* [133] sonicate 0.5–1 g of dry ground polluted sediment for 1 hr in 40 ml of 0.1*M* hydrochloric acid. Unpolluted samples are agitated ultrasonically in doubly distilled water before hydride generation determination. Seidel *et al.* [134] use a 'mild' 0.3*M* hydrochloric acid leach to extract tin and organotin compounds from sediments. An acidified solvent has also been utilized by Hattori *et al.* [135]. The sediments are digested in methanolic hydrochloric acid and the digest is extracted with benzene. The extract is cleaned up on an activated silica gel column, and hydrogenated with a solution of $NaBH_4$ in ethanol.

A comparison of reagent efficiencies for leaching sediments in water was made by Donard *et al.* [80]: 0.5-g samples of dried and ground sediments were digested with 5 ml of 5*M* hydrochloric acid, 5*M* nitric acid, or 0.1*M* sodium hydroxide, then sonicated for 1 hr and finally shaken for 12 hr in the dark. The leachates were diluted to 50 ml with doubly distilled water and centrifuged. An aliquot of the supernatant solution was adjusted to pH 2 before analysis by hydride generation. Best results were obtained with the alkaline digestion solution. However, the comparison between the different techniques, in terms of both efficiency and reproducibility, may call into question in general the validity of data obtained with hydride generation. One of the main problems is the potential dismutation and disproportionation of methyltin compounds or/and reaction with naturally occurring sulphide in the digestion liquor. A slightly different approach has been proposed by Gilmour *et al.* [136], based on direct hydride generation from buffered sediment samples. The recovery from sulphide-spiked sediments is good for mono- and dimethyltin, but the low recovery of trimethyltin (60%) suggests a potential loss by formation of volatile tetramethyltin [137, 138].

Some of the problems mentioned above appear to be avoided in the extraction of butyltin compounds from sediments. Butyltins in sediments are mainly determined by a solvent extraction method. The dried sediments are refluxed for 2 hr with 0.25 g of tropolone in 100 ml of benzene. The mixture is filtered before pentylation [26]. The $Bu_xPe_{(4-x)}Sn$ compounds are then purified by passage through a silica gel column and concentrated prior to analysis [85]. Finally, recent developments indicate that butyltin compounds in sediments can be determined at the low pg level by the hydride generation technique. Extraction from freeze-dried homogenized sediments is done in centrifuge tubes after addition of enough calcium chloride and hydrochloric acid to give a final concentration of 2.5*M* for both, in a final volume of 10 ml. Each sample is then shaken for 15 hr and filtered through a 0.4-μm polycarbonate filter. Extraction of sediments spiked with methyltin and butyltin compounds gave an average recovery ranging from 70 to 103% for all alkyltin species [139, 140]. In recent work by Müller [50] capillary GC/FPD has been applied for the detection of butyl-, cyclohexyl- and phenyltins in both sediments and sewage sludges. The sample (1–20 g) is spiked with an internal standard and acidified with hydrochloric acid to pH 2–3. The slurry is extracted with a 0.25% solution of tropolone in diethyl ether. The organic extract is then filtered and dried, and the volume is reduced by evaporation at room temperature before derivatization as described earlier.

7.4.4 Biological samples

Surprisingly enough, even fewer methods have been developed for the determination of organotins in biological matrices, although this objective was the initial motive for the research in this field, undertaken on account of the high toxicity of the organotin compounds. This paucity of methods reveals the difficulty of applying these analytical techniques to such complex systems as biological matrices.

Organotin compounds were initially determined by hydride generation in algal tissues and marine invertebrates by Seidel *et al.* [134]. A 10-g sample was rinsed with doubly distilled water, then digested in 60 ml of 0.2*M* acetic acid. After homogenization, the sample was centrifuged and the supernatant liquid decanted for analysis. Tugrul *et al.* [133] crushed in a grinder the soft parts of limpet and fish tissues (0.5–2 g) in 50 ml of 0.04*M* hydrochloric acid. Other procedures have been adapted to algal material. Fresh algal tissues were homogenized in 60 ml of 0.02*M* sorbitol/0.01*M* Tris-acetate buffer solution [141]. Analyses of seaweed tissues were performed by Donard *et al.* [142]. After cleaning and rinsing in artificial seawater, the plant material was blotted dry, and 0.2 g of tissue was frozen in liquid nitrogen, then crushed with a mortar and pestle, and homogenized in a Teflon tissue-grinder with 5 ml of doubly distilled water. The final mixture was digested in 5 ml of 1*M* sodium hydroxide for 48 hr and then acidified to pH 2 with 5*M* nitric acid before analysis by hydride generation. Biological tissues were in general first purged without addition of $NaBH_4$, to detect any volatile Me_4Sn. The addition of an antifoaming agent or adjustment of the bubbler to just above the sample surface is required, owing to the excessive foaming produced in the borohydride reaction.

Maher [143] has reported on the use of the hydride/QFAAS system for analysis of marine biological specimens.

The sample pretreatment for the determination of butyltins in the tissues of oligochaetes by Grignard derivatization differs slightly from that described for sediments. One gram of oligochaetes is kept in 1–2 ml of hydrochloric acid for 2 hr at room temperature. The final mixture is diluted fivefold with distilled water and extracted as for water samples [26].

Very little has so far been published concerning the determination of organotin compounds in biological matrices by HPLC/GFAAS. The digestion of fish and crustaceous tissues is under current development. Several attempts have been made so far. The solubilization of dry tissues with a 'Lumatom' solution†. (3 hr digestion at

† Lumatom® is a commercial tissue solubilizer (H. Kürner, Herderstrasse 2, D-8200 Rosenheim, FRG).

50°C) and dilution with toluene is a very effective method of preparing biological samples for the determination of total tin by GFAAS. To date, however, the separation of butyltin species by gel filtration has been unsuccessful. Another digestion procedure for dry tissues, based on use of 0.04*M* hydrochloric acid followed by extraction with 1% toluene solution of tropolone is also very effective for the determination of total tin by GFAAS, but the speciation of butyltins by HPLC/GFAAS is not possible. The failures originate from the high solubility of organotin compounds in the lipid phase of the samples and from deleterious effects of this large amount of lipids on the HPLC column [144].

7.5 TYPICAL ENVIRONMENTAL CONCENTRATIONS

Concentration ranges for inorganic germanium and organogermanium compounds in waters are given in Table 7.2. The distribution of inorganic tin, methyltin and butyltin species in various matrices are given in Tables 7.3 and 7.4. In comparison, few environmental data obtained with the HPLC/GFAAS system have been published. Butyltins have been determined in anti-fouling organometal polymer sandblasting grits at the μg/g level [60]. Results obtained for sewage sludges with the GC/FPD system illustrate the considerable concentration of butyl- or phenyltins in these samples. The concentrations detected were in the μg/g range for butyl- and phenyltins in some treatment plants [50].

7.6 CONCLUSIONS

The analytical techniques developed for the speciation and determination of organotin and organogermanium compounds can still be considered as recent. In general, many of them have achieved excellent separation and sufficient sensitivity for the determination of these species at the pg level in water samples. Hydride generation or derivatization with a Grignard reagent after extraction has been successfully applied to a wide range of matrices. Though separation of triorganotins by HPLC seems to be well developed for simple liquid matrices, the sequential aspect of the detection mode affects the sensitivity and limits applications of these techniques to contaminated samples.

None of these techniques can completely substitute for another, and they are essentially complementary rather than competitive. Hydride generation methods are very efficient for the determination of volatile compounds containing few carbon atoms, such as methyl- and butyltins in water samples. Derivatization with a Grignard reagent requires more sample pretreatment steps but seems more appropriate for higher alkyl compounds (butyltins) and to the determination of organotins in solid matrices. Although not yet completely developed, HPLC/GFAAS systems should allow in the future the separation and determination of longer alkyl chain compounds in complex matrices, and the relatively low cost of these methods should promote their routine application to the speciation and determination of organotin compounds in the environment. However, these determinations are based only on the presence of the metal atom and on the retention time, and cannot substitute for an accurate identification of a new molecule by gas chromatography/mass spectro-

Table 7.2 — Typical environmental concentration of germanium species (as Ge).

Sample location	Technique	D.L.*(ng)	Ge	$MeGe^{3+}$	Me_2Ge^{2+}	Me_3Ge^+	References
			Water samples (ng/l.)				
Freshwaters							
Yukon (USA)	GC/GFAAS	0.58–2.90	3.6–5.2	ND	ND	ND	[28]
Rhine (Germany	GC/GFAAS	0.58–2.90	0.7–12	ND	ND	ND	[28]
Estuarine waters							
Peace River and Charlotte Harbor (USA)	GC/GFAAS	0.11–0.23	0.7–7.1	0–18.6	0–10.0	0	[30]
Ochlockonee River and Bay Estuary	GC/GFAAS	0.11–0.23	2.4–5.7	0–19.5	0–12.0	0	[30]
Tejo River and Estuary (Portugal)	GC/GFAAS	0.11–0.23	0.6–7.4	0–19.6	0–6.6	tr	[30]
Sea-waters							
Atlantic, Sargasso Sea	GC/GFAAS	0.11–0.23	0.4–1.8	21.9–24.5	10.0–11.5	0	[30]
Bering Sea (GEOSECS 219)	GC/GFAAS	0.11–0.23	2.2–12.3	18.2–20.3	8.0	0	[30]
Baltic Sea (10 m depth)	GC/GFAAS	0.58–2.90	1.2–3.3	3.8–5.7	2.6–3	nd	[132]
Baltic Sea (210 m depth)	GC/GFAAS	0.58–2.90	29.7–37.5	6.8–7.5	4.1–5.1	nd	[132]

*Range of absolute detection limits for the different inorganic germanium and organogermanium compounds recalculated to 3 times the standard deviation of the blank when possible and for 100 ml sample size.
tr = trace; ND = not determined; nd = not detected.

Table 7.3 — Typical environmental concentrations of tin and methyltin species (as Sn).

Sample location	Technique	D.L.*(ng)	TRISn†	$MeSn^{3+}$	Me_2Sn^{2+}	Me_3Sn^+	Me_4Sn	References
				Water samples (ng/l.)				
Rainwaters								
Tampa (USA)	GC/FPD	0.007–0.02	2.7–41	0.7–22	nd–4.8	nd–1.1	ND	[39]
La Jolla (USA)	GC/QFAAS	0.15	1–3	nd–10	nd	nd	ND	[133]
Freshwaters								
Lake Carroll (USA)	GC/FPD	0.007–0.02	3.3–7.7	nd	0.6–1	3.3–3.6	ND	[39]
Lake Michigan (USA)	GC/QFAAS	0.4–2.0	84–490	6–18	nd–63	nd	ND	[40]
Lake Ontario (USA)	GC/QFAAS	0.1	130–980	350–1220	100–400	nd	ND	[49]
Lake Superior (USA)	GC/FPD	2.0	ND	150–230	30	nd–50	ND	[145]
Estuarine waters								
Baltimore Harbor (USA)	GC/FPD	1.3–5.3	200–300	nd	50–100	10–20	50–100	[45]
Great Bay microlayer (USA)	GC/QFAAS	0.03–0.9	373–1338	182–661	nd–196	132–415	nd	[43]
Great Bay 1 m depth (USA)	GC/QFAAS	0.03–0.9	30–646	51–414	nd–35	70–508	nd	[43]
Great Bay porewater (USA)	GC/QFAAS	0.03–0.9	52–360	320–690	20–58	10–40	nd	[43]
San Diego Bay (USA)	GC/QFAAS	0.4–2	6–38	2–8	15–45	nd	nd	[40]
Hillsborough Bay (USA)	GC/FPD	0.007–0.02	nd–13	nd–0.7	0.9–2.4	0.3–0.7	nd	[39]
Sea-waters								
Pacific Ocean	GC/QFAAS	0.4–2	0.3–0.8	nd	nd	nd	nd	[40]
Gulf of Mexico	GC/FPD	0.007–0.02	2.2–62	nd–15	0.7–3.2	nd–1	nd	[39]
				Sediments (μg/g dry weight)				
Bay sediments (Turkey)	GC/QFAAS	0.15	0.5–11	nd–0.011	nd–0.013	0.0001–0.020	nd	[133]
Lake Geneva (Switzerland)	GC/QFAAS	0.03–0.9	0.200–0.280	0.042–0.056	0.033–0.039	0.543–0.552	nd	[80]
Great Bay (USA)	GC/QFAAS	0.03–0.9	0.430–0.600	nd–0.080	nd–0.049	nd	nd	[140]
				Algae (ng/g wet weight)§				
San Diego (USA)‡	GC/QFAAS	0.4–2	1.4–115	ND	nd–0.2	ND	nd–1.1	[134]
Mission Bay (USA)	GC/QFAAS	0.75–3	0.4–2.0	nd–0.4	nd–0.2	nd	nd	[141]
Great Bay (USA)	GC/QFAAS	0.03–0.9	183–395	30–71	13.52	nd–3	nd–6	[142]
				Fish (ng/g dry weight)				
Mediterranean Sea	GC/QFAAS	0.15	86–260	0.8–27	2.6–2.9	1.2–1.3	nd	[133]

ND = not determined; nd = not detected
*See Table 7.2 for definition
§All species
‡Assuming a dry weight/wet weight ratio of 0.09.

Table 7.4 — Typical environmental concentrations of tin and butyltin species (as Sn).

Sample location	Technique	D.L. (ng)*	TRISn†	n-$BuSn^{3+}$	n-Bu_2Sn^{2+}	n-Bu_3Sn^+	n-Bu_4Sn	References
				Water samples (ng/l.)				
Freshwaters								
Rhine (Germany)	CCF/FPD	0.1	ND	ND	ND	3–10	ND	[46]
Lake Superior microlayer (Canada)	GC/FPD	1	11600–111000	nd	nd–710	nd–680	ND	[145]
Lake Superior (Canada)	FC/FPD	1	nd–730	nd	nd–560	nd–20	ND	[145]
Detroit River (Canada)	GC/FPD	1	nd–2943	nd–114.7	nd–7.13	nd–69.6	ND	[86]
Estuarine waters								
Chesapeake Bay microlayer (USA)	GC/FPD	0.1–2.2	ND	nd–165	91–150	tr–1872	nd–395	[48]
Chesapeake Bay (USA)	GC/FPD	0.1–2.2	ND	nd	nd–109	tr–900	nd–4200	[48]
Tejo Estuary (Portugal)	GC/FPD	0.02	0.9–3	0.02–1.2	ND	ND	ND	[146]
				Sediments (μg/g dry weight)				
Mission Bay (USA)	GC/QFAAS	0.4–2	0.13–2.64	nd–0.011	ND	ND	ND	[134]
Ontario (Canada)	GC/FPD	1	0.08–1.74	nd–0.58	nd–0.35	nd–0.54	ND	[26]
Great Bay (USA)	GC/QFAAS	0.03–0.9	0.430–0.600	0.003–0.030	nd–0.015	0.012–0.044	ND	[140]
				Algae (ng/g wet weight)				
Great Bay (USA)	GC/QFAAS	0.03–0.9	183–395	nd–6	nd	nd–50	ND	[142]

ND = not determined; nd = not detected; tr = trace.
*See Table 7.2 for definition.
†See Table 7.3 for definition.

metry. Nonetheless, these techniques make a novel and important contribution in more detailed investigations of tin and germanium cycles in ecosystems. However, considerable efforts have still to be made to improve their determination in samples such as sediments and biological tissues.

Acknowledgements — One of the authors (O. F. X. Donard) is thankful to Professor J. H. Weber from the University of New Hampshire, under whose direction many of the analytical techniques for organotin compounds were developed. The help and comments of Professor J. P. Quintard were most appreciated, along with the understanding and support of Professors M. Ewald, M. Astruc and C. Belin.

REFERENCES

[1] J. A. J. Thompson, M. G. Sheffer, R. C. Pierce, Y. K. Chau, J. J. Cooney, W. R. Cullen and R. J. Maguire, *Organotin Compounds in the Aquatic Environment: Scientific Criteria for Assessing their Effects on Environmental Quality*, NRCC/CNRC, Ottawa, 1985.
[2] Organisation Mondial de la Santé, *Etain et organostanniques: Mise au point préliminaire*. Organisation Genève, 1980.
[3] P. J. Craig, *Environ. Technol. Lett.*, 1980, **1**, 225.
[4] S. J. Blunden and A. H. Chapman, *Environ. Technol. Lett.*, 1982, **3**, 267.
[5] S. J. Blunden, P. A. Cusack and R. Hill, *The Industrial Uses of Tin Chemicals*, Royal Society of Chemistry, London, 1985.
[6] M. A. Tompkins, *Ph.D. Thesis*, University of South Florida, Tampa, 1977.
[7] P. N. Froelich, Jr. and M. O. Andreae, *Science*, 1981, **213**, 205.
[8] B. L. Lewis, P. N. Froelich and M. O. Andreae, *Nature*, 1985, **313**, 303.
[9] P. Smith and L. Smith, *Chem. Brit.*, 1975, **11**, 208.
[10] *Chem. Brit.*, 1982, **18**, 850.
[11] C. O. Evans and S. Karpel, *Organotin Compounds in Modern Technology*, Elsevier, Amsterdam, 1985.
[12] R. B. Laughlin and O. Linden, Jr., *Ambio*, 1985, **14**, 88.
[13] C. Alzieu, M. Héral, Y. Thibaud, M. J. Dardignac and M. Feuillet, *Rev. Trav. Inst. Pêches Marit.*, 1982, **45**, 101.
[14] C. Alzieu, Y. Thibaud, M. Héral and B. Boutier, *Rev. Trav. Inst. Pêches Marit.*, 1980, **44**, 305.
[15] C. Alzieu and Y. Thibaud, *Bull. Acad. Nat Med.*, 1983, **167**, 473.
[16] M. Astruc and R. Pinel, *L'étain dans l'environnement, Rapport CNRS-PIREN — ASP Dechets et Environnement*, 1982.
[17] H. Vrijhof, *Sci. Total Environ.*, 1985, **43**, 221.
[18] H. Plum, *Inf. Chim.*, 1982, **220**, 135.
[19] S. J. Blunden, L. A. Hobbs and P. J. Smith, in *Environmental Chemistry*, H. J. M. Bowen (ed.), pp. 49–77. Royal Society of Chemistry, London, 1984.
[20] P. J. Craig and S. Rapsomanikis, *Environ. Sci. Technol.*, 1985, **19**, 726.
[21] S. Rapsomanikis and J. H. Weber, *Environ. Sci. Technol.*, 1985, **18**, 352.
[22] S. Rapsomanikis and J. H. Weber, in *Organometallic Compounds in the Environment*, P. J. Craig (ed.), pp. 279–303. Longmans, Harlow, 1986.
[23] O. F. X. Donard and J. H. Weber, *Environ. Sci. Technol.*, 1985, **19**, 1104.
[24] L. Randall and J. H. Weber, *Sci. Total Environ.*, 1986, **57**, 191.
[25] F. E. Brinkman, J. A. Jackson, W. R. Blair, G. J. Olson and W. P. Iverson, in *Trace Metals in Seawater*, C. S. Wong, E. Boyle, K. W. Bruland, J. D. Burton and E. D. Goldberg (eds.), pp. 39–72. Plenum Press, New York.
[26] R. J. Maguire, *Environ. Sci. Technol.*, 1984, **18**, 291.
[27] J. T. Byrd and M. O. Andreae, *Science*, 1982, **218**, 565.
[28] P. N. Froelich, G. A. Hambrick, M. O. Andreae, R. A. Mortlock and J. M. Edmond, *J. Geophys. Res.*, 1985, **90**, 1133.
[29] P. N. Froelich, L. W. Kaul, J. T. Byrd, M. O. Andreae and K. K. Roe, *Est. Coast. Shelf Sci.*, 1985, **20**, 239.
[30] G. A. Hambrick III, P. N. Froelich, Jr., M. O. Andreae and B. L. Lewis, *Anal. Chem.*, 1984, **56**, 421.
[31] A. H. Chapman, *Anal. Proc.*, 1983, **20**, 210.
[32] S. J. de Mora, C. N. Hewitt and R. M. Harrison, *Anal. Proc.*, 1984, **21**, 415.
[33] G. Weber, *Z. Anal. Chem.*, 1985, **321**, 217.
[34] F. J. Fernandez, *At. Abs. Newsl.*, 1977, **16**, 33.
[35] J. C. Van Loon, *Anal. Chem.*, 1979, **51**, 1139A.

[36] L. Ebdon, R. W. Ward and D. A. Leathard, *Anal. Proc.*, 1982, **19**, 110.
[37] R. M. Harrison, C. N. Hewitt and S. J. de Mora, *Trends Anal. Chem.*, 1985, **4**, 8.
[38] G. E. Parris, W. R. Blair and F. E. Brinckman, *Anal. Chem.*, 1977, **49**, 378.
[39] R. S. Braman and M. A. Tompkins, *Anal. Chem.*, 1979, **51**, 12.
[40] V. F. Hodge, S. L. Seidel and E. D. Goldberg, *Anal. Chem.*, 1979, **51**, 1256.
[41] M. O. Andreae and P. N. Froelich, Jr., *Anal. Chem.*, 1981, **53**, 287.
[42] M. O. Andreae and J. T. Byrd, *Anal. Chim. Acta*, 1984, **156**, 147.
[43] O. F. X. Donard, S. Rapsomanikis and J. H. Weber, *Anal. Chem.*, 1986, **58**, 772.
[44] L. Randall, O. F. X. Donard and J. H. Weber, *Anal. Chim. Acta*, 1986, **184**, 197.
[45] J. A. A. Jackson, W. R. Blair, F. E. Brinckman and W. P. Iverson, *Environ. Sci. Technol.*, 1982, **16**, 110.
[46] M. D. Müller, *Z. Anal. Chem.*, 1984, **317**, 32.
[47] R. J. Maguire and H. Huneault, *J. Chromatog.*, 1981, **209**, 458.
[48] C. L. Matthias, J. M. Bellama, G. J. Olson and F. E. Brinckman, *Environ. Sci. Technol.*, 1986, **20**, 609.
[49] Y. K. Chau, P. T. S. Wong and G. A. Bengert, *Anal. Chem.*, 1982, **54**, 246.
[50] M. D. Müller, *Anal. Chem.*, 1987, **59**, 617.
[51] C. Lattard and J. L. Rocca, *Analusis*, 1983, **11**, 457.
[52] L. Ebdon, S. J. Hill and P. Jones, *Analyst*, 1985, **110**, 515.
[53] D. T. Burns, F. Glockling and M. Harriott, *Analyst*, 1981, **106**, 921.
[54] A. Y. Cantillo and D. A. Segar, *Heavy Metals in the Environment, Toronto* 1975, pp. 183–204. CEP Consultants, Edinburgh.
[55] T. M. Vickrey, M. S. Buren and H. E. Howell, *Anal. Lett.*, 1978, **12**, 1075.
[56] T. M. Vickrey, H. E. Howell, G. V. Harrison and G. J. Ramelow, *Anal. Chem.*, 1986, **52**, 1743.
[57] F. E. Brinckman, W. R. Blair, K. L. Jewett and W. P. Iverson, *J. Chromatog. Sci.*, 1977, **15**, 493.
[58] F. E. Brinckman, K. L. Jewett, W. P. Iverson, K. J. Irgolic, K. C. Ehrhardt and R. A. Stockton, *J. Chromatog.*, 1980, **191**, 31.
[59] E. J. Parks, F. E. Brinckman and W. R. Blair, *J. Chromatog.*, 1979, **185**, 563.
[60] K. L. Jewett and F. E. Brinckman, *J. Chromatog. Sci.*, 1981, **19**, 583.
[61] R. Pinel, M. Z. Benabdallah, A. Astruc, M. Potin-Gautier and M. Astruc, *Analusis*, 1984, **12**, 344.
[62] D. C. Reamer, C. Veillon and P. T. Tokousbalides, *Anal. Chem.*, 1981, **53**, 245.
[63] O. Donard and Ph. Pedemay, *Anal. Chim. Acta*, 1983, **153**, 301.
[64] S. Rapsomanikis, O. F. X. Donard and J. H. Weber, *Anal. Chem.*, 1986, **58**, 35.
[65] J. P. Quintard and M. Peyrere, *Revs. Si, Ge, Sn, Pb Compounds*, 1980, **4**, 151.
[66] A. Woollins and W. R. Cullen, *Analyst*, 1984, **109**, 1527.
[67] Y. K. Chau, P. T. S. Wong and P. D. Goulden, *Anal. Chim. Acta*, 1976, **85**, 421.
[68] G. J. Olson, F. E. Brinckman and J. A. Jackson, *Int. J. Environ. Anal. Chem.*, 1983, **15**, 249.
[69] W. A. Aue and H. H. Hill, Jr., *J. Chromatog.*, 1972, **70**, 158.
[70] W. A. Aue and C. G. Flinn, *J. Chromatog.*, 1977, **142**, 145.
[71] B. W. Wright, M. L. Lee and G. M. Booth, *J. High Resol. Chromatog. Chromatog. Commun.*, 1979, **2**, 489.
[72] T. Nakahara, *Prog. Anal. Atom. Spectrosc.*, 1983, **6**, 263.
[73] C. J. Soderquist and D. G. Crosby, *Anal. Chem.*, 1978, **50**, 1435.
[74] R. S. Braman and M. A. Tompkins, *Anal. Chem.*, 1978, **50**, 1088.
[75] S. Tugrul, T. I. Balkas, E. D. Goldberg and I. Salihoglu, *VIth Workshop on Marine Pollution of the Mediterranean*, pp. 497–504. ICSEM/IOC/UNEP, Cannes, 1982.
[76] M. O. Andreae, in *Trace Metals in Seawaters*, C. S. Wong, E. Boyle, K. W. Bruland, J. D. Burton and E. D. Goldberg (eds.), pp. 1–19. Plenum Press, New York, 1983.
[77] R. Pinel, I. G. Gandjar, M. Z. Benadallah, A. Astruc and M. Astruc, *Analusis*, 1984, **12**, 404.
[78] J. Dědina, *Anal. Chem.*, 1982, **54**, 2097.
[79] F. D. Pierce and H. R. Brown, *Anal. Chem.*, 1977, **49**, 1417.
[80] O. F. X. Donard, L. Randall, S. Rapsomanikis and J. H. Weber, *Int. J. Environ. Anal. Chem.*, 1986, **27**, 55.
[81] P. J. Craig and S. Rapsomanikis, *Inorg. Chim. Acta*, 1983, **80**, L19.
[82] H. A. Meinema, T. Burger-Wiersma, G. Versluis-de Haan and E. C. Gevers, *Environ. Sci. Technol.*, 1978, **12**, 288.
[83] Y. K. Chau and P. T. S. Wong, *Proc. DOE/NBS Workshop on Environmental Speciation and Monitoring Needs*, pp. 65–80. NBS, Gaithersburg, MD., 1981.
[84] R. J. Maguire, J. H. Carey and E. J. Hale, *J. Agric. Food Chem.*, 1983, **31**, 1060.
[85] R. J. Maguire and R. J. Tkacz, *J. Agric. Food Chem.*, 1985, **33**, 947.
[86] R. J. Maguire, R. J. Tkacz and D. L. Sartor, *J. Great Lakes Res.*, 1985, **11**, 320.
[87] Y. Arakawa, O. Wada, T. H. Yu and H. Iwai, *J. Chromatog.*, 1981, **207**, 237.

[88] D. T. Burns, F. Glockling and M. Harriott, *J. Chromatog.*, 1980, **200**, 305.
[89] G. K. H. Tam, G. Lacroix and J. F. Lawrence, *J. Chromatog.*, 1983, **259**, 350.
[90] F. H. Pollard, G. Nickless and D. J. Cooke, *J. Chromatog.*, 1964, **13**, 48.
[91] L. Ebdon, R. W. Ward and D. A. Leathard, *Analyst*, 1982, **107**, 129.
[92] B. Welz and M. Melcher, *Analyst*, 1983, **108**, 213.
[93] J. Dědina and I. Rubeška, *Spectrochim. Acta*, 1980, **35B**, 119.
[94] J. Fazakas, *Talanta*, 1984, **31**, 573.
[95] R. M. Dagnall, K. C. Thompson and T. S. West, *Analyst*, 1968, **93**, 518.
[96] S. Kapila and C. R. Vogt, *J. Chromatog. Sci.*, 1980, **18**, 144.
[97] W. A. Aue and C. G. Flinn, *Anal. Chem.*, 1980, **52**, 1537.
[98] R. J. Maguire and R. J. Tkacz, *J. Chromatog.*, 1983, **268**, 99.
[99] Y. Arakawa, O. Wada, T. H. Yu and H. Iwai, *J. Chromatog.*, 1981, **216**, 209.
[100] D. R. Jones IV and S. E. Manahan, *Anal. Chem.*, 1976, **48**, 502.
[101] M & T, *Speciation of butyltins in fish tissues by solvent separation and AAS–graphite furnace: Standard Test Methods*, AA-33, M & T Chemical Inc., Rahway, New Jersey, 1979.
[102] J. J. Cleary and A. R. D. Stebbing, *Mar. Pollut. Bul.*, 1985, **16**, 350.
[103] L. E. Hallas and J. J. Cooney, *Appl. Environ. Microbiol.*, 1981, **41**, 446.
[104] G. Neubert and H. Andreas, *Z. Anal. Chem.*, 1976, **280**, 31.
[105] C. S. Weiss, K. L. Jewett, F. E. Brinckman and R. H. Fish, in *Proc. Workshop on Environmental Speciation and Monitoring Needs for Trace Metal-Containing Substances from Energy-Related Processes, Gaithersburg, Maryland*, 1981, pp. 197–210. N.B.S., Washington, D.C., 1981.
[106] T. H. Yu and Y. Arakawa, *J. Chromatog.*, 1983, **258**, 189.
[107] W. Langseth, *Talanta*, 1984, **31**, 975.
[108] W. Langseth, *J. Chromatog.*, 1984, **315**, 351.
[109] L. Jirackova-Audoin, D. Ranceze and J. Verdu, *Analusis*, 1985, **13**, 59.
[110] E. Lundberg, B. Bergmark and W. Frech, *Anal. Chim. Acta*, 1982, **142**, 129.
[111] P. Hocquellet and N. Labeyrie, *Analusis*, 1975, **3**, 505.
[112] P. Hocquellet, *Rev. Franc. Corps Gras*,1984, **31**, 117.
[113] P. Hocquellet, *At. Spectrosc.*, 1985, **6**, 69.
[114] P. Hocquellet and N. Labeyrie, *At. Abs. Newsl.*, 1977, **16**, 124.
[115] H. Fritzsche, W. Wegscheider and G. Knapp, *Talanta*, 1979, **26**, 219.
[116] D. T. Burns, D. Dadgar and M. Harriott, *Analyst*, 1984, **109**, 1099.
[117] Y. Thibaud, *Rev. Trav. Inst. Pêches Marit.*, 1982, **44**, 349.
[118] T. M. Vickrey, H. E. Howell and M. T. Paradise, *Anal. Chem.*, 1979, **51**, 1880.
[119] T. M. Vickrey, G. V. Harrison and G. J. Ramelow, *Anal. Chem.*, 1981, **53**, 1573.
[120] V. Zatka, *Anal. Chem.*, 1978, **50**, 538.
[121] L. Zhou, T. T. Chao and A. L. Meier, *Talanta*, 1984, **31**, 73.
[122] M. Tominaga and Y. Umezaki, *Anal. Chim. Acta*, 1979, **110**, 55.
[123] H. L. Trachman, A. J. Tyberg and P. D. Branigan, *Anal. Chem.*, 1977, **49**, 1090.
[124] K. Ohta and M. Suzuki, *Anal. Chim. Acta*, 1979, **107**, 245.
[125] M. L. Kaiser, S. R. Koirtyohann and E. J. Hinderberger, *Spectrochim. Acta.*, 1981, **36B**, 773.
[126] E. Pruszkowska, D. C. Manning, G. R. Carnrick and W. Slavin, *Atom. Spectrosc.*, 1983, **4**, 87.
[127] R. Pinel, M. Z. Benabdallah, A. Astruc and M. Astruc, *Anal. Chim. Acta*, 1986, **181**, 187.
[128] E. J. Parks, W. R. Blair and F. E. Brinckman, *Talanta*, 1985, **32**, 633.
[129] J. R. Knechtel and J. L. Fraser, *Analyst*, 1978, **103**, 104.
[130] I. B. Peetre and B. E. F. Smith, *Mikrochim. Acta*, 1974, **2**, 301.
[131] J. T. Byrd, *Ph.D. Thesis*, Florida State University, 1984.
[132] O. M. Andreae and P. N. Froelich, Jr., *Tellus*, 1984, **36B**, 101.
[133] S. Tugrul, T. I. Balkas and E. D. Goldberg, *Mar. Poll. Bull.*, 1983, **14**, 297.
[134] S. L. Seidel, V. F. Hodge, and E. D. Goldberg, *Thalas. Jugoslav.*, 1980, **16**, 209.
[135] Y. Hattori, A. Kobayashi, S. Takemoto, K. Takami, Y. Kuge, A. Sugimae and M. Nakamoto, *J. Chromatog.*, 1984, **315**, 341.
[136] C. C. Gilmour, J. H. Tuttle and J. C. Means, *Anal. Chem.*, 1986, **58**, 1848.
[137] P. J. Craig and S. Rapsomanikis, *J. Chem. Soc. Chem. Commun.*, 1982, 114.
[138] P. J. Craig and S. Rapsomanikis, *Environ. Technol. Lett.*, 1984, **5**, 407.
[139] L. Randall, J. S. Han and J. H. Weber, *Environ. Technol. Lett.*, 1986, **7**, 571.
[140] J. H. Weber, O. F. X. Donard, L. Randall and J. S. Han, in *Oceans* '86, Institute of Electrical and Electronic Engineers, Piscataway NJ, 1986.
[141] I. Ishii, *Bull. Japan. Soc. Sci. Fish.*, 1982, **48**, 1609.
[142] O. F. X. Donard, F. T. Short and J. H. Weber, *Can. J. Fish. Aqua. Sci.*, 1986, **44**, 140.
[143] W. Maher, *Anal. Chim. Acta*, 1982, **138,** 365.
[144] R. Pinel, M. Z. Benabdallah, M. Astruc and J. P. Quintard, *Fifth International Conderence on the*

Organometallic and Coordination Chemistry of Germanium, Tin and Lead, Padua (Italy), 1986, p. C41.

[145] R. J. Maguire, Y. K. Chau, G. A. Bengert, E. L. Hale, P. T. S. Wong and O. Kramar, *Environ. Sci. Technol.*, 1982, **16**, 698.

[146] M. O. Andreae, J. T. Byrd and P. N. Froelich, Jr., *Environ. Sci. Technol.*, 1983, **17**, 731.

8

Lead

M. Radojević
Department of Chemistry, University of Manchester Institute of Science and Technology, PO Box 88, Manchester M60 1QD, England

8.1 INTRODUCTION

Environmental pollution by organic lead is almost entirely due to the manufacture and use of tetra-alkyllead (R_4Pb) compounds as petrol additives, and the toxicity of these species is well documented [1]. Varying proportions of the five R_4Pb compounds (tetramethylated, tetraethyllead and their mixed derivaties) are added to petrol to increase the octane rating of fuels for high-compression internal combustion engines. A possible natural source of organic lead may be the environmental methylation of inorganic lead [2]. In the environment, R_4Pb compounds decompose to inorganic lead with trialkyllead (R_3Pb^+) and dialkyllead (R_2Pb^{2+}) compounds as fairly persistent intermediates [3]. There is evidence to suggest the existence of a biogeochemical cycle of organic lead [4]. Organolead compounds which may be encountered in the environment are listed in Table 8.1 and sensitive, species-specific

Table 8.1 — Alkyllead compounds encountered in environmental samples.

R_4Pb	R_3Pb^+	R_2Pb^{2+}	$[RPb^{3+}]$*
Me_4Pb	Me_3Pb^+	Me_2Pb^{2+}	$[MePb^{3+}]$
Me_3EtPb	Me_2EtPb^+	$MeEtPb^{2+}$	$[EtPb^{3+}]$
Me_2Et_2Pb	$MeEt_2Pb^+$	Et_2Pb^{2+}	
$MeEt_3Pb$	Et_3Pb^+		
Et_4Pb			

* Tentatively identified in chromatograms; presence in the environment is improbable.

analytical techniques are required for their determination in a variety of environmental samples, Physicochemical speciation studies of organolead compounds in environmental media are necessary for assessing the impact of man-made pollution and in investigation of the environmental methylation of lead. Concern over lead pollution has resulted in even more stringent limits on the organolead content of petrol in many countries, and environmental monitoring of these species would enable an investigation into the effect of control measures to be made.

Many of the methods capable of measuring organic lead are either not specific or not sensitive enough for environmental application, are generally cumbersome, and may involve several analysis steps in order to yield complete information on the various species [5]. The volatility of alkyllead compounds permits their separation by chromatography, but the use of conventional detectors, such as flame ionization, is precluded by the interference from hydrocarbons present at higher concentrations. Interfacing chromatography with element-specific atomic-absorption spectrometry (AAS) provides one of the most sensitive, specific, and versatile analytical techniques for organolead compounds. Since most environmental laboratories are equipped with some form of chromatographic and AAS equipment the setting up of a chromatography/AAS system can be fairly inexpensive and straightforward.

This chapter is a critical review of chromatography/AAS methods employed for organic lead in environmental analysis, together with the sampling, extraction and derivatization techniques.

8.2 CHROMATOGRAPHIC/ATOMIC SPECTROMETRY TECHNIQUES FOR LEAD DETERMINATION

Experimental chromatography/atomic spectrometry systems which have been applied for the determination of alkyllead compounds are summarized in Table 8.2.

Table 8.2 — Detectors for chromatography employed for lead determination. GC — gas chromatography, HPLC — high-performance liquid chromatography, AAS — atomic-absorption spectrometry, AES — atomic emission spectrometry, FAAS — flame AAS, FAFS — flame atomic fluorescence spectrometry, ETA — electrothermal atomization, SFD — silica furnace detector, ICP — inductively coupled plasma, MPD — microwave plasma detection.

GC	HPLC
FAAS	FAAS
FAFS	ETA–AAS
ETA–AAS	ICP–AES
SFD–FAAS	
SFD–ETA–AAS	
MPD–AES	
ICP–AES	

The following requirements have to be considered when selecting a system for environmental analysis: high specificity because of the complex nature of the samples, high sensitivity because of the low levels in the environment, simplicity of interfacing, and reasonable cost. Of the systems listed in Table 8.2, those which satisfy these requirements have been employed for environmental measurements. If chromatography and atomic-absorption spectrometry are already used in the laboratory, as in the case with most environmental laboratories, a system for organolead analysis can be set up without the purchase of additional instruments. Any necessary modifications to the existing equipment are simple to make, and specialized detector cells are easy to construct. Detectors for chromatography employed in organometallic analysis have been discussed in Chapters 2–5 and those systems which have been specifically applied to lead compounds will be described here. Derivatization techniques used in the determination of R_3Pb^+ and R_2Pb^{2+} compounds, which are less volatile than R_4Pb species, will also be outlined.

8.2.1 High-performance liquid chromatography/atomic spectrometry

This technique has not been widely used for organolead analysis and the reported studies are summarized in Table 8.3. HPLC has been interfaced with various atomic spectrometry detectors and these have been dealt with in Chapter 6. The effluent from the HPLC can readily be introduced into the nebulizer of an FAAS instrument but a major problem with this technique is the need for a higher efficiency of nebulization to obtain optimum sensitivity. HPLC interfaced to ETA–AAS does not suffer from this disadvantage but the analysis may be exceedingly slow because the graphite-furnace AAS instrument accepts only discrete liquid droplets, at a rate of a single drop every two minutes. The effluent from the HPLC can be stored in an autosampler or the chromatograph may be operated in a stopped-flow mode. Electrothermal atomization has been achieved by using GFAAS with post-column digestion [10] and by using a high-temperature furnace in the optical path of a Zeeman atomic-absorption spectrometer [9]. Complete speciation of the five R_4Pb compounds was achieved in less than 20 min by HPLC/ICP–AES [12], and in about 4 min with reversed-phase HPLC/FAAS [11]. Reversed phase HPLC/ICP–AES has been employed for the speciation of R_4Pb compounds, and ion-exchange HPLC/ICP–AES in analysis for R_3Pb^+ species. Nebulization presents a problem with ICP–AES detection and the plasma exhibits a low tolerance for organic solvents. Two different nebulizers, a number of different columns, and several mobile phases were investigated by Ibrahim *et al.* [14]. Several mobile phases were also investigated in a study employing HPLC/FAAS [13]. Et_2PbCl_2, Me_3PbOAc, Et_3PbCl and $(C_6H_5)_3PbOAc$ (where OAc is acetate) were successfully separated and separation of Et_4Pb could be achieved by switching to a different solvent.

From the detection limits quoted in Table 8.3 it is apparent that at present HPLC/atomic spectrometry techniques are not sufficiently sensitive for general environmental analysis and they have only been successfully applied to the determination of R_4Pb compounds in petrol. In a comparative study it was found that HPLC/ICP–AES is less sensitive to R_3Pb^+ species than to R_4Pb compounds [14]. Furthermore, HPLC/ICP–AES is not readily available in most environmental laboratories and a major disadvantage of this technique is its cost.

8.2.2 Gas chromatography/plasma atomic emission spectrometry

The microwave-induced plasma is suitable for the analysis of gases but not of liquids, because of the high excitation temperature compared with the gas temperature. MPD has therefore been employed as a detector for gas chromatography and the analytical systems which have been applied to organolead analysis are summarized in Table 8.4. MPD is considered to be the cheapest of all the analytical plasmas, but it is not readily available in most environmental laboratories and has not been used

Table 8.3 — HPLC coupled with atomic spectrometry detectors (NR — not reported, GFAAS — graphite furnace AAS, ZAAS — Zeeman AAS).

Detector	Compound	Detection limit (Pb, ng)	Column	Mobile phase	Flow- (ml/min)	Application	Reference
FAAS	Me_4Pb Et_4Pb	NR NR	ODS	3:2 v/v H_2O/CH_3OH	1	Gasoline	[6]
GFAAS	Ph_6Pb_2	42	Lichrosorb SI-100 10 μm silica	Hexane–methylene chloride (95:5)	0.33	NR	[7]
ICP	Pb	(a)	NR	75% Ethanol	1.4	Gasoline	[8]
ZAAS	Me_4Pb Et_4Pb Ph_4Pb	NR NR NR	Hitachi	Methanol	0.67	NBS gasoline standard	[9]
GFAAS	Me_4Pb Me_3EtPb Me_2Et_2Pb $MeEt_3Pb$ Et_4Pb	NR NR NR NR NR	Lichrosorb 10 μm C-18 ODS	80:20 CH_3OH/H_2O for 28 ml followed by a step gradient to 100% methanol	0.5	NR	[10]
FAAS	Me_4Pb Me_3EtPb Me_2Et_2Pb $MeEt_3Pb$ Et_4Pb	(~10) (~10) (~10) (~10) (~10)	μ Bondopak C_{18}	Acetonitrile– water (70:30)	3	Gasoline, waste lubricating oils, leachates of potentially hazardous wastes	[11]
ICP	Me_4Pb Et_4Pb	2 11	10 μm C_2	Butanol–ethanol– water (15:35:50)	1	Gasoline	[12]
FAAS	Et_2PbCl_2 Me_3PbOAc Et_3PbCl Ph_3PbOAc	NR NR NR NR	Whatman Partisil- 10 Reversed phase C-18, ODS-2	55 Acetonitrile– 45 water — 0.1*M* NaCl	NR	Laboratory experiments	[13]
	Et_4Pb	NR		80 Acetonitrile– 20 methanol			
ICP	Me_4Pb Et_4Pb	2 11	C_2 Partisil-10 SCX	n-Butanol–ethanol– water (15:35:50)	1	NR	[14]
	Me_3PbCl Et_3PbOAc Pr_3PbOAc Bu_3PbOAc	NR NR NR NR		n-Butanol–ethanol– water (15:35:50) with 0.5*M* ammonium acetate	1.5		

(a) d.l. of 10.9 μg/l. Pb^{2+} in aqueous solution.

Table 8.4 — Gas chromatography coupled with plasma emission spectrometry. MPD — microwave plasma detection, ICP — inductively coupled plasma, DCP — direct-current plasma.

Detector	Compound	Detection limit (Pb, ng)	Column	Column T (°C)	Carrier flow (ml/min)	Injector T (°C)	Interface T (°C)	Application	Reference
MPD	Et_4Pb	NR	2.5% Dexsil 300 on Chromosorb 750	170	50 (He)	170	170	NR	[15]
MPD	Me_4Pb Me_3EtPb Me_2Et_2Pb $MeEt_3Pb$ Et_4Pb	0.006 0.010 0.023 0.035 0.040	3% OV-1 on Chromosorb W	80	22 (Ar, 1% H_2)	130	NR	Air	[16]
ICP	Me_4Pb Et_4Pb	NR NR	SP 1000	150	30 (N_2)	NR	200	Gasoline	[17]
DCP	Et_4Pb Et_3PrPb Et_2Pr_2Pb $EtPr_3Pb$ Pr_4Pb	NR NR NR NR NR	5% OV-101 on Chromosorb 750, and 3% OV-210 on Ultrabond 20M	From 80 at 8°C/min	40 (He)	210	220	Laboratory studies	[18]
MPD	Me_3PbCl, Et_3PbCl	(a)	SP-2100 fused silica WCOT capillary column, and OV-100 SCOT glass capillary column	140–160	1 (He)	180	180	Tap water	[19]
MPD	Me_3BuPb, Et_3BuPb	(b)	SP-2100 fused silica WCOT capillary column	Room T for 1 min, then 70–270 at 20°C/min	1 (He)	NR	250	Industrial plant effluent	[20]

(a) d.l. of 10–30 μg/l. Et_3PbCl in aqueous solution.
(b) d.l. of 35 μg/l. Me_3PbCl and 5.6 μg/l. Et_3PbCl in aqueous solution.

extensively in field studies. In this technique compounds eluted from the GC are directed into a microwave discharge sustained in helium or argon. There is a limit to the amount of organic material which can be introduced into the plasma per unit time, as in most cases the injected solvent will extinguish the plasma, and/or result in carbon deposition on the walls of the atom cell. Hence, interfacing is not as straightforward as with some of the other techniques and the interface generally incorporates a high-temperature valve inside an oven, which permits venting of the solvent away from the plasma. Absolute detection limits for the R_4Pb compounds with GC/MPD are low enough to permit environmental analysis and compare favourably with detection limits based on GC/AAS, which are shown in Tables 8.5–8.8. Me_3PbCl and Et_3PbCl have been determined by using GC/MPD after extraction of aqueous solutions with 50 ml of benzene in the presence of sodium chloride and at pH between 5 and 7, and low-temperature evaporation of the extract to between 0.25 and 4 ml [19]. The complete procedure took over 2 hr, and an efficiency of 50% was found for the recovery of Et_3PbCl. The efficiency for Me_3PbCl was suspected to be even lower. A detection limit in the range 10–30 μg/l was reported for Et_3PbCl in tapwater. GC/MPD has also been applied to the determination of R_3Pb^+ compounds after extraction into benzene and butylation with n-butylmagnesium chloride [20]. The resulting R_3BuPb derivatives were enriched on a precolumn trap containing Tenax-GC. The solvent could be volatilized from the trap while the higher boiling point R_3BuPb species were retained. This precolumn enrichment trap was inserted into a specially adapted injector port and the trapped R_3BuPb compounds were purged from the Tenax into the GC/MPD system for analysis. This technique has been applied to an industrial plant effluent. In another study approximately 1% of hydrogen was added to the argon carrier gas to prevent formation of lead deposits on the walls of the silica capillary, and the technique was applied to R_4Pb compounds in petrol and air samples [16].

8.2.3 Gas chromatography/atomic-absorption spectrometry

GC/AAS has been the most widely used chromatography/atomic spectrometry combination for the specific determination of alkyllead compounds [42]. The instrumentation is relatively inexpensive and readily available in most environmental laboratories, and the interfacing is relatively simple as the GC effluent can be readily introduced into the atomic-absorption spectrometer. Most of the experimental systems employ commercially available gas chromatographs with a heated injection port and a temperature-programmable oven containing a packed GC column. Custom-built GC columns have also been interfaced to AAS detectors and these are generally employed for the cold-trapping of volatile organolead compounds, which are then thermally desorbed into the AAS.

The first GC/AAS system employed flame atomic-absorption spectrometry (FAAS) with an air–acetylene flame and the interfacing was achieved by introducing the GC effluent directly into the AAS nebulizer [21]. Dilution effects can be reduced by introducing the GC effluent directly into the burner head [23] and this interface results in decreased peak broadening and enhanced sensitivity [28]. More recently, flameless electrothermal atomization atomic-absorption spectrometry (ETA–AAS) techniques have been utilized. These have used either graphite-furnace atomic-absorption spectrometry (GFAAS) or custom-built electrothermal detector cells.

Table 8.5 — Gas chromatography coupled with conventional flame AAS.

Compound	Detection limit (ng Pb)	Column	Column T (°C)	Carrier flow (ml/min)	Injector T (°C)	Transfer line T (°C)	Detector T (°C)	Application	Reference
Me_4Pb, Et_4Pb	NR NR	10% Apiezon M on Chromosorb R	100–200 at 20°C/min	40 (N_2)	190	NR	NR	Gasoline	[21]
Me_4Pb, Me_3EtPb, Me_2Et_2Pb, $MeEt_3Pb$, Et_4Pb	NR NR NR NR NR	25% TCP on Chromosorb WAW	110	40 (N_2)	NR	NR	NR	Gasoline	[22]
Me_4Pb Et_4Pb	20 20	10% PEG 20M (Carbowax) on Porasil	130	120 (H_2)	NR	Room T	NR	Gasoline	[23]
Me_4Pb Me_3EtPb Me_2Et_2Pb $MeEt_3Pb$ Et_4Pb	80 80 80 80 80	3% OV-1 on Chromosorb W	40 for 2 min then to 90 at 5°C/min	65 (N_2)	NR	Room T	NR	Gasoline	[24]
Me_4Pb Et_4Pb	17 81	20% SE-52 on Chromosorb W	125	90 (Ar)	125	~130	NR	Gasoline	[25]
Me_4Pb Et_4Pb	10–20 10–20	Poropak Q	200	50 (H_2)	NR	Room T	NR	Gasoline	[26]
Me_3EtPb	1.5	3% OV-101 on Chromosorb W	50–200 at 40°C/min	140 (N_2)	NR	NR	2400	NR	[27]
Me_4Pb Et_4Pb	1 2	5% Carbowax 20M on Chromosorb 750	175	25 (N_2)	NR	NR	NR	NR	[28]
Me_4Pb Me_3EtPb Me_2Et_2Pb, $MeEt_3Pb$, Et_4Pb	20 20 20 20 20	20% 1.2.3-tris-(2-cyanoethoxy)-propane on Chromosorb P coated with 1% KOH	90	10–200 (N_2)	85	NR	NR	Gasoline	[29]

Table 8.6 — Gas chromatography coupled with conventional electrothermal AAS.

Compound	Detection limit (Pb, ng)	Column	Column T (°C)	Carrier flow (m/min)	Injector T (°C)	Transfer line T (°C)	Detector T (°C)	Application	Reference
(i) *Graphite Furnace*									
Me_3EtPb	10	4% SE-301 6% OV-210 on Gas chrom Q	150	50 (Ar)	NR	Rom T	1700	Gasoline	[30]
Me_4Pb Et_4Pb	0.12 1.1	20% SE-52 on Chromosorb W	125	90 (Ar)	125	130	1300	Gasoline	[25]
Me_3EtPb	0.03	3% OV-101 on Chromosorb W	50–200 at 40°C/min	140 (N_2)	NR	NR	1500	NR	[27]
Me_4Pb Me_3EtPb Me_2Et_2Pb $MeEt_3Pb$ Et_4Pb	0.04 0.06 0.06 0.07 0.09	3% OV-101 on Gaschrom Q	50–175 at 20°C/min	30 (Ar)	150	200	2000	Gasoline	[31]
Me_4Pb Me_3EtPb Me_2Et_2Pb $MeEt_3Pb$ Et_4Pb	0.04 0.04 0.04 0.04 0.04	3% OV-101 on Chromosorb W	90–200 at 40°C/min	140 (N_2)	NR	80	1500	Gasoline, car exhaust, air, laboratory studies	[32]
(ii) *Carbon Rod Atomizer*									
Me_4Pb	0.037	Tricresyl phosphate on Chromosorb W	95	150 (Ar)	100	95	2000	Gasoline	[33]
Me_4Pb Me_3EtPb Me_2Et_2Pb $MeEt_3Pb$ Et_4Pb	0.1 0.1 0.1 0.1 0.1	20% TCP on Chromosorb W	100	30 (Ar)	125	100	2000	Gasoline	[34]
Me_3EtPb	0.07	3% OV-101 on Chromosorb W	50–200 on 40°C/min	140 (N_2)	NR	NR	1500	NR	[27]

Table 8.7 — Gas chromatography coupled with silica furnace flame AAS.

Compound	Detection limit (Pb, ng)	Column	Column T (°C)	Carrier flow (ml/min)	Injector T (°C)	Transfer line T (°C)	Detector T (°C)	Application	Reference
Me_3EtPb	0.09	3% OV-101 on Chromosorb W	50–200 at 40°C/min	140 (N_2)	NR	NR	950	NR	[27]
Me_4Pb Et_4Pb	0.017 0.017	5% Carbowax 20M on Chromosorb 750	159	64 (N_2)	NR	NR	NR	Gasoline, simulated arson samples	[28]
Me_4Pb Me_3EtPb Me_2Et_2Pb $MeEt_3Pb$ Et_4Pb	0.023 0.026 0.031 0.038 0.053	1,2,3-tris-(2-cyanoethoxy)-propane on Chromosorb W	45–145 at 20°C/min	30 (He)	NR	Room T	950	Air, water, fish, laboratory studies	[35]
Me_4Pb Et_4Pb	0.02 0.03	3% OV-101 on Gaschrom Q	35–135 at 20°C/min	165 (He)	NR	150	950	Air	[36]
Me_4Pb Me_3EtPb Me_2Et_2Pb $MeEt_3Pb$ Et_4Pb	0.012 0.017 0.021 0.020 0.025	3% OV-101 on Gaschrom Q	40–160 20°C/min	167 (He)	150	160	950	Rainwater, motorway run-off water, street dust, atmospheric aerosol	[37]
Me_4Pb Me_3EtPb Me_2Et_2Pb $MeEt_3Pb$ Et_4Pb Me_3BuPb Me_2Bu_2Pb Et_3BuPb Et_2Bu_2Pb	0.04 0.05 0.05 0.07 0.07 0.05 0.08 0.09 0.10	10% OV-101 on Chromosorb W	50–150 at 10°C/min	40 (Ar)	150	140	NR	Rainwater	[38]

Table 8.8 — Gas chromatography coupled with silica furnace electrothermal AAS.

Compound	Detection limit (Pb, ng)	Column	Column T (°C)	Carrier flow (ml/min)	Injector T (°C)	Transfer line T (°C)	Detector T (°C)	Application	Reference
Me_4Pb	0.1	3% OV-101 on Chromosorb W	50 for 2 min then to 150 at 15°C/min	70 (N_2)	150	Room T	1000	NR	[24]
Me_3EtPb	0.1								
Me_2Et_2Pb	0.1								
$MeEt_3Pb$	0.1								
Et_4Pb	0.1								
Me_3EtPb	0.1	3% OV-101 on Chromosorb W	50–200 at 40°C/min	140 (N_2)	NR	NR	1000	NR	[27]
Me_4Pb	5	Poropak Q	235	150 (N_2)	NR	Room T	980	EPA control samples	[39]
Me_3EtPb	0.0087	10% SP-2100 on Chromosorb HP	−90 to 60 in 5 min	102 (He)	NR	Room T	950	NR	[40]
Me_2Et_2Pb	0.0105								
Me_4Pb	0.01	10% OV-101 on Chromosorb W	80–180 at 20°C/min	120 (N_2)	150	120	950	Environmental samples, laboratory studies	[41]
Me_3EtPb	0.01								
Me_2Et_2Pb	0.02								
$MeEt_3Pb$	0.03								
Et_4Pb	0.03								
Me_3PrPb	0.02								
Me_2Pr_2Pb	0.03								
Et_3PrPb	0.05								
Et_2Pr_2Pb	0.08								
Pr_4Pb	0.27								
Me_3BuPb	0.03								
Me_2Bu_2Pb	0.04								
Et_3BuPb	0.05								
Et_2Bu_2Pb	0.11								

Various detector cells have also been employed in GC/FAAS and these offer considerable improvement over conventional flame detectors.

GC/AAS systems which have been employed for the determination of R_4Pb compounds are summarized in Tables 8.5–8.8 together with the detection limits, relevant experimental conditions, and applications. Glass and stainless-steel columns have been used and the transfer lines are generally heated. GC ovens have been operated in isothermal and temperature-programmed modes. The greatest sensitivity has been obtained with systems employing a silica furnace detector (SFD) which has been either electrothermally heated or suspended in an air–acetylene flame above the burner head and in the path of the light-beam. Ebdon *et al.* [28] used several different detector cells as well as the conventional flame technique for the determination of R_4Pb compounds by GC/AAS. Best results were obtained when the GC effluent was mixed with hydrogen and introduced through a hole in the side of a ceramic detector tube manufactured from recrystallized alumina. More recent studies have confirmed that a hydrogen diffusion flame burnt inside the tube greatly improves the atomization efficiency [40,41,43]. Forsyth and Marshall [43] ascribed this effect to hydrogen radicals resulting from atomization of the hydrogen in the furnace. They proposed that these radicals are involved in the formation and atomization of lead hydride. They found that in the absence of hydrogen an appreciable portion of the lead which was not atomized when alkyllead entered the furnace was deposited on the walls of the silica tube. If hydrogen was then admitted to the furnace these deposits were rapidly revolatilized as lead hydride. The air–acetylene flame served mainly to heat the ceramic tube, which was used to increase the atom residence times. Optimum dimensions of the detector tubes have to be found for each particular AAS instrument. T-shaped silica tubes have been wrapped with heating wire to serve as electrothermal detector cells. The development of AAS detectors for organolead analysis is discussed in Chapter 2. GC/AAS systems have been successfully applied in environmental analysis, when equipped with a silica furnace detector. GC/SFD–AAS is the most sensitive and therefore the most appropriate technique for environmental studies [44,45] and for this reason the GC/SFD–ETA–AAS system will be discussed in greater detail.

Figure 8.1 illustrates the GC/AAS system at Essex University, which has been employed for the complete speciation of alkyllead in environmental samples. The system consists of three main instrumental components: a Perkin-Elmer F17 gas chromatograph, a Perkin-Elmer 305 atomic-absorption spectrometer and a GN concentrator. The gas chromatograph was interfaced to the electrochemically heated silica-furnace detector cell, and the GN concentrator could be interfaced to the gas chromatograph by a Swagelok fitting on the end of the interface tube. A heated injector port could also be interfaced to the GN concentrator for injecting volatile R_4Pb standards dissolved in n-hexane. The GC/AAS system was optimized in the usual way [36] with R_4Pb and alkylated R_3Pb^+ standards, and the operating conditions selected for environmental analysis are given in Table 8.9. The change in instrument response to alkyllead species was noted when the hydrogen and carrier gas flow-rates, the GC temperature programme and the temperature of the injection port and transfer line were varied. The conditions selected for routine analysis were those which gave the highest sensitivity with best peak resolution. The reproducibility of the system was studied by injecting seven aliquots of a standard containing 1 ng

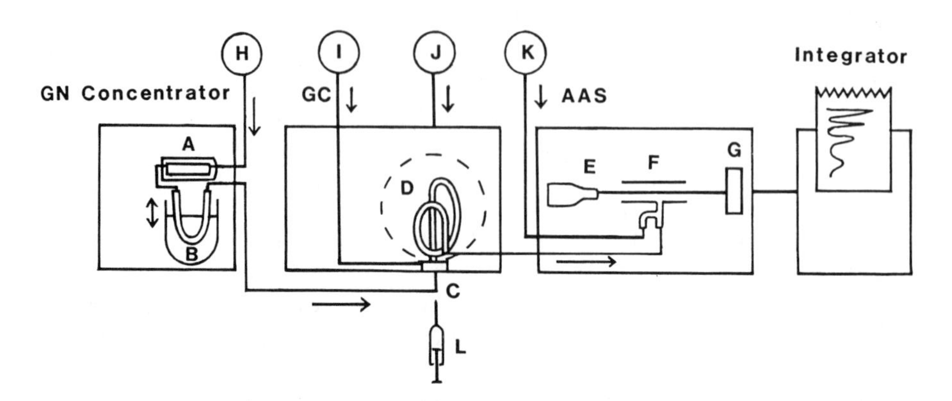

Fig. 8.1 — GC/AAS system at Essex University. A, Adsorption tube in oven. B, Cryogenic trap. C, Injection port. D, GC column inside oven. E, Pb hollow cathode lamp. F, Electrothermally heated silica tube. G, Monochromator. H, Helium. I, Nitrogen. J, Air. K, Hydrogen. L, Manual injection. Details of instrumentation and operating conditions are given in Table 8.9.

Table 8.9 — Operating conditions of GC/AAS system at Essex University.

Perkin-Elmer F17 gas chromatograph
- Column: silanized 1 m long glass column (6 mm o.d. × 2 mm i.d.) packed with 10% OV-101 on Chromosorb W (80–100 mesh)
- Injector temperature: 150°C
- Temperature programme of oven: 80–180°C at 20°C/min
- Carrier gas flow-rate: 120 ml/min

Perkin-Elmer 305 atomic-absorption spectrophotometer
- Hollow-cathode lamp: 8 mA; 283.3 nm Pb line
- Slit setting: 3
- Scale expansion: 3
- Deuterium arc background corrector

Electrochemical detector cell
- Silica tube 16 cm long (12 mm diameter) wrapped with 22 gauge nichrome wire, insulated with a porcelain sleeve and glass fibre, controlled by a Variac transformer at 40 V and 300 W.
- Temperature inside cell: 950°C
- Hydrogen flow-rate: 80 ml/min

GN concentrator (GN Instrumentation Consultancy)
- Packing: ground glass (250/500 μm)
- Helium flow-rate: 187 ml/min

Transfer lines: PTFE tubing (1.6 mm i.d.) heated to 120°C

Rikadenki Mitsui chart recroder:
- Scale expansion: 10 mV
- Chart speed: 2 cm/min

Shimadzu integrator

of lead, for each R_4Pb compound, and recording the peak heights. The relative standard deviations were between 7 and 12%. Calibration data were obtained by measurement of a series of standards containing alkyllead at between 0.1 and 1 ng (as Pb). A linear response was found. One and two metre long columns were used and were silanized as follows. The columns were filled with a 5% v/v solution of dichlorodimethylsilane, $(CH_3)_2SiCl_2$, and left for 24 hr, after which they were emptied out and rinsed with methanol. The columns were conditioned by heating at 180°C for 2 hr. A more sophisticated method was employed by Rapsomanikis *et al.* [40] who optimized their custom-built GC/AAS system by means of a computer program based on a Simplex algorithm.

A typical GC/AAS trace of R_4Pb and ionic propylated alkyllead compounds is shown in Fig. 8.2. Detection limits were estimated to be in the range from 0.01 ng of Pb for Me_4Pb to 0.08 ng of Pb for Et_2Pr_2Pb, for peak-height measurements and definition of the detection limit as 3 times the standard deviation of the baseline noise, divided by the sensitivity. For the five R_4Pb compounds (R = Me, Et or their combinations), it was found that equal quantities of lead gave equal peak areas (±4%) and peak areas were employed when analysing environmental samples. Both helium and nitrogen were employed as GC carrier gases and no difference in system performance was noted. When analysing some polluted samples is was necessary to hold the final temperature of the GC oven for 8 min in order to elute alkylated inorganic Pb^{2+} completely. Signals for n-hexane (the solvent) and other organic compounds may appear on the GC/AAS trace, and these may be eliminated by use of the background corrector. When environmental samples were analysed the background corrector was switched on. Up to 50 μl of n-hexane could be tolerated, and gave no interference with the alkyllead peaks. Longer columns could be used to improve peak separation, but this resulted in longer retention and analysis times and longer periods of column clean-up to remove Pr_4Pb. Environmental samples were quantified by comparison with external R_4Pb standards, internal standardization and standard-addition. Standard-addition is the most reliable method for positive identification of peaks, from the increase in peak height when a particular standard is added to the sample extract. The experimental system at Essex University has been applied to environmental analysis, viz. analysis of rainwater [46], atmospheric gases and aerosols [47], surface and potable waters [48] and dust, sediment and soil samples [49].

8.2.4 Derivatization techniques

Trialkyllead (R_3Pb^+) compounds may be separated by gas chromatography and detected with electron-capture [50], flame-ionization [51], and microwave-plasma detectors [52, 53] but with a sensitivity inadequate for environmental analysis. Estes *et al.* [19] have determined Me_3PbCl and Et_3PbCl by GC/MPD and report detection limits in the range 10–30 μg/l, for Et_3PbCl. Robinson *et al.* [53] have determined Et_3PbCl by direct injection of aqueous solutions into a GC/AAS system, and Chau and Wong [52] analysed the extract by GC/AAS. They reported that the sensitivity was much less for the ionic alkyllead species then for R_4Pb compounds. It is, however, possible to perform sensitive analysis for R_3Pb^+ and R_2Pb^{2+} species by GC/AAS after derivatization to render the compounds more volatile. Derivatization techniques employed so far have consisted of: (a) hydride generation, (b) extraction

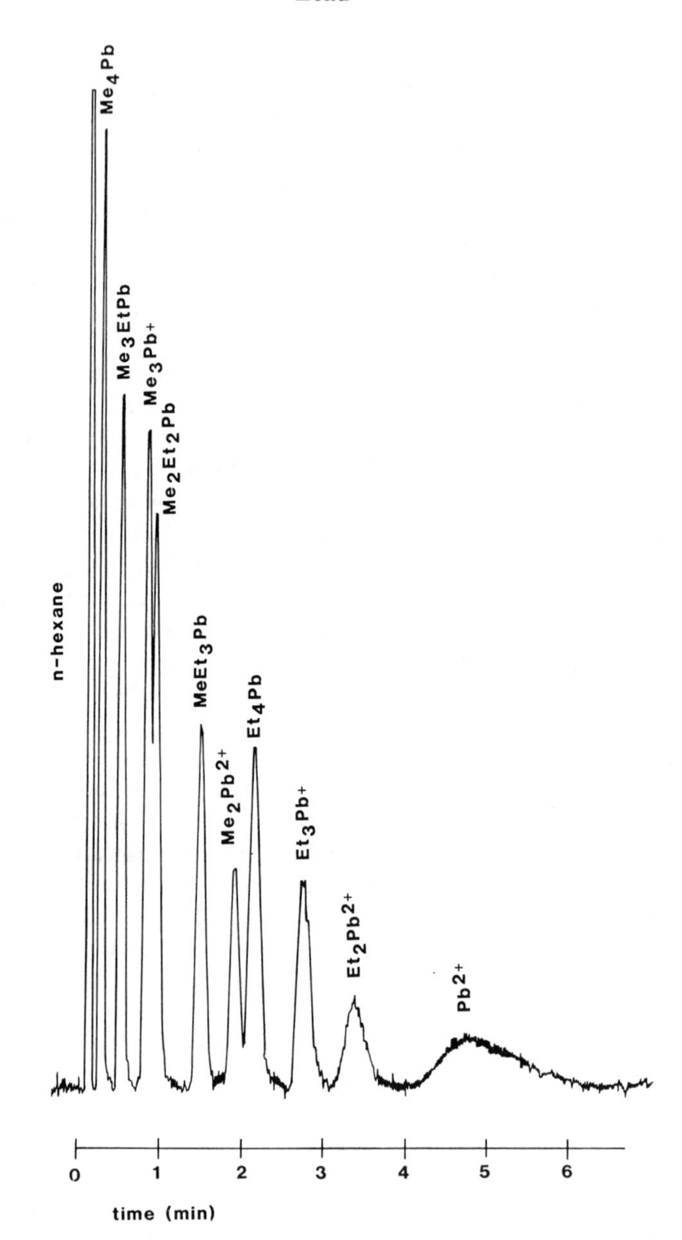

Fig. 8.2 — GC/AAS trace of propylated R_4Pb, R_3Pb^+ and R_2Pb^{2+} compounds without background correction.

into an organic solvent and derivatization with an alkylating agent, and (c) *in situ* alkylation in the aqueous sample, followed by head-space analysis.

Yamauchi *et al.* [54] have developed a hydride generation/SFD–ETA–AAS technique for the determination of Et_3Pb^+, Et_2Pb^{2+} and Pb^{2+} in urine. Samples and standards were reacted with sodium borohydride in a reaction vessel and the gaseous lead hydrides collected on silica wool in a U-trap cooled to liquid-nitrogen temperature. The trap was then warmed to room temperature and the hydrides were passed into a silica cell (heated electrothermcally to 1000°C) placed in the path of the light-beam of an atomic-absorption spectrometer. Recoveries of spikes added to urine were >90% for all the species and detection limits of 5 ng of Pb for Et_3Pb^+ and Et_2Pb^{2+} and 0.1 μg for Pb^{2+} were reported. However, these species could not be determined simultaneously, as different sample preparation techniques were required for each compound.

The most widely used technique for the preparation of R_3Pb^+ and R_2Pb^{2+} species for GC/AAS analysis has been extraction into an organic solvent, followed by alkylation to the volatile R_4Pb form by use of Grignard reagents: n-butylmagnesium chloride [44], propylmagnesium chloride [41] and phenylmagnesium chloride [43]. Alkyllead ions are generally extracted from aqueous solution into an organic solvent in the presence of sodium chloride and sodium diethyldithiocarbamate (NaDDTC) and alkylated by addition of the Grignard reagent to the extracts. Recoveries of ionic alkyllead species are summarized in Table 8.10 and represent the extraction/alkylation efficiences. R_4Pb compounds are unaffected by the derivatization.

Table 8.10 — Recoveries (extraction/alkylation efficiencies) in %. Number of replicates is given in brackets.

Reference	Me_3Pb^+	Et_3Pb^+	Me_2Pb^{2+}	Et_2Pb^{2+}	Pb^{2+}
(i) *Butylation*					
[18]	95±8(NR)	95±8(NR)	—	—	—
[44]*	100±7(2)	98±6(2)	98±5(2)	97±6(2)	93±5(2)
[38]†	105±10(10)	102±7(10)	90±8(10)	108±8(10)	—
[37]	90(1)	100(1)	30(1)	15(1)	—
[55](a)*	90±3(3)	100±5(3)	29±4(3)	—	—
(b)*	103±7(3)	105±7)3)	45±12(3)	—	—
[41]†	92±12(5)	95±9(5)	32±3(3)	18±4(3)	—
(ii) *Propylation*					
[41]†	104±5(5)	100±8(5)	95±10(5)	98±12(5)	6(1)

(a)Method of Harrison and Radojević [37].
(b)Method of Chakraborti *et al.* [38].
* Mean and average deviation from mean.
† Mean and standard deviation from mean.

High recoveries of R_3Pb^+ species by the butylation method have been found by all workers; however, there is disagreement regarding the recovery of R_2Pb^{2+} compounds. Chau *et al.* [44] and Chakraborti *et al.* [38] both found almost quantitative recoveries of R_2Pb^{2+} species, but in three other studies low recoveries were found for similar procedures [37,41,55]. Chau *et al.* [44] extracted the ionic alkyllead species into benzene, whereas Chakraborti *et al.* [38] employed pentane as the solvent, and evaporated it prior to the butylation. One ml of n-butylmagnesium chloride reagent was added and the alkylated species were extracted into 250 μl of nonane for analysis. This method is not suitable for the simultaneous determination of R_4Pb, R_3Pb^+ and R_2Pb^{2+} species in one analysis because the volatile R_4Pb

compounds are lost during the rotary evaporation. In the studies which found inefficient extraction/butylation of R_2Pb^{2+} species, n-hexane was employed as the solvent. High recoveries were, however, found for the R_2Pb^{2+} species when propylation rather than butylation was used [41]. Propylation results in R_4Pb compounds which are more volatile than the corresponding butylated species and it is therefore the preferred method for gas chromatographic separation. Radojević *et al.* [41] compared the propylation and butylation techniques using the same extraction procedure and analytical instrumentation, and from their results it would appear that the low recovery of R_2Pb^{2+} species with the butylation method may be due to less efficient alkylation of these compounds than with the propylation reagent, because the extraction efficiency was the same in all tests. R_3Pb^+ and R_2Pb^{2+} compounds have also been phenylated to the R_4Pb form and the derivatives determined by GC/SFD–ETA–AAS but this technique has not been applied to environmental samples [43].

Chau *et al.* [44] and Estes *et al.* [20] both recommend that aqueous solutions be adjusted to pH 7 before extraction. Chakraborti *et al.* [38] investigated the effect of the solution pH on the extraction of R_3Pb^+ and R_2Pb^{2+} species and found that the most efficient extraction of all the ions takes place at pH 9. Harrison and Radojević [37] found that varying the pH between 4.3 and 7 had no significant effect on the extraction of ionic alkyllead species in the presence of sodium chloride and NaDDTC. Inorganic Pb^{2+} is also extracted and alkylated to the R_4Pb form and with many environmental samples this may result in an intense peak near the end of the chromatographic analysis, which may require a long elution time before the instrument response returns to the normal base line. This is more of a problem with the butylation than the propylation technique because Bu_4Pb is considerably less volatile than Pr_4Pb, and therefore takes longer to be eluted. The coextraction of inorganic Pb^{2+} ions may be prevented by adding EDTA to the solutions before the extraction [38].

Radojević *et al.* [41] determined alkyllead ions according to the following procedure. Five ml of 0.5*M* NaDDTC, 5 g of sodium chloride and 5 ml of n-hexane were added to 1 litre of solution containing the ionic alkyllead compounds either alone or in combination with R_4Pb and Pb^{2+}. The mixture was shaken for 30 min. NaDDTC forms a complex with Pb(II) which may be extracted by organic solvents [56] and this also applies to R_3Pb^+ and R_2Pb^{2+} ions. The organic and aqueous phases are separated and the hexane phase is transferred to a small glass vial. Then 0.5 ml of Grignard reagent (either 2*M* propylmagnesium chloride in diethyl ether or 2.18*M* n-butylmagnesium chloride in tetrahydrofuran) is added and the mixture is gently shaken for 8 min. This results in conversion of the ionic alkyllead species into the tetra-alkyllead form:

$$R_3Pb^+ + R^1MgCl \rightarrow R_3R^1Pb + MgCl^+$$

$$R_2Pb^{2+} + 2R^1MgCl \rightarrow R_2R^1_2Pb + 2MgCl^+$$

where R may be methyl (CH_3) or ethyl (C_2H_5), and R^1 may be propyl (C_3H_7) or butyl (C_4H_9). The extract is washed with 5 ml of 0.5*M* sulphuric acid to destroy any excess of Grignard reagent. The phases are separated and the extract is dried with anhydrous sodium sulphate. Up to 50 μl of the extract may be injected into the GC/AAS system for analysis. Absolute detection limits for this technique [41] are given in Table 8.8. The extract may also be concentrated by purging with nitrogen in the case of environmental samples containing particularly low concentrations of these species. Concentration efficiencies increase with decreasing volatility of the alkyllead species, from 50% for Me_4Pb to 81% for Et_2Pr_2Pb. By use of the propylation technique, concentrating the extract, and analysing with GC/SFD–ETA–AAS, the following detection limits for alkyllead ions in aqueous solution could be achieved (ng/l. Pb): Me_3Pb^+ (0.3), Et_3Pb^+ (0.6), Me_2Pb^{2+} (0.4) and Et_2Pb^{2+} (1.0) and these are superior to those reported by other workers. Somewhat poorer detection limits may be obtained by using the butylation procedure and otherwise identical analytical techniques and detection systems [41]. For analysis by butylation followed by GC/MPD–AES, detection limits of 35 and 5.6 μg/l, for Me_3PbCl and Et_3PbCl have been reported [20]. Chau *et al.* [44] reported detection limits of 0.1 μg/l. Pb for the individual ionic alkyllead species, by butylation followed by GC/SFD–ETA–AAS, and an absolute detection limit of 0.1 ng of Pb. For the butylation followed by GC/SFD–FAAS, Chakraborti *et al.* [38] reported detection limits of 1.25 ng/l. Pb for Me_3Pb^+ and 2.5 for Et_2Pb^{2+}. Absolute detection limits of 0.05 and 0.1 ng of Pb were quoted for these two species respectively. With the butylation technique care should be taken in interpretation of the peaks due to Me_2Bu_2Pb and Et_3BuPb, which are poorly separated. With the propylation technique this is not a problem. The propylation technique offers better recoveries of R_2Pb^{2+} species, a shorter analysis time and better resolution of some peaks, and the ionic alkyllead species may be determined simultaneously with R_4Pb compounds, as shown in Fig. 8.2. The derivatization techniques described here have been employed extensively in the analysis of environmental samples.

A promising technique which has not yet been fully developed for environmental analysis involves alkylation of the R_3Pb^+ and R_2Pb^{2+} compounds to the volatile R_4Pb form *in situ* in aqueous solutions. Me_3Pb^+ and Me_2Pb^{2+} have been ethylated in aqueous solution with sodium tetraethylborate ($NaBEt_4$) and the volatile products determined by head-space analysis with a custom-built GC/SFD–ETA–AAS system [40]. Solutions (50 ml) containing ionic alkyllead species were placed in 120-ml Hypo-vials, the pH was adjusted to 4.1 and the vials were capped. Three ml of 0.43% $NaBEt_4$ solution were injected into each vial, the solutions were shaken for 15 min, and then purged for 8.7 min with helium (102 ml/min flow-rate). The volatile R_4Pb species were trapped by 10% SP-2100 on Chromosorb HP kept at −190°C (liquid nitrogen), after passing through an empty glass water-trap kept at −78°C (dry-ice/acetone). The trap was then heated to 60°C electrochemically to desorb the tetraalkyllead compounds into an electrothermally heated silica furnace placed in an AAS. Detection limits of 0.18 and 0.21 ng/l. Pb for Me_3Pb^+ and Me_2Pb^{2+} respectively were reported [40]. As this procedure at present involves ethylation, only the methylated species may be determined, as the ethyllead ions together with Pb^{2+} will be converted into Et_4Pb. For this approach to be of use in environmental analysis, *in situ* butylation or propylation would have to be developed.

8.3 SAMPLING AND EXTRACTION PROCEDURES

GC/AAS may be employed for the analysis of gases or solvent extracts. Gaseous R_4Pb compounds in the atmosphere may be collected and concentrated on solid adsorbents and then thermally desorbed into the GC/AAS system for analysis. Gaseous R_3Pb^+ and R_2Pb^{2+} compounds are collected in water, and aqueous samples are extracted with an organic solvent (n-hexane or benzene). These extracts may be alkylated and analysed by injecting aliquots into the GC/AAS system. Sediment samples may either be extracted with a solvent or the volatile R_4Pb compounds may be thermally volatilized from the sample onto a solid adsorbent and desorbed into the GC/AAS. If necessary, the extracts may be concentrated.

8.3.1 Air samples

Gas-phase organolead compounds in air are separated from particulate matter by filtration. Sampling and analysis techniques for vapour phase R_4Pb are summarized in Table 8.11 and these have generally involved trapping the R_4Pb species from the

Table 8.11 — Sampling and analysis techniques for vapour phase R_4Pb compounds in air, and detection limits.

Adsorbent material	Temperature of trap (°C)	Maximum sampling volume (l.)	Detection limit of individual R_4Pb species (ng/m^3; Pb)	Detector	Reference
10% Apiezon L on Universal support (60/80 mesh)	−196	70	<100	FAAS	[57]
3% OV-1	−80	—	—	GC/FAAS	[58]
10% SE-52 on Chromosorb P	−80	2200	0.5	GC/MPD	[16]
3% OV-101 on Chromosorb W (80–100 mesh)	−72	70	0.5	GC/GFAAS	[32]
5% Apiezon L on Chromosorb WAW (30/80 mesh)	−190	200	1	GC/GFAAS	[59]
Glass beads (4 mm dia.)	−130	400	0.1–0.3	GC/GFAAS	[60]
Porapak Q	Ambient	89	0.25–0.37	GC/SFD–FAAS	[36]

filtered air stream with a solid adsorbent held at low temperature, followed by thermal desorption into the detector system. Detection limits for the reported techniques are also given in Table 8.11.

Harrison *et al.* [57] filtered air through a 0.22 μm filter and trapped the gas phase alkyllead compounds by means of a stainless-steel tube filled with 10% Apiezon L on silanized acid-washed Universal support and cooled with liquid nitrogen. After sampling, the adsorption tube was attached to the combustion air stream of a flame atomic-absorption spectrometer and heated to 130°C to desorb the volatile alkyllead compounds. Nitrogen was passed through the tube to elute the analytes into the atomic-absorption spectrometer for Pb determination. This method is not specific for the individual R_4Pb species but measures the total "volatile" organolead concentration. A small heating coil was positioned before the adsorption tube during sampling, to ensure that all condensation of water or other compounds from the air occurred within the adsorption tube. This method was developed further by Rohbock *et al.* [59]. They removed the water vapour from the filtered air by passing the air stream through an empty glass tube held at −78°C (dry-ice/methanol mixture) and positioned upstream from the sampling trap. In the laboratory the adsorption tubes were heated at a fast rate to 130°C and the desorbed alkyllead compounds were flushed into either an ETA–AAS or a GC/ETA–AAS system for analysis. The authors claim that the cold-trap used for removing water vapour was completely by-passed by the lead alkyls, but in other studies it was found that a fraction of the R_4Pb species is retained by similar traps held at −78°C [60].

For species-specific GC/MPD–AES determination of R_4Pb compounds Reamer *et al.* [16] sampled air by passing it through a 0.2 μm filter and then through Teflon adsorption tubes cooled with a dry-ice/methanol mixture. The R_4Pb compounds were removed from the adsorbent by freeze-drying. The adsorbent was transferred into a 50 ml round-bottomed flask and dried for 12 hr to remove the trapped R_4Pb and water. The evolved vapours were trapped at liquid-nitrogen temperature, extracted into hexane and analysed by GC/MPD–AES. Collection efficiencies for Et_4Pb were found to be between 84 and 100% and a breakthrough volume of 2.2 m^3 was determined. The air samples in this study generally had volumes of about 0.6 m^3. This method is fairly lengthy, and impracticable, and the analytical system is not readily available in environmental laboratories.

Chau *et al.* [58] sampled air by passing it though a 0.45 μm filter and a 25 cm long U-shaped glass column packed with 3% OV-1 held at −80°C (dry-ice/methanol mixture), at a flow-rate of 130–150 ml/min. After completion of sampling the column was connected to a four-way valve between the carrier gas inlet and the injector port of the gas chromatograph and warmed to about 50°C and the R_4Pb compounds were swept into the GC/FAAS system by the carrier gas for species-specific determination. Columns containing the adsorbed air samples could be stored for at least four days in dry-ice. The detector was a conventional flame AAS of modest sensitivity (detection limit of about 0.08 μg of Pb for the individual R_4Pb species) making the method unsuitable for the analysis of ambient air.

A similar sampling procedure was employed by Radziuk *et al.* [32]. Air was sampled with a U-shaped Teflon-lined aluminium tube packed with GC packing and cooled with a dry-ice/methanol mixture. Moisture was removed by condensation from the air stream in a glass U-tube held at −15°C (sodium chloride in crushed ice) placed upstream of the sampling trap. After sampling the tube was removed and stored in dry-ice until analysis. In the laboratory the trap was attached to a 4-way valve between the carrier gas inlet and the injector port. The trap was immersed in boiling water and after 1 min the volatile R_4Pb species were introduced into a sensitive GC/ETA–AAS system for analysis. Motor-vehicle exhaust fumes and ambient air were analysed by this technique.

De Jonghe and co-workers sampled R_4Pb compounds in air by cold-trapping on glass beads [60–63]. Air was passed through a 0.4 μm filter and a Pyrex tube filled with glass beads and cooled with a liquid-nitrogen/ethanol mixture. The tube was then warmed to 60°C and the R_4Pb compounds were transferred into an absorption tube packed with 0.2 g of OV-101 on Gas-Chrom Q and held at liquid-nitrogen temperature (−196°C). The tube was connected to a 4-way valve between the carrier gas inlet and the injection port and the R_4Pb compounds were swept into the GC/ETA–AAS system by the carrier gas. The effect of varying the temperature of the cold-trap between −21 and −196°C, on the collection of R_4Pb compounds, was investigated [60]. Quantitative trapping was achieved at temperatures of −120°C and lower. Sampling rates of up to 10 l./min were possible with this technique and up to 400 litres of air could be sampled.

A similar cold-trapping technique has been developed for the simultaneous determination of vapour phase R_4Pb, R_3Pb^+ and R_2Pb^{2+} compounds [64]. After filtration through a 0.45 μm filter, air was passed through a 20 cm long U-shaped glass tube (1 cm diameter) containing 2 mm glass beads. The tube was immersed in a liquid-nitrogen/ethanol slush bath at −130°C for the duration of sampling. After sampling, the tube was removed from the slush and 10 ml of demineralized doubly distilled water and 2.5 ml of n-hexane were added. The tube was allowed to warm to room temperature. The beads and liquid were transferred to a glass tube and 1.5 ml of NaDDTC solution and 1.5 g of sodium chloride were added. The tube was shaken for 30 min, the phases were separated, and the organic fraction was butylated or propylated according to the procedure described in the previous section. Extracts were analysed by GC/SFD–FAAS and GC/SFD–ETA–AAS. Quantitative trapping of all volatile alkyllead compounds (R_4Pb, R_3Pb^+ and R_2Pb^{2+}) was obtained by this technique. A major problem with cold-traps is that water vapour condenses and freezes out in the trap, resulting in a decrease in the air flow-rate with time and restricting the sample volume. Furthermore there are practical difficulties in using cold-trapping techniques in the field, which precludes their use in remote areas.

Nielsen *et al.* [65] evaluated the use of several solid adsorbents for sampling R_4Pb and concluded that Porapak QS and N were the most suitable. The adsorbent (3.5 g) was packed into stainless-steel tubes (50 cm × 6.3 mm) and analysis was by GC/MS. A similar method for sampling and determination of R_4Pb in air was described by Hewitt and Harrison [36]. R_4Pb compounds were passed through a ferrous sulphate filter to remove atmospheric ozone and sampled on Porapak Q contained inside a stainless-steel tube at room temperature. Ozone was removed to avoid the decomposition of R_4Pb on the adsorption medium. Eighty litres of air could be collected at various sampling intervals (8–24 hr) and the tubes stored at −10°C prior to analysis. In the laboratory, R_4Pb compounds were thermally desorbed into the GC/SFD–FAAS system by means of a two-stage GN concentrator (GN Instrumentation Consultancy). The sample was transferred from the sampling tube into a glass-lined stainless-steel U-shaped tube (15.2 cm × 0.2 cm i.d.) containing 4% Apiezon M on Chromosorb P (60/80 mesh) The sampling tube was heated to 150°C and helium was used to sweep the R_4Pb compounds into the U-shaped cold-trap held at liquid-nitrogen temperature. After 5 min the trap was raised from the coolant and flash-heated to 90°C. The R_4Pb compounds were desorbed and swept into the GC/AAS system for analysis. A collection efficiency of 100% was found, provided the breakthrough volume of 89 litres for 0.45 g of Porapak Q was not exceeded. No sample loss was observed from capped tubes stored at −10°C for up to 14 days. The ozone filter had only a slight effect on the analysis; it removed less than 2% of Me_4Pb and Et_4Pb when these were present at 2 µg/m^3 concentration (as lead) in a standard atmosphere generated in the laboratory. This technique has been employed extensively in field measurements of R_4Pb compounds [36, 66, 67]. By a similar technique a breakthrough volume of 200 litres was found [47].

Total vapour phase organolead concentrations in air are generally determined by a chemical method involving the sampling of filtered air with a solution of iodine monochloride [68]. Vapour-phase organolead compounds are converted into R_2Pb^{2+}, which is selectively extracted and determined as Pb^{2+} by ETA–AAS. Simultaneous measurements of the total vapour-phase organolead concentrations and R_4Pb alone by using GC/AAS have revealed the presence of vapour-phase organolead compounds other than R_4Pb in the atmosphere [36,66,68a]. These are believed to be R_3Pb^+ and R_2Pb^{2+} species, intermediates in the gas-phase decomposition of R_4Pb [66] but until recently no specific method was available for the determination of these species in air. Cold-trapping on glass beads, followed by extraction, derivatization and GC/AAS analysis is a specific technique for these species but it is not sensitive enough for environmental applications [64]. A sensitive and specific technique for vapour phase R_3Pb^+ and R_2Pb^{2+} compounds, involving absorption in water, has been developed [64]. Filtered (0.45 µm) air is sampled with two bubblers connected in series and each containing 80 ml of demineralized doubly distilled water. After sampling is completed, the contents of the two bubblers are combined and transferred to a glass bottle and 5 ml of 0.5*M* NaDDTC, 5 g of sodium chloride and 5 ml of n-hexane are added. The solution is shaken for 30 min and the phases are separated. The extract is butylated or propylated and analysis performed by GC/AAS. The technique has been calibrated by using standard vapour generators in the laboratory. A slight interference from R_4Pb compounds was found, but less than 4% of Me_4Pb and less than 1% of Et_4Pb were recovered in the form of R_3Pb^+. This method has been applied successfully in field studies at urban and rural sites [47].

Organolead compounds in the atmosphere may be present in particulate matter as well as in the vapour phase. De Jonghe *et al.* [62] have developed a sensitive technique for the determination of particle-associated R_4Pb, involving the leaching of R_4Pb compounds from the filter into an organic solvent and analysis by GC/AAS. Quantitative recovery of R_4Pb compounds from spiked samples was found. A technique for the simultaneous determination of R_4Pb, R_3Pb^+ and R_2Pb^{2+} compounds in atmospheric aerosols has been described [66]. Particulate matter was sampled onto GF/A filter papers with a Hi-Vol sampler. After sample collection the filter papers were rolled up and placed inside tubes, and 30 ml of water, 1.5 ml of 0.5*M* NaDDTC, 1.5 g of sodium chloride and 1.5 ml of n-hexane were added. The tubes were shaken for 30 min, and the organic layers were separated, derivatized with a butylating Grignard reagent and analysed by GC/SFD–FAAS.

A scheme for the total speciation of alkyllead in the atmosphere which has been applied by Allen *et al.* [47] is shown in Table 8.12. Gas-phase R_4Pb compounds were

Table 8.12 — Sampling and analysis techniques for the complete physicochemical speciation of atmospheric alkyllead by GC/AAS and determination of inorganic lead [47].

Sample	Compounds determined	Sampling rate	Sampling method	Analytical technique
Vapour	R_4Pb	100 ml/min	Stainless-steel tubes packed with Poropak Q	Thermal desorption/ GC/AAS
Vapour	R_3Pb^+, R_2Pb^{2+}	1 l./min	Two water bubblers in series	Extraction-propylation/GC/AAS
Aerosol	R_4Pb, R_3Pb^+, R_2Pb^{2+}	1 m^3/min	Hi-Vol filtration through GF/A	Extraction/propylation/GCAAS
Aerosol	Pb^{2+}	1 l./min	Membrane filter 0.45 μm	HNO_3 extraction/ GFAAS

sampled by using stainless-steel tubes (8.2 cm long × 0.5 cm i.d.) packed with Poropak Q, at ambient temperature. The tubes were plugged at both ends with silanized glass wool. A filter for removing atmospheric ozone, consisting of a piece of Teflon tubing (5 cm × 0.5 cm i.d.) packed with 0.5 g of ferrous sulphate crystals, was attached to the upstream end of the tube, and preceded by a 0.45 μm filter. A GN concentrator was used for the thermal desorption of R_4Pb compounds into the GC/AAS system. The sampling tube was heated to 160°C with the carrier gas flowing at a rate of 180 ml/min through the tube, and the R_4Pb species were transferred into a glass-lined stainless-steel U-tube packed with 0.5 g of ground glass (250/500 μm), which was found to perform better than several other packings which were investigated. The U-tube was kept at liquid-nitrogen temperature during the transfer. After the completion of this procedure the U-tube was flash-heated to 175°C and the R_4Pb compounds were swept into the GC/AAS system. Prior to sampling, each tube was thermally cleaned by flushing with helium for 10 min at 170°C and a breakthrough volume of 200 litres was found for this technique. Gas-phase R_3Pb^+ and R_2Pb^{2+} compounds were collected in water, and atmospheric aerosols were sampled on GF/A filters with a Hi-Vol sampler. Sampling, extraction and derivatization techniques similar to those just described were employed. In all cases the organolead compounds were determined by GC/SFD–ETA–AAS and inorganic lead was determined by analysing the 0.45 μm filter by ETA–AAS. Collection efficiencies and detection limits are shown in Tables 8.13 and 8.14 respectively.

Vapour phase R_4Pb compounds have been determined at a variety of sites by GC/AAS [32,36,60,62,66] and GC/MPD-AES has been employed in one study [16]. Concentrations have been found to fall off with distance from sources. Highest concentrations have been observed near motor cars, petrol stations and inside tunnels, and lowest concentrations have been measured in rural air. In urban air, the contribution of alkyllead compounds to total lead lies within the range 1–10%. Alkyllead compounds in atmospheric aerosols are generally present at concentrations $<1\%$ of gas-phase R_4Pb. The reported measurements of alkyllead in the atmosphere by species-specific techniques have recently been reviewed [42].

8.3.2 Aqueous samples

Many organic solvents can be used to extract R_4Pb compounds from aqueous samples, but those which are suitable for environmental analysis must fulfil two

Table 8.13 — Extraction efficiencies of ionic alkyllead compounds from environmental samples (analysis was by propylation/GC/ETA–AAS [41]).

Sample	Recovery (%)				Reference
	Me_3Pb^+	Et_3Pb^+	Me_2Pb^{2+}	Et_2Pb^{2+}	
Air(a)	78	70	66	46	[64]
Atmospheric aerosol	89	100	101	104	[69]
Unfiltered aqueous	102	100	98	102	[69]
Dust	112	100	71	100	[46]
Soil	71	68	54	50	[46]
Sediment	108	85	85	54	[46]
Milli-Q water(b)	104	100	95	98	[41]

(a) Collection efficiency.
(b) Included for comparison.

Table 8.14 — Detection limits for alkyllead compounds in environmental samples analysed with the GC/AAS system at Essex University. Extraction and concentration efficiencies are taken into account.

Compound	Detection limits for Pb			
	Aqueous (a) sample (ng/l.)	Air (b) (ng/m^3)	Aerosol (b) (pg/m^3)	Dust, sediment (c) (ng/g)
Me_4Pb	0.2	0.14	0.19	0.02
Me_3EtPb	0.1	0.14	0.13	0.02
Me_2Et_2Pb	0.2	0.28	0.22	0.04
$MeEt_3Pb$	0.3	0.49	0.30	0.06
Et_4Pb	0.3	0.83	0.30	0.06
Me_3Pb^+	0.3	0.34	0.27	0.02
Et_3Pb^+	0.6	0.76	0.54	0.06
Me_2Pb^{2+}	0.4	0.53	0.32	0.04
Et_2Pb^{2+}	1.0	1.96	0.86	0.10

(a) 1-litre sample.
(b) 24-hr sample.
(c) 10-g sample.

requirements. They must be capable of extracting R_4Pb with high efficiency, and the peak due to their non-specific absorption must not interfere with the R_4Pb peaks in GC/AAS analysis. Benzene and n-hexane have both been employed for extracting alkyllead compounds from environmental samples. Generally, workers have added 5–50 ml of solvent to 100–2000 ml of aqueous sample in glass bottles, and shaken these for 30 min. The organic and aqueous phases are separated and the extracts analysed by GC/AAS. Extractions may be done in the presence of sodium chloride and NaDDTC and the extracts derivatized if the complete speciation of R_4Pb,

R_3Pb^+ and R_2Pb^{2+} species is required. For the determination of only ionic alkyllead species in rainwater, Chakraborti *et al.* [38] have employed a method involving extraction into pentane, reduction to dryness with a rotary evaporator, butylation and extraction into a small volume (250 μl) of nonane. Up to 20 μl of the nonane extract was introduced into the GC/SFD–FAAS system for analysis. Alternatively, ionic alkyllead species in aqueous samples may be derivatized *in situ* to the R_4Pb form and determined by head-space analysis. The volatile R_4Pb compounds are purged from the solution and onto a solid adsorbent for subsequent analysis. This method eliminates many potential sources of sample contamination, by minimizing sample handling. Fifty-ml samples are placed in 120 ml Hypo-vials and these are sealed on site. Detection limits for this technique are comparable to those obtained with concentration of solvent extracts. R_4Pb compounds can be determined by head-space analysis followed by GC/AAS without the addition of any reagent. So far, no suitable technique exists for determination of R_3Pb^+ and R_2Pb^{2+} species. A reagent has been employed to ethylate Me_3Pb^+ and Me_2Pb^{2+} *in situ* [40]. As the ethylation results in Me_3EtPb and Me_2Et_2Pb species, R_4Pb compounds present in the sample may interfere. Therefore, it is necessary to make an independent analysis for R_4Pb compounds, in vials to which no derivatizing reagent is added. These techniques have been successfully employed in environmental analysis [48] but more development work is required before all the organolead compounds may be speciated in a single analysis. In the future this method may offer an elegant and convenient alternative to the extraction/derivatization/GC/AAS method.

Estes *et al.* [19] extracted 500 ml samples of tap water with 50 ml of benzene in the presence of sodium chloride and evaporated the extract to between 0.25 and 4 ml under reduced pressure. Portions (2–5 μl) of this extract were analysed by GC/MPD–AES. An efficiency of only 50% for Et_3Pb^+ was found and an even lower efficiency was suspected for Me_3Pb^+. This method was applied to R_3Pb^+ spikes in tap water. Chau *et al.* [69] found an average efficiency of 88.9% for extraction of R_4Pb species from lake water into hexane and reported a detection limit of 0.5 μg/l. for Pb. Cruz *et al.* [70] analysed for R_4Pb compounds by volatilizing them from aqueous samples at 150°C and trapping the total amount on a small segment of cooled chromatographic column. The organolead compounds were then volatilized into the GC/AAS. Efficiencies of particular extraction/analysis procedures are evaluated by analysing environmental samples which have been spiked with known quantities of alkyllead compounds. Samples selected for these tests should contain naturally occurring organolead species at levels low enough not to interfere with the test results.

Radojević and Harrison [48] extracted unfiltered samples (500–2000 ml) with 5–50 ml of n-hexane in the presence of 50 g of sodium chloride per litre of solution and 5 ml of 0.5*M* NaDDTC, shaking the mixture for 30 min on a mechanical shaker. The solvent and aqueous phases were separated and 0.5 ml of propylmagnesium chloride in diethyl ether or n-butylmagnesium chloride in tetrahydrofuran was added to the extract if R_3Pb^+ and R_2Pb^{2+} compounds were also to be determined. The mixture was shaken gently for 8 min, washed with dilute sulphuric acid and dried with anhydrous sodium sulphate. When required, the extracts were concentrated by means of a stream of nitrogen. Efficiencies of these procedures are given elsewhere [48] and typical detection limits are given in Table 8.14. The extraction was done in the sampling bottles because of adsorption of R_4Pb species on the walls of vessels [37]. Samples were not filtered, because of the possible adsorption of R_4Pb compounds on suspended particles [71]. The presence of suspended matter in aqueous samples considerably reduces the recovery of R_4Pb compounds from the near quantitative level obtained for clean water, as illustrated by the results of laboratory tests shown in Table 8.15. An efficiency of 60% was therefore employed

Table 8.15 — Recovery of R_4Pb spikes from aqueous samples [69].

Sample	Recovery (%)				
	Me_4Pb	Me_3EtPb	Me_2Et_2Pb	$MeEt_3Pb$	Et_4Pb
Demineralized doubly distilled water (a)	95(±9)	108(±16)	115(±13)	119(±4)	100(±6)
Filtered motorway run-off water (a)	96(±5)	100(±8)	101(±6)	106(±9)	102(±3)
Tap water	80	102	119	124	110
Rain water (unfiltered)	58	58	59	61	62
River water (unfiltered)	62	61	61	61	62

(a) Mean and standard deviation of four determinations for each sample.

for correcting the results of environmental analysis of unfiltered samples for R_4Pb. Furthermore, it is desirable to keep the time period between sampling and extraction to a minimum because R_4Pb species may be converted fairly rapidly into the R_3Pb^+ form in solution [3]. No such problems were encountered with R_3Pb^+ and R_2Pb^{2+} species, which are water-soluble and fairly stable if the solutions are stored in the dark. It is recommended to employ dark bottles for all sampling and to add the solvent to the sample on site; the samples may be further processed in the laboratory. In the work of Radojević and Harrison [48] extraction was completed on the same day as sampling and the extracts were analysed either on the same or the next day. Extraction efficiencies for R_3Pb^+ and R_2Pb^{2+} compounds in aqueous samples are given in Table 8.13. Rainwater samples have been collected by using bulk funnel-in-bottle samplers. Motorway run-off samples have been collected by using an auto-sampler, and storm profiles of alkyllead compounds determined [72].

8.3.3 Sediment, dust and soil samples

Chau *et al.* [69a] extracted river and lake sediments by vigorously shaking 5 g of wet sediment with 5 ml of 0.1*M* EDTA and 5 ml of hexane in a capped test-tube for 2 hr. The mixture was centrifuged for 10 min and a 5–10 μl portion of the hexane extract was injected into the GC/AAS system for analysis. An average extraction efficiency of 84 ± 9% was found for R_4Pb compounds (Me_4Pb, Et_4Pb and the mixed alkyl species) and a detection limit of 0.01 μg/g. No alkyllead species were found in environmental samples analysed in the study. Chau *et al.* [73] heated 1 g of sediment in a glass U-tube at 100°C for 20 min to volatilize the R_4Pb compounds. During the heating, air was passed through the U-tube into an empty U-tube, which served as a moisture trap, and a U-tube packed with 3% OV-1 Chromosorb W, which served as a sample trap. Both tubes were immersed in a dry-ice/methanol bath for the cold-trapping of water vapour in the first tube and volatilized R_4Pb compounds in the second, and an air flow-rate of 70 ml/min was employed. The sample trap was removed from the sampling line and connected to the GC/AAS system for determination of R_4Pb compounds. A detection limit of 0.1 ng/g Pb was reported. Cruz *et al.* [70] have employed a similar method and quote a detection limit of 0.5 ng/g for R_4Pb (as Pb). Neither group of workers found any R_4Pb species in the environmental sediment samples analysed.

All the aforementioned techniques are suitable for the determination of R_4Pb compounds only. Chau *et al.* [45] were the first to achieve complete speciation of R_4Pb, R_3Pb^+ and R_2Pb^{2+} species in river sediments by using GC/AAS. Dried (1–2 g) or wet (5 g) sediment samples were extracted in capped vials with 3 ml of benzene after addition of 10 ml of water, 6 g of sodium chloride, 1 g of potassium iodide, 2 g of sodium benzoate, 3 ml of 0.5*M* NaDDTC and 2 g of coarse glass beads. The vials were shaken for 2 hr and aliquots (1 ml) of the benzene extract were butylated and analysed by GC/SFD–ETA–AAS. Recoveries of R_3Pb^+ compounds are given in Table 8.16; a detection limit of 15 ng/g Pb was reported. Recoveries of R_4Pb

Table 8.16 — Recoveries of R_3Pb^+ and R_2Pb^{2+} compounds from solid samples, by butylation GC/AAS.

References	Sample	Recovery (%)			
		Me_3Pb^+	Et_3Pb^+	Me_2Pb^{2+}	Et_2Pb^{2+}
[45]	Sediment	111±4	94±4	113±14	93±15
[37]	Dust	58	48	—	—
[74]	Dust	93±2	89±2	87±5	105±5

compounds were not reported. Two of the sediments from the St. Lawrence River, Canada, contained R_4Pb compounds at concentrations of 330–1300 ng/g (as Pb) and R_3Pb^+ and R_2Pb^{2+} species at concentrations from below the detection limit to 200 ng/g (Pb).

Radojević and Harrison [49] extracted sediment, dust and soil samples according to the following procedure. A 10 g sample was placed in a glass-stoppered tube together with 100 ml of 'Milli-Q' water, 5 g of sodium chloride, 5 ml of 0.5*M* NaDDTC and 5 ml of n-hexane. The tube was shaken for 30 min and the phases were separated. The extract was alkylated with either a butylating or propylating reagent and concentrated by evaporation in a Kuderna–Danish apparatus or by a stream of nitrogen, if necessary. For R_4Pb compounds an extraction efficiency of 55% was found and the extraction efficiencies of R_3Pb^+ and R_2Pb^{2+} species are given in Table 8.13. With some samples, problems of low recoveries of the extract were encountered. The problem was eliminated by slightly modifying the technique. A 10 g sample was shaken for 30 min in 100 ml of water containing 5 g of sodium chloride and filtered with a 0.45 μm filter. The filtrate was extracted with n-hexane and the extract propylated as already described. This procedure was efficient for R_3Pb^+ and R_2Pb^{2+} species but the water-insoluble R_4Pb compounds were not recovered. Since hexane may extract and concentrate many organic compounds present in sediment, dust and soil samples, and since these may produce non-specific absorption if present in sufficient quantities, background correction is recommended in the analysis by GC/AAS.

8.3.4 Biological samples

Chau *et al.* [69a] homogenized fish tissue and extracted 2 g of homogenate into 5 ml of hexane with 5 ml of 0.1*M* EDTA present in the tubes. The tubes were shaken vigorously for 2 hr in a reciprocating shaker and centrifuged to facilitate phase separation. Portions (5–10 μl) of the hexane extract were analysed by GC/AAS for R_4Pb compounds. The extraction efficiencies for R_4Pb compounds from fish are given in Table 8.17 and a detection limit of 0.025 μg/g (as Pb) was reported [69a]. Chau *et al.* [73] and Cruz *et al.* [70] analysed fish homogenate by volatilizing the R_4Pb compounds from the sample and collecting them on GC packing in a U-tube. The sampling line and the procedure adopted were the same as described in the previous section for sediment samples. The R_4Pb species were desorbed from the packing into the GC/AAS system for analysis. Detection limits of 0.1 and 0.5 ng/g Pb have been reported by Chau *et al.* [73] and Cruz *et al.* [70] respectively for R_4Pb. These techniques have been applied successfully to environmental samples. Chau *et al.* [69a] measured a concentration of 0.26 μg/g (as Pb) of Me_4Pb in one sample of fillet out of some 50 fish samples analysed for R_4Pb. In all the other samples R_4Pb compounds were below the detection limit. Chau *et al.* [73] found R_4Pb compounds in 17 out of 107 fish samples analysed. Me_4Pb, Me_3EtPb, Me_2Pb, $MeEt_3Pb$ and Et_4Pb were observed at concentrations <10 ng/g (as Pb). Cruz *et al.* [70] reported the presence of the five R_4Pb species in four samples of fish at <5 ng/g Pb levels. The concentrations given above for biological samples all refer to the wet weight of sample.

Chau *et al.* [45] were the first to report simultaneous measurements of R_4Pb, R_3Pb^+ and R_2Pb^{2+} species in fish samples by GC/AAS. Two g of a paste prepared by homogenizing the fish tissue were digested in 5 ml of tetramethylammonium hydroxide (TMAH) solution in a water-bath at 60°C for 1–2 hr until the tissue was completely dissolved to give a pale yellow solution. After cooling, the solution was neutralized to pH 6–8 with 6*M* hydrochloric acid and the mixture was extracted with 3 ml of benzene by shaking on a mechanical shaker for 2 hr. Two g of sodium chloride and 3 ml of 0.5*M* NaDDTC were present in the solutions. The mixture was centrifuged and a measured amount (1 ml) of the benzene extract was butylated by addition of 0.2 ml of 0.9*M* n-butylmagnesium chloride in tetrahydrofuran and mixing for 10 min. The mixture was washed with 2 ml of 0.5*M* sulphuric acid and the phases were separated. The extract was dried with anhydrous sodium sulphate and 10–20 μl portions were analysed by GC/SFD–ETA–AAS. A detection limit of 7.5 ng/g Pb was achieved for alkyllead compounds in biological samples (fish and aquatic weeds) and the extraction efficiencies for R_3Pb^+ and R_2Pb^{2+} compounds are given in Table 8.17. R_4Pb extraction efficiencies were not reported for this technique. R_4Pb, R_3Pb^+ and R_2Pb^{2+} species were identified in carp, pike, white sucker and small mouth bass samples taken from the St. Lawrence River, Canada. Measured lead concentrations were in the range 917–7600 ng/g for R_4Pb and 200–7900 ng/g for R_3Pb^+ and R_2Pb^{2+}. Alkyllead concentrations were also determined in two macrophyte samples and lead concentrations of 68–21700 ng/g for R_4Pb and 130–660 ng/g for R_3Pb^+ and R_2Pb^{2+} were reported. A somewhat different method employing tetrabutylammonium hydroxide and propylation has been used to analyse homogenized fish samples [69]. Algae were processed in a blender with sea-water, extracted with n-hexane in the presence of sodium chloride and NaDDTC, propylated, and analysed by GC/SFD–ETA–AAS. Yamauchi *et al.* [54] have determined Et_3Pb^+, Et_2Pb^{2+} and inorganic Pb^{2+} in urine by hydride generation followed by ETA–AAS. Recoveries of 99.7±5.6% for Et_3Pb^+, 97.3±7.5% for Et_2Pb^{2+} and 91.4±2.1% for Pb^{2+} were reported. These compounds were found in the urine of a patient suffering from acute tetraethyllead poisoning and Et_2Pb^{2+} made up about 50% of the total lead in the urine 21 days after exposure. Et_3Pb^+ concentrations (expressed as lead) of between <1 and 13 ng/l. and Et_2Pb^{2+} concentrations of between <5 and 177 ng/l. were observed in the period between 21 and 29 days after exposure [75].

8.4 ENVIRONMENTAL CONCENTRATIONS AND PATHWAYS

GC/AAS has been the most widely used chromatography/atomic spectrometry technique for environmental analysis of alkyllead compounds. Typical GC/AAS traces for some environmental samples are shown in Fig. 8.3, illustrating the occurrence of R_4Pb, R_3Pb^+ and R_2Pb^{2+} species. To date, over 250 environmental samples have been analysed and the results are reported elsewhere [42,46–49,76]. A typical cross-section of results is shown in Table 8.18. R_3Pb^+ compounds containing mixed alkyl groups (Me_2EtPb^+ and $MeEt_2Pb^+$) and what appear to be monoalkyl lead compounds ($MePb^{3+}$ and $EtPb^{3+}$) have also been observed and quantified in many environmental samples, and their presence has been noted by other workers [38]. These compounds have been tentatively identified by the relationships of molecular weights of alkyllead species to retention time. However, complete identification is not at present possible, in the absence of standards of these compounds. A recent study suggests that the existence of RPb^{3+} species may be due to an analytical artefact [77]. R_4Pb compounds have been measured in numerous gasoline samples by chromatography/AAS techniques, and GC/AAS has been applied to environmental samples only fairly recently. A comprehensive review of concentrations and pathways of alkyllead compounds in the environment has been published [42].

Environmental pathways of organic lead are shown in Fig. 8.4. Alkyllead compounds are released into the environment from motor vehicles, accidental spillages, effluents from R_4Pb manufacture, evaporation at petrol stations and other minor sources such as motor boats, chain-saws and lawn mowers. A natural source of organic lead has been proposed [2, 78–80] and is believed to involve environmental methylation of inorganic lead. Most of the volatile R_4Pb compounds are released into the atmosphere where they may decompose by reaction with hydroxyl radicals under the influence of sunlight to give the more stable vapour-phase trialkyllead and dialkyllead species. These decomposition products are more soluble than the R_4Pb species in water and are readily scavenged by cloud and rain droplets. A minor

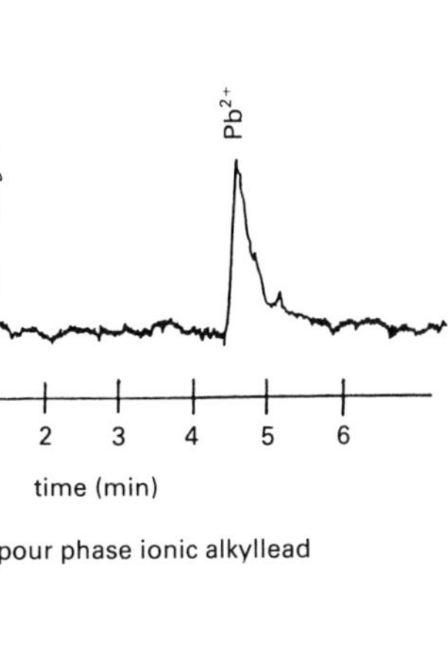

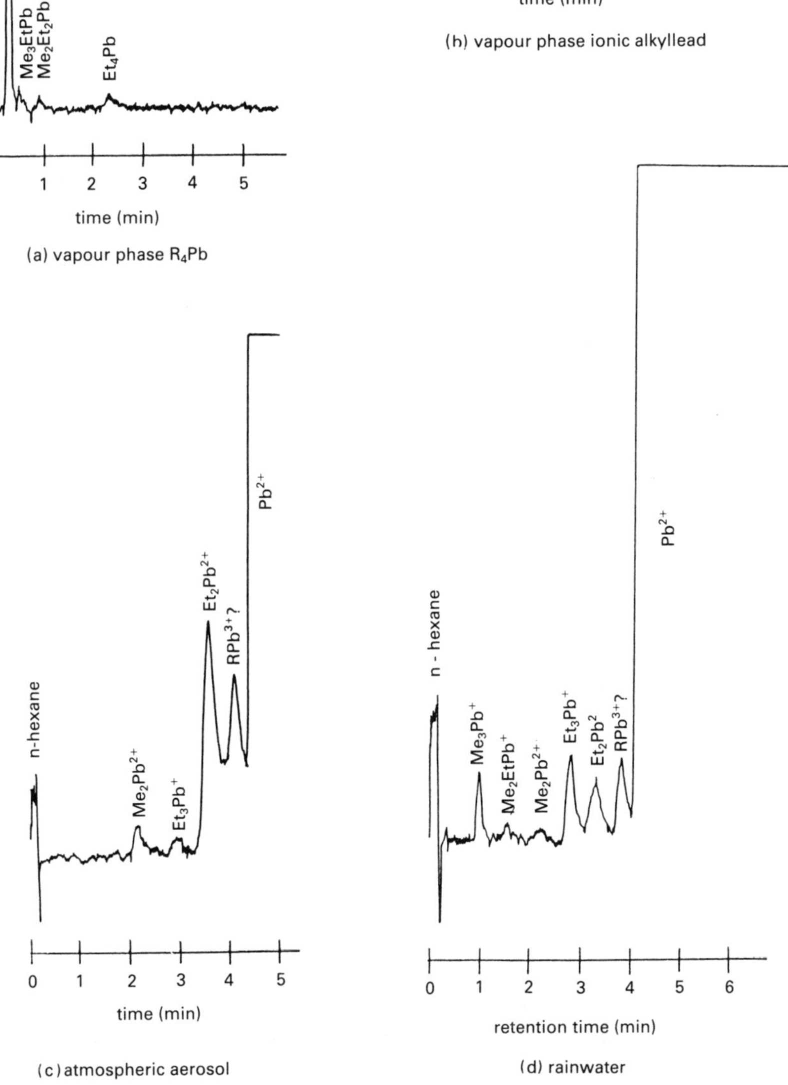

Fig. 8.3 — GC/AAS trace of environmental samples with background correction. All except (a) are propylated extracts.

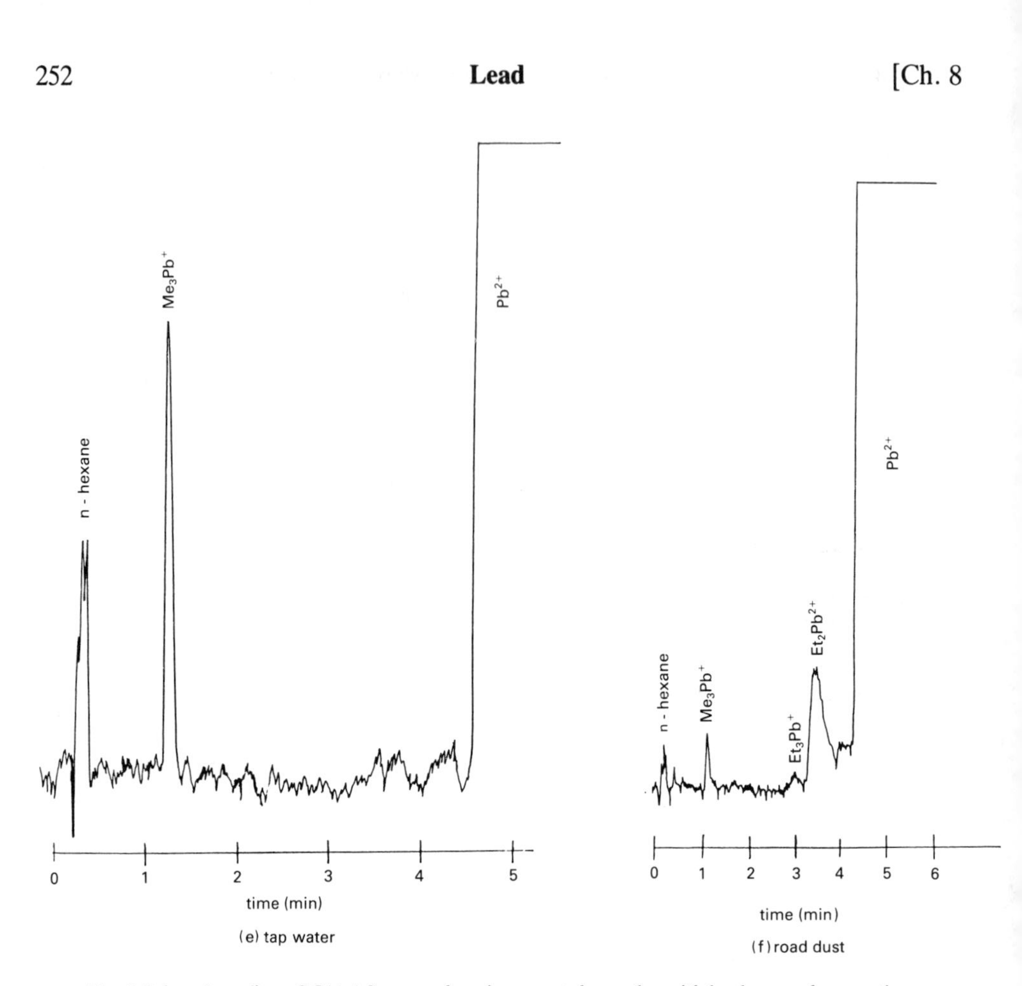

Fig. 8.3 (continued) — GC/AAS trace of environmental samples with background correction. All except (a) are propylated extracts.

fraction of organic lead in the atmosphere is adsorbed on aerosols. Organolead compounds are removed from the atmosphere by dry and wet deposition and rainwater appears to be an important source of organic lead in surface waters. Other sources of alkyllead species in surface waters are direct discharges from alkyllead manufacture, accidental spillages and the probable natural methylation of inorganic lead. Sediments and sea algae have been shown to produce organolead compounds [55,69]. Organolead compounds may be adsorbed on suspended particles in natural waters and bioaccumulated by fish and other organisms. The chemical cycle of organic lead is also indicated in Fig. 8.4. Ionic alkyllead species, R_3Pb^+ and R_2Pb^{2+}, are fairly persistent intermediates in the environmental decomposition of R_4Pb. Monoalkyllead (RPb^{3+}) compounds have also been postulated as intermediates in the decomposition of organic lead to Pb^{2+} but they are considered aa being very unstable, decomposing spontaneously and instantaneously to inorganic Pb^{2+} [81]. Organolead species decompose very slowly in the dark and the decomposition reactions are accelerated by solar radiation [3,71,82]. Compounds containing methyl

Table 8.17 — Recoveries of alkyllead compounds from fish.

Reference	Recovery (%)									Number of
	Me_4Pb	Me_3EtPb	Me_2Et_2Pb	$MeEt_3Pb$	Et_4Pb	Me_3Pb^+	Et_3Pb^+	Me_2Pb^{2+}	Et_2Pb^{2+}	replicates
[69a]	72.2±8	72.3±5	76.2±5	75.2±9	75.3±8	—	—	—	—	4
[70]	>98	>98	>98	>98	>98	—	—	—	—	—
[45]	—	—	—	—	—	86±15	92±7	71±18	101±20	2

Table 8.18 — Concentrations of alkyllead compounds (expressed as lead) in selected environmental samples [69].

Sample	Me_4Pb	Me_3EtPb	Me_2Et_2Pb	$MeEt_3Pb$	Et_4Pb	Me_3Pb^+	Me_2EtPb^+	$MeEt_2Pb^+$	Et_3Pb^+	Me_2Pb^{2+}	Et_2Pb^{2+}	RPb^{3+} (c)	Pb^{2+}
Air (vapour) ng/m^3 (a)	1.1	—(b)	—	—	—	0.5	—	—	—	—	—	—	
Aerosol pg/m^3 (a)	—	—	—	—	—	0.1	—	—	—	0.2	0.7	—	65000
Rainwater ng/l. (a)	—	—	—	—	—	71.8	1.0	—	—	14.2	28.0	8.0	6000
Snow ng/l.	—	—	—	—	—	20.0	—	—	10.0	—	20.0	30.0	22000
River water ng/l.	—	—	—	—	—	0.8	—	—	—	—	—	—	3000
Tap water ng/l.	—	—	—	—	—	0.7	—	—	—	—	—	—	<1280
Road surface water ng/l.	—	—	—	—	—	499	57	—	129	108	1132	436	10000
Road dust ng/g	0.2	—	—	—	0.4	1.2	0.1	—	2.3	0.4	—	—	ND
Intertidal sediment ng/g	—	—	—	—	—	0.1	—	—	0.2	0.1	—	—	ND
Fish ng/g	—	—	—	—	—	3.1	—	—	—	—	—	—	ND
Seaweed ng/g	—	—	—	—	—	0.3	—	—	—	—	—	—	ND

(a) Simultaneous 24-hr samples.
(b) Below detection limit.
(c) Probable artefact of analysis and not present in environment.
ND — Not determined.

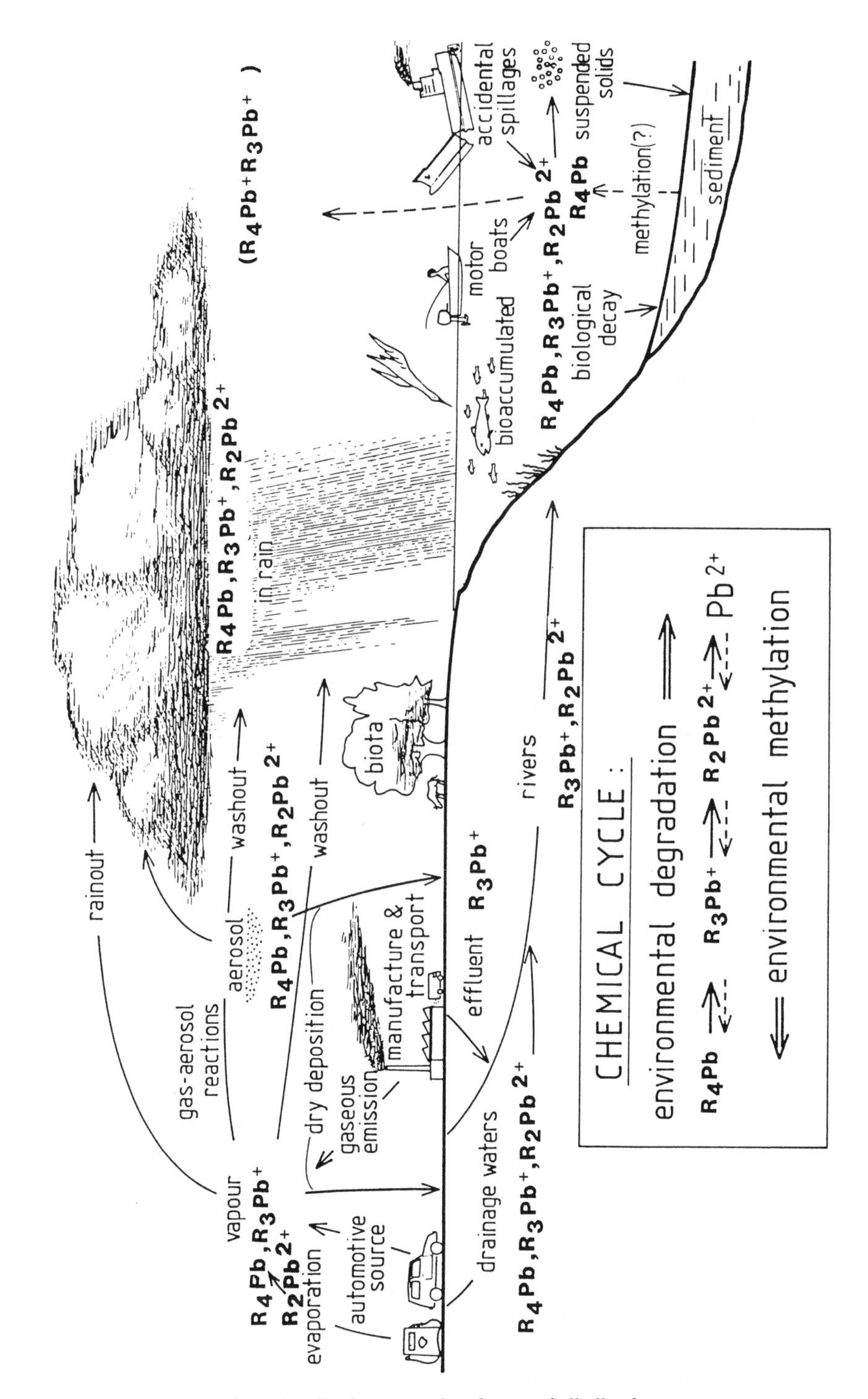

Fig. 8.4 — Environmental pathways of alkyllead.

groups are more stable to photochemical decomposition than those with ethyl groups and the degradation reactions may be influenced by impurities present in environmental media. The mechanisms of the environmental alkylation reactions are still not understood but they are inefficient and appear to involve biological mediation. The biogeochemical cycle of organic lead interacts with the well established cycle of inorganic lead [83] through environmental decomposition and alkylation reactions, sink and source mechanisms respectively of organic lead in the environment.

REFERENCES

[1] P. Grandjean and T. Nielsen, *Residue Rev.*, 1979, **72**, 98.
[2] R. M. Harrison and D. P. H. Laxen, *Nature*, 1978, **275**, 738.
[3] R. M. Harrison, C. N. Hewitt and M. Radojević, *Int. Conf. Chemicals in the Environment, Lisbon, Portugal*, p. 110. Selper, London, 1986.
[4] C. N. Hewitt and R. M. Harrison, in *Organometallic Compounds in the Environment*, P. J. Craig (ed.), p. 92. Longmans, Harlow, 1986.
[5] F. G. Noden in *Lead in the Marine Environment*, M. Branica and Z. Konrad (eds.), pp. 83–91. Pergamon Press, Oxford, 1980.
[6] C. Botre, F. Cacace and R. Cozzani, *Anal. Lett.*, 1976, **9**, 825.
[7] F. E. Brinckman, W. R. Blair, K. L. Jewett and W. P. Iverson, *J. Chromatog. Sci.*, 1977, **15**, 493.
[8] C. H. Gast, J. C. Kraak, H. Poppe and F. J. M. J. Maessen, *J. Chromatog.*, 1979, **185**, 549.
[9] H. Koizumi, R. D. McLaughlin and T. Hadeishi, *Anal. Chem.*, 1979, **51**, 387.
[10] T. M. Vickrey, H. E. Howell, C. V. Harrison and G. J. Ramelow, *Anal. Chem.*, 1980, **52**, 1743.
[11] J. D. Messman and T. C. Rains, *Anal. Chem.*, 1981, **53**, 1632.
[12] M. Ibrahim, T. W. Gilbert and A. Caruso, *J. Chromatog. Sci.*, 1984, **22**, 111.
[13] J. W. Robinson and E. D. Boothe, *Spectrosc. Lett.*, 1984, **17**, 689.
[14] M. Ibrahim, W. Nisamaneepong, D. L. Haas and J. A. Caruso, *Spectrochim. Acta*, 1985, **40B**, 367.
[15] B. D. Quimby, P. C. Uden and R. M. Barnes, *Anal. Chem.*, 1978, **50**, 2112.
[16] D. C. Reamer, W. H. Zoller and T. C. O'Haver, *Anal. Chem.*, 1978, **50**, 1449.
[17] D. Sommer and K. Ohls, *Z. Anal. Chem.*, 1979, **295**, 337.
[18] S. A. Estes, C. A. Poirier, P. C. Uden and R. M. Barnes, *J. Chromatog.*, 1980, **196**, 265.
[19] S. A. Estes, P. C. Uden and R. M. Barnes, *Anal. Chem.*, 1981, **53**, 1336.
[20] S. A. Estes, P. C. Uden and R. M. Barnes, *Anal. Chem.*, 1982, **54**, 2402.
[21] B. Kolb, G. Kemmner, F. H. Schleser and E. Wiedeking, *Z. Anal. Chem.*, 1966, **221**, 166.
[22] T. Katou and R. Nakagawa, *Bull. Inst. Environ. Sci. Technol.*, 1974, **1**, 19.
[23] D. T. Coker, *Anal. Chem.*, 1975, **47**, 386.
[24] Y. K. Chau, P. T. S. Wong and P. D. Goulden, *Anal. Chim. Acta*, 1976, **85**, 421.
[25] R. Bye, P. E. Paus, R. Solberg and Y. Thomassen, *At. Abs. Newsl.*, 1978, **17**, 131.
[26] D. T. Coker, *Ann. Occup. Hyg.*, 1978, **21**, 33.
[27] B. Radziuk, Y. Thomassen, L. R. P. Butler, J. C. Van Loon and Y. K. Chau, *Anal. Chim. Acta*, 1979, **108**, 31.
[28] L. Ebdon, R. W. Ward and D. A. Leathard, *Analyst*, 1982, **107**, 129.
[29] P. R. Ballinger and I. M. Whittemore, *Proc. Am. Chem. Soc. Division Petroleum Chemistry, Meeting, Atlantic City*, 1968, pp. 133–138.
[30] D. A. Segar, *Anal. Lett.*, 1974, **7**, 85.
[31] W. De Jonghe, D. Chakraborti and F. C. Adams, *Anal. Chim. Acta*, 1980, **115**, 89.
[32] B. Radziuk, Y. Thomassen, J. C. Van Loon and Y. K. Chau, *Anal. Chim. Acta*, 1979, **105**, 255.
[33] J. W. Robinson, L. E. Vidaurreta, D. K. Wolcott, J. P. Goodbread and E. Kiesel, *Spectrosc. Lett.*, 1975, **8**, 491.
[34] J. W. Robinson, E. L. Kiesel, J. P. Goodbread, R. Bliss and R. Marshall, *Anal. Chim. Acta*, 1977, **92**, 321.
[35] S. J. de Mora, C. N. Hewitt and R. M. Harrison, *Anal. Proc.*, 1984, **21**, 415.
[36] C. N. Hewitt and R. M. Harrison, *Anal. Chim. Acta*, 1985, **167**, 277.
[37] R. M. Harrison and M. Radojević, *Environ. Technol. Lett.*, 1985, **6**, 129.
[38] D. Chakraborti, W. R. A. De Jonghe, W. E. Van Mol, R. J. A. Van Cleuvenbergen and F. C. Adams, *Anal. Chem.*, 1984, **56**, 2692.
[39] T. W. Brueggemeyer and J. A. Caruso, *Anal. Chem.*, 1982, **54**, 872.
[40] S. Rapsomanikis, O. F. X. Donard and J. H. Weber, *Anal. Chem.*, 1986, **58**, 35.
[41] M. Radojević, A. Allen, S. Rapsomanikis and R. M. Harrison, *Anal. Chem.*, 1986, **58**, 658.

[42] M. Radojević and R. M. Harrison, *Sci. Total Environ.*, 1987, **59**, 157.
[43] D. S. Forsyth and W. D. Marshall, *Anal. Chem.*, 1985, **57**, 1299.
[44] Y. K. Chau, P. T. S. Wong and O. Kramar, *Anal. Chim. Acta*, 1983, **146**, 211.
[45] Y. K. Chau, P. T. S. Wong, G. A. Bengert and J. L. Dunn, *Anal. Chem.*, 1984, **56**, 271.
[46] M. Radojević and R. M. Harrison, *Atmos. Environ.*, 1987, **21**, 2403.
[47] A. G. Allen, M. Radojević and R. M. Harrison, *Environ. Sci. Technol.*, 1988, **22**, 517.
[48] M. Radojević and R. M. Harrison, *Environ. Technol. Lett.*, 1986, **7**, 519.
[49] M. Radojević and R. M. Harrison, *Environ. Technol. Lett.*, 1986, **7**, 525.
[50] K. Hayakawa, *Japan. J. Hyg.*, 1971, **26**, 377.
[51] W. De Jonghe and F. C. Adams, *Z. Anal. Chem.*, 1983, **314**, 552.
[52] Y. K. Chau and P. T. S. Wong, in *Trace Element Speciation in Surface Waters and its Ecological Implications*, G. C. Leppard (ed.), pp. 87–102. Plenum Press, New York, 1983.
[53] J. W. Robinson, E. L. Kiesel and I. A. L. Rhodes, *J. Environ. Sci. Health*, 1979, **A14**, 65.
[54] H. Yamauchi, F. Arai and Y. Yamamura, *Indust. Health*, 1981, **19**, 115.
[55] C. N. Hewitt, *Ph.D. Thesis*, University of Lancaster, 1985.
[56] J. Starý, *The Solvent Extraction of Metal Chelates*, pp. 156–161. Pergamon Press, Oxford, 1964.
[57] R. M. Harrison, R. Perry and D. H. Slater, *Atmos. Environ.*, 1974, **8**, 1187.
[58] Y. K. Chau, P. T. S. Wong and H. Saitoh, *J. Chromatog. Sci.*, 1976, **14**, 162.
[59] E. Rohbock, H. W. Georgi and J. Miller, *Atmos. Environ.*, 1980, **14**, 89.
[60] W. R. A. De Jonghe, D. Chakraborti and F. C. Adams, *Anal. Chem.*, 1980, **52**, 1974.
[61] D. Chakraborti, S. G. Jiang, P. Surkijn, W. De Jonghe and F. C. Adams, *Anal. Proc.*, 1981, **18**, 347.
[62] W. R. A. De Jonghe, D. Chakraborti and F. C. Adams, *Environ. Sci. Technol.*, 1981, **15**, 1217.
[63] S. G. Jiang, D. Chakraborti, W. De Jonghe and F. C. Adams, *Z. Anal. Chem.*, 1981, **305**, 177.
[64] C. N. Hewitt, R. M. Harrison and M. Radojević, *Anal. Chim. Acta*, 1986, **188**, 229.
[65] T. Nielsen, H. Egsgaard, E. Larsen and G. Schroll, *Anal. Chim. Acta*, 1981, **124**, 1.
[66] R. M. Harrison, M. Radojević and C. N. Hewitt, *Sci. Total Environ.*, 1985, **44**, 235.
[67] R. M. Harrison, C. N. Hewitt and M. Radojević, in *Proc. Int. Conf. Heavy Metals in the Environment*, Vol. 1, pp. 171–173. C. E. P. Consultants, Edinburgh, 1985.
[68] J. Birch, R. M. Harrison and D. P. H. Laxen, *Sci. Total Environ.*, 1980, **14**, 31.
[68a] W. R. A. De Jonghe and F. C. Adams, *Talanta*, 1982, **29**, 1057.
[69] M. Radojević, A. Allen, S. Rapsomanikis and R. M. Harrison, unpublished data.
[69a] Y. K. Chau, P. T. S. Wong, G. A. Bengert and O. Kramar, *Anal. Chem.*, 1979, **51**, 186.
[70] R. B. Cruz, C. Lorouso, S. George, Y. Thomassen, J. D. Kinrade, L. R. P. Butler, L. Lye and J. C. Van Loon, *Spectrochim. Acta*, 1980, **35B**, 775.
[71] A. W. P. Jarvie, R. N. Markall and H. R. Potter, *Environ. Res.*, 1981, **25**, 241.
[72] R. M. Harrison, M. Radojević and S. J. Wilson, *Sci. Total Environ.*, 1986, **50**, 129.
[73] Y. K. Chau, P. T. S. Wong, O. Kramar, G. A. Bengert, R. B. Cruz, J. O. Kinrade, J. Lye and J. C. Van Loon, *Bull. Environ. Contam. Toxicol.*, 1980, **24**, 265.
[74] D. Chakraborti, R. Van Cleuvenbergen and F. C. Adams, in *Chemicals in the Environment, Proc. Int. Conf. Lisbon*, pp. 298–304. Selper, London, 1986.
[75] Y. Yamamura, F. Arai and H. Yamauchi, *Indust. Health*, 1981, **19**, 125.
[76] M. Radojević, A. Allen and C. N. Hewitt, *Proc. Int. Conf. Heavy Metals in the Environment*, Vol. 1, pp. 256–258. C. E. P. Consultants, Edinburgh, 1987.
[77] R. J. A. Van Cleuvenbergen, D. Chakraborti and F. C. Adams, *Anal. Chim. Acta*, 1986, **182**, 239.
[78] F. Huber, U. Schmidt and H. Kirchmann, in *Organometals and Organometalloids: Occurrence and Fate in the Environment*, F. E. Brinckman and J. M. Bellama (eds.), pp. 65–81. ACS, Washington DC, 1978.
[79] A. W. P. Jarvie, R. N. Markall and H. R. Potter, *Nature*, 1975, **255**, 217.
[80] J. A. J. Thompson and J. A. Crerar, *Marine Pollut. Bull.*, 1980, **11**, 251.
[81] H. Shapiro and F. W. Frey, *The Organic Compounds of Lead*, Interscience, New York, 1968.
[82] W. R. A. De Jonghe, W. E. Van Mol and F. C. Adams, *Anal. Chem.*, 1983, **55**, 1050.
[83] J. O. Nriagu, *Properties and the biogeochemical cycle of lead*, in *The Biogeochemistry of Lead in the Environment*, J. O. Nriagu (ed.), Elsevier, Amsterdam, 1978.

9

Arsenic and antimony

S. C. Apte
Water Research Centre, Medmenham Laboratory, Henley Road, Marlow, Buckinghamshire, England
A. G. Howard and A. T. Campbell
The University, Southampton, England

9.1 INTRODUCTION

The breakthroughs over the past 15 years in our understanding of environmental metalloid chemistry have relied heavily on the application of sensitive analytical techniques that couple the separation power of chromatography with the selectivity of detection by atomic spectrometry. The aim of this chapter is to review combination techniques that are applicable to the determination of arsenic and antimony species in the environment and to provide additional information regarding sample collection, storage and pretreatment procedures. Information is also given on various detectors and chromatographic separations that have not, to date, been utilized in combination. It is hoped that this information may be of use to the reader who is attempting to improve existing techniques or assemble previously untried coupled methods. Many of the techniques described may also be of use in identifying previously unknown metalloid-containing biomolecules.

There has been comparatively little work on the determination of antimony species, mainly because of the low environmental concentrations of this element. Many techniques applicable to analysis for arsenic, however, may also be applicable to antimony with some modification, and improvement in detection limit. It is therefore appropriate that the two elements are dealt with together in the same chapter.

A detailed review of the environmental chemistry of these elements is beyond the scope of this chapter. A brief background will therefore be given which includes environmental concentrations for each element in waters, sediments, air and tissues, based on data obtained in the last 15 years. This is intended to help the reader match the most appropriate combination technique to the application of interest.

9.2 ENVIRONMENTAL CHEMISTRY

9.2.1 Arsenic

Most arsenic derived from man-made sources is released as a by-product of mining and metal refining processes or burning fossil fuels. Industrial uses are relatively small and include applications as herbicides (arsenate, monomethyl- and dimethyl-arsenic salts) primarily in the cotton industry, as wood preservatives (chromated copper arsenite), and as additives in glass. Despite the growing use of gallium arsenide in semiconductors, the actual gross consumption of arsenic by the electronics industry is in fact quite small.

The environmental chemistry of arsenic was initially of interest owing to the toxic nature of certain arsenic species to man and the alarmingly high arsenic concentrations found in some marine organisms [1]. Recent work, however, has revealed a rich environmental chemistry that involves conversion between oxidation states, formation of organometalloid species, and metabolism by living organisms. Perhaps the most significant discoveries regarding its environmental chemistry are as follows.

(i) Reduction and biomethylation of inorganic arsenic by fungi [2], bacteria [2], algae [3] and mammals, including man [4,5], resulting in the presence of mono-, dimethyl- and trimethylated arsenic compounds in the environment [6–8].
(ii) Biological concentration of arsenic in the marine food chain, and the resulting conversion of inorganic arsenic into compounds which are believed to be non-toxic, such as arseno-sugars, arsenobetaine and arsenocholine.

Arsenic compounds that are known to be naturally present in the environment are illustrated in Fig. 9.1; these include arsenite, arsenate, monomethylarsonate (MMA), dimethylarsinate (DMA), arsenobetaine (ASB), arsenocholine (ASC) and three arseno-sugars that have been isolated from seaweeds [9]. This is by no means the complete inventory of arsenic-containing biomolecules; there are several compounds, including arsenic-containing lipids, for which the structures have yet to be elucidated [7].

The toxicity of inorganic arsenicals is infamous. The element's historical link with homicide, however, obscures a wide variation in toxicity between various arsenic species. The most toxic arsenical known is arsine (rat LD50 3 mg/kg [7]), followed by arsenite (rat LD50 20–60 mg/kg [10]), and then arsenate. The toxicity of MMA and DMA is lower by a factor of about 1000, and arsenobetaine, the predominant arsenic-containing compound in many higher marine organisms, has been demonstrated to be non-toxic when fed to the rat [11] and other organisms [12]. When ingested by higher organisms, including man [4], arsenobetaine is excreted in unchanged form in the urine [11]. Arsenic(III), MMA, and DMA are believed to exert their toxicity by inhibiting thiol-containing enzymes [13], whereas arsenate toxicity results from inhibition of oxidative phosphorylation [14].

Until recently, arsenic was believed to be a non-essential element, but evidence for its essentiality has been presented, arising from deprivation studies in a range of mammals [15]. Arsenic is thought to be involved in arginine, manganese and zinc metabolism and stimulates haemoglobin production.

The concentrations of arsenic in the terrestrial environment are influenced by the underlying geology (high arsenic concentrations are found in igneous regions) and

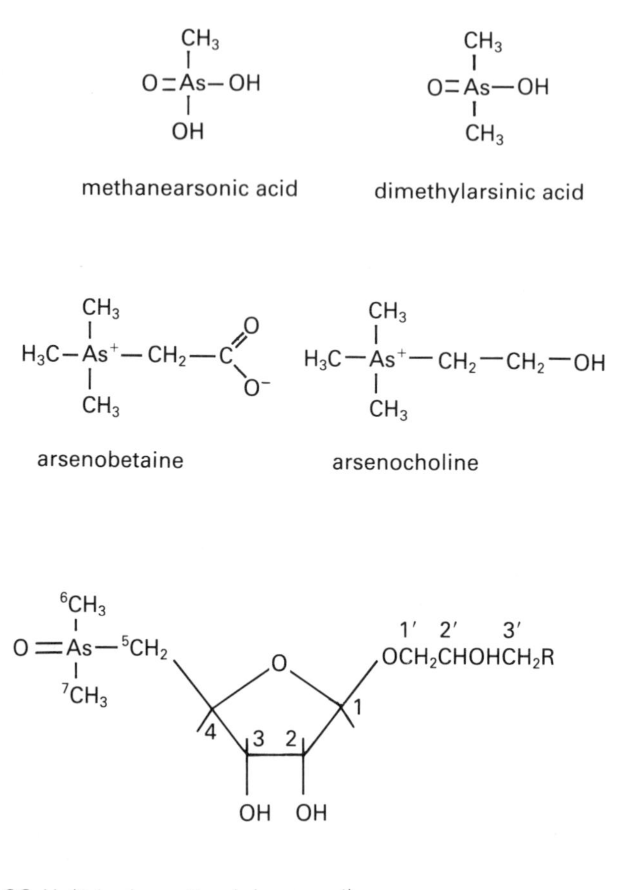

Fig. 9.1 — The structure of organoarsenic compounds identified in environmental samples.

span a wide range for rocks, soils, vegetation and freshwaters. For instance, arsenic concentrations may range from 1 to 50 ppm in soil [16], from 0.01 to 5 ppm in plant foliage in uncontaminated soils [16] and from 0.2 to 100 μg/l. in freshwaters. In contrast, arsenic concentrations in most ocean waters are fairly constant, ranging between 1 and 1.5 μg/l. Arsenic in oxygenated waters is predicted by thermodynamic calculations to be present predominantly as arsenate, but significant quantities of arsenite, MMA and DMA may also be found (see Table 9.1). Biologically mediated processes are primarily responsible for the presence of the latter two species: aquatic algae and bacteria reduce and biomethylate arsenate, releasing dimethylarsinate (DMA) and monomethylarsonate (MMA) into the water column [7]. An additional source of arsenite may be from the pore waters surrounding anoxic sediments.

For a detailed discussion of arsenic chemistry, including use, environmental chemistry, biochemistry, toxicity and carcinogenicity, the recent symposium proceedings on arsenic [5] and the review by Andreae [7] are recommended. A table of typical arsenic concentrations, relating mainly to the marine environment, is given in Table 9.1.

9.2.2 Antimony

Antimony compounds find limited application in industry in alloys, glasses, fire retardants, ceramics, enamels, semiconductors and dyestuffs. Phenylantimony compounds have minor uses as trypanicides and were at one time used in the treatment of syphilis [25]. Very little information is available on the environmental chemistry of antimony, probably as a result of its low environmental concentrations and the lack of sufficiently sensitive analytical techniques. Inorganic antimony compounds are very toxic (comparable to arsenic); their toxic effects are thought to result from irreversible binding to thiol-containing enzymes [26].

Most of our knowledge of the chemical speciation of antimony is restricted to waters. Four distinct species have been identified in natural waters by using hydride generation cold-trap procedures: antimony(V), antimony(III), 'methylantimony' and 'dimethylantimony', which are assumed to be present as their oxo-acids [27]. Antimony(V) is the predominant species in oxygenated waters [27–29], but levels of antimony(III) far higher than predicted by thermodynamic calculations, together with methylated antimonials, have been observed (see Table 9.2). In the oceans, methylantimony species may comprise about 10% of the total dissolved antimony, with the monomethyl species being predominant [34]. There are no reports, to date, of methylantimonials being found in biological tissues, though antimony(III) has been detected in algae [31], which suggests that algae are capable of reducing antimony(V). Biomethylation of antimony by terrestrial organisms has not, to date, been satisfactorily demonstrated [25].

The environmental chemistry of organoantimony species has been reviewed in detail by Craig [25]. Typical environmental concentrations of antimony species are given in Table 9.2.

9.3 SAMPLE COLLECTION, STORAGE AND STABILIZATION

9.3.1 General considerations

When storing samples for metalloid speciation analysis, it is important to stabilize them against changes in oxidation state or changes induced by microbial activity, as well as against losses by volatilization and adsorption. For both arsenic and antimony, contamination is rarely problematic, as long as normal trace metal working procedures are adopted.

9.3.2 Waters

9.3.2.1 Arsenic

Few losses in total arsenic have been observed during storage of fresh and saline waters at natural pH, in acid-washed glass, PTFE or polyethylene containers [35–37]. Acidification of samples does not appear necessary, but is advisable as it has the additional effect of arresting biological activity. Cheam and Agemian [35] found that natural water samples were stable for at least 125 days if acidified to 0.2% v/v sulphuric acid. Nitric acid should be avoided for preservation purposes if analysis

Table 9.1 — Typical arsenic species concentrations in the environment (expressed as As).

		Units	As(III)	As(V)	MMA	DMA	References
Water							
Pacific Ocean (S. California)	1 m depth	μg/l.	0.023	1.24	0.007	0.117	[17]
	55 m depth		0.415	1.10	<0.006	0.014	[17]
Baltic Sea 57°20′ N 20°30′ E	10 m depth	μg/l.	0.108	0.234	0.032	0.209	[18]
	235 m depth		0.794	0.208	0.003	0.006	[18]
Beaulieu estuary (Hants UK)	date: 1/6/82	μg/l.	0.03	0.69	0.09	0.20	[8]
	3/2/82		<0.02	1.08	<0.02	<0.02	[8]
Tamar estuary (Devon UK)	date: 18/7/83	μg/l.	0.41	5.31	0.29	0.81	[19]
Sediment interstitial water			13.7	15.9	0.44	0.56	[19]
Rainwater (La Jolla, California)	date: 11/9/76	μg/l.	<0.002	0.094	<0.002	<0.002	[20]
Human urine		μg/l.	1.9	3.9	1.8	15.0	[6]
Sediments and soils							
			Total As	Leachable As	As(V)	As(III)	
Freshwater reservoir sediment		(μg/g dry weight)	39.3	18.6	15.6	3.0	[21]
Uncontaminated soils		(μg/g dry weight)	1–20				[16]

Air						
Approximate tropospheric concentration overland (northern hemisphere)	ng/m^3	1.3 total As				[22]
Marine biological tissues						
		Inorganic As			Organic As	
Eastern common crayfish	μg/g fresh weight	0.30			35.5	[23]
Mud crab		0.05			1.9	[23]
King prawns		0.02			13.1	[23]
Macroalgae		Inorganic As		MMA	DMA	
Pelagophycus porra	ng/g wet weight	3.45		5.36	10.6	[17]
Ligartina spinosa		8.53		2.31	4.88	[17]
Eastern N. Pacific phytoplankton		As(III)	As(V)	MMA	DMA	
Sample 1	ng/g dry weight	63	30	5.9	490	[18]
Sample 2	ng/g dry weight	27	429	27	740	[18]
		Total As	arsenobetaine	arsenocholine		
Herring	μg/g fresh weight	1.1	0.98	—		[24]
Haddock		6.0	4.7	—		[24]
Shrimp		7.2	0.58	5.2		[24]

Table 9.2 — Typical antimony species concentrations in the environment (expressed as Sb).

	Units	Sb(V)	Sb(III)	MMS	DMS	References
Waters						
Gulf of Mexico	ng/l.	149.0	4.4	5.3	3.2	[27]
Mississippi River		148.0	<0.3	2.3	ND	[27]
River Rhine (Oppenheim)		231.0	0.4	1.2	ND	[27]
Loch Ewe, NW Scotland		142	3	NA	NA	[29]
Baltic Sea 57°20′ N 20°03′ E						
10 m depth		68.2	1.0	3.9	<0.7	[18]
235 m depth		1.2	13.6	3.2	<0.7	[18]
Air						
Aerosol concentration 55 m above sea level			Total Sb			
English Channel	ng/m^3		<0.06–2.7			[30]
Northern Atlantic			<0.02–0.4			[30]
Marine biological tissues						
Algae		Total Sb	Sb(III)		Sb(V)	
Phytoplankton Eastern N. Pacific						
Sample 1	ng/g dry weight	54	43		11	[18]
Sample 2		64	18		46	[18]
Sargassum sp.	ng/g dry weight	100–210				[31]
Enteromorpha sp.		210–240				[31]
Ecklonia radiata		94				[32]
Mollusca						
Littorina littoralis	ng/g dry weight	49–95				[33]
Nucella lapillus		7–14				[33]
Fish						
Marone labrax	ng/g dry weight	10				[33]
Platichthys flesus		6				[33]

MMS=monomethylantimony; DFMS=dimethylantimony; NA=not analysed for; ND=not detectable.

includes hydride generation, owing to potential interferences caused by this acid. Further details of procedures that are applicable to the storage of samples for total inorganic arsenic analysis can be found in the review by Gunn [37].

Methylated arsenic species are stable for months if water samples containing them are frozen [21] or preserved by the addition of 4 ml of concentrated hydrochloric acid per litre of sample [21,38]. If left unpreserved for long periods, the concentrations of these species can change unpredictably [38], presumably as a result of bacterial activity.

A detailed storage experiment has been conducted on natural waters containing low μg/l. concentrations of arsenic species [21]. Whilst total inorganic arsenic and methylated arsenic concentrations were stable for over 30 days for samples stored frozen or at room temperature in polyethylene containers, the preservation of arsenic(III) was more difficult to achieve. To prevent losses by oxidation, the only satisfactory method was rapid freezing in liquid nitrogen and storage at below −80°C. Storage at higher temperatures, e.g. −18°C, caused marked changes in concentration. At much higher concentrations (1–4 μg/ml), arsenic(III) appears to be more stable. Satisfactory stabilization for up to 6 weeks has been achieved by acidification to pH 2 and refrigeration at near 0°C [39].

9.3.2.2 Antimony

For presevation of total inorganic antimony, filtered samples have been satisfactorily stabilized by acidification with nitric acid to pH 1.6 [40] or by addition of hydrochloric acid [41]. Andreae *et al.* [27] found storage of estuarine water in linear polyethylene bottles in the dark at room temperature for two days acceptable for the preservation of inorganic and methylated antimony species. For prolonged storage, rapid freezing in liquid nitrogen and storage at below −30°C is the advised method [38].

9.3.3 Sediments

There is little information available on the storage of sediments. Changes in overall redox conditions are to be avoided (especially with anoxic sediments) and it is also advisable to arrest biological activity. Crecelius *et al.* [21] looked at the effects of storage for 30 days on the concentrations of arsenic species in reservoir sediments which had been frozen at −18°C or stored at between 0 and 4°C. With both treatments, small changes were observed in arsenate and arsenite concentrations, and methylated arsenic concentrations dropped by between 20 and 30%. It was therefore recommended that sediment samples should be analysed as soon as possible after collection.

9.3.4 Biological tissues

There is no information available that is unique to arsenic and antimony. Preservation by deep freezing is advisable for long-term storage to minimize the risk of chemical or biological degradation.

9.3.5 Air

Apparatus for the sampling of air for particulate and vapour forms of inorganic arsenic is described by Walsh *et al.* [22]. For the selective and quantitative sampling of the atmosphere for arsines, Braman [42] has used columns containing small glass beads coated with either gold or silver. The trapped arsenic species were eluted as their oxidation products with potassium hydroxide solution.

9.4 SAMPLE PREPARATION PROCEDURES

9.4.1 Digestion prior to total elemental analysis

9.4.1.1 Waters

Organoarsenic compounds, that would otherwise not be detectable by hydride generation procedures, have been successfully decomposed by base hydrolysis [43] and automated acid persulphate digestion [37]. UV oxidation has also been employed [45,46] though this may not completely degrade methylarsenic species, if present.

9.4.1.2 Solid samples

If arsenic is initially present in an inorganic form, then a number of mixed-acid digests may be employed [47–49]; care must be taken to keep the mixture oxidizing to ensure that arsenic is in its quinquevalent state, as this avoids losses of arsenic trichloride (b.p 130°C) by volatilization. For the total digestion of siliceous materials such as rocks and sediments, hydrofluoric acid should be used in combination with various oxidizing acids [49,50]. Problems may otherwise be encountered with volatilization of arsenic trifluoride [50].

Difficulties have been encountered with the incomplete digestion of dimethylarsenic species in biological materials, resulting in low recoveries. Dimethylarsinic acid (cacodylic acid) is a highly stable compound that is unaffected by boiling in concentrated nitric acid for several hours [7]. Neither nitric/sulphuric or sulphuric/peroxide acid digests are completely effective [51–53]. Complete mineralization has been achieved by using a vanadium pentoxide/sulphuric/nitric acid digest [51] or dry ashing at 500°C with magnesium nitrate/magnesium oxide [7]. Slightly lower recoveries are reported, however, with the latter method for lobster (−13.1%) and shellfish (−6.3%) tissues [51].

For the determination of total inorganic antimony in biological samples, various mixed-acid digests are reported to be satisfactory [32,47,54]. In addition, low-temperature oxygen ashing [55] has been used for the digestion of algal samples [31].

9.4.2 Speciation digests

In the absence of suitable non-invasive methods of analysis, information on the chemical speciation of arsenic and antimony in biological tissues and other solids can only be obtained after the release of detectable metalloid-containing fragments from sample matrices. Many so-called speciation methods are capable, at best, of determining molecular fragments, i.e. the analyte atom and its nearest neighbour atoms or functional groups. It is therefore necessary to leach these detectable fragments chemically from the matrix, which may involve hydrolysis of larger parent molecules. As the complete inventory of arsenic- and antimony-containing biomolecules is not known, it is virtually impossible to test the effectiveness of such procedures.

Tissue homogenization, followed by extraction with solvents such as water, methanol [56] and chloroform/methanol [57], has been used in arsenic speciation studies. As methylated arsenic species are quite stable, it is possible to determine MMA and DMA together with total inorganic arsenic in samples after leaching with warm (60–70°C) [52,58], or cold [59,60] concentrated hydrochloric acid. Such methods have been used in the determination of DMA and MMA in sediments and biological tissues. Such treatments undoubtedly destroy some of the larger arsenic-containing biomolecules. Several more complex separation schemes are described for the isolation of arsenic-containing biomolecules such as arsenobetaine [61,62], arsenocholine [56,63] and arseno-sugars [9]. These typically involve steps such as solvent extraction, TLC, and column chromatography.

For the determination of maximum bioavailable arsenic and antimony in aerosol particulates, Austin and Millward [30] leached filters with 3% v/v nitric acid in a sonic bath for 45 min. A selective leach for the determination of arsenic speciation in sediments has recently been evaluated with freshwater reservoir sediments [21]. With an aqueous solution at pH 2.3, arsenic(III) alone is selectively extracted from dried sediments, whereas at pH 11.9 arsenic(V), MMA and DMA are extracted.

Little information is available regarding the extraction of antimony species. Kantin [31] extracted soluble antimony(III) and antimony(V) from algae by homogenization with 0.2*M* acetic acid followed by sonication and centrifugation (5000 rpm for 10 min).

9.5 PRECONCENTRATION PROCEDURES

Preconcentration techniques may prove to be a useful adjunct to combination techniques to improve detection limits. Co-precipitation with iron or lanthanum hydroxide may be used to concentrate both inorganic antimony [32,64,65] and arsenic species [48,65]. Antimony(III) and antimony(V) have also been preconcentrated from natural waters by co-precipitation with hydrous zirconium oxide [66]. The selective extraction of arsenic(III) and antimony(III) from solution into a suitable organic solvent is possible after complexation with ammonium pyrollidinedithiocarbamate at pH between 2 and 5 [67–69], or, for arsenic(III), complexation with ammonium *sec*-butyldithiophosphate [70]. Extraction of arsenic(III), MMA and DMA as their tri-iodides into organic solvents, may also be achieved from highly acidic solutions in the presence of iodide [59,71]. Norin *et al.* [56] report the selective extraction of arsenocholine from biological extracts in the presence of arsenobetaine, as its ion-pair with hexanitrodiphenylamine. By use of radiolabelled standards the extraction efficiency was shown to be 97%. Arsenobetaine was hardly extracted at all. Inorganic arsenic has been separated from organically bound arsenic by distillation as arsenic trichloride: marine crustacea samples were acidified with hydrochloric acid/hydrobromic acid prior to distillation [24].

Preconcentration of arsenic species on both cation- [72,73] and anion- [74] exchange columns may also be applicable. Both inorganic arsenic and antimony may be removed from sea-water by passage through a hydrous zirconium oxide column [32,75]. The principal application of this method has been in preparing sea-water low in arsenic and antimony for use in preparing standards [32,75]. Howard *et al.* [76] describe the synthesis of a silica gel with attached mercapto functional groups, that is capable of selectively preconcentrating arsenite from natural water samples in the presence of other arsenic species.

9.6 HYDRIDE GENERATION

9.6.1 Introduction

Derivatization by hydride formation and subsequent detection in the gas phase by atomic spectrometry has had an immense impact on the determination of arsenic and antimony at trace levels, allowing the simple and reliable determination of metalloids at sub-μg/l. levels in aqueous samples. As a component of a combination technique, hydride generation may serve two roles: to derivatize samples prior to GC separation, or as a post-column derivatization step in liquid chromatography that enhances sensitivity and improves detection limits.

Strictly speaking, hydride generation cannot be used to deduce solution speciation; information can only be obtained about the oxidation state of the metalloid, and, when cold trapping or gas chromatography is used, the degree of alkylation. Many workers make the mistake of claiming to detect arsenite and arsenate species by using hydride generation techniques when in fact they mean arsenic(III) and arsenic(V). This distinction is especially important when analysing biological samples. It is important to note that certain naturally occurring arsenic compounds, for instance arsenobetaine, are not 'hydride-reducible species', i.e. they do not react with sodium borohydride to form volatile hydrides. The limitations of hydride generation derivatization procedures should be fully taken into account, especially when biological samples are dealt with.

For comprehensive details of hydride generation and its application to arsenic determination the report by Gunn [37] and the more general review by Nakahara [77] are recommended.

9.6.2 Apparatus

Derivatization by hydride generation requires adjustment of sample pH by the addition of acid or a buffer and the addition of a hydride transfer reagent. Hydrides are rapidly evolved, and can be efficiently stripped from solution by an inert carrier gas. Two distinct types of hydride generation apparatus have been employed in combination techniques.

(i) A discrete volume reactor, where the sample is held in a chamber and measured volumes of reagents are added. Examples are illustrated in Figs. 9.2 and 9.3.
(ii) On-line continuous-flow hydride generation apparatus which can be constructed from standard 'AutoAnalyzer' components with the inclusion of a purpose-built gas-liquid separator (e.g. [44,82,83]).

Both types of generator can be coupled to a gas chromatograph or cold trap, but only the on-line continuous-flow systems are suitable for post-column derivatization in liquid chromatgraphy.

The continuous-flow methods may be preferred on the grounds of improved reproducibility and less manual sample manipulation. There is also evidence to suggest that they are less prone to chemical interference (see Section 9.6.5). They also allow the incorporation of other automated steps such as UV photolysis [46] and acid persulphate digestion [44]. The sensitivity and detection limits of on-line methods that have no preconcentration step are governed largely by sample flowrate and the efficiency of the gas–liquid separator. The discrete-volume methods are

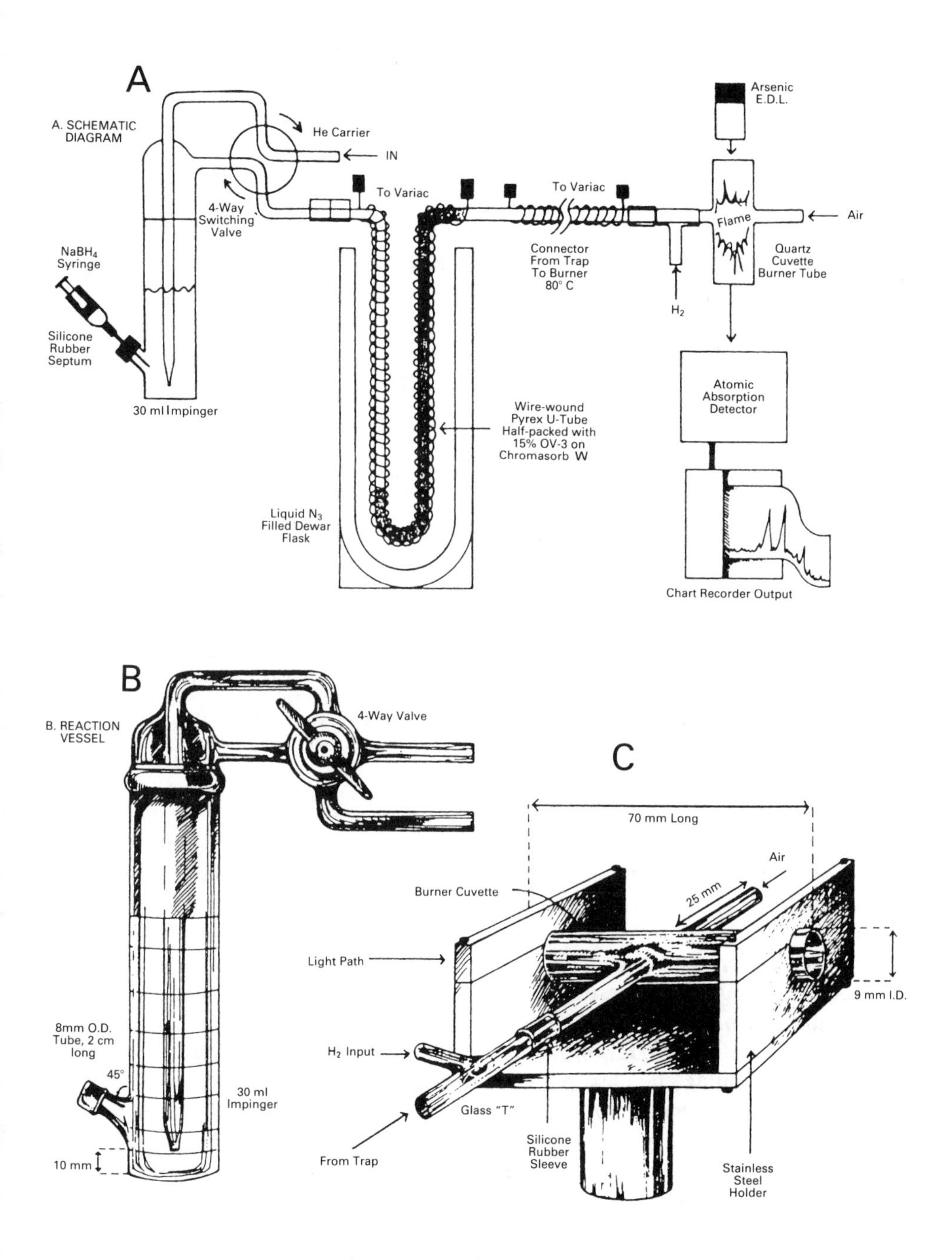

Fig. 9.2 — Hydride-generation/cold-trap AAS apparatus for the determination of arsenic species in aqueous samples. (A) Schematic diagram of the system. (B) Detail of the discrete volume hydride generator. (C) Silica cuvette burner. The metal holder has a cylindrical base that fits a standard AAS burner socket. Gas flow-rates: hydrogen 250 ml/min, air 150 ml/min, helium carrier 50–200 ml/min. (Reproduced from [21] by permission of the copyright holders).

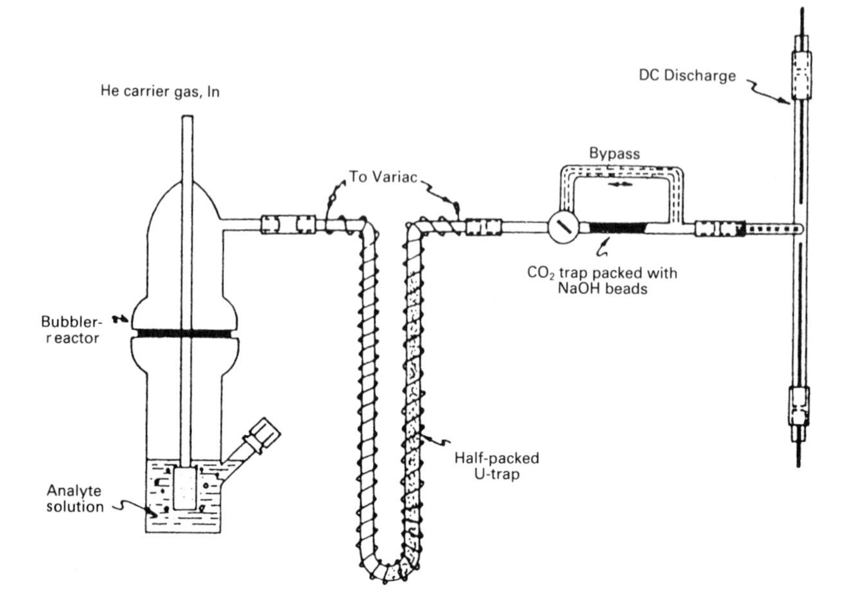

Fig. 9.3 — Hydride-generation/cold-trap/helium flow discharge AES detector apparatus. This system has been used by Braman and co-workers to determine arsenic species [78] and inorganic antimony [79] in environmental samples. Details of detector construction and operation may be found in [80,81]. (Reproduced by permission from R. S. Braman, D. L. Johnson, C. C. Foreback, J. M. Ammons and J. L. Bricker, *Anal. Chem.*, 1977, **49,** 621. Copyright 1977, American Chemical Society).

capable of handling large sample volumes (typically 50–250 ml) and are the methods of choice if very low detection limits are required. These hydride generators may also be automated [84], but this is far more complicated.

9.6.3 Generation conditions

By adopting suitable generation conditions, it is possible to generate hydrides quantitatively from arsenic(III), arsenic(V), DMA, MMA, trimethylarsine oxide, antimony(V), antimony(III), monomethylstibonic acid and dimethylstibonic acid. The two determining chemical parameters are reaction pH and choice of hydride transfer reagent. In early work Zn/H^+ was used as the hydride transfer reagent [37,85,86]. This is no longer popular because the generation times are long (between 60 sec and 30 min [37,85,86] and the efficiencies are typically 20% or less [88]. Sodium borohydride in tablet, powder, or more popularly solution form (1–10%) is the most widely used reagent. Its advantages include speed, efficiency and effectiveness over a wide pH range. At optimum pH, the reaction is essentially instantaneous and is reported to be more than 90% efficient [89–91]. Dilute sodium borohydride solutions readily hydrolyse, resulting in hydrogen production. Solutions may be stabilized by addition of sodium hydroxide, which facilitates storage of the solution [92].

If total inorganic arsenic or tervalent inorganic antimony is to be determined, then the sample hydrogen-ion concentration is adjusted to between 0.1 and 6*M* (actual optimum concentrations vary greatly between methods). Hydrochloric acid is the preferred reagent; high concentrations of oxidizing acids e.g. nitric, sulphuric and perchloric acids, are not recommended, as they can cause interferences [93–95].

Several workers have studied the effect of pH on hydride generation yields of both arsenic [78,82,96,97] and antimony species [27,98,99]. Slightly different responses are found between systems, but generalizations can be made. Arsenic(III) gives a near constant response from pH$<$0 to pH 10, whereas arsenic(V) gives similar response to arsenic(III) under acid conditions, but no response at pH $>$5. DMA and MMA give optimal hydride yield under mildly acidic conditions [20,82].A similar pattern in behaviour is found for antimony species. A constant yield of stibine is obtained from Sb(III) over the pH range 1–7 [27,98,99], whereas the yield from antimony(V) is effectively zero at pH $>$2. At optimum pH, the hydride yield from antimony(V) is at best only 40% of that for Sb(III). Equalization of response may be obtained by the addition of an auxiliary reductant such as potassium iodide [27,98,99]. Methylantimony species are reported to form hydrides under mildly acidic conditions and can be determined along with antimony(III) at pH between 2 and 3 [27]. The variation of hydride yields from differing oxidation states with pH has been exploited to obtain information on oxidation states in water samples (e.g. [8,17,21]). If the sample is suitably buffered, both arsenic(III) [20,82,96] and antimony(III) [27,98,99] may be selectively determined in the presence of their higher oxidation states, which do not form hydrides at the selected pH.

The differences in results reported for hydride yields from various arsenic species by different workers [82,97,100,101] are due partially to variations in apparatus and generation conditions. Much of the confusion results from failure to take into account all the factors that influence hydride generation yield. These include reaction pH, sodium borohydride concentration [102], efficiency of gas–liquid transfer, reagent contact time and reaction kinetics. Additionally, hydrogen evolved during the reaction may affect the signal response of the detector. True optimization of hydride generation conditions can only really be achieved by using a multivariate optimization approach such as a simplex method [103], which must be applied for each hydride-forming species. Adequate results may be obtained, however, by using a univariate approach to optimization. It is crucial to optimize the hydride yield experimentally for the specific system in use.

9.6.4 Reagent blanks

Blanks in continuous flow systems contribute to the background signal. This generally only presents a problem if blank levels are very high, and may result in increased baseline noise and a reduced linear range. For discrete-volume systems, especially those involving cold-trapping preconcentration, inorganic arsenic and antimony blanks can be significant. Acids usually contribute little to blanks and may be purified efficiently by distillation. Sodium borohydride is the major blank source for both inorganic arsenic and antimony, the concentration of which may vary greatly between batches of the reagent. The easiest approach to blank control is to screen several batches of the reagent and select a suitable ‘low metalloid batch’ which is then bought in bulk. Alternatively, clean-up procedures such as filtration, recrystallization [96] or removal by hydride formation [104] may be adopted.

9.6.5 Interferences

Hydride generation techniques for the interference-free determination of arsenic species in fresh and saline waters and sediment extracts have been reported by many workers. Several interference studies using low-level hydride-based speciation methods have confirmed interference-free operation in analysis of natural waters [27,82,98]. There are, however, cases such as the analysis of biological samples, where interferences may be encountered. Extensive studies of interferences, primarily affecting inorganic arsenic, have been made [44,82,105] and the following are known to be potentially interfering substances affecting hydride generation procedures [37]: oxidizing acids; transition metals; other hydride-forming elements.

It is important to realize, however, that the severity of an interference very much depends on the type of system in use, and the relative proportion of analyte to interfering substance. For instance, a recent comparison study of interferences in a batch hydride-generation system and in two continuous-flow systems (coupled with ICP–AES) [106] showed great differences between the systems with regard to susceptibility to interference. Interferences were markedly less (or even absent) with the continuous-flow systems. It is recommended that interferences are evaluated for the hydride system in use, and when analysing new sample matrices it is advisable to check for interferences by using the method of standard additions.

The masking of transition metal interferences encountered during inorganic arsenic determination has been reviewed in detail by Gunn [37]. Many of these interferents may be effectively masked by the addition of EDTA [82,107,108], thiosemicarbazide [109] and 1,10-phenanthroline [109] (both more effective than EDTA under acid conditions), thiocyanate [110] and citrate [96]. A detailed study of masking procedures for use with a hydride-generation/cold-trap/AAS method for arsenic speciation has been made [82]. Nitric acid and nitrite interfere by the formation of volatile nitrogen oxide species that cause gas-phase interference [95]. This can be overcome by the addition of sulphamic acid [95] or suitable diazotization reagents such as sulphanilamide [111]. A potentially severe interference which may affect low-level cold-trap speciation methods, namely, the presence of variable quantities of residual free chlorine in batches of concentrated hydrochloric acid, has been noted by Andreae [38]. This interference can be removed by purging acid solutions with an inert gas prior to use.

9.7 DETECTION SYSTEMS

9.7.1 Introduction

The various atomic spectrometry detectors available for arsenic and antimony determination will now be discussed in isolation from chromatographic separations. Distinction is made between detection systems suitable for liquid and gaseous phase sample-introduction. It is difficult to compare the relative merits of detectors, especially for gaseous sample introduction, as detection limits are very much dependent on the nature of the sample introduction, peak shape and mode of measurement (peak area or peak height). There are also variations in the definition of detection limit. The detection limits quoted therefore serve only as an approximate comparison of detectors. It should also be borne in mind that the coupling of any detector to chromatographic apparatus may result in poorer performance than that otained by direct sample introduction.

9.7.2 Atomic-absorption spectrometry

The principle resonance lines for arsenic are at 189.0, 193.7 and 197.2 nm, all of which lie in the vacuum ultraviolet range of the electromagnetic spectrum. This presents problems of source attenuation as a result of low reflectance of mirrors, and absorption by the atmosphere, flames and lenses. Thus poor linear ranges, low sensitivity, and, owing to high background noise, poor detection limits are frequent problems. The lines at 189.0 and 193.7 nm are of approximately equal sensitivity, whereas that at 197.2 nm is less sensitive by a factor of about 1.7 [112]. This line may be of practical use, however, as it has a greater linear range [21] and the signal-to-noise ratio may also be better.

Improvements in AAS performance may be obtained by fitting a photomultiplier tube giving a more sensitive response to ultraviolet radiation [82], or by using a more intense radiation source than the hollow-cathode lamp (HCL). Radiofrequency-excited electrodeless discharge lamps (EDLs) are reported to give 5–10 times more intense radiation than do hollow-cathode lamps [113] and give better detection limits (owing to improved signal-to-noise ratios) as well as extended linear range. Several workers claim that arsenic EDLs give about double the sensitivity of HCLs [113–115]. This can only be the case if the EDL emission linewidths are narrower than those from the HCL.

Principal resonance lines for antimony are found at 217.6, 206.8, and 231.12 nm, of which that at 217.6 nm is the most sensitive. A spectral bandpass of less than 0.2 nm is required to isolate this line from two non-absorbing lines at 217.9 and 217.0 nm. If this is not done, drastic reductions in sensitivity and linear range will be observed. Hollow-cathode lamps are reasonably satisfactory sources, though Andreae *et al.* [27] report improvement by a factor of three in linear range by use of an antimony EDL.

9.7.3 Atomic-emission spectrometry

The principal emission lines for arsenic lie in the vacuum ultraviolet range (e.g. 197.3 nm) and are therefore not often employed, owing to limitations in the performance of most spectrometers. These lines may be of practical use if high-performance spectrometers having an evacuated optical system are available. As a result, the lines at 228.8 nm for arsenic and at 252.5 nm for antimony are widely used. Flame-based excitation is not employed, as there is insufficient excitation energy and consequently the sensitivity is poor. More energetic sources such as plasmas and electrical discharges are required.

9.7.4 Liquid sample introduction

9.7.4.1 Flame AAS

For arsenic, flame AAS offers poor detection limits for aqueous samples. Virtually all premixed flames (air–acetylene, air–hydrogen, air–propane) absorb a high degree of source radiation (about 60% at 193.7 nm [87]. This results in noise-limited detection limits which are typically in the low ppm range (Table 9.3). There are few reported interferences in the air–acetylene flame [119] and the nitrous oxide–acetylene [120] flame, though the sensitivity is lower in the latter. A twofold improvement in sensitivity has been obtained by shielding the nitrous oxide–acetylene flame with nitrogen [120]. An argon–hydrogen flame, that burns by air diffusion, absorbs far less source radiation (about 15% [115]) than do premixed gas flames. The much lower flame temperatures of between 300 and 500°C [121,122], however, result in greater interferences, especially for liquid samples, where desolvation and dissociation can present problems. Signal response may also vary with oxidation state in this flame. For instance, it is reported that arsenic(III) standards give 2–3 times the response of arsenic(V) standards [123]. This type of flame is best suited for atomization of gaseous species such as hydrides.

For antimony determination, both air–acetylene and air–propane flames are applicable. Few interferences have been encountered when using the air–acetylene flame. Typical detection limits (217.6 nm line) are 0.05–0.1 μg/ml [112].

9.7.4.2 Graphite-furnace AAS

This detector is inherently sensitive, but is limited to small liquid sample volumes of typically 10–50 μl. Absolute detection limits for arsenic are around 25 pg [126], which corresponds to a sample concentration detection limit of about 0.5 μg/l. Arsenic is stable, on thermal pretreatment (ashing), up to 600°C and this may be extended up to 1100°C, irrespective of furnace tube type, by adding a nickel [124] or lanthanum [125] matrix modifier.

There is little available information on GFAAS determination of antimony. Absolute detection limits are around 20 pg [117]. An optimum atomization temperature of 1800°C is reported [115]. Antimony may be ashed at 1000°C without loss if ammonium dichromate is used as matrix modifier [115].

9.7.4.3 Atomic-emission spectrometry

A commercially available triple-electrode direct current arc plasma, equipped with a high-resolution echelle monochromator, has been evaluated for the analysis of aqueous arsenic(V) samples [126]. The emission line at 193.696 nm gives a limit of detection of 0.063 μg/ml and linear response up to 100 μg/ml. Enhancement by certain acids and chromium is noted. Although little used for arsenic and antimony determination, this system has the appeal of operation in a simultaneous multielement mode.

With ICP–AES, detection limits of around 15–50 μg/l. [87,127] for arsenic are obtainable, which is comparable to the flame AAS detection limits for an argon–hydrogen flame (see Table 9.3). ICP–AES however, has the advantage of being far less prone to interference. The plasma may have a tendency to be extinguished if high concentrations of volatile organic solvents are introduced (e.g. certain HPLC eluents). This can be avoided to a certain extent by operating at higher radiofrequency powers or by cooling the nebulizer [128].

9.7.5 Gaseous sample introduction

9.7.5.1 Atomic-absorption spectrometry

Hydride generation may be used in conjunction with flame AAS by introduction of

Table 9.3 — A comparison of atomic emission and atomic absorption detectors for arsenic and antimony determination: aqueous sample introduction.

		Detection limit (μg/ml)	Reference
Arsenic			
AAS			
Argon–hydrogen flame	EDL source	0.02	[115]
Air–acetylene flame	EDL source	0.15	[115]
	HCL source	2.3	[115]
GFAAS		0.0005	[116]
DC-arc AES		10	[87]
DC plasma AES		0.1	[87]
ICP-AES		0.05	[87]
Antimony			
AAS, air–acetylene flame		0.050	[112]
GFAAS		0.001	[117]
ICP–AES		0.2	[118]

HCL=hollow-cathode lamp; EDL=electrodeless discharge lamp.

the hydrides through the nebulizer assembly [129,130], but this combination is rather insensitive. Fiorino *et al.* [84] have described modifications to a conventional burner assembly, for hydride introduction. The nebulizer is removed, and the hydride is introduced into the burner chamber by a nitrogen-carrier/hydrogen mixture. The flame, which burns by air entrainment, is shielded by borosilicate glass plates and fused-silica end-windows. Absolute detection limits (with an EDL source) were 3 ng for arsenic and 2 ng for antimony.

The most widely used hydride atomizer consists of a heated fused-silica tube aligned in the spectrometer light-path, and into which an inert carrier gas containing the hydride is introduced. The tube is heated either by an air–acetylene flame [82,96,131–133] or electrothermally [134–136] to about 800–1000°C. Electrothermal heating may be achieved by wrapping heating wire around the tube, or by housing the tube in a small removable furnance. Electrothermal heating is preferred as it allows better control over atomization temperature, gives even heating and eliminates noise from flame flicker. Some tube atomizers have fused-silica end-windows to prevent diffusion flames from burning at the tube ends, which can give rise to increased background absorption [37]. Hydrogen may be introduced into the tube [137] to increase the sensitivity.

Heated fused-silica cells have several advantages over flame atomization. The atomization process is more efficient, atoms are concentrated into a smaller volume, and the residence time of atoms in the light-path is longer. Absolute detection limits of 0.05 ng for antimony [98] and between 0.04 and 0.12 ng for arsines [19] have been reported. An alternative sensitive atomization cell is the fused-silica cuvette burner [20] (see Fig. 9.2), which consists of a small hydrogen–helium diffusion flame burning in a small area of a fused-silica tube. No additional form of heating is required. Air and the helium-carrier/hydrogen mixture are supplied from opposite sides of the burner. Detection limits obtained with this cell are reported in Table 9.4. This atomizer is very sensitive, but less robust than the heated silica tube.

The atomization mechanism of hydrides in silica tubes has been the subject of

Table 9.4—Limits of detection (ng) obtained with various detectors coupled with hydride generation and cold trapping [34].

Species	Silica tube AAS	Graphite-furnace AAS	Electron-capture detector
As(III)	0.03	0.09	—
MMA	0.03	0.09	0.4
DMA	0.05	0.15	0.2
Sb(III)	0.04	0.12	—

much attention. In a cool hydrogen-rich flame (e.g. the silica cuvette burner), H and OH radicals are believed to be the operative species. Similarly, atomization in the heated tubes is thought not to proceed by straight thermal decomposition, but by the same radical mechanism [138]. This is because a temperature greater than 1800°C is required for the thermal decomposition [139]. The source of hydrogen for the free radical atomization process is believed to be the hydride generation reaction itself. Welz and Melcher [138] have drawn attention to the importance of the silica tube surface in the atomization process. In new tubes, a surface film may be present that gives rise to both poor sensitivity and poor precision. This has to be removed by thermal pretreatment, or by etching with hydrofluoric acid, before optimum performance is attained.

GFAAS can be coupled with ease to hydride generation [27,40,140,141] but this combination is less popular. A potential drawback is that the tube has to be sustained at the atomization temperature for quite long times (e.g. >1 min) during each determination, resulting in poor tube life. Parris *et al.* [139] have studied the optimum conditions required for arsine atomization. Bare uncoated graphite tubes were preferred, at a temperature of 1800°C. Sensitivity was greatly increased by introducing hydrogen into the furnace, which suggests that hydrogen radicals are also involved during the hydride atomization process. For antimony determination, Andreae *et al.* [27] report the optimum atomization temperature as 2200°C. A comparison of absolute detection limits for GFAAS and other detectors can be found in Table 9.4.

9.7.5.2 Atomic-emission spectrometry

Braman and Dynako [80] have described a DC arc excitation helium-glow detector contained within a fused-silica tube. This is suitable for the determination of gas-phase species only and has been coupled with hydride generation for the determination of both arsenic and antimony species (Fig. 9.3) [78,79,81]. By use of the most sensitive emission lines at 228.8 nm (As) and 252.5 nm (Sb), absolute detection limits of 1.0 ng (As) and 0.5 ng (Sb) can be obtained. Water vapour and carbon dioxide must both be removed from the gas stream, as they cause molecular interferences. Feldman [91] has described improvements to the detector, including modifications to the electrodes which make it more robust, reproducible and sensitive. These modifications resulted in a tenfold improvement in the detection limit for arsenic species.

Hydrides of arsenic, antimony and bismuth have been simultaneously determined by hydride generation coupled to ICP–AES. The hydrides may be introduced into the plasma in an argon carrier stream [142]. Typical detection limits of 1 μg/l. for both elements have been reported [142], with no mutual interference [143].

Microwave-induced plasmas (MIP) have a low energy density and hence are easily disturbed, if not extinguished, by high solvent concentrations. Even when gas-phase introduction is used most GC systems are equipped with a valve system allowing the solvent peak to be vented to the atmosphere, thus preventing plasma overload and possible fouling of the plasma containment capillary. Most MIP systems are also susceptible to molecular interferences caused by water vapour and carbon dioxide, and these compounds must either be eliminated by selective trapping or separated chromatographically. Mulligan *et al.* [144] compared the responses when various microwave cavities were used in the determination of arsenic and antimony as their hydrides. They tested the Beenakker, 3/4-wave Evenson, 1/4-wave Evenson and Broida cavities, operated at atmospheric pressure with helium/argon or pure helium plasmas (3/4-wave Evenson and Beenakker only). All cavities had similar absolute detection limits for each element, of between 20 and 40 ng, and the response was linear over two orders of magnitude. A pure helium plasma, sustained in a Beenakker cavity was favoured, however, as it was easiest to tune, more reproducible, and more stable in the presence of foreign matter. Spectral background interferences were observed for arsenic and consituted 15–45% of the signal obtained for 0.5 μg of arsenic with the argon/helium plasma, compared to a much higher level of 60–75% with the pure helium plasma. Much lower detection limits (20 pg of As, 50 pg of Sb) have been attained by using a reduced pressure (5–10 mmHg) argon plasma [145]. Signal response for arsenic compounds was found to be relatively independent of the molecular structure [146].

The relative line intensities for arsenic in a pure helium plasma sustained in a Beenakker cavity, and measured with a high-resolution echelle spectrometer have been compared [147] (see Table 9.5). The echelle monochromator was preferred as its high resolution improved the selectivity and signal-to-noise ratio.

Table 9.5 — Relative line intensities for arsenic determination by microwave plasma AES microwave-induced helium plasma sustained at atmospheric pressure in a Beenakker cavity).

Wavelength (nm)	Relative emission intensity
193.696	90
200.334	190
228.812	750
234.984	1000

9.7.5.3 Miscellaneous detectors

Theoretically, atomic-fluorescence spectrometry (AFS) offers improvements in detection limits over AAS whilst retaining the advantage of selectivity. There has been little application of this technique, the major obstacle being that of finding stable and sufficiently intense excitation sources.

Dispersive AFS coupled with hydride introduction has been reported [148,149] to reduce detection limits by a factor of 2–30, compared to those obtained by AAS. Non-dispersive AFS has been employed for the determination of both arsenic and antimony [150,151]. This technique is potentially attractive, as arsenic and antimony have several lines in the vacuum ultraviolet, and using a solar blind photomultiplier eliminates the need for a monochromator. With an EDL source and an argon–hydrogen entrained air flame atomization cell, the detection limits were 0.05 ng for arsenic and 0.1 ng for antimony, and the calibration curves were linear over 4 decades [151].

Several detectors have been described which detect molecular emission. A possible drawback to their use is the lack of selectivity, and the increased susceptibility to interferences owing to the broadness of the emission bands. Two systems, however, have the attraction of being non-flame devices. Fujiwara *et al.* [152] used ozone chemiluminescence (detection limits 0.15 ng of As, 110 ng of Sb), and Tao *et al.* [153] determined arsenic and antimony by ultraviolet-stimulated photoluminescence by excitation at 210 nm. The broad band emission (250–700 nm) was monitored in a modified conventional fluorimeter at 325 nm. Limits of detection were 0.6 ng for arsenic and 9.0 ng for antimony.

9.8 HYDRIDE-GENERATION COLD-TRAP METHODS

9.8.1 Applications

These combination techniques involve the on-line coupling of a hydride generator and cold trap with detection by atomic spectrometry. The cold trap functions as both a preconcentrator when immersed in liquid nitrogen, and as a chromatographic column when the trap is warmed. This dual function is advantageous as it eliminates the need for separation by conventional gas chromatography. Such methods are dealt with in a separate section as they involve the use of purpose-built apparatus and because of their relative importance within the field of trace metalloid analysis. By cold-trap techniques, inorganic arsenic, MMA and DMA, as well as the corresponding antimony species, have been routinely determined in waters and extracts of sediments and biological tissues. Detection limits, which are largely dependent on the sample volume and detection system used, are typically in the low or sub ng/l. range. This makes the cold-trap techniques the most sensitive currently available for metalloid species determination. A comparison of cold-trap methods and their detection limits is given in Table 9.6.

The first combination method of this type was reported by Braman *et al.* [78] who determined inorganic and methylated arsenic species (Fig. 9.3). The hydrides are generated in a reactor at pH 1, purged from solution by helium as carrier gas and trapped in a glass U-tube half-packed with glass beads. On warming of the trap by electrical heating, the hydrides are eluted in order of their boiling points (Table 9.7) into a DC arc helium-glow AES detector (see Section 7.5.2). By use of a slightly modified version of this apparatus [79], inorganic antimony concentrations in natural waters have also been determined. Various improvements have been made to this system, including semi-automation and improvement of the detector [91]. This has resulted in improvements in both detection limits and reproducibility.

Cold-trap methodology was extended significantly by Andreae [20,27,34,38] who coupled it to AAS equipped with a sensitive silica cuvette burner, thus transforming the technique into one applicable in most analytical laboratories. With this system,

Table 9.6 — Comparison of hydride-generation cold-trap combination techniques for determination of arsenic and antimony species.

Species determined	System details	Detection limits	Precision	References
As(III), As(V), MMA, DMA, TMA	Discrete volume hydride generator. Cold trap half packed with glass beads. DC arc–helium glow detector	4–20 ng/l. (50 ml sample)	5–11% rsd at 5 μg/l.	[78]
As(III), As(V), MMA, DMA, TMA	Generally improved version of apparatus above	20 ng/l. (5 ml sample)	no data given	[91]
As(III), As(V), MMA, DMA	Discrete volume hydride generator. Trap half packed with 15% OV-3 on Chromosorb W. Silica cuvette AAS detector	0.3 ng/l. (100 ml sample)	At the μg/l. level about 4% rsd for inorganic species. Slightly worse for methylated species	[38]
As(III), As(V), MMA, DMA	Similar apparatus to above [38], with slight improvements	3–12 ng/l. (30 ml sample)	At the low μg/l. level: As(III), As(V); MMA 3–7% rsd; DMA 8–11% rsd	[21]
As(III), As(V), MMA, DMA	Continuous flow hydride generator. Narrow bore cold trap packed with glass beads. Heated silica tube AAS detector	~20 ng/l. (30 ml sample)	At 1 μg/l. level: 3–5% rsd for all species	[82]
Sb(III), Sb(V), MMS, DMS	Apparatus identical to that for As, but slightly different generation conditions used	~0.5 ng/l. (100–250 ml sample)	At the ng/l. level: Sb(III), Sb(V) 1–3% rsd; MMS, DMS 10–20% rsd	[38]

MMS=monomethylantimony; DMS=dimethylantimony; rsd=relative standard deivation.

Table 9.7 — Boiling points of various hydrides.

	B.P (°C)		B.P (°C)
arsine	−55	stibine	−17
monomethylarsine	2	monomethylstibine	41
dimethylarsine	35.6	dimethylstibine	60.7
trimethylarsine	70	trimethylstibine	79.4
phenylarsine	148		

reliable data have been obtained for metalloid speciation in natural waters. With sample volumes of about 100 ml, sub-ng/l. detection limits are attainable for both arsenic and antimony species (see Table 9.6). This system has undergone several stages of development and the current state-of-the-art apparatus for arsenic speciation is depicted in Fig. 9.2. The same apparatus may be used for antimony speciation analysis if the hydride generation conditions are modified. Detailed instructions are available in the literature on the setting up of this system for both arsenic [21,38] and antimony speciation analysis [38]. Typical chromatograms obtained for both arsenic and antimony species with this type of system are illustrated in Figs. 9.4 and 9.5 respectively. Peak area measurement is preferred to peak height as it gives better precision. One drawback of this system, however, is low sample throughput, typically 2 per hour.

An alternative cold-trap system for the determination of arsenic speciation is described by Howard *et al.* [19,52,82]; the major difference is the use of an automated continuous-flow hydride generator and a small-bore cold trap packed with glass beads (Fig. 9.6). Atomization is achieved in a silica tube heated in an air–acetylene flame or an electric furnace. The latter method is preferred as it is less noisy and more reproducible [19]. As there is little interaction of the hydrides with the packing material used, the hydrides are eluted within 1 min without additional heating. The chromatography is reproducible as long as room temperature is not subject to severe fluctuations. A typical chromatogram obtained with the system is reproduced in Fig. 9.7; adequate analytical performance (Table 9.6) is attained with peak height measurement. Semi-automation of hydride generation allows ease of operation, improved precision and higher sample throughput (typically 7 per hour), but, as a result of the relatively low sample volume (<10 ml), the limits of detection are about 20 ng/l. This is sufficient, however, for the determination of arsenic species in most estuarine and coastal waters.

9.8.2 Design considerations

Several practical problems may be encountered when combining hydride generation with cold-trapping. These result from back-pressure and clogging of the trap with water carried over from the hydride generator. Water condensation may be avoided by adding a drying stage after the hydride generation. Chemical drying with sodium hydroxide pellets [82], calcium chloride [140], or 85% phosphoric acid [91] have proved effective. In some cases, chemical agents are reported to adsorb some of the hydrides so a physical drying trap such as a U-tube immersed in an ice-bath [20] may be preferred. Alternatively, a design of cold trap that has the capacity to accommo-

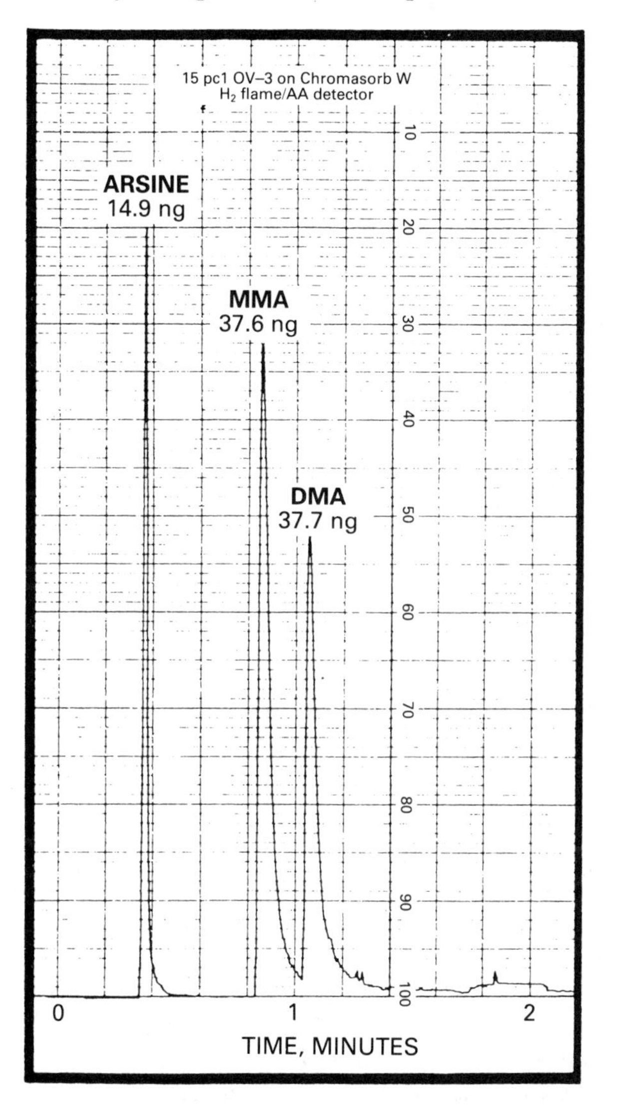

Fig. 9.4 — Typical chromatogram of arsenic standards, obtained with the hydride-generation/cold-trap AAS system described by Crecelius *et al.* (see Fig. 9.2). Retention times are arsine 24 sec, monomethylarsine 53 sec, dimethylarsine 66 sec. (Reproduced from [21] by permission of the copyright holders).

date ice build-up has been applied (e.g. [21,38,79]). These traps are dried out by heating after every determination.

Ideally, cold-trap packings should be capable of quantitative preconcentration, should not irreversibly retain hydrides, should give adequate chromatographic separation on warming and be resistant to the effects of water vapour and condensation. Several trap packings have been advocated, including glass beads, 200 μm–3 mm in diameter (e.g. [78,82,85,140]), silanized glass wool [20] and PTFE shavings [154,155]. There has been much disagreement between workers regarding the

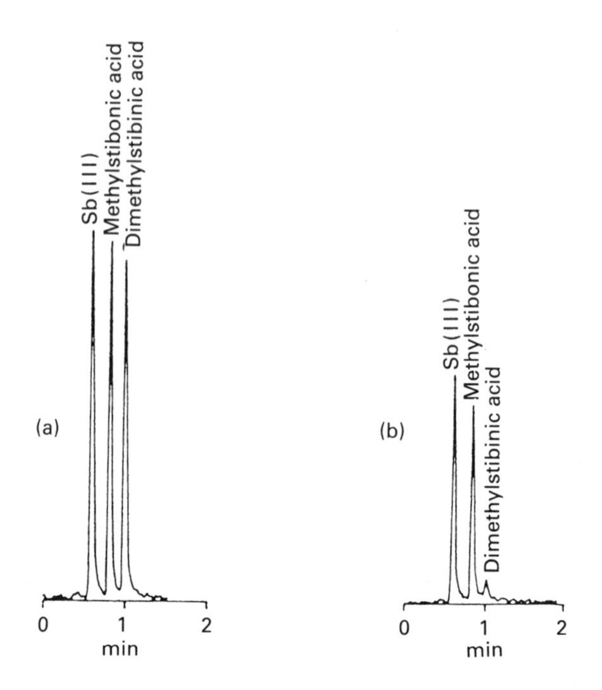

Fig. 9.5 — Typical chromatogram of antimony species, obtained with the hydride-generation/cold-trap AAS system described by Andreae *et al.* [27,34,38]. (a) Stibines produced by sodium borohydride reduction of 2ng of Sb each as Sb(III), methylstibonic acid and dimethylstibinic acid. (b) Analysis of a sample of seawater (100 ml) from the open Gulf of Mexico, 12 March, 1981.

relative merits of the various packings. A very effective cold trap consists of a U-tube packed with 3% OV-3 on Chromosorb W, which is first conditioned by silylation and heating [21,38]. To elute the hydrides, the trap must be heated to above room temperature. This is achieved by electrical heating, which gives a reproducible temperature programme. Howard and co-workers [19,82] have found that silylated glass beads packed in a narrow bore U-tube of special design (see Fig. 9.6) give acceptable separations, and recent work [19] has shown that this trap, packed with hydrofluoric acid-etched glass beads (40 mesh), appears to give the best separation of arsenic species (Fig. 9.7).

For the reproducible operation of cold-trap apparatus, it is important to ensure that the trap is kept reasonably free from water. This is best achieved by thorough drying of the trap by heating between determinations. To avoid adsorption losses and peak tailing, it is advisable to keep transfer lines as short as possible. In addition, Crecelius *et al.* [21] recommend that the transfer line between trap and detector is heated to about 80°C. This avoids water condensation which may otherwise affect signal response, especially that for dimethylarsine. It is also advisable to condition the apparatus prior to starting analysis, by running a few high-concentration standards.

There is little information available regarding the analysis of higher boiling-point hydride derivatives, although there are reports of trimethylarsenic species determination by hydride generation (see Table 9.6). Poor detection limits may, however, be expected for higher molecular-weight hydrides, owing to lower volatility and resultant peak broadening.

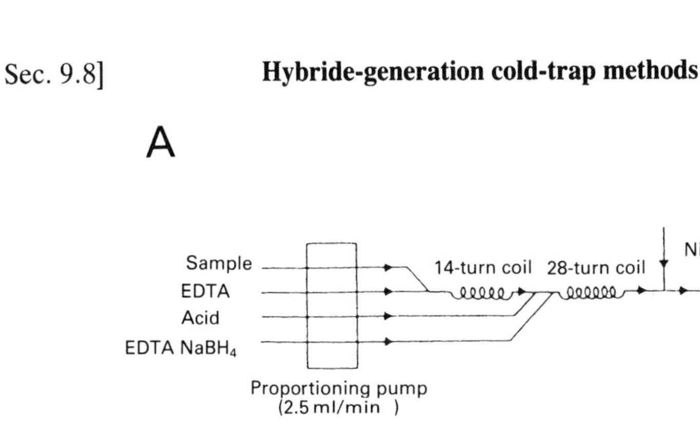

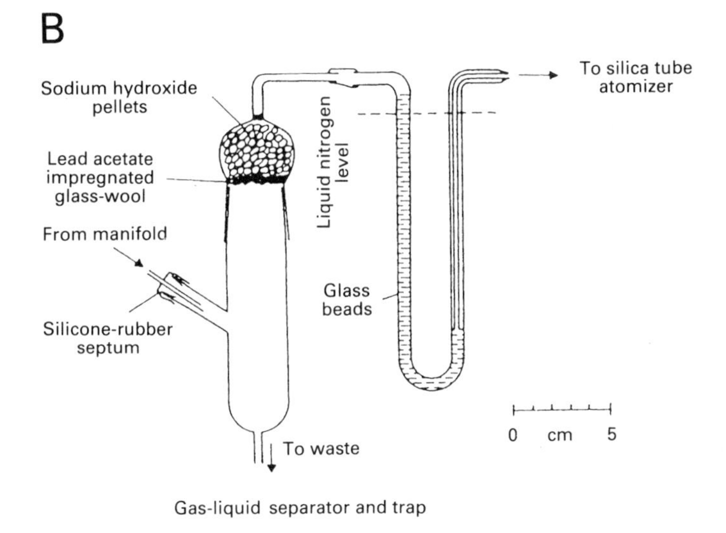

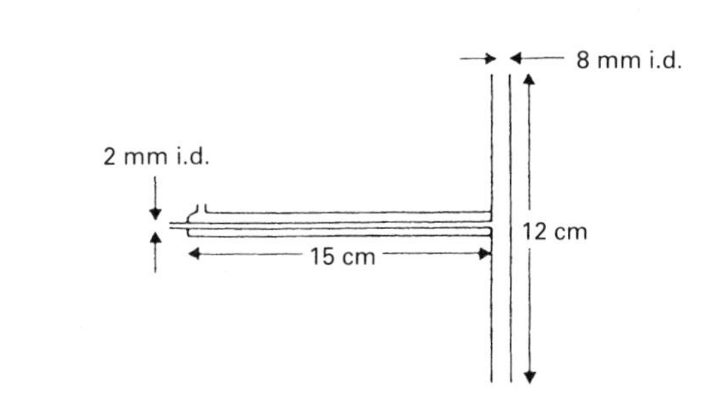

Fig. 9.6 — Details of continuous-flow hydride-generation/cold-trap/AAS system for the determination of arsenic speciation. (a) Schematic diagram of the continuous flow hydride generation system (the EDTA reagent is optional, depending on whether interferences are encountered). (b) Gas–liquid separator and cold trap. (c) Silica tube atomizer which is positioned in the AAS light-path. The tube may be heated electrically or by an air-acetylene flame. (Reproduced by permission from A. G. Howard and M. H. Arbab–Zavar, *Analyst,* 1981, **106,** 213. Copyright 1981, Royal Society of Chemistry).

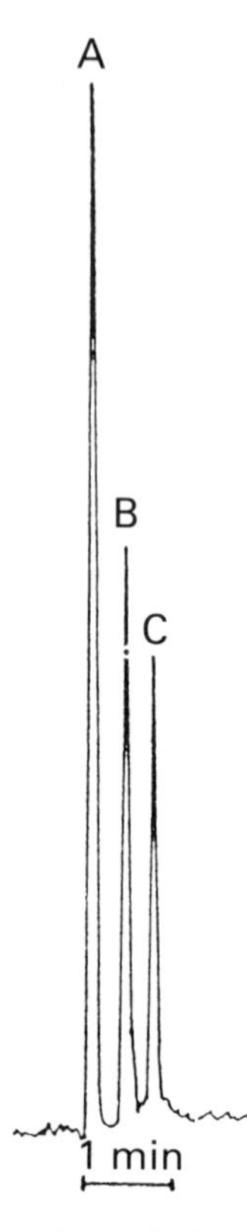

Fig 9.7 — Typical chromatogram of arsenic standards, obtained by using the apparatus shown in Fig. 9.6. The peaks (1 ng, as arsenic) are for (a) arsine, (b) monomethylarsine, (c) dimethylarsine. The arsine peak includes a reagent blank contribution of 0.3 ng.

9.9 COUPLED GAS CHROMATOGRAPHY METHODS

9.9.1 General considerations

There have been few reliable combination GC methods reported for arsenic species. This is due mainly to the lack of suitable derivatization procedures. No GC combination methods have been reported specifically for antimony species, although inorganic antimony has been determined as its hydride [156–157] and its triphenylstibine derivative [145]. Most GC separations have been done on packed columns, and to date there has been negligible use of capillary columns.

9.9.2 Derivatization and gas chromatographic separation

It is desirable that any derivatization step should be simple, applicable to a wide range of arsenic species and form high yields of derivatives that are stable on storage. There are few, if any, methods that meet these criteria. Unless it is directly coupled to GC (for instance by a cold trap), hydride generation is unsuitable, as the derivatives are generally too volatile to be concentrated by procedures such as solvent extraction. Nevertheless, dimethylarsine has been extracted into toluene with satisfactory recovery [146]; however, this method could not be used for monomethylarsine, as losses were too great. Both monomethylarsine and dimethylarsine have been successfully preconcentrated by gas purging from solution and trapping in cold toluene (−5°C). The authors reported, however, that rearrangement reactions took place in the toluene extracts, which makes the validity of this method questionable.

Both arsine and stibine are amenable to GC separation. The thermal stability of arsine and stibine has been investigated and no degradation at a temperature of 40°C was detected over a period of 4 min [159].Owing to the low boiling points, low-temperature separations, often on rather crude columns, have been used. For instance, adequate separations have been attained on packed columns containing 10% silicone gum rubber on Porapak Q (50–80 mesh), maintained at 16°C [160], silica gel (80–100 mesh) at 45°C [161], and plastic tubes packed with Chromosorb 102 (60–80 mesh) maintained at room temperature [157].

Several derivatization methods have been reported for arsenic species. These include the formation of the dithiocarbamates of MMA, DMA and inorganic arsenic [162], trimethylsilyl derivatives [163], and iododimethylarsine formed by derivatization of DMA [164,165]. None of these methods, however, has proved satisfactory, owing to poor derivative stability or degradation of derivatives on the column. Phenylation of inorganic arsenic and antimony under anhydrous conditions, with phenylmagnesium bromide in benzene or ether [145,166] to form triphenyl-arsine and triphenylstibine has been reported. Prior preconcentration from aqueous samples was achieved by co-crystallization with thionalide [145] or extraction with dithiocarbamate followed by evaporation to dryness [166].

Arsenic(III) and MMA may be derivatized quite simply in aqueous media by complexation with 2,3-dimercaptopropanol [British Anti-Lewisite (BAL)], to form stable five-membered ring structures which are extractable into benzene [167]. The derivatives were separated on a 2-m column packed with 20% SE-30 on 80/100 Chromosorb W AW DMCS. Detection by sulphur-specific flame photometry yielded detection limits of 0.02 ng of As(III) and 0.04 ng of MMA. Unfortunately, DMA does not react with BAL. This method has been used for the interference-free determination of inorganic arsenic in marine waters, tissues and sediments [168]. By separation on a 1-m glass column packed with 5% OV-101 on Chromosorb W and electron capture detection a theoretical detection limit of 10 pg (two times the baseline noise) was attained, but in practice this was limited by a reagent blank of around 160 pg [168].

Inorganic arsenic, MMA and DMA react in aqueous media with thioglycollic acid methyl ester (TGM) [169] to form derivatives extractable into cyclohexane. TGM also reduces arsenic(V) to arsenic(III), so differentiation between oxidation states cannot be made. EDTA was added to the solutions to mask possible transition metal interfernces. No interferences were found when analysing blood or urine samples. Derivatives (5 ppm concentration) were reported to be stable for more than 4 days. Dix *et al.* [170] have satisfactorily separated these derivatives on a wide bore 10-m non-polar capillary column, the separation being complete within 15–20 min.

9.9.3 Applications of coupled GC methods

9.9.3.1 GC–AAS

Except for cold-trap methods, there are virtually no couplings of GC apparatus to

AAS reported for the determination of arsenic or antimony species. Norin *et al.* [56] describe a pyrolysis GC—AAS system for the determination of arsenic compounds in purified biological tissue extracts. A commercial GC and pyrolyser unit were coupled to a silica cuvette burner of the type described by Andreae [20]. After sample pyrolysis at 400°C for 5 sec and separation on a 2-m packed column (10% UCC W 9821 on Chromosorb W AW 80/100 mesh), an arsenic-specific chromatogram was obtained which contained a maximum of two peaks. These were identified by fast atom bombardment mass spectrometry as trimethylarsine and dimethylvinylarsine. With this system, it was possible to differentiate arsenobetaine, which on pyrolysis gave a single peak (trimethylarsine), from arsenocholine and acetylarsenocholine, which both gave two peaks.

9.9.3.2 GC-AES (including multielement methods)

Talmi and Norvell [145] describe a coupled GC–MIP system operated at low pressure, for the determination of arsenic and antimony as their triphenyl derivatives (formed by Grignard reaction) and of several alkylarsines dissolved in cold toluene. Detection wavelengths were the 228.8 nm arsenic and 259.8 nm antimony lines. This system was very sensitive; absolute detection limits were 20 pg for As, 50 pg for Sb, but sample detection limits were not as impressive (low μg/l. range) owing to the small injection volumes used. With a packed column (2.5% silicone XE-60 on Chromosorb G) coupled with an MIP employing an atmospheric helium plasma sustained in a Beenakker cavity, DMA and MMA have been determined as their thioglycollic methyl ester derivatives [171]. Detection was by monitoring the arsenic emission at 228.8 nm, or the more intense sulphur emission at 545.4 nm. The absolute detection limit with the sulphur line was 80 pg.

A number of papers have been produced on multielement hydride generation procedures which couple the same basic hydride generation GC system to various atomic spectrometry detectors. The complex system (Fig. 9.8) involves semi-automated hydride generation, cold-trapping in a condensation U-tube packed with glass helices, and separation on a Chromosorb 102 column at room temperature. The detectors used include MIP–AES [157,158], ICP–AES [155] and AAS [172). For the MIP system with an Evenson 1/4-wave cavity and helium–argon plasma, the response was linear over 2 orders of magnitude, and detection limits of 7 ng of As and 10 ng of Sb were attained (sample volume 20 ml). Inorganic As, Se, Sn and Sb have been determined in extracts of whole blood, enriched flour and NBS orchard leaves, following suitable pretreatment procedures [158]. There are no reports of this system being used for routine sample analysis.

9.10 HPLC COMBINATION METHODS

9.10.1 General considerations

There are several papers devoted to the subject of HPLC coupled with atomic spectrometry and its application to determining arsenic speciation in waters [173–176], soil extracts [177,178], biological tissues [62,179–181] and oil shale process waters [182]. The species determined include arsenite, arsenate, MMA, DMA, arsenobetaine, arsenocholine and phenylarsonic acid. There are no details, however, of HPLC combination methods for the determination of antimony species.

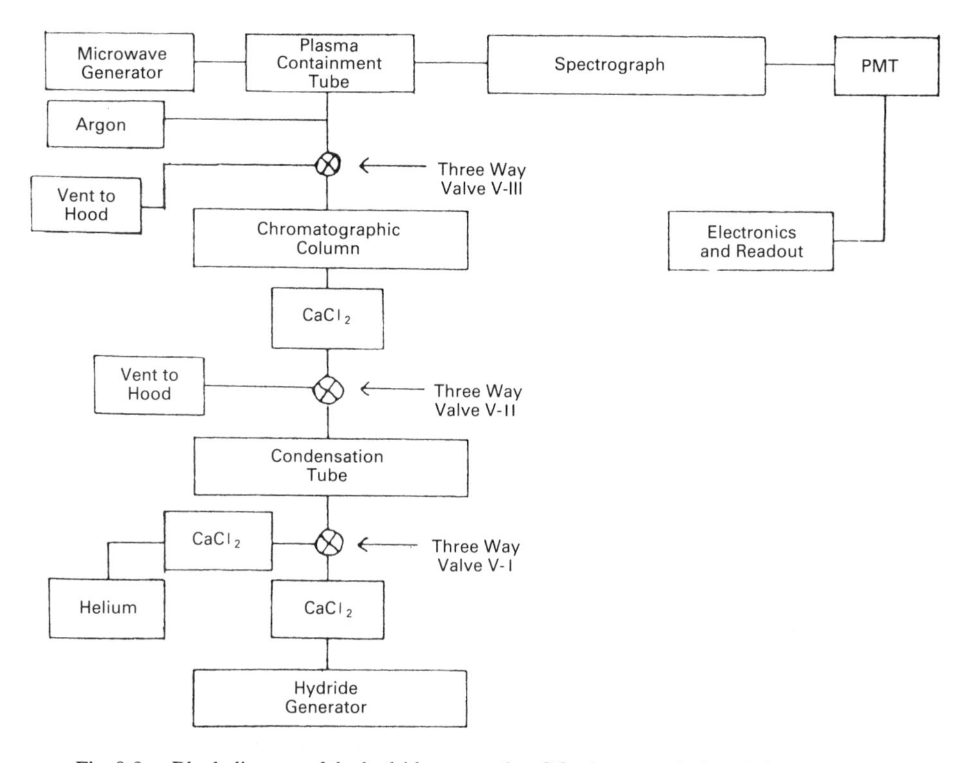

Fig. 9.8 — Block diagram of the hydride-generation GC microwave-induced plasma system for the determination of inorganic As, Sb, Se and Sn. (Reproduced by permission from W. B. Robbins and J. A. Caruso, *J. Chromatog. Sci.*, 1979, **17**, 360. Copyright 1979, Preston Publications, Inc.).

A comparison of selected coupled HPLC methods, including their detection limits, may be found in Table 9.8.

Sample preconcentration is a potentially important requirement for HPLC methods, to enhance detection limits and remove interferents. Detection limits of between 10 and 200 μg/l. are currently achieved by HPLC combination methods and may be adequate for biological samples, but for natural waters where typical DMA concentrations may be less than 0.2 μg/l., preconcentration by at least 100–500-fold is required. Practically, this precludes HPLC from water analysis, except for highly contaminated waters. There has been only one report on sample preconcentration prior to HPLC determination of arsenic species. Tye *et al.* [173] preconcentrated arsenate, DMA and MMA from acidified solutions on a precolumn packed with Zipax pellicular anion-exchange resin. This method however, had the serious drawback of not preconcentrating arsenite.

9.10.2 HPLC separation of arsenic species

The polarity of inorganic and most organoarsenic species makes them amenable to both ion-exchange HPLC and reversed-phase HPLC separations. For application in biological studies, it would be desirable to separate arsenite, arsenate, MMA, DMA, and trimethylarsenic species such as arsenobetaine and arsenocholine in one run. No

Table 9.8 — Comparison of HPLC combination methods for the determination of arsenic species.

Combination	Species determined	Matrix	Chromatography	Performance	Reference
RP-HPLC ZAA GFAAS	ASB, ASC, arsenite/arsenate	standards only	C18 μ-Bondapak column; eluent: water, acetic acid, acetonitrile mixture with ion-pairing reagent.	no data given RT 28 min	[183]
RP-HPLC ZAA GFAAS	MMA, DMA arsenite, arsenate	synthetic river water	C-18 Lichrosorb RP-1 column; eluent switching with 2 eluents	lod 150 μg/l. RT 50 min precision 3–8% at 2.6 μg/ml	[174]
RP-HPLC ICP–AES	MMA, DMA, arsenite, arsenate, phenylarsonate	standards only	Hamilton PRP column; eluent switching with 3 eluents	lod 130 μg/l. RT 27 min	[127]
RP–HPLC ICP–AES	ASB, ASC MMA, DMA	biological tissue extracts	Hamilton PRP column; eluent: 0.05*M* Na heptanesulphonate in 2.5% acetic acid solution	lod 75 μg/l. RT 10 min	[62]
IE–HPLC ICP–AES	ASB, DMA, MMA, arsenate, arsenite	biological tissue extracts	Nucleosil strong anion-exchange column; eluent: 0.025*M* phosphate buffer	lod about 150 μg/l. RT 15 min	[179]
IE–HPLC ICP–AES	DMA, MMA, arsenite, arsenate,	biological tissue extracts	Aminex A27 anion-exchange column; gradient elution: water to 0.5*M* ammonium carbonate solution; column and injector heated to 45°C	lod 120 μg/l. RT 20 min	[180]

IE-HPLC HYDRIDE AAS	DMA, MMA, arsenite, arsenate, *p*-aminophenylarsonate	standards only	Dionex pellicular low-capacity anion-exchange column; carbonate/borate eluent	lod 10 μg/l. precision 11% rsd at 20 μg/l. RT 10 min	[184]
IE-HPLC HYDRIDE AAS	DMA, MMA arsenite, arsenate	soil waters, drinking waters	BAX 10 strong anion-exchange column; ammonium carbonate eluent	lod about 20 μg/l. RT 4 min	[173]
IE-HPLC HYDRIDE AAS	DMA, MMA, arsenite, arsenate	seaweed extracts	Nucleosil 10SB anion-exchange column; phosphate buffer eluent	lod about 40 μg/l. RT 25 min	[181]
IE-HPLC GFAAS	DMA, MMA arsenite, arsenate	soil extracts	Dionex pellicular anion-exchange column; gradient eluetion with methanol, water, ammonium carbonate solutions	lod 0.5 μg/g (in soils) RT 36 min	[178]
IE-HPLC GFAAS	arsenite, arsenate DMA, MMA, phenylarsonate	oil shale process waters	Dionex pellicular anion-exchange column, gradient elution with water, methanol and ammonium carbonate solution	lod 5–20 μg/l. RT 50 min	[182]

RT=run time (excluding column conditioning time); lod=limit of detection; IE=ion-exchange; RP=reversed phase; ZAA=Zeeman effect background correction AAS; ASB=arsenobetaine; ASC=arsenocholine.

system is yet available that achieves this comprehensive separation. For waters and sediment or soil extracts, separation of arsenite, arsenate, MMA and DMA may be adequate. It is important to realize that even with the advantages of element-specific detection, biological samples and soil extracts may require extensive sample preparation and purification prior to analysis [e.g. 62,178] in order to avoid interferences occurring at the separation stage. Clean-up methods include TLC, solvent extraction and column chromatography.

For arsenite, arsenate, MMA and DMA, some of the most impressive separations have been obtained on strong anion-exchange and low-capacity pellicular anion-exchange resins (e.g. Figs. 9.9, 9.10a and 9.11b) with water, methanol and

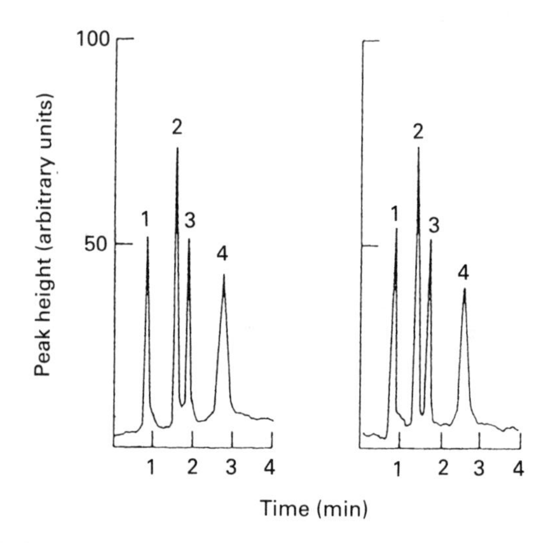

Fig. 9.9 — HPLC hydride-generation AAS. Duplicate chromatograms of 10 ng arsenic standards (injection volume 100 μl). The peaks are for (1) arsenite, (2) dimethylarsinite, (3) monomethylarsonate, (4) arsenate. Separation was achieved on a strong anion-exchange column with 0.1*M* ammonium carbonate eluent. (Reproduced by permission from C. T. Tye, S. J. Haswell, P. O'Neill and K. C. C. Bancroft, *Anal. Chim. Acta,* 1985, **169,** 195. Copyright 1985, Elsevier Science Publishers).

ammonium carbonate solution as eluents. For higher biomolecules, the best separations are depicted in Fig. 9.11(a and c). A rapid separation of several arsenic species has been described by Spall *et al.* [180]. On an anion-exchange column with an elution gradient from water to 0.5*M* ammonium carbonate, arsenate, arsenite, DMA, MMA , phenylarsonic acid and triphenylarsine oxide were separated within 15 min (Fig. 9.11b). The column was held at 45°C to improve resolution. Low *et al.* [185] describe a complex column-switching technique which involves both ion-exchange and reversed-phase columns coupled with ICP–AES detection. Arsenate, arsenite, MMA, DMA and arsenobetaine were separated.

9.10.3 HPLC–AAS

Owing to the insensitivity of flame AAS, its direct coupling to HPLC has rarely been used. Morita *et al.* [179] found HPLC–AAS with an argon–hydrogen flame atom cell

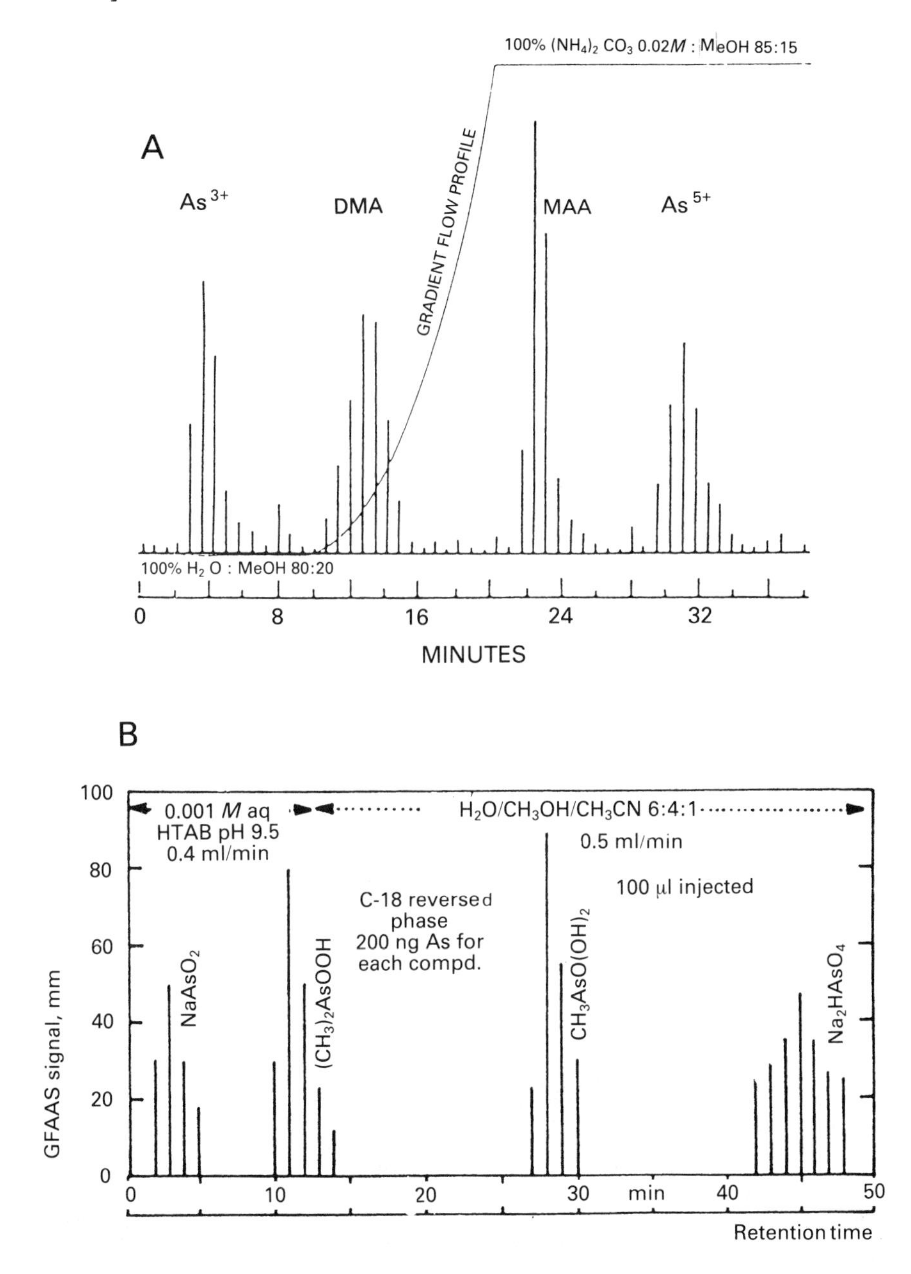

Fig. 9.10 — Comparison of HPLC–GFAAS chromatograms for the separation of arsenate, arsenite, monomethylarsonate and dimethylarsinite. (A) Ion-exchange HPLC. Separation on a low-capacity pellicular anion-exchange column by gradient elution, 20 μl of standard mixture (30 ng of each compound). Vertical axis is the integrated peak area. Maximum response corresponds to 0.3 absorbance. (Reproduced by permission from A. E. Woolson and N. Aharonson, *J. Assoc. Off. Anal. Chem*., 1980, **63,** 523. Copyright 1980, Association of Official Analytical Chemists). (B) Reversed-phase HPLC. Chromatographic conditions as indicated on the diagram. (Reproduced from [174] by permission of the copyright holders, CEP Consultants).

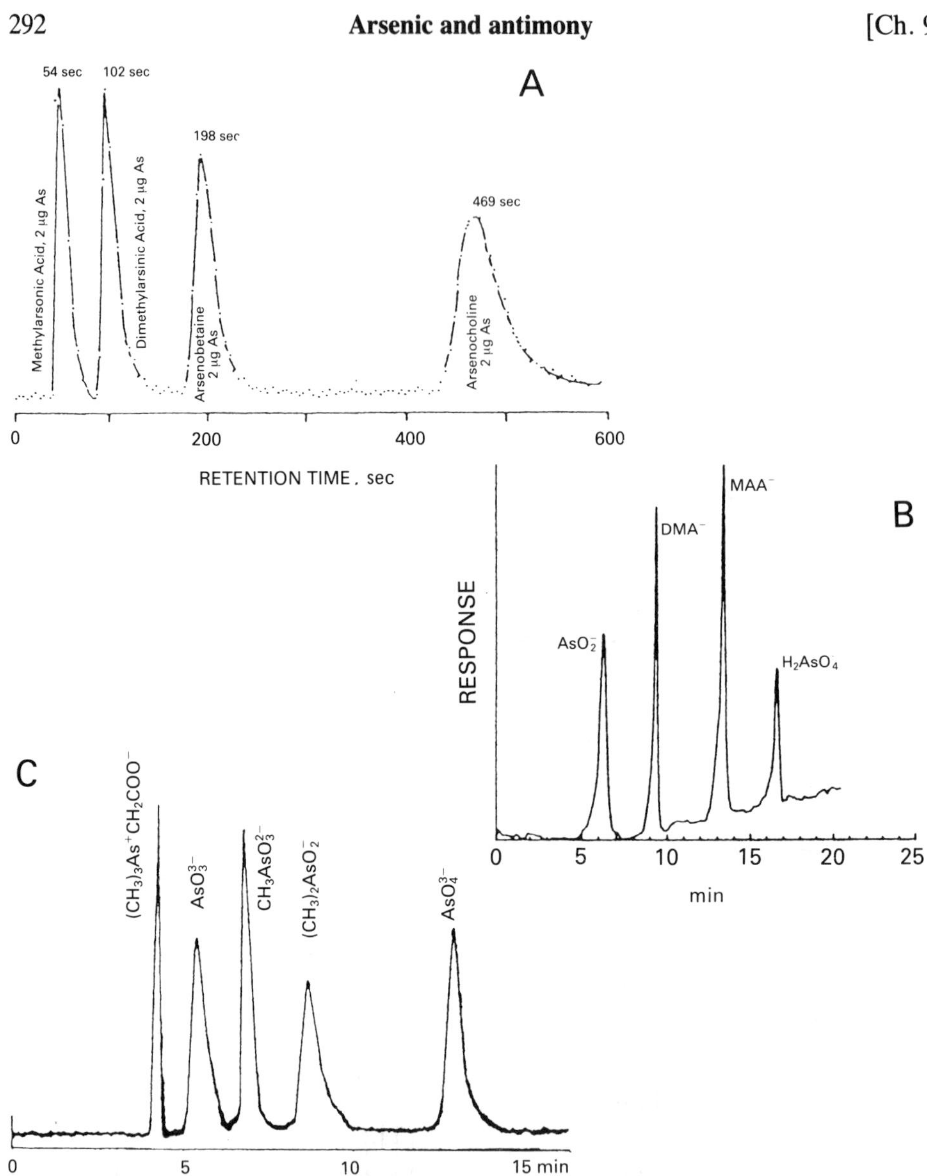

Fig. 9.11 — Comparison of HPLC–ICP–AES chromatograms of arsenicals found in biological tissues. (A) Reversed-phase separation of arsenocholine, arsenobetaine, methylarsonate and dimethylarsinite. Column, Hamilton PRP-1; mobile phase, 0.05*M* sodium heptane sulphonate in 21.5% aqueous acetic acid. Injection volume, 100 µl. (Reproduced by permission from K. A. Francesconi, P. Micks, R. A. Stockton and K. J. Irgolic, *Chemosphere,* 1985, **14,** 1443. Copyright 1985, Pergamon Press). (B) Separation of arsenate, arsenite, methylarsonate and dimethylarsinite. Aminex A-27 Radial PAK anion-exchange column with a linear elution gradient from water to 0.5*M* ammonium carbonate in 15 min followed by isocratic flow for 5 min. (Reproduced by permission from W. D. Spall, J. Glynn, J. L. Andersen, J. G. Valdez and L. R. Gurley, *Anal. Chem.,* 1986, **58,** 1340. Copyright 1986, American Chemical Society). (C) Separation of arsenate, arsenite, methylarsonate, dimethylarsinite and arsenobetaine. Nucleosil strong anion-exchange column, 0.025*M* phosphate buffer eluent, 350 ng (as arsenic) of each standard injected. (Reproduced by permission from M. Morita, T. Uehiro and K. Fuwa, *Anal. Chem.,* 1981, **53,** 1806. Copyright 1981, American Chemical Society.)

had only a twentieth of the sensitivity of HPLC coupled to an ICP–AES. No limit of detection was reported. Post-column on-line hydride generation can be used to give a substantial gain in sensitivity and improvement in detection limit [173,176,181,184] (see Fig. 9.9 and Table 9.8), but peak broadening due to dilution and dead-volume effects can be a drawback. The hydride-coupled systems are also limited to the determination of arsenic species that form volatile hydrides. The determination of non-hydride-reducible species such as arsenobetaine might be possible if some form of on-line digestion step were applied prior to hydride generation.

The physical means of coupling GFAAS to HPLC are described in Chapter 6. As GFAAS is a discrete sampling technique, with a minimum of 45–60 sec cycle time between determinations [183,186], an element-specific chromatogram consisting of discontinuous pulses is obtained (Fig. 9.10). Most significantly, the discontinuous nature of the detector signal and its relatively slow cycle time place limits on the chromatographic separation. Broad peaks are required for reliable determination of species. Measurement can be achieved equally effectively by peak area determination or more simply, by summing individual GFAAS signals [175]. It is important to select an eluent that does not interfere with the GFAAS determination or affect sensitivity. This is particularly important when using gradient elution or solvent-switching methods. An eluent that requires a minimum of thermal pretreatment and gives interference-free determination is desirable, hence the application of ammonium carbonate solution and water as eluents [177,182].

HPLC–GFAAS systems for the determination of arsenic species have been developed for both ion-exchange [175,177,182,186] and reversed-phase columns [62,174,175,183] (Table 9.7). Eluents are generally more complex with reversed-phase separations and may result in a high background molecular absorption, requiring the use of a Zeeman-effect background-correction sytem [187]. In contrast, ion-exchange HPLC has been successfully coupled to AA spectrometers having only deuterium lamp background-correction. Both systems have been evaluated for the determination of arsenic species in soil extracts and waters [175] and were found comparable in terms of accuracy, but detection limits with the anion-exchange sytem were an order of magnitude lower. Detection limits, which largely depend on the type of spectrometer used, range between 20 [182] and 150 μg/l. [174]. Two typical HPLC–GFAAS separations are presented in Fig. 9.10.

9.10.4 HPLC–AES

There is only one report of coupling HPLC to a direct current plasma spectrometer (DCP). Morita *et al.* [179] found DCP detection had only a fifth of the sensitivity of ICP–AES. No limit of detection was given.

ICP–AES has proved to be the most effective arsenic-specific liquid chromatography detector to date, as it is reasonably interference-free and allows continuous monitoring of the eluent stream. This must, however, be balanced against the high cost of ICP–AES instrumentation. Coupling is straightforward, and may be simply achieved by a PTFE tube connection and adjustment of the HPLC flow-rate to match the nebulizer uptake rate. With conventional ICP–AES apparatus, detection limits of between 75 and 150 μg/l. [62,127,179,180] have been reported for various arsenic species, and both reversed-phase and ion-exchange separations (see Table 9.7). Typical chromatograms are illustrated in Fig. 9.11.

9.11 STANDARDIZATION AND ACCURACY

Primary inorganic standards are available for both arsenic and antimony. Methylated arsenic standards are commercially available but may be impure. It is therefore advisable to check their purity by speciation analysis, and their total arsenic concentration by flame AAS. If necessary, purification may be achieved by column chromatography [72]. Antimonates have variable stoichiometries, so it is advisable to check at least the total antimony concentration by cross-calibration by flame AAS against antimony(III) primary standards.

There are no commercially available standards of higher arsenic biomolecules. Arsenobetaine and arsenocholine have to be synthesized by literature procedures [12,56,61]. The same is true for organoantimony compounds. The synthesis of methylantimony species has been described by Meinema and Noltes [188], but there is now some doubt about the authenticity of the structures reported.

There are at present no reference materials available for speciation measurements, which makes assessment of analytical accuracy difficult. Several standard reference materials are, however, available for waters, sediments and tissues, that are certified in terms of total inorganic arsenic and antimony content. These should be used in the absence of better alternatives. Some doubts have been expressed over the arsenic reference value for Bowen's Kale; two independent groups [52,189] have reported values around 0.27 μg/g, compared to the certified value of 0.14 μg/g.

Crecelius *et al.* [21] have noted that dilute arsenite standards can be oxidized spontaneously if prepared in demineralized water, probably as a result of trace oxidants present in solution. As a consequence they preferred to prepare standards in river water having low arsenite concentrations. A similar problem has been reported by Andreae [38] for preparation of dilute antimony(III) standards. This can be overcome by adding ascorbic acid, or by preparing the standards in natural waters. In many laboratories, distilled water may prove satisfactory for low-level standard preparation.

9.12 CONCLUSIONS AND FUTURE DEVELOPMENTS

Element-specific detection offers better selectivity than do conventional GC and HPLC detectors and in some cases better sensitivity. Conventional gas chromatographic methods have so far failed to make a significant impact on arsenic and antimony speciation analysis, primarily because of the absence of satisfactory derivatization procedures. Combination methods that involve hydride generation with cold trapping are powerful techniques for the speciation analysis of aqueous samples containing arsenic and antimony at sub ng/l. concentrations. These techniques are typically 1000-fold more sensitive than the coupled liquid chromatography techniques presently available. The latter techniques, however, come into their own for the identification of biologically important molecules. There is still room for further work in the development of more rapid HPLC separations, and methods for the determination of arseno-sugars and as yet unidentified arsenic biomolecules. HPLC methods have not so far been applied to antimony speciation analysis. Progress on this front will require the development of suitable preconcentration steps and improvements in detector sensitivity.

Many of the coupled methods described in this chapter require lengthy analysis times of more than 30 min per sample (i.e. no more than 16 determinations in an average working day). This considerably limits sample throughput and leaves little scope for accurate calibration and checking for instrumental drift. The balance between acceptable analytical performance and sample throughput should be taken into account when selecting methods.

REFERENCES

[1] A. C. Chapman, *Analyst,* 1926, **51,** 548.
[2] F. Challenger, *Chem. Rev.,* 1945, **362,** 315.
[3] M. O. Andreae and D. D. Klumpp, *Environ. Sci. Technol.,* 1979, **13,** 738.
[4] E. A. Crecelius, *Environ. Health Persp.,* 1978, **19,** 147.
[5] S. A. Peoples, in *Arsenic: Industrial, Biomedical Environmental Perspectives; Proc. As Symp. Gaithersburg Maryland,* W. H. Lederer and R. J. Fensterheim (eds.), Van Nostrand Reinhold, New York, 1983.
[6] R. S. Braman and C. C. Foreback, *Science,* 1973, **182,** 1247.
[7] M. O. Andreae, in, *Organometallic Compounds in the Environment: Principles and Reactions,* P. J. Craig (Ed.), Chapter 5, Longmans, Harlow, 1986.
[8] A. G. Howard, M. H. Arbab-Zavar and S. C. Apte, *Est. Coastal Shelf Sci.,* 1984, **19,** 493.
[9] J. S. Edmonds and K. A. Francesconi, *Nature,* 1981, **289,** 602.
[10] S. A. Peoples, in *Arsenical Pesticides,* E. A. Woolson (ed.), Chapter 1. American Chemical Society, Washington D.C., 1975.
[11] T. Kaise, S. Watanabe and K. Itoh, *Chemosphere,* 1985, **14,** 1327.
[12] J. R. Cannon, J. S. Edmonds, K. A. Francesconi, C. L. Raston, J. B. Saunders, B. W. Skelton and A. H. White, *Aust. J. Chem.,* 1981, **34,** 787.
[13] F. C. Knowles and A. A. Benson, *Trends Biochem. Sci.,* 1983, **8,** 178.
[14] M. J. Gresser, *J. Biol. Chem.,* 1981, **256,** 5981.
[15] E. O. Uthus, W. E. Cornatzer and F. H. Nielsen, in *Arsenic: Industrial, Biomedical Environmental Perspectives; Proc. As Symp. Gaithersburg, Maryland,* W. H. Lederer and R. J. Fensterheim (eds.), Van Nostrand Reinhold, New York, 1983.
[16] R. D. Wauchope, in *Arsenic: Industrial, Biomedical Environmental Perspectives; Proc. As Symp. Gaithersburg, Maryland,* W. H. Lederer and R. J. Fensterheim (eds.), Van Nostrand Reinhold, Princeton, 1983.
[17] M. O. Andreae, *Deep Sea Res.,* 1978, **25,** 391.
[18] M. O. Andreae and P. N. Froelich, *Tellus,* 1984, **36B,** 101.
[19] S. C. Apte, *Ph.D Thesis,* University of Southampton, 1985.
[20] M. O. Andreae, *Anal. Chem.,* 1977, **49,** 820.
[21] E. A. Crecelius, N. S. Bloom, C. E. Cowan and E. A. Jenne, *Speciation of Selenium and Arsenic in Natural Waters and Sediments,* Vol. 2, *Arsenic Speciation,* EPRI, Battelle Northwest Laboratories, Washington, 1986.
[22] P. R. Walsh, R. A. Duce and J. L. Fasching, *J. Geophys. Res.,* 1979, **84,** 1710.
[23] J. Flanjak, *J. Sci. Food Agric.,* 1982, **33,** 579.
[24] J. F. Lawrence, P. Michalik, G. Tam and H. B. S. Conacher, *J. Agric. Food Chem.,* 1986, **34,** 315.
[25] P. J. Craig, in *Organometallic Compounds in the Environment: Principles and Reactions,* P. J. Craig (ed.), Chapter 10. Longmans, Harlow, 1986.
[26] E. Berman, *Toxic Metals and their Analysis,* Heyden, London, 1980.
[27] M. O. Andreae, J. F. Asmodé, P. Foster and L. Van't dack, *Anal. Chem.,* 1981, **53,** 1766.
[28] K. K. Bertine and D. S. Lee, in *Trace Metals in Seawater,* Vol. 4, C. S. Wong, E. Boyle, K. W. Bruland, J. D. Burton and E. D. Goldberg (eds.), Plenum Press, New York, 1983.
[29] S. C. Apte, A. G. Howard, R. J. Morris and M. J. McCartney, *Mar. Chem.,* 1986, **20,** 119.
[30] L. S. Austin and G. E. Millward, *Cont. Shelf Res.,* 1986, **6,** 459.
[31] R. Kantin, *Limnol. Oceanog.,* 1983, **28,** 165.
[32] W. Maher, *Anal. Lett.,* 1986, **19,** 295.
[33] T. M. Leatherland and J. D. Burton, *J. Mar. Biol. Assoc. UK,* 1974, **54,** 457.
[34] M. O. Andreae, in *Trace Metals in Seawater,* Vol. 4, C. S. Wong, E. Boyle, K. W. Bruland, J. D. Burton and E. D. Goldberg (eds.), Plenum, Press, New York, 1983.
[35] V. Cheam and H. Agemian, *Analyst,* 1980, **105,** 737.
[36] R. Massee, F. M. J. M. J. Maessen and J. J. M. de Goeij, *Anal. Chim. Acta,* 1981, **127,** 181.
[37] A. M. Gunn, *Research Centre Medmenham, Tech. Rept.,* TR 169, 1981.

[38] M. O. Andreae, in *Methods of Seawater Analysis,* K. Grasshof, M. Ehrhardt and K. Kremling (eds.), pp. 168–173. Verlag-Chemie, Weinheim, 1982.
[39] J. Aggett and M. R. Kriegman, *Analyst,* 1987, **112,** 153.
[40] R. E. Sturgeon, S. N. Willie and S. S. Berman, *Anal. Chem.,* 1985, **57,** 6.
[41] M. Thompson, B. Pahlavanpour and L. T. Thorne, *Water Res.,* 1981, **15,** 407.
[42] R. L. Braman, in *Arsenical Pesticides,* E. A. Woolson (ed.), Chapter 8. American Chemical Society, Washington D.C., 1975.
[43] M. O. Andreae, in *Arsenic: Industrial, Biomedical Environmental Perspectives: Proc. As Symp. Gaithersburg, Maryland,* W. H. Lederer and R. J. Fensterheim (eds.), Van Nostrand Reinhold, Princeton, 1983.
[44] A. M. Gunn, *Water Research Centre, Medmenham, Techn . Rept.,* TR 191, 1983.
[45] G. E. Stringer and M. Atterp, *Anal. Chem.,* 1979, **51,** 731.
[46] M. Fishman and R. Spencer, *Anal. Chem.,* 1977, **49,** 1599.
[47] T. T. Gorsuch, *Analyst,* 1959, **84,** 135.
[48] J. E. Portman and J. P. Riley, *Anal. Chim. Acta,* 1964, **31,** 509.
[49] R. Bock, *Decomposition Methods in Analytical Chemistry,* Int. Textbook Co., Glasgow, 1979.
[50] S. Bajo, *Anal. Chem.,* 1978, **50,** 649.
[51] J. F. Uthe, H. C. Freeman, J. R. Johnston and P. Michalik, *J. Assoc. Off. Anal. Chem.,* 1974, **57,** 1363.
[52] M. H. Arbab-Zavar, *Ph.D. Thesis,* University of Southampton, 1982.
[53] D. D. Nygaard and J. H. Lowry, *Anal. Chem.,* 1982, **54,** 803.
[54] Standing Committee of Analysts, *Antimony in Effluents and Raw, Potable and Seawaters by Spectrophotometry using Crystal Violet; Tentative Method,* Department of the Environment, HMSO, London, 1982.
[55] C. E. Gleit and W. D. Holland, *Anal. Chem.,* 1962, **34,** 1454.
[56] H. Norin, R. Rhyage, A. Christakopoulos and M. Sandstrom, *Chemosphere,* 1983, **12,** 299.
[57] D. W. Klumpp and P. J. Peterson, *Mar. Biol.,* 1981, **62,** 297.
[58] A. W. Fitchett, E. H. Daughtrey and P. Mushak, *Anal. Chim. Acta,* 1975, **79,** 93.
[59] A. Yasui, C. Tsutsumi and S. Toda, *Agric. Biol. Chem.,* 1978, **42,** 2139.
[60] A. Shinagawa, K. Shiomi, H. Yamanaka and T. Kikuchi, *Bull. Jap. Soc. Sci. Fish.,* 1983, **49,** 75.
[61] J. S. Edmonds, K. A. Francesconi, J. R. Cannon, C. L.Raston, B. W. Skelton and A. H. White, *Tetrahedron Lett.,* 1977, **18,** 1543.
[62] K. A. Francesconi, P. Micks, R. A. Stockton and K. J. Irgolic, *Chemosphere,* 1985, **14,** 1443.
[63] J. F. Lawrence, P. Michalik, G. Tam and H. B. S. Conacher, *J. Agric. Food Chem.,* 1986, **34,** 315.
[64] E. H. DeCarlo, H. Zeitlin and Q. Fernando, *Anal. Chem.,* 1981, **53,** 1104.
[65] W. Reichel and B. G. Bleakley, *Anal. Chem.,* 1974, **46,** 59.
[66] A. H. Abu-Hillal and J. P. Riley, *Anal. Chim. Acta,* 1981, **131,** 175.
[67] K. Subramanian and J.C. Meranger, *Anal. Chim. Acta,* 1981, **124,** 131.
[68] P. Puttemans and D. L. Massart, *Anal. Chim. Acta,* 1982, **141,** 225.
[69] T. Kamada and Y. Yamamoto, *Talanta,* 1977, **24,** 330.
[70] D. Chakraborti, F. Adams and K. J. Irgolic, *Z. Anal. Chem.,* 1986, **323,** 340.
[71] A. R. Byrne, *Anal. Chim. Acta,* 1972, **59,** 91.
[72] E. A. Dietz and M. A. Perez, *Anal. Chem.,* 1976, **48,** 1088.
[73] J. Persson and K. Irgum, *Anal. Chim. Acta,* 1982, **138,** 111.
[74] S. S. Sandhu and P. Nelson, *Environ. Sci. Technol.,* 1979, **13,** 476.
[75] M. G. Haywood and J. P. Riley, *Anal. Chim. Acta,* 1976, **85,** 219.
[76] A. G. Howard, M. Volkan and D. Y. Ataman, *Analyst,* 1987, **112,** 159.
[77] T. Nakahara, *Prog. Anal. Atom. Spectrosc.,* 1983, **6,** 163.
[78] R. S. Braman, D. L. Johnson, C. C. Foreback, J. M. Ammons and J. L. Bricker, *Anal. Chem.,* 1977, **49,** 621.
[79] R. S. Braman and M. A. Tompkins, *Anal. Chem.,* 1978, **50,** 1088.
[80] R. S. Braman and A. Dynako, *Anal. Chem.,* 1968, **40,** 95.
[81] R. S. Braman, L. L. Justen and C. C. Foreback, *Anal. Chem.,* 1972, **44,** 2195.
[82] A. G. Howard and M. H. Arbab-Zavar, *Analyst,* 1981, **106,** 213.
[83] P. N. Vijan, A. C. Rayner, D. Sturgis and G. R. Wood, *Anal. Chim. Acta,* 1976, **82,** 329.
[84] J. A. Fiorino, J. W. Jones and S. G. Capar, *Anal. Chem.,* 1976, **48,** 120.
[85] W. Holak, *Anal. Chem.,* 1969, **41,** 1712.
[86] S. Terashima, *Anal. Chim. Acta,* 1976, **86,** 43.
[87] R. R. Brooks, D. E. Ryan and H. Zhang, *Anal. Chim. Acta,* 1981, **131,** 1.
[88] F. J. Schmidt and J. L. Royer, *Anal. Lett.,* 1973, **6,** 17.
[89] I. May and L. P. Greenland, *Anal. Chem.,* 1977, **49,** 2376.
[90] M. Inhat and H. J. Miller, *J. Assoc. Off. Anal. Chem.,* 1977, **60,** 813.

[91] C. Feldman, *Anal. Chem.,* 1979, **51,** 664.
[92] J. R. Knechtel and J. L. Fraser, *Analyst,* 1978, **103,** 104.
[93] H. K. Kang and J. L. Valentine, *Anal. Chem.,* 1977, **49,** 1829.
[94] P. N. Vijan and G. R. Wood, *Talanta,* 1976, **23,** 89.
[95] R. M. Brown, R. C. Fry, J. L. Moyers, S. J. Northway, M. B. Denton and G. S. Wilson, *Anal. Chem.,* 1981, **53,** 1560.
[96] J. Aggett and A. C. Aspell, *Analyst,* 1976, **101,** 341.
[97] R. K. Anderson, M. Thompson and E. Culbard, *Analyst,* 1986, **111,** 1143.
[98] S. C. Apte and A. G. Howard, *J. Anal. Atom. Spectrom.,* 1986, **1,** 221.
[99] M. Yamamoto, K. Urata and Y. Yamamoto, *Spectrochim. Acta,* 1981, **36B,** 671.
[100] M. H. Arbab-Zavar and A. G. Howard, *Analyst,* 1980, **105,** 744.
[101] T. A. Hinners, *Analyst,* 1980, **105,** 751.
[102] S. C. Apte and A. M. Gunn, Water Research Centre, Medmenham, unpublished results, 1986.
[103] D. Betteridge, A. P. Wade and A. G. Howard, *Talanta,* 1985, **32,** 709.
[104] C. Cremsini, M. Dall'Aglio and E. Ghiara, in *Proc. Int. Conf. Heavy Metals in the Environment, London,* p. 341. CEP Consultants, Edinburgh, 1979.
[105] A. E. Smith, *Analyst,* 1975, **100,** 300.
[106] J. W. Hershey and P. N. Keliher, *Spectrochim. Acta,* 1986, **41B,** 713.
[107] G. E. M. Aslin, *J. Geochem. Explor.,* 1976, **6,** 321.
[108] R. Belcher, S. L. Bogdanski, E. Henden and A. Townshend, *Analyst,* 1975, **100,** 522.
[109] G. F. Kirkbright and M. Taddia, *Anal. Chim. Acta,* 1978, **100,** 145.
[110] J. Guimont, M. Pichette and N. Rheaume, *At. Abs. Newsl.,* 1977, **16,** 53.
[111] G. A. Cutter, *Anal. Chim. Acta,* 1983, **149,** 391.
[112] K. C. Thompson and R. J. Reynolds, *Atomic Absorption, Fluorescence and Flame Emission Spectroscopy,* 2nd Ed., Griffin, London, 1978.
[113] W. B. Barnett, *At. Abs. Newsl.,* 1973, **12,** 142.
[114] K. G. Brodie, *Am. Lab.,* 1977, **9,** No. 3, 73.
[115] B. Welz, *Atomic Absorption Spectroscopy,* Verlag-Chemie, Weinheim, 1985.
[116] J. W. Owens and E. S. Gladney, *At. Abs. Newsl.,* 1976, **15,** 47.
[117] W. B. Barnett and E. A. McLaughlin, *Anal. Chim. Acta,* 1975, **80,** 285.
[118] V. A. Fassel and R. N. Knisely, *Anal. Chem.,* 1974, **46,** 1110A.
[119] K. E. Smith and C. W. Frank, *Appl. Spectrosc.,* 1968, **22,** 765.
[120] G. F. Kirkbright and L. Ranson, *Anal. Chem.,* 1971, **43,** 1238.
[121] A. Ando, M. Suzuki, K. Fuwa and B. L. Vallee, *Anal. Chem.,* 1969, **41,** 1974.
[122] M. Verlinden, H. Deelstra and P. Adrianessens, *Talanta,* 1981, **28,** 637.
[123] R. Kaszerman and K. Theurer, *At. Abs. Newsl.,* 1976, **15,** 129.
[124] R. D. Ediger, *At. Abs. Newsl.,* 1975, **14,** 127.
[125] J. Korečková, W. Frech, E. Lundberg, J. Å. Perrsson and A. Cedergren, *Anal. Chim. Acta,* 1981, **130,** 267.
[126] K. Smolander and M. Kaupinnen, *Analyst,* 1986, **111,** 1029.
[127] K. J. Irgolic, R. A. Stockton, D. Chakraborti and W. Beyer, *Spectrochim. Acta,* 1983, **38B,** 437.
[128] K. J. Irgolic, *Sci. Total Environ.,* 1987, **64,** 61.
[129] J. R. Castillo, J. Lanaja, M. C. Martinez and J. Aznárez, *Analyst,* 1982, **107,** 1488.
[130] C. J. Peacock and S. C. Singh, *Analyst,* 1981, **106,** 931.
[131] K. C. Thompson and D. R. Thomerson, *Analyst,* 1974, **99,** 595.
[132] D. E. Fleming and G. A. Taylor, *Analyst,* 1978, **103,** 101.
[133] W. H. Evans, F. J. Jackson and D. Dellar, *Analyst,* 1979, **104,** 16.
[134] P. N. Vijan and G. R. Wood, *At. Abs. Newsl.,* 1974, **13,** 33.
[135] J. R. S. Broughton and C. W. Fuller, *Proc. Anal. Div. Chem. Soc.,* 1977, **14,** 112.
[136] R. C. Chu, G. P. Barron and P. A. W. Baumgarner, *Anal. Chem.,* 1972, **44,** 1476.
[137] Y. K. Chau, P. T. S. Wong and P. D. Goulden, *Anal. Chem.,* 1975, **47,** 2279.
[138] B. Welz and M. Melcher, *Analyst,* 1983, **108,** 213.
[139] G. E. Parris, W. R. Blair and F. E. Brinckman, *Anal. Chem.,* 1977, **49,** 378.
[140] M. McDaniel, A. D. Shendrikar, K. D. Reizner and P. W. West, *Anal. Chem.,* 1976, **48,** 2240.
[141] G. Drasch, L. V. Meyer and G. Kauert, *Z. Anal. Chem.,* 1980, **304,** 141.
[142] M. Thompson, B. Pahlavanpour, S. J. Walton and G. F. Kirkbright, *Analyst,* 1978, **103,** 568.
[143] M. Thompson, B. Pahlavanpour, S. J. Walton and G. F. Kirkbright, *Analyst,* 1978, **103,** 705.
[144] K. J. Mulligan, M. H. Hahn, J. A. Carusol and F. L. Fricke, *Anal. Chem.,* 1979, **51,** 1935.
[145] Y. Talmi and V. E. Norvell, *Anal. Chem.,* 1975, **47,** 1510.
[146] Y. Talmi and D. T. Bostick, *Anal. Chem.,* 1975, **47,** 2145.
[147] K. B. Olsen, D. S. Sklarew and J. C. Evans, *Spectrochim. Acta,* 1985, **40B,** 357.
[148] K. C. Thompson, *Analyst,* 1975, **100,** 307.

[149] L. C. Ebdon, J. R. Wilkinson and K. W. Jackson, *Anal. Chim. Acta,* 1982, **136,** 191.
[150] K. Tsuji and K. Kuga, *Anal. Chim. Acta,* 1974, **72,** 85.
[151] K. Tsuji and K. Kuga, *Anal. Chim. Acta,* 1978, **97,** 51.
[152] K. Fujiwara, Y. Watanabe, K. Fuwa and J. D. Winefordner, *Anal. Chem.,* 1982, **54,** 125.
[153] H. Tao, A. Miyazaki, K. Bansho and Y. Umezaki, *Anal. Chem.,* 1984, **56,** 181.
[154] S. C. Apte and A. G. Howard, *J.Anal. Atom. Spectrom.,* 1986, **1,** 379.
[155] M. A. Eckhoff, J. P. McCarthy and J. P. Caruso, *Anal. Chem.,* 1982, **54,** 165.
[156] W. B. Robbins and J. A. Caruso, *J. Chromatog. Sci.,* 1979, **17,** 360.
[157] F. A. Fricke, W. D. Robbins and J. A. Caruso, *J. Assoc. Off. Anal. Chem.,* 1978, **61,** 1118.
[158] W. B. Robbins, J. A. Caruso and F. L. Fricke, *Analyst,* 1979, **104,** 35.
[159] K. Fujita and T. Takadu, *Talanta,* 1986, **33,** 203.
[160] R. Belcher, S. L. Bogdanski, E. Henden and A. Townshend, *Anal. Chim. Acta,* 1977, **92,** 33.
[161] R. K. Skogerboe and A. P. Bejmuk, *Anal. Chim. Acta,* 1977, **94,** 297.
[162] E. H. Daughtrey, A. W. Fitchett and P. Mushak, *Anal. Chim. Acta,* 1975, **79,** 199.
[163] F. T. Henry and T. M. Thorpe, *J. Chromatog.,* 1978, **166,** 577.
[164] C. J. Soderquist, D. G. Crosby and J. B. Bowers, *Anal. Chem.,* 1974, **46,** 155.
[165] S. Fukui, T. Hirayama, M. Nohara and Y. Sakagami, *Talanta,* 1981, **28,** 402.
[166] G. Schwedt and H. A. Rüssel, *Chromatographia,* 1972, **5**, 242.
[167] S. Fukui, T. Hirayama, M. Nohara and Y. Sakagami, *Talanta,* 1983, **30,** 89.
[168] K. W. M. Siu, S. Y. Roberts and S. S. Berman, *Chromatographia,* 1984, **19,** 398.
[169] B. Beckerman, *Anal. Chim. Acta,* 1982, **135,** 77.
[170] K. Dix, C. J. Cappon and T. Y. Toribara, *J. Chromatog. Sci.,* 1987, **25,** 164.
[171] H. Haraguchi and A. Takatsu, *Spectrochim. Acta,* 1987, **42B,** 235.
[172] M. H. Hahn, K. J. Mulligan, M. E. Jackson and J.A. Caruso, *Anal. Chim. Acta,* 1980, **118,** 115.
[173] C. T. Tye, S. J. Haswell, P. O'Neill and K. C. C. Bancroft, *Anal. Chim. Acta,* 1985, **169,** 195.
[174] D. Chakraborti and K. J. Irgolic, in *Proc. Int. Conf. Heavy Metals in the Environment, Athens,* T. D. Lekkas (ed.), Vol. 2, p. 484. CEP Consultants, Endinburgh, 1985.
[175] F. E. Brinckman, K. L. Jewett, W. P. Iverson, K. J. Irgolic, K. C. Ehrhardt and R. A. Stockton, *J. Chromatog.,* 1980, **191,** 31.
[176] D. S. Bushee, I. S. Krull, P. R. Demko and S. B. Smith, *J. Liq. Chromatog.,* 1984, **7,** 861.
[177] E. A. Woolson and N. Aharonson, *J. Assoc. Off. Anal. Chem.,* 1980, **63,** 523.
[178] R. Iadevaia, N. Aharonson and E. A. Woolson, *J. Assoc. Off. Anal. Chem.,* 1980, **63,** 742.
[179] M. Morita, T. Uehiro and K. Fuwa, *Anal. Chem.,* 1981, **53,** 1806.
[180] W. D. Spall, J. G. Lynn, J. L. Andersen, J. G. Valdez and L. R. Gurley, *Anal. Chem.,* 1986, **58,** 1340.
[181] T. Maitani, S. Uchiyama and Y. Saito, *J. Chromatog.,* 1987, **391,** 161.
[182] R. H. Fish, F. E. Brinckman and K. L. Jewett, *Environ. Sci. Technol.,* 1982, **16,** 174.
[183] R. A. Stockton and K. J. Irgolic, *Int. J. Environ. Anal. Chem.,* 1979, **6,** 313.
[184] G. R. Ricci, L. S. Shepard, G. Colovos and N. E. Hester, *Anal. Chem.,* 1981, **53,** 610.
[185] G. K. C. Low, G. E. Batley and S. J. Buchanan, *J. Chromatog.,* 1986, **368,** 423.
[186] F. E. Brinckman, W. R. Blair, K. L. Jewett and W. P. Iverson, *J. Chromatog. Sci.,* 1977, **15,** 493.
[187] L. C. Ebdon, S. Hill and R. W. Ward, *Analyst,* 1987, **112,** 1.
[188] H. A. Meinema and J. G. Noltes, *J. Organomet. Chem.,* 1972, **36,** 313.
[189] B. Pahlavanpour, M. Thompson and L. Thorne, *Analyst,* 1981, **106,** 467.

10

Mercury

S. Rapsomanikis
Chemical Engineering Department, Aristotelian University of Thessaloniki, Thessaloniki, Greece

10.1 OCCURRENCE OF MERCURY IN THE ENVIRONMENT

Mercury is used in a variety of products and industrial processes but in recent years its use has dwindled [1] (see also Tables 10.1 and 10.2). Concern over the effects of mercury on entering the environment has been the reason for limiting its use. There have been several accidents, with fatalities from misuse or spillage of mercury compounds, in the last 20 years [2,3]. In Minamata (Japan) the effluent of a chemical plant containing methylmercury entered the local bay and fish bioconcentrated the pollutant mercury compound. Fatalities occurred after consumption of the contaminated fish. In Iraq (1971–1972) consumption of seed sprayed with fungicidal methylmercury compounds resulted in 459 fatalities.

Problems associated with detecting methylmercury in fish or birds came to light in the late 1960s because the elevated concentrations found could not be attributed to methylmercury spillage or misuse [4,5]. It was later demonstrated that mercury can be methylated in the environment and accumulated in the tissue of fish or birds [6,7]. Upon entering the environment, mercury in any form and from a variety of sources (Table 10.1) can be transformed into its toxic methyl derivatives (Fig. 10.1). The chemical pathways for the methylation of mercury in the environment have been recently reviewed [8]. The need to speciate and hence understand the biogeochemical cycling of mercury has led to the search for sensitive analytical methods. Analysis for total mercury in any matrix involves a digestion procedure, to transform the organometallic or otherwise bound mercury species into Hg^{2+}, followed by reduction by tin(II) or sodium borohydride and determination of the resultant elemental mercury by atomic spectrometry. Organic mercury is determined after extraction into organic solvents, stripping into aqueous solution for cleaning purposes, re-extraction into an organic solvent, and chromatography. Electron-capture detection is usually used in gas chromatography (absolute limits of detection about 50–100 pg)

Table 10.1 — Use of organomercury compounds. (Reproduced by permission from P. J. Craig, *Organometallic Compounds in the Environment,* Copyright 1986, Longmans).

Compound	Use	Comments
CH_3HgX	Agricultural seed dressing, fungicide	Banned Sweden 1966, USA 1970, as seed disinfectant. Not used today in Europe or USA. Used in laboratories
C_2H_5HgX	Cereal seed treatment	Banned USA, Canada 1970. Used in UK
RHgX	Catalyst for urethane, vinyl acetate production	
C_6H_5HgX	Seed dressings, fungicide, slimicide, general bactericide. For pulp, paper, paints	Banned as slimicide USA 1970. Banned as rice seed dressing Japan 1970. Used in UK
p-$CH_3C_6H_5HgX$	Spermicide	
$ROCH_2CH_2HgX$	Seed dressings, fungicides	Banned Japan 1968. Used in UK
$ClCH_2CH(OCH_3)CH_2HgX$	Fungicide, pesticide, preservative	
Thiomersal	Antiseptic, C_2H_5Hg derivative	
Mercurochrome	Antiseptic, organomercury fluorescein derivative	
Mersalyl	Diuretic, methoxyalkyl derivative, $RCH_2CH(OCH_3)CH_2HgX$	Little used today. R= *o*–$COOHCH_2OC_6H_4CONH$–
Chlormerodrin	Diuretic, methoxyalkyl derivative, $NH_2CONHCH_2CH(OCH_3)CH_2HgCl$	Little used today
Mercarbolide	*o*-HOC_6H_4HgCl	*o*-Chloromercuriphenol
Mercurophen	*o*-NO_2-*p*-$ONaC_6H_3HgOH$	
Mercurophylline	Diuretic	

X=anionic group; wide range of X known, e.g. OAc^-, PO_4^{3-}, Cl^-, $NHC(NH)NHCN^-$, etc.

but the use of extremely pure solvents is necessary, together with tedious cleaning procedures, to avoid co-elution of electron-capturing species with the organomercury compounds [9]. Inorganic mercury cannot be determined by gas chromatography. However, its concentration in the media under investigation can be derived from the difference between the total and organic mercury content.

Techniques for the detection and speciation of mercury-containing compounds in environmental media usually involve the coupling of chromatography (gas or liquid) with atomic spectrometry (atomic absorption, emission and fluorescence). Extensive cleaning procedures can hence be avoided and unambiguous identification and quantification of organomercury compounds achieved. In this chapter such techniques will be reviewed, their applications critically evaluated and future developments assessed.

10.2 SPECIATION BY GC–AAS

The volatility of mercury metal simplifies its detection by atomic spectrometry. Problems with nebulization and atomization in classical flame atomic-absorption spectrometry are avoided by chemical reduction of the mercury species to the element and transfer of the cold metal vapour into the optical path from the excitation source. The pioneering work of Hatch and Ott [10] describes an experimental set-up utilizing a reduction vessel and a fused-silica cell. Samples are digested

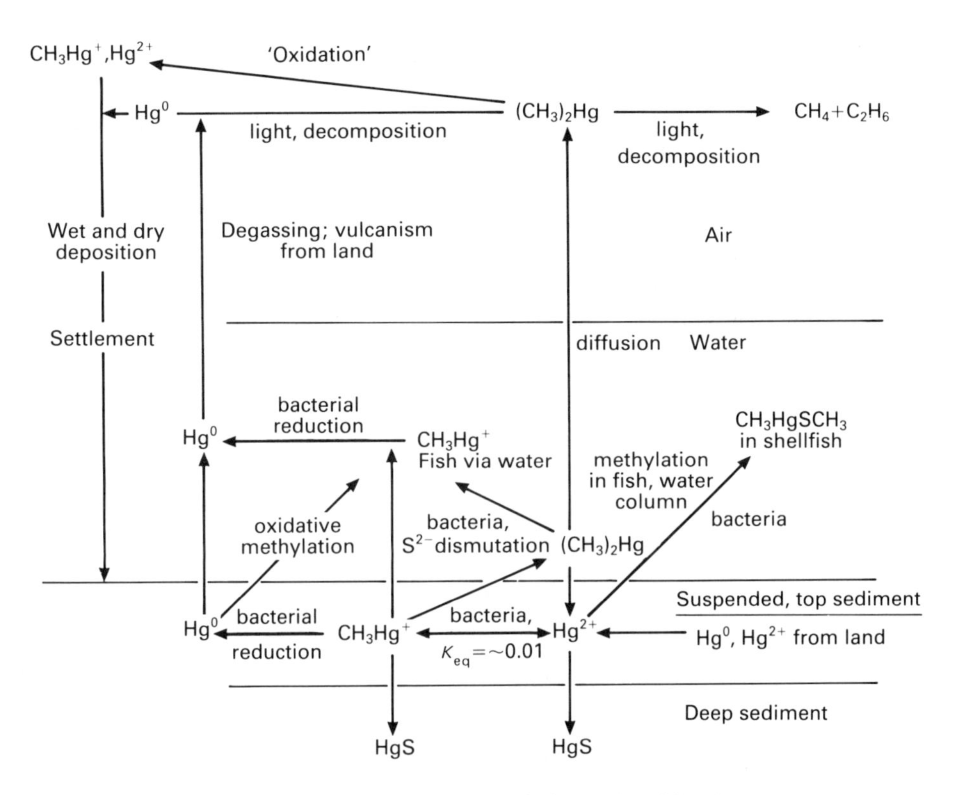

Fig. 10.1 — The biogeochemical cycle of mercury. (Reproduced by permission from P. J. Craig, *Organometallic Compounds in the Environment*, Copyright 1986, Longmans).

Table 10.2 — Mercury consumption by use (tonnes). (Reproduced by permission from Bureau of Mines and Minerals Yearbook, 1984).

	1959, USA	1968, USA	1978, USA	1984, USA
Agriculture	110	118	21 (1975)	—
Amalgamation	9	9	<0.5 (1975)	—
Catalytic	33	66	29 (1975)	12
Dental	95	106	18	49
Electrical/batteries	426	677	619	1170
Chlor-alkali	201	602	385	253
Laboratory	38	69	14	7.5
Instruments	351	275	120	98
Paint	121	369	309	160
Paper/pulp	150	14	<0.5	—
Pharmaceuticals	59	15	15 (1975)	—
Metal for inventory/other	298	298	216	48

Total world production 1984=6000 tonnes. Total USA usage 1984=1884 tonnes.

with acid and reduction is effected with stannous sulphate. After drying, the cold vapour is swept into a long-path cylindrical cell placed in the beam of light from a mercury hollow-cathode lamp. The absolute sensitivity is 20 ng for 1% absorption (absorbance 0.0044) and the limit of detection is 1.0 ng/ml (S/N=2). The principle of this method has been applied widely for the determination of mercury and it has also been exploited commercially [11] (MAS-50 Mercury Analyzer, Perkin-Elmer Corp., U.S.A.). In a large number of cases the cold vapour detector has been used for the detection of organomercurials in the effluent from a gas chromatograph. Longbottom and Dressman [12] used the MAS-50 instrument to detect the organomercurials in the GC effluent after it had passed through the flame ionization detector. The non-ionic organomercurials (dimethyl-, diethyl-, dipropyl- and dibutylmercury) were decomposed to give mercury metal as they were eluted, and after 'on-line' cooling and drying the metal vapour was swept into the analyser (Fig. 10.2). A glass column,

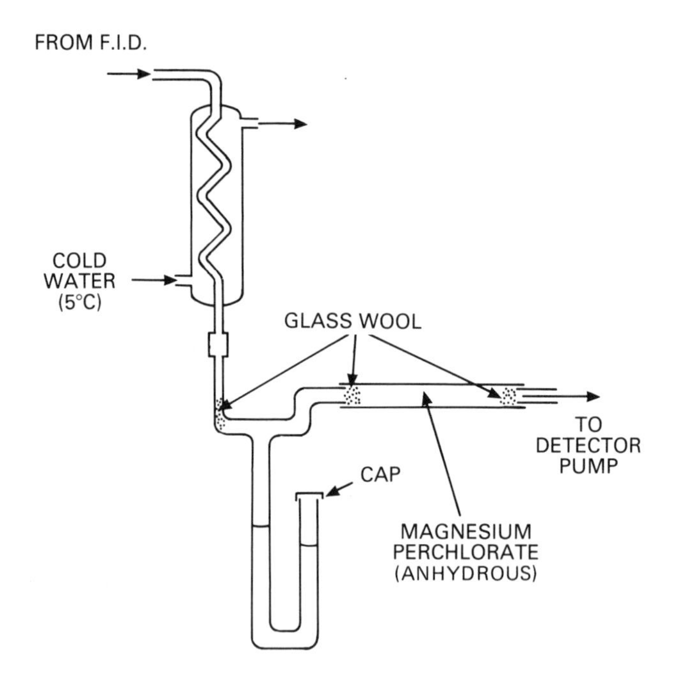

Fig. 10.2 — Condenser and dryer assembly for the transfer line of the GC-MAS-50 analyser. (Reproduced by permission from R. Dumarey, R. Heindryckx and R. Dams, *Atom. Spectrosc.*, 1981, **2**, 51. Copyright 1981, Perkin-Elmer).

180 cm long, packed with 5% DC-200 and 3% QF-1 on Gas-Chrom Q was used for the separation of the organomercurials. Separation and detection of ionic organomercurials was not attempted and the limit of detection for dimethylmercury was reported to be 20 pg.

In a similar experimental set-up, Dumarey *et al.* [13] built an interface line which could accommodate the fused-silica GC column threaded through it (Fig. 10.3). The

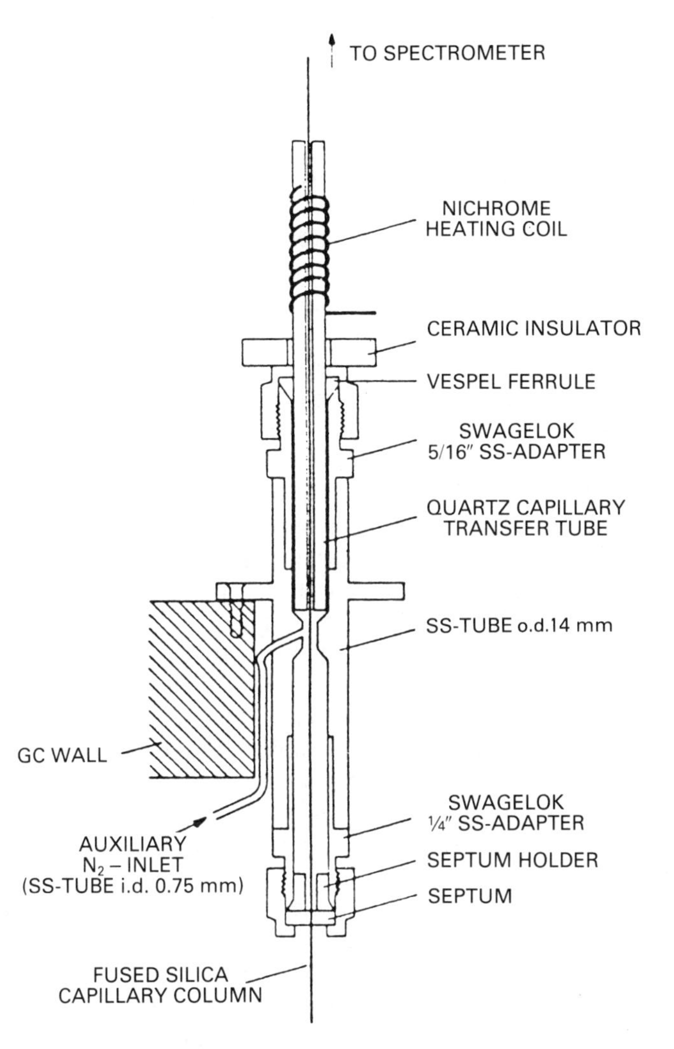

Fig. 10.3 — The GC–AAS interface. (Reproduced by permission from R. Dumarey, R. Dams and P. Sandra, *J. High. Resol. Chromatog. Chromatog. Commun.* 1982, **5**, 687. Copyright 1982, Hüttig).

organomercurials were decomposed when passing through the line, part of which was heated to 600°C and the rest was kept at 200°C. With GC temperature programming, Me_2Hg, Et_2Hg, MeHgCl and EtHgCl could be chromatographed and detected with limits of detection of 100 pg.

10.2.1 Biological samples

The instrumental arrangement above [13] was very similar to the one used by Bye and Paus [14] to determine alkylmercury compounds in fish tissues. Homogenized tissue samples were acid-digested, extracted with toluene and mercury was stripped into an aqueous phase as organomercury thiosulphate complexes. After addition of potassium iodide, the organomercury iodides were extracted with benzene and chromatographed on a column packed with 10% SP-2300 on Chromosorb W, 80–100 mesh. The GC effluent passed through a fused-silica tube at 600°C where the organomercurials were thermally decomposed to elemental mercury before being detected by cold-vapour AAS. An absolute limit of detection of 3.5 ng as CH_3Hg^+ was obtained, with a limit of detection of 0.3 ppm in the original sample. Although recoveries were not quoted, the analyses were done by the standard-addition method and the relative standard deviation of the method was 2.2%. However, the chromatographic separations were not ideal, having tailing peaks and long retention times.

Gonzalez and Ross [15] used the flame of the FPD (flame photometric detector) of a GC, to decompose the eluted organomercurials thermally before the effluent gas was dried and swept into the fused-silica cell of the atomic-absorption instrument (the FPD of the GC alone would not respond to mercury compounds). They found that this configuration had low sensitivity and produced broad tailing peaks. On replacing the FPD flame by a 31×0.8 cm fused-silica furnace held at 780°C, using an oxygen flow of 25 ml/min and reducing the length of the transfer lines by using the MAS-50 instrument, they observed a marked improvement. A 180 cm long, 6 mm o.d. column, packed with 5% HIEFF-2AP on Chromosorb WHP 80–100 mesh was used. No stripping procedures were used for fish tissues digested with 6*M* hydrochloric acid. The benzene extract was injected directly into the GC and the range of linearity extended to 45 ng of mercury as methylmercury. The absolute sensitivity was 25 pg and the recoveries were comparable to those of the Westöö method [4].

For the speciation of mercuric compounds in human urine and sweat samples, Robinson and Wu [16] constructed a GC–AAS interface with the graphite-furnace T-cell mounted on top of the GC. The optics of the atomic-spectrometry system were built around the T-cell, which was kept at 1500°C (Fig. 10.4). For the GC separation of inorganic and organic mercuric chlorides a Teflon column 60 cm long and 3 mm i.d. packed with 5% DEGS on Chromosorb W AWDMCS was used. Samples required no pretreatment and injections of aqueous samples resulted in an absolute sensitivity of 100 pg. However, in both urine and sweat samples no organomercurials were detected, and no record of the chromatograms of the inorganic mercury species detected was presented. It appears there is a large component of non-volatile mercury in the samples, which could not be chromatographed. Its presence was established by comparing inorganic mercury values obtained by GC–AAS with total mercury estimations.

10.2.2 Air samples

Speciation of mercury and organomercury compounds in air is important in establishing their concentration fluxes and deducing their biogeochemical cycles in the environment. They are usually collected on chromatographic support media and/or silver and gold filaments, thermally desorbed, and detected by GC–AAS.

Bzenzinska *et al.* [17] used Tenax GC and glass chips for absorbing mercury compounds after sampling for 15–30 hr at 2 l./min. Breakthrough volumes for the compounds collected at ambient temperature were not reported, nor was the study of collection efficiencies explicit. A study of a number of columns and separating conditions proved that Tenax was the most suitable solid support for the separation

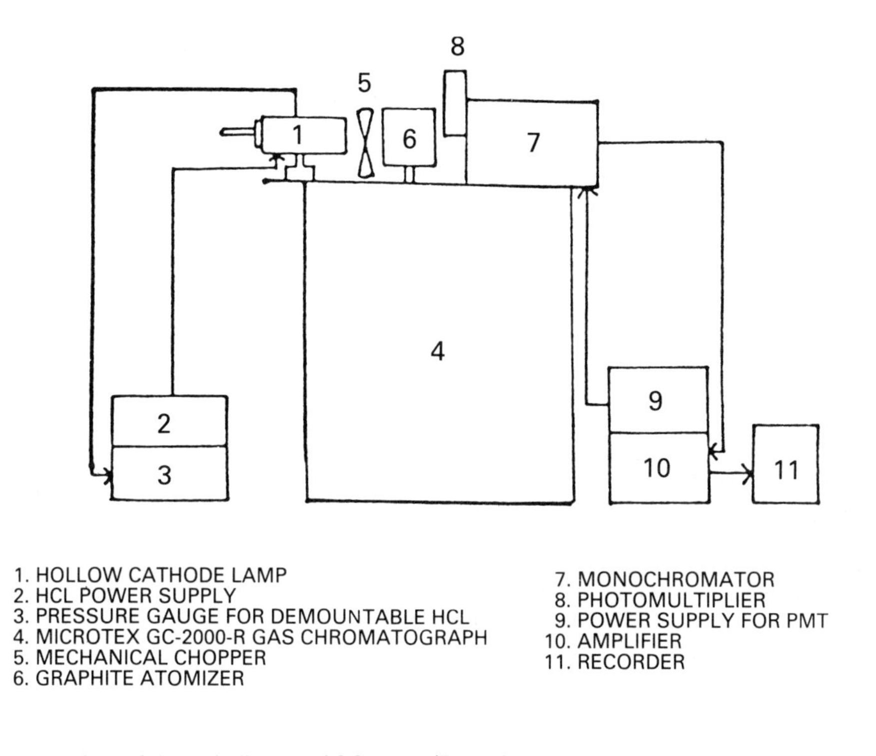

Fig. 10.4 — Schematic diagram of GC–AAS. (Reproduced by permission from J. W. Robinson and J. C. Wu, *Spectrosc. Lett.*, 1985, **18**, 47. Copyright 1985, Marcel Dekker).

of elemental mercury, methylmercury and dimethylmercury. However, in analysis of atmospheric samples the determination of elemental mercury appeared troublesome and was abandoned. After thermal desorption the collected mercury compounds were trapped in 0.1 ml of cooled benzene, and 10-μl aliquots were taken for analysis on a 2 m long 2 mm i.d. glass column packed with Tenax, with temperature programming. The effluent was thermally decomposed at 900°C (Fig. 10.5) and the mercury detected in a silica T-cell of the AAS instrument. Absolute limits of detection for methylmercury and dimethylmercury compounds were 5 and 2 ng respectively.

Schroeder and Jackson [18,19] directed their efforts towards the efficient collection of elemental mercury, inorganic mercury, methylmercury and dimethylmercury in a sampling train containing four different sorbing materials (Fig. 10.6). By reversal of the gas flow and raising of the temperature, the mercury species were swept through to an atomic-absorption or atomic-fluorescence detector. For the study of the properties of each sorbent, each mercury compound was injected into a stream of mercury-free air and sorbed on the sorbent under investigation. The amount that passed through was collected on gold filaments and determined by heating these to 450°C and sweeping the metal vapour into an atomic-absorption detector. Subsequently the material collected by the sorbent was thermally desorbed and deter-

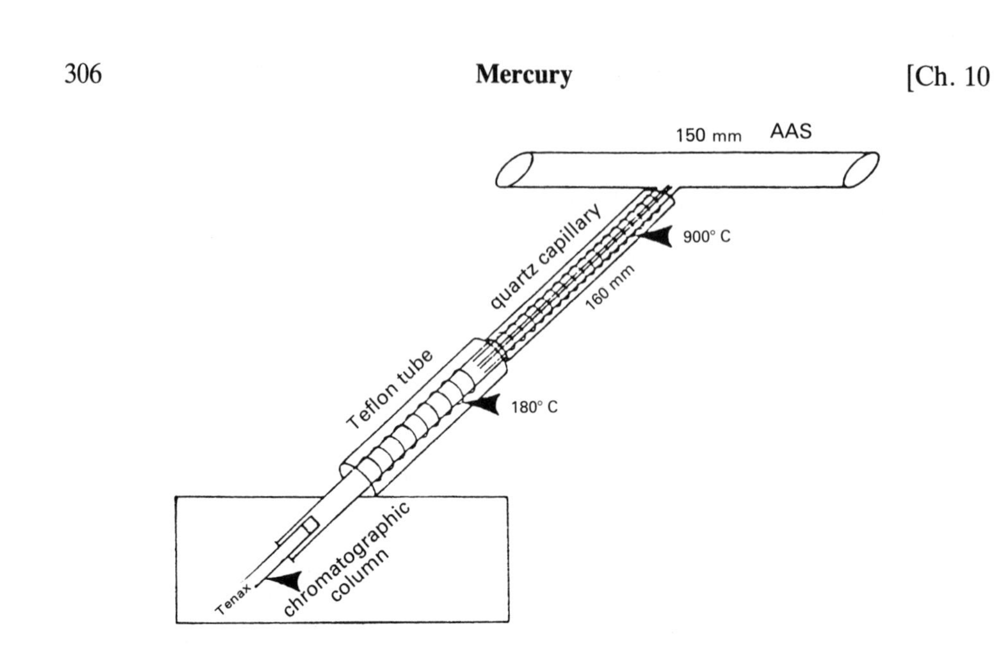

Fig. 10.5 — The GC–AAS interface. (Reproduced by permission from A. Bzezinska, J. Van Loon, D. Williams, K. Oguma, K. Fuwa and I. H. Harguchi, *Spectrochim. Acta*, 1983, **38B,** 1339. Copyright 1983, Pergamon Press).

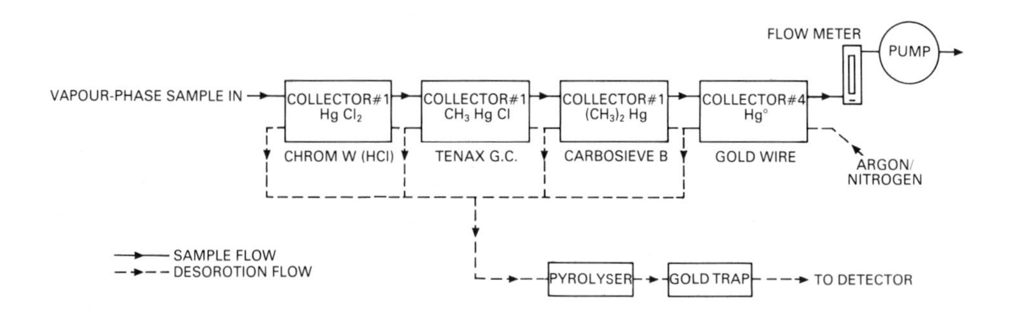

Fig. 10.6 — Schematic diagram of the atmospheric mercury speciation train. (Reproduced by permission from W. H. Schroeder and R. Jackson, *Chemosphere*, 1984, **9,** 1041. Copyright 1984, Pergamon Press.

mined in a similar manner. Retention volumes for the mercury species collected on a particular sorbent were not quoted, and gas-phase standard mixtures containing mercuric chloride and methylmercuric chloride vapour are difficult to envisage. Experimental details for making standards for any of the compounds under investigation were not described, nor were sensitivities or limits of detection. It is hence not possible to establish the practical application of the technique and the possible interferences when analysing environmental samples.

10.3 SPECIATION BY GC–AES

Atomic emission in plasmas is the most commonly used method of detection of organomercury compounds after they have been eluted from the GC column, presumably because the separate thermal decomposition step used in GC–AAS is not needed.

Grossman *et al.* [20] studied the selectivity of a microwave discharge detector, described by McCormack *et al.* [21], coupled to a GC. The selectivity ratios of a large number of organic compounds relative to dimethylmercury were satisfactory, demonstrating the usefulness of the detection system. A stainless-steel column, 300 cm long and 6 mm i.d., packed with 20% Carbowax 20M on 60/80 Chromosorb P was used with argon as carrier gas. The sensitivity for dimethylmercury (signal to give 0.5 mV response) was 0.26 ng of mercury (as dimethylmercury).

Quimby *et al.* [22] described the coupling of a GC with a microwave medium plasma maintained at atmospheric pressure (Fig. 10.7). The Beenakker TM_{010}

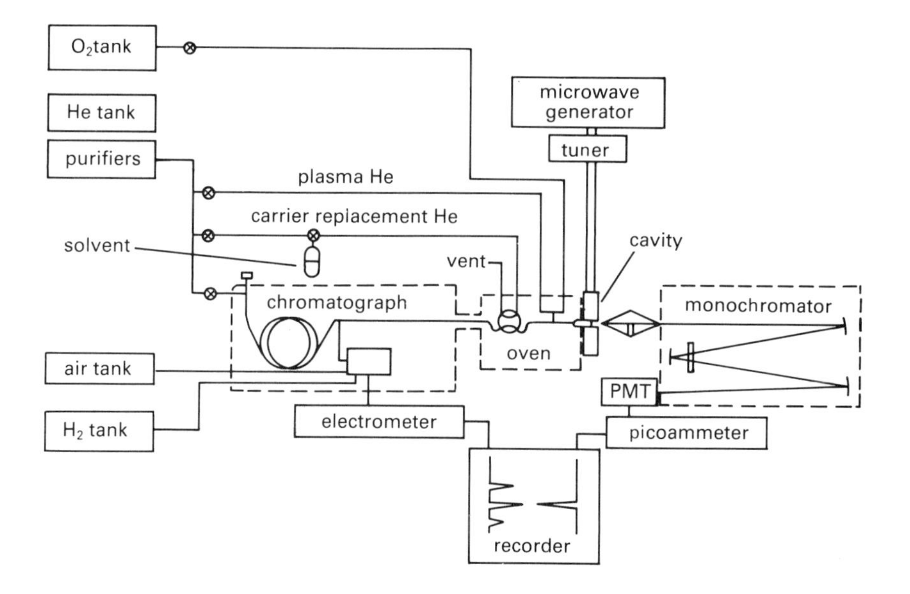

Fig. 10.7 — Block diagram of the GC–MPD (microwave plasma detector). (Reproduced by permission from B. D. Quimby, P. C. Uden and R. M. Barnes, *Anal. Chem.*, 1978, **50,** 2112. Copyright 1978, American Chemical Society).

resonant cavity is viewed axially at the brightest zone and can also be used for the detection of a number of other elements. Diphenylmercury was determined by using a stainless-steel column, 90 cm long and 3 mm o.d, packed with 5% OV-17 on 100/120 mesh Chromosorb 750. For a flow-rate of 50 ml/min through the cavity, the detection limit of mercury as diphenylmercury was 1.0 pg/sec (S/N=2). Organomercury compounds can hence be selectively determined by this method, with a precision of ca. 5%.

Rice *et al.* [23] described the use of the atmospheric-pressure active-nitrogen afterglow (APAN) as a detector for gas chromatography (Fig. 10.8). The detector

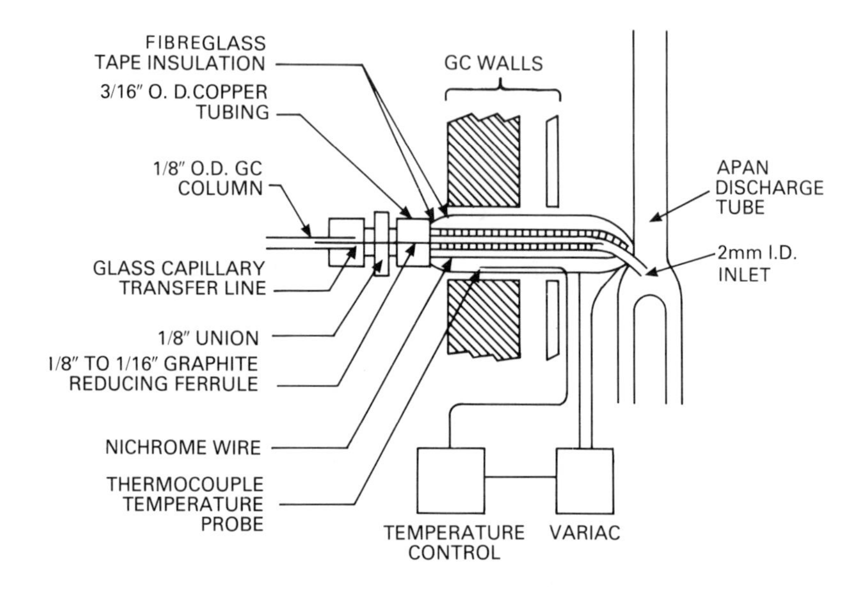

Fig. 10.8 — Interface between GC and the APAN (atmospheric-pressure active-nitrogen) detector. (Reproduced by permission from G. W. Rice, J. R. Richard, A. P. D'Silva and V. A.Fassel, *Anal. Chem.*, 1981, **53**, 1519. Copyright 1981, American Chemical Society).

was more sensitive to organomercury compounds than to organotins or organoleads. For all the organometallic compounds tested, a 120 cm 3 mm o.d glass column was used, packed with 4% SE-30/6% SP-2401 on 100–120 mesh Supelcoport. The limit of detection for mercury as dimethylmercury was 2 pg (S/N=3), with a higher sensitivity than that for the other organometallic compounds. The separation of a number of organomercurials was demonstrated [$(CH_3)_2Hg$, $(C_2H_5)_2Hg$, $(i\text{-}C_3H_7)_2Hg$, $(C_3H_7)_2Hg$, $(C_4H_9)_2Hg$, $(C_6H_{13})_2Hg$ and $(C_6H_5)_2Hg$]. However, the usefulness of the technique for the separation and determination of ionic organomercury compounds was not established.

A low-pressure helium microwave plasma detector coupled to a capillary GC has been described by Olsen *et al.* (Fig. 10.9) [24]. The apparatus is a modification of the one described by Quimby *et al.* [22]. The organomercury compounds were separated in a fused-silica capillary column, 30 m long, coated with a 1.04 μm thick film of DB-5. Injections were splitless, hence helium make-up gas was needed to sustain the plasma. Diaryl, dialkyl and alkylmercuric chlorides could be detected by the MES detector with limits of detection of ca. 100 pg. Use of hydrogen as scavenger gas in the MES reduced the tailing of the organomercury peaks. Comparison of the low-

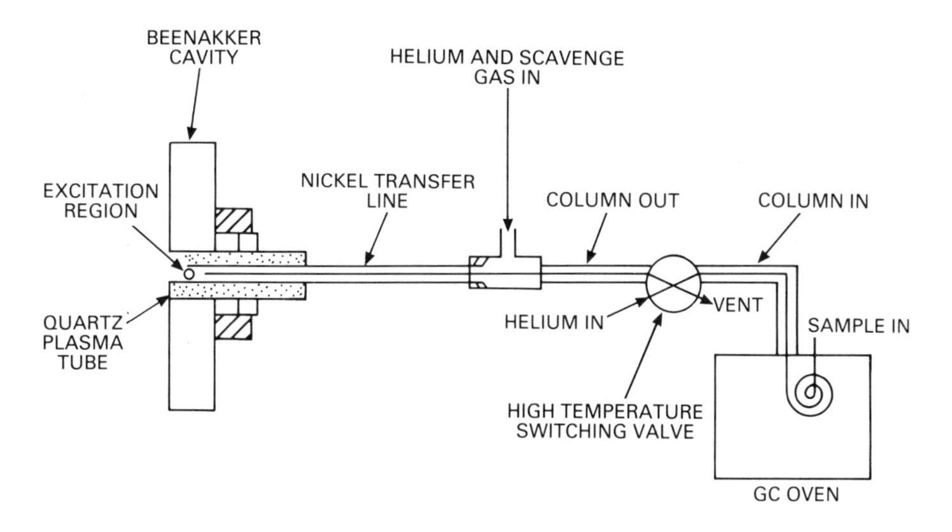

Fig. 10.9 — Interface between GC and the atmospheric-pressure MPD. (Reproduced by permission from K. B. Olsen, D. S. Sklarew and J. C. Evans, *Spectrochim. Acta*, 1985, **40B,** 357. Copyright 1985, Pergamon Press).

pressure MES detector with the atmospheric-pressure MES revealed that the latter was more sensitive to dimethylmercury by two orders of magnitude (limit of detection ca. 1 pg) under the same chromatographic conditions.

10.3.1 Water and sediment samples

In a comprehensive study of the microwave emission spectrometric detector coupled to GC, Talmi [25] described extraction procedures and calculated extraction efficiencies, using different digestion agents, volume of extractant and derivatizing halides for the determination of methylmercuric ion. The apparatus was similar to that described by McCormack *et al.* [21]. Aqueous samples containing CH_3HgI can be extracted into 2 ml of benzene and after drying, chromatographed in a 90 cm long 3 mm i.d. column, packed with 4% FFAP on 80–100 mesh Gas-Chrom Q. Water samples can also be injected directly into a column of the same dimensions packed with 1% FFAP graphitized carbon, 80–100 mesh. The absolute detection limit is 0.5 pg of mercury as CH_3HgX (or C_2H_5HgX; X=Cl, Br or I) and 2 pg as $(CH_3)_2$ Hg.

Using the same apparatus as above, Talmi and Norvell [26] employed a benzene solution containing 0.5% Alamine 336-S to extract CH_3HgCl from spiked spring water and synthetic sea-water. The samples had to be adjusted to pH 9 before extraction. It was postulated that an $R_3N{\cdot}CH_3HgCl$ complex was formed rather than an $R_4N^+(CH_3HgCl)^-$ ion-associate. The complex was broken at the injection port, and CH_3HgCl was eluted before the amines. Chromatographic conditions were similar to those described above, but the packing was 6% FFAP on 80% 100 mesh Gas-Chrom Q. The absolute limit of detection established was 10 pg with a detection limit of 2.5 ng/l.

The APAN detector was used by Rice *et al.* [27] to detect alkylmercuric chlorides and dimethylmercury in water and sediments. The aqueous samples were acidified with 6*M* hydrochloric acid, extracted with two 20-ml portions of dichloromethane, and the extract was evaporated to 1.0 ml before analysis. Samples analysed for dimethylmercury were not acidified prior to extraction. Sediments (1.0 g) were acidified and extracted with dichloromethane. The whole sample was filtered under suction, and the organic layer was separated and concentrated to 1.0 ml prior to analysis. Gas chromatographic conditions were similar to those [23] described on p. 000. The limits of detection were 2 pg for dimethylmercury, 5 pg for diethylmercury, 10 pg for dipropylmercury, 20 pg for dibutylmercury and 50 pg for methylmercuric chloride. Recoveries for methylmercury compounds from sediments and water were ca. 94%.

Chiba *et al.* [28] described a procedure for extraction of methylmercuric chloride from sea-water. The atmospheric-pressure helium microwave plasma, coupled to a GC, was very similar to the one described by Quimby *et al.* [22]. The sample was acidified and extracted with three portions of benzene. The combined organic phase was shaken with a solution containing 1% L-cysteine and 0.8% sodium acetate. The aqueous phase was acidified and re-extracted with benzene. A 500-fold concentration was achieved in this manner and the alkylmercury chlorides were chromatographed in a column 1 m long, 3 mm i.d, packed with 15% DEGS on 80–100 mesh Chromosorb W. Absolute limits of detection (S/N=2) were 0.02 pg/sec for methylmercuric and ethylmercuric chloride, and 0.03 pg/sec for dimethylmercury (80 ml/min gas flow through the plasma cavity).

10.3.2 Biological samples

An original publication by Bache and Lisk [29] described the application of the apparatus devised by McCormack *et al.* [21] to determination of alkylmercurials in fish samples. The extraction method was the one described by Westöö [4] and the amounts of alkylmercurials needed to give a signal of 0.5 mV were: dimethylmercury and dimethylmercuric chloride, 0.6 ng; methylmercuric dicyandiamide 0.8 ng; phenylmercuric acetate 8.8 ng and methylmercuric dithizonate 0.7 ng. A 180 cm long and 4 mm i.d. column, packed with 10% OV-17 and 10% QF-1 on 80–100 mesh Gas-Chrom Q, was used for all compounds except dimethylmercury, for which a 60 cm glass column, 4 mm i.d., packed with 60–80 mesh Chromosorb 101, was used. The instrumentation described by Rice *et al.* [27] was used for the determination of organomercury compounds in fish and urine. Homogenized fish tissue (1.0 g) was washed with acetone and methylene chloride before being shaken with 6*M* hydrochloric acid. The mixture was extracted with two portions of methylene chloride and the extracts were combined and concentrated to 1.0 ml. Urine samples were acidified with concentrated hydrochloric acid (10 drops per 25 ml of sample) and extracted with methylene chloride, and the extracts were each concentrated to 1.0 ml. The chromatographic conditions were the same as for the water and sediment samples ([27]; Section 10.3.1) and recoveries of up to 96% were recorded. For the determination of methylmercuric chloride from fish tissues and insects Talmi's set-up ([25]; Section 10.3.1) was used with an extraction procedure similar to the one just described, but with benzene instead of methylene chloride, and smaller amounts of sample (0.2–0.5 g). McCormack's GC–MES system [21] was used by Decadt *et al.* [30] in combination with semi-automatic head-space analysis for the determination of methylmercury in biological samples. The tissue of birds (liver or kidney) was lyophilized, then homogenized in 20 ml of water, and 0.5 ml of the aqueous sample was placed in the reaction vial. Addition of sulphuric acid and potassium bromide ensured the generation of methylmercuric bromide, which was converted into the iodide by addition of iodoacetic acid and sodium thiosulphate. The vials were immediately capped and the head-space analysed by using a semi-automatic head-space sampler (Perkin-Elmer, HS-6). A Teflon column packed with 10% AT-100 on Chromosorb W AW (80–100 mesh) was used, and gave a 3 pg absolute limit of detection.

10.3.3 Air samples

Air samples are usually collected in trains of absorbers containing a variety of sorbents for the selective collection of the different mercury species. Braman and Johnson [31] used a sampling train which contained (a) a glass wool filter for the removal of some particulate forms of mercury; (b) Chromosorb-W, 45–60 mesh, coated with 3% SE-30, for the removal of $HgCl_2$ and particulates that pass through the prefilter; (c) Chromosorb-W, 45–60 mesh, treated with 0.05*M* sodium hydroxide, for the removal of CH_3HgCl; (d) silvered glass beads for the removal of elemental Hg; (e) gold-coated glass beads for the removal of $(CH_3)_2Hg$.

The detector was a microwave discharge tube with a heated chamber (Fig. 10.10).

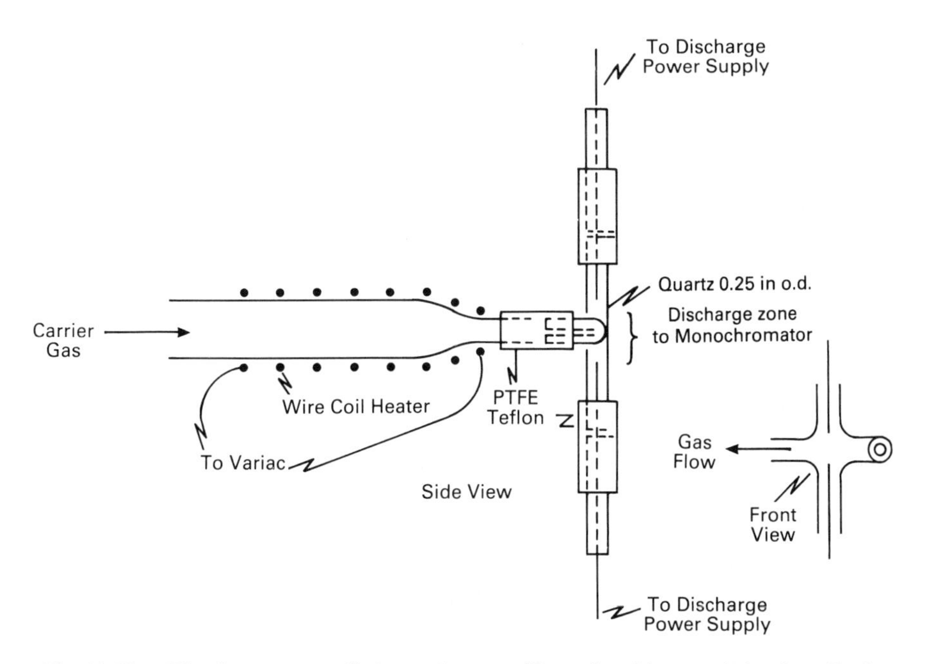

Fig. 10.10 — The direct current discharge detector. (Reproduced by permission from R. S. Braman and D. L. Johnson, *Environ. Sci. Technol.*, 1974, **8**, 996, Copyright 1974, American Chemical Society).

Although, strictly speaking, the apparatus is not a GC–AES assembly (each part of the train was subjected to analysis separately, with desorption by heating), it is a very useful method of analysis for organomercurials in the atmosphere. The absolute limit of detection is 10 pg, and the concentration limit of detection depends on the volume of air sampled.

An instrumental set-up described by Gay *et al.* [32] used a direct current plasma emission detector on a gas chromatograph to determine either total mercury or to speciate inorganic mercury, methylmercury and dimethylmercury in the atmosphere. Total mercury was determined by trapping on silver and gold, and direct desorption into the plasma, and speciation was effected by trapping in U-tubes containing 4% FFAP on 80–100 mesh Supelcoport. After desorption of analyte from these traps at 125°C, the effluent was passed through a 180 cm long 6 mm o.d. column packed with 4% FFAP on 80–100 mesh Supelcoport. Elemental mercury, dimethylmercury and methylmercuric ion could be separated; the limit of detection was 5 pg Hg (S/N=2).

The system described by Talmi [25] was used by Ballantine and Zoller [33] to determine methylmercury compounds and total mercury in the atmosphere. Methylmercuric compounds were collected in 4 cm long tubes packed with Chromosorb 101, then thermally desorbed into tubes 9 cm in length packed with 5% FFAP on Gas Chrom Q, and finally thermally desorbed into the gas chromatographic column, 1 m long, 6 mm o.d., packed with 5% FFAP on Gas Chrom Q. Dimethylmercury compounds were collected in a cold-trap containing Chromosorb 101, thermally desorbed into another trap with the same packing, and finally desorbed into the chromatographic column (same dimensions as above, packed with Chromosorb 101) at temperatures much lower than those for monoethylmercuric ion. Total mercury was determined by using a trap with gold-coated beads, desorbing into a similar trap and sweeping through an empty column. Retention volumes for methylmercury compounds on a number of packing materials were calculated, and the detection limits were the same as the ones quoted by Talmi [25].

10.4 SPECIATION BY HPLC–AAS AND HPLC–AES

For determination of organomercury compounds by GC–atomic spectrometry, derivatization steps and clean-up procedures are needed. The presence of other organometallic species in the same environmental samples may promote dealkylation or alkyl transfer reactions during these steps and/or during volatilization of the species in the gas chromatograph. In theory, HPLC coupled with atomic spectrometry should effect separation of organomercurials after simple extraction procedures, maintain the integrity of the original sample and provide adequate specific sensitivity.

Approaches to coupling HPLC with atomic spectrometry have been reviewed in Chapter 6. The discrete sampling of the HPLC effluent by an autosampler and injection into a graphite furnace has been used for the determination of organomercurials by Brinckman *et al.* [34]. A Lichrosorb C_8 column was used with an initial eluent consisting of a 94:6 v/v mixture of 0.01*M* ammonium acetate and methanol containing 25 μg/g 2-mercaptoethanol, with a gradient change to methanol containing 25 μg/g 2-mercaptoethanol, at the rate of 10% methanol/min. CH_3Hg^+, $C_2H_5Hg^+$ and n-$C_3H_7Hg^+$ were separated and the detection limits were 33, 41 and 66 ng respectively.

Gost *et al.* [35] described the connection of an HPLC instrument to an inductively coupled argon plasma through the nebulizer of the latter. CH_3Hg^+, $(CH_3)_2Hg$, $C_2H_5Hg^+$ and $C_3H_7Hg^+$ were separated but no limits of detection were reported.

Fujita and Takabatake [36] described the separation of inorganic, methyl and ethylmercuric compounds on an ODS (LS-410) resin in an HPLC column and the construction of an in-line post-column reaction vessel (Fig. 10.11). The effluent was

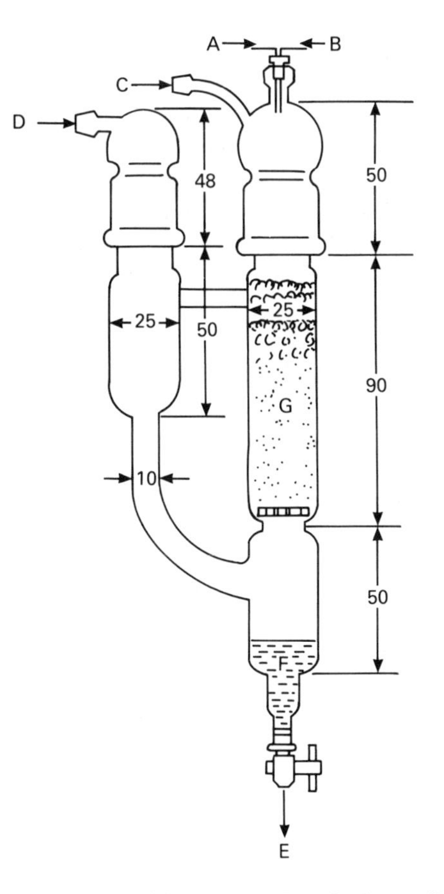

Fig. 10.11 — Reducing vessel: (A) inlet of the HPLC effluent; (B) reducing solution inlet; (C) air flow inlet; (D) Hg vapour outlet; (E) waste outlet; (F) waste; (G) 2 g of glass wool; dimensions in mm. (Reproduced by permission from M. Fujita and E. Takabatake, *Anal. Chem.*, 1983, **55**, 444. Copyright 1983, American Chemical Society).

continuously treated with sodium borohydride and each mercury compound was reduced to elemental mercury which was swept to an atomic-absorption spectrometer for cold-vapour detection. No absolute limits of detection were quoted but the sensitivity of the method is believed to be in the low ng range.

Robinson and Booth [37] described the use of HPLC–AAS instrumentation for the separation and determination of $HgCl_2$, CH_3HgCl, p-HOC_6H_4HgCl, C_2H_5HgCl, C_6H_5HgCl and $(C_6H_5)_2Hg$. The HPLC column was a Whatman Partisil-10, reversed-phase C-18, ODS-2, 25 cm long. Its outlet was connected to the nebulizer of the atomic-absorption spectrometer, operated in the flame mode. The HPLC separation was performed with a 55:45 v/v acetonitrile–0.1*M* sodium chloride or bromide mixture. The mercury compounds were well separated. Detection limits were not quoted, however, and seem likely to be high (in the low μg range).

Krull *et al.* [38] described the coupling of HPLC with ICP by post-column treatment of the effluent with sodium borohydride in two mixing chambers, passage through a gas/solvent separator, and introduction of the mercury vapour into the ICP. The HPLC was operated in the gradient mode with a mobile phase consisting of 0.06*M* ammonium acetate and 0.005% v/v 2-mercaptoethanol in acetonitrile–water mixture, with gradient increase of the acetonitrile concentration from 15 to 75% v/v. Two 15-cm columns packed with 5 μm C_{18} stationary phase were used in series. Absolute limits of detection for $HgCl_2$, CH_3HgCl, C_2H_5HgCl and $(CH_3)_2Hg$ ranged from 5 to 12 ng.

10.4.1 Biological samples

Holak [39] described the coupling of HPLC with cold-vapour atomic-absorption (Fig. 10.12) for the separation and detection of mercury compounds in fish. A Zorbax ODS (5 μm) 25 cm long column was used with 60:40 v/v methanol/0.05*M* ammonium acetate, containing 0.01% 2-mercaptoethanol. The eluent was burned in a copper tube and the elemental mercury produced was swept to the cold-vapour cell of the AAS instrument. A fine suspension of fish tissue was obtained by blending a known amount in water. This was then mixed with equal amounts of diatomaceous earth (Celite 545, acid-washed), transferred into a 22 mm i.d. chromatographic column, and eluted with chloroform. The 20 ml of chloroform extract collected were extracted with 2 ml of 0.01*M* sodium thiosulphate. Methylmercuric thiosulphate was thus formed in the aqueous phase and did not need further treatment before determination. The detection limit (S/N=2) was 0.6 ng, with 93–106% recovery of spikes.

10.5 CONCLUSIONS

Applications of the methods reviewed in this chapter are collected in Table 10.3 together with limits of detection and range of concentrations of alkylmercury compounds in environmental and biological samples. Choosing a method on the basis of the limit of detection is not an easy task, however, because detection limits are not and cannot be standardized by a common definition. Two analytical committees have tried to remedy this by introducing guidelines for reporting sensitivities and limits of detection [40,41].

Gas chromatography/cold-vapour AAS is a sensitive enough method for deter-

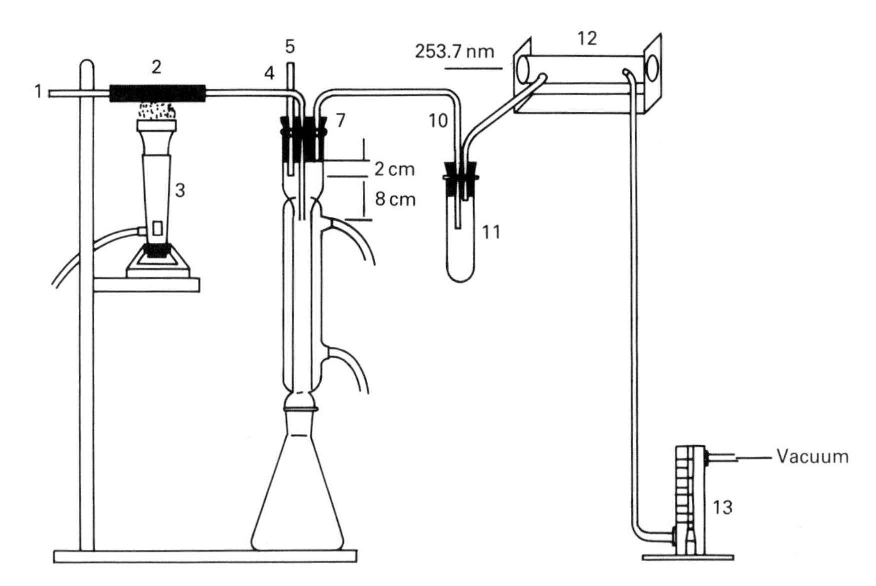

Fig. 10.12 — Atomic-absorption detection of mercury vapour generated from the HPLC effluent. (1) HPLC effluent; (2) copper tube; (3)Méker burner; (4), (5), (6) narrow-bore HPLC tubing; (7) rubber stopper; (8) condenser; (9) Erlenmeyer flask; (10) narrow-bore PTFE tubing; (11) solvent trap; (12) atom cell, 10 cm with silica windows; (13) flowmeter. (Reproduced by permission from W. Holak, *Analyst*, 1982, **107,** 1457. Copyright 1982, Royal Society of Chemistry).

mining organomercurials in environmental samples, provided these are converted into elemental mercury before the cold-vapour cell is reached. Detection of organomercury compounds by decomposition on a heated silica cell similar to the one used for determination of organotins and organoleads is troublesome because of ionization of mercury. Preconcentration of the sample may be advantageous provided care is taken not to lose volatile organomercurials or alter the integrity of the sample.

Gas chromatography coupled to the microwave induced plasma has been the most widely used technique for the determination of organomercurials because of its sensitivity and selectivity. However, the use of microwave generation of plasmas, as described in Section 10.3, needs expert handling. Commercially available systems are safe and versatile.

Systems utilizing HPLC coupled to AAS or AES detectors are still underdeveloped, with limits of detection inferior to those of the gas chromatographic techniques. Their advantage lies in the preservation of the integrity of the organomercury complex if simple extraction techniques are used.

For the organomercury compounds found in the environment, it is not possible to use techniques analogous to those (derivatization for formation of volatile species, cold-trapping and analysis of the whole samples by GC–AAS or GC–AES) used for organolead, organoarsenic and organotin compounds. Hence extraction with organic solvents, stripping for clean-up, and re-extraction into the solvent of choice, are essential. The problems of contamination with mercury by spillage, biological

Table 10.3 — Analysis of environmental and biological samples by chromatography–atomic spectrometry methods.

Method[a]	Absolute limit of detection (ng)	Sample	Range or limit of concentration	Reference
GC–CVAA	3.5[b] ($MeHg^+$)	Fish ($MeHg^+$)	2.2 μg/g	[14]
GC–GFAA	0.10[c] (Hg^{2+})	Urine (Hg^{2+})	0–3.3 μg/g	[16]
		Sweat (Hg^{2+})	0–0.6 μg/g	[16]
GC–CVAA	3.5[c] (Me_2Hg)	Air (Me_2Hg)	0–1.3 ng/m^3	[17]
	14.0[c] ($MeHg^+$)	Air ($MeHg^+$)	0.002–30 ng/m^3	[17]
GC–MES	0.0005[b] ($MeHg^+$)	Fish ($MeHg^+$)	0.009–0.042 μg/g	[25]
	0.002[b] (Me_2Hg)	Insect ($MeHg^+$)	0.034–0.108 μg/g	[25]
GC–APAN	0.050[d] ($MeHg^+$)	Fish ($MeHg^+$)	0.34–10.2 μg/g	[27]
GC–MES	0.36[e] ($MeHg^+$)	Sea-water ($MeHg^+$)	0.085 ng/g	[28]
GC–MES	0.6[f] ($MeHg^+$)	Fish ($MeHg^+$)	0.18–0.20 μg/g	[29]
GC–MES	0.003[e] ($MeHg^+$)	Bird liver ($MeHg^+$)	0.5–2.6 μg/g	[30]
		Bird kidney ($MeHg^+$)	0.72–1.57 μg/g	[30]
AST–DCP	0.05[b] (Hg)	Air (Hg)	37–888 ng/m^3	[31]
		Air (Hg^{2+})	0–43 ng/m^3	[31]
		Air ($MeHg^+$)	0–36 ng/m^3	[31]
		Air (Me_2Hg)	0–12 ng/m^3	[31]
GC–MES	0.05[b] (Hg)	Air ($MeHg^+$)	2.1–7.3 ng/m^3 *day*	[33]
		Air ($MeHg^+$)	0.74–4.7 ng/m^3 *night*	[33]
HPLC–CVAA	0.6[e] ($MeHg^+$)	Fish ($MeHg^+$)	0–1.01 μg/g	[39]

[a] GC–CVAA: Gas chromatography–cold vapour atomic absorption
GC–GFAA: Gas chromatography–graphite furnace atomic absorption
GC–MES; Gas chromatography–microwave emission spectrometry
GC–APAN: Gas chromatography–atmospheric-pressure active-nitrogen
AST–DCP: Air sampling train–direct current plasma
HLPC–MES: High-performance liquid chromatography–microwave emission spectrometry
[b] Procedure of calculating limit of detection was not defined
[c] Absolute sensitivity; amount to produce 0.0044 absorbance
[d] 3 times standard deviation of baseline noise
[e] 2 times standard deviation of baseline noise
[f] For 0.5 mV signal

transformations of mercury compounds, and cycling of mercury in the environment, have not ceased to exist, although interest in the development of methods for determining the metal and its compounds has waned in recent years. The electron-capture detector is not specific enough for the detection of 'unclean' samples obtained by using impure solvents. Construction of interfaces and/or mercury-specific detectors to suit particular needs is usually left to individual laboratories. It is hoped that the present review will help with such a task.

REFERENCES

[1] P. J. Craig in *Organometallic Compounds in the Environment,* P. J. Craig (ed.), Chapter 2, Longmans, Harlow, 1986.
[2] W. Smith and A. Smith (eds.), *Minamata,* Holt Rinehart and Winston, New York, 1975.
[3] F. Bakir, S. F. Damliyi, L. Amin-Zaki, M. Murtadha, A. Khalidi, N. Y. Al-Ravi, S. Tikriti, H. I. Dhahir, T. W. Clarkson, J. C. Smith and R. A. Doherty, *Science,* 1973, **181**, 230.
[4] G. Westöö, *Acta Chem. Scand.,* 1966, **20**, 2131.
[5] K. Noren and G. Westöö, *Var Foeda,* 1967, **19**, 13.
[6] S. Jensen and A. Jernelov, *Nature,* 1969, **223**, 753.
[7] J. M. Wood, F. S. Kennedy and C. G. Rosen, *Nature,* 1968, **220**, 173.

[8] S. Rapsomanikis and J. H. Weber, in *Organometallic Compounds in the Environment,* P. J. Craig (ed.), Chapter 8, Longman, Harlow, 1986.
[9] J. A. Rodriguez-Vasquez, *Talanta,* 1978, **25**, 299.
[10] W. R. Hatch and W. L. Ott, *Anal. Chem.*, 1968, **40**, 2085.
[11] R. Dumarey, R. Heindryckx and R. Dams, *Atom. Spectrosc.*, 1981, **2**, 51.
[12] J. E. Longbottom and R. C. Dressman, *Chromatog. Newslett.*, 1973, **2**, 17.
[13] R. Dumarey, R. Dams and P. Sandra, *J. High Resol. Chromatog. Chomatog. Commun.*, 1982, **5**, 687.
[14] R. Bye and P. E. Paus, *Anal. Chim. Acta,* 1979, **107**, 169.
[15] J. G. Gonzalez and R. T. Ross, *Anal. Lett.*, 1972, **5**, 683.
[16] J. W. Robinson and J. C. Wu, *Spectrosc. Lett.*, 1985, **18**, 47.
[17] A. Bzezinska, J. Van Loon, D. Williams, K. Oguma, K. Fuwa and I. H. Haraguchi, *Spectrochim. Acta,* 1983, **38B**, 1339.
[18] W. H. Schroeder and R. Jackson, *Chemosphere,* 1984, **9**, 1041.
[19] W. H. Schroeder and R. Jackson, *Intern. J. Environ. Anal. Chem.*, 1985, **22**, 447.
[20] W. E. L. Grossman, J. Eng and Y. C. Tong, *Anal. Chim. Acta,* 1972, **60**, 447.
[21] A. J. McCormack, S. C. Tong and W. D. Cooke, *Anal. Chem.*, 1965, **37**, 1470.
[22] B. D. Quimby, P. C. Uden and R. M. Barnes, *Anal. Chem.*, 1978, **50**, 2112.
[23] G. W. Rice, J. J. Richard, A. P. D'Silva and V. A. Fassel, *Anal. Chem.*, 1981, **53**, 1519.
[24] K. B. Olsen, D. S. Sklarew and J. C. Evans, *Spectrochim. Acta,* 1985, **40B**, 357.
[25] Y. Talmi, *Anal. Chim. Acta,* 1975, **74**, 107.
[26] Y. Talmi and V. E. Norvell, *Anal. Chim. Acta,* 1976, **85**, 203.
[27] G. W. Rice, J. R. Richard, A. P. D'Silva and V. A. Fassel, *J. Assoc. Off. Anal. Chem.*, 1982, **65**, 14.
[28] K. Chiba, K. Yoshida, K. Tanabe, H. Haraguchi and K. Fuwa, *Anal. Chem.*, 1983, **55**, 450.
[29] C. A. Bache and D. J. Lisk, *Anal. Chem.*, 1971, **43**, 950.
[30] G. Decadt, W. Baeyens, D. Bradley and L. Goeyens, *Anal. Chem.*, 1985, **57**, 2788.
[31] R. S. Braman and D. L. Johnson, *Environ. Sci. Technol.*, 1974, **8**, 996.
[32] D. D. Gay, K. O. Wirtz, L. C. Fortmann and H. L. Kelly, *Spectra 2000,* 1979, **7**, 72.
[33] D. S. Ballantine, Jr. and W. H. Zoller, *Anal. Chem.*, 1984, **56**, 1288.
[34] F. E. Brinckman, W. R. Blair, K. L. Jewett and W. P. Iverson, *J. Chromatog. Sci.*, 1977, **15**, 493.
[35] C. H. Gast, J. C. Kraak, H. Poppe and F. J. M. J. Maessen, *J. Chromatog.*, 1979, **185**, 549.
[36] M. Fujita and E. Takabatake, *Anal. Chem.*, 1983, **55**, 454.
[37] J. W. Robinson and E. D. Boothe, *Spectrosc. Lett.*, 1984, **17**, 673.
[38] I. S. Krull, D. S. Bushee, R. G. Schleicher and S. B. Smith Jr., *Analyst,* 1982, **107**, 1457.
[39] W. Holak, *Analyst,* 1982, **107**, 1457.
[40] L. H. Keith, W. Crummett, J. Deegan Jr., R. A. Libby, J. K. Taylor and G. Wentler, *Anal. Chem.*, 1983, **55**, 2210.
[41] Analytical Methods Committee, *Analyst,* 1987, **112**, 199.

11

Selenium combination techniques

Alan G. Howard
Chemistry Department, The University, Southampton, Hampshire, England

11.1 INTRODUCTION

Ranking seventeenth in crustal abundance, selenium is one of the minor but biologically essential elements [1]. It has found a variety of industrial applications, in electronic rectifiers and photoelectric cells, the production of stainless steel and chromium plating solutions, and in the manufacture of glass and ceramics, pigments and some pharmaceutical preparations.

Its abundance in the environment is variable; for example the selenium content of soils ranges from 0.1 ppm in selenium-deficient areas of New Zealand [2] to 1200 ppm in a toxic region of Ireland [3]. This wide concentration range is also reflected in the concentrations found in non-saline waters, from 0.1 μg/l. to extremes of 9 mg/l. [4] but most lie between 0.2 and 10 μg/l. On the basis of thermodynamic predictions, selenium would be expected to be present in natural waters as selenate, and to a very minor extent, selenite. As a result of biological transformations, however, methylated species such as dimethylselenide, dimethyldiselenide and dimethylselenone are also found [5]. These have been shown to be produced by micro-organisms in sewage sludge and soil [6] and are to be found in some natural waters. They have not, as yet, been positively identified as being present in sea-water. The marine geochemistry of selenium has been discussed by Cutter and Bruland [7] in terms of three distinct dissoved selenium species: selenite, selenate and organic selenide. The organic selenide fraction has not been characterized, but might be expected to contain seleno-amino-acids and their derivatives. In surface waters, organic selenide makes up about 80% of the total dissolved selenium, the highest levels being coincident with the primary productivity maximum. In deep ocean waters selenite and selenate are the dominant species, with no evidence of organic selenide.

In addition to the organoselenium species mentioned previously, a number of selenium compounds have been found in living organisms. These are often selenium analogues of sulphur-containing biochemicals and include selenocoenzyme A, selenobiotin, *Se*-methylselenocysteine, *Se*-methylselenomethionine, selenocystathione, the trimethylselenonium ion, selenohomocystine, selenocystine and selenomethionine. A high proportion of the selenium which is present in plants and animals is incorporated in protein, both as selenoamino-acids and selenotrisulphide-linked.

For more general reviews on the occurrence of selenium, its biological consequences and determination in a variety of matrices, the reader is referred to the articles by Rosenfeld and Beath [8], Zingaro [9], Bem [10], Raptis *et al.* [11], Odom [12], Fishbein [13], Shamberger [14], Adriano [15] and Craig [16].

In this chapter the combination techniques are broken down into component parts, sample storage, derivatization, separation and detection being considered separately. In this way it is possible to isolate the properties of the system components that can be employed in the construction of both novel and extant combination systems. Examples of practical systems and their applications are then described.

11.2 SAMPLE COLLECTION AND STORAGE

The considerations that must be applied to the collection of samples for selenium determination are similar to those for most other trace element work. What is rather unusual about selenium is the need to consider volatile forms of the analyte. Once a sample has been obtained, particular regard must be paid to the possibility of biomethylation during storage, and any losses by this route must be minimized. Gaseous samples for determination of the volatile alkylselenide species can be collected by cryogenic trapping of alkylselenide species from air [17–19], by using the apparatus shown in Fig. 11.1.

Almost quantitative recovery of dimethylselenide, dimethyldiselenide and diethylselenide can be obtained, whether the trap is filled with glass beads or glass wool (silanized or untreated) [18].

Historically, inadequacies in the storage of trace element samples have led to gross analytical errors. Nowhere is the problem more acute than in the analysis of water, especially if selenium speciation is to be preserved. Early work in the area favoured the use of Pyrex glass for the storage of both acidified and unacidified water samples. More recently, however, there has been a move back to the use of polyethylene, probably as a result of improved production techniques and sample storage procedures. Shendrikar and West [20] studied the loss of selenite by adsorption on the walls of a variety of containers, and concluded that losses were minimal on Pyrex. Such bottles, with caps lined with aluminium foil, have been found to be suitable for the preservation of selenite in non-acidified natural water samples [21]. Whilst selenite samples could be preserved for 2 weeks, 50% of the selenite was lost within 7 days. Polyethylene bottles were no better, a 59% loss being experienced in 5 days. Rapid freezing (dry-ice) of spiked samples in polyethylene bottles, and subsequent storage in a deep freeze for 3 days, led to a 12% loss of selenite and a 48% loss of selenate. Quick freezing of samples in liquid nitrogen gave only minimal losses and, once frozen, samples could be stored in a freezer. Acidification can be helpful in the preservation of water samples, but this must not be overdone if speciation is to be maintained. Both Se(IV) and Se(VI) speciation is preserved by storing acidified ($1M$ H^+) samples in Teflon-capped Pyrex glass [22]. Of

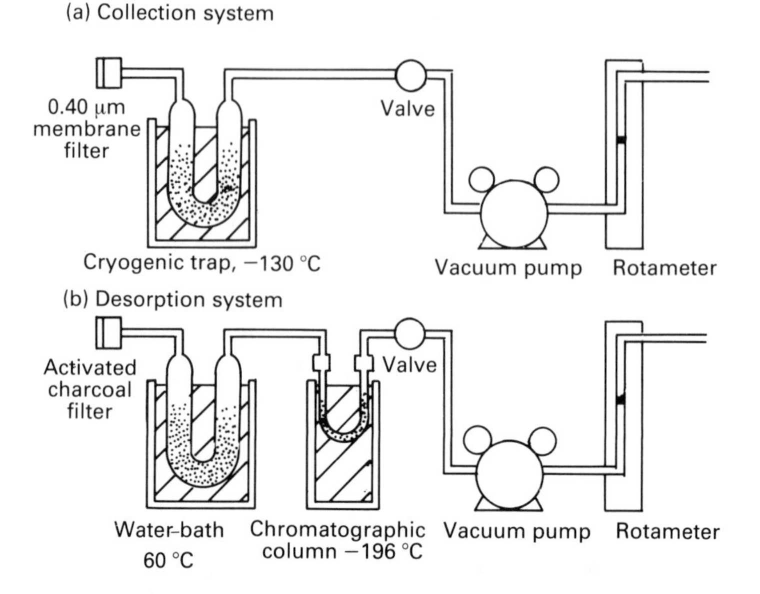

Fig. 11.1 — Systems for the collection (a) and desorption (b) of alkylselenium species present in air. (Reproduced with permission from S. Jiang, H. Robberecht and F. Adams, *Atmos. Environ.*, 1983, **17**, 111. Copyright 1983, Pergamon Press.)

the available methods, freezing is generally preferred as it minimizes contamination and unforeseen speciation changes. The storage of samples containing methylselenium species is difficult and immediate analysis is to be recommended if the volatile components are not to be completely lost [21].

11.3 PRETREATMENT

If analyses for selenium are to be carried out without regard for speciation, any number of digestion procedures can be employed to simplify the sample matrix provided that due attention is paid to the potential volatility of the analyte. Selenium chlorides are volatile during wet ashing and the use of concentrated hydrochloric acid is therefore not to be advised. Biological materials can be digested with a nitric–perchloric acid mixture at 210°C without loss of selenium [23] but perchloric acid is to be avoided if possible, owing to the explosion hazard associated with its use. An alternative procedure which has been shown to be satisfactory is to use a nitric/sulphuric acid mixture in digestion apparatus fitted with a condenser system [24]. Organoselenium species in water can be converted into inorganic selenium by ultraviolet irradiation of a sample containing borax and hydrogen peroxide [25]. Whilst most digestion procedures for the determination of total selenium are designed to maintain the element as selenate, the determination methods almost without exception necessitate conversion of the selenium into a single chemical species; this is normally selenite. A number of techniques can be employed to reduce selenate to selenite but one of the most reliable is digestion of the sample with 6*M* hydrochloric acid for 4 hr at 60°C.

Sample pretreatment for selenium speciation is more difficult as it is necessary to use procedures which are both quantitative and preserve speciation. Speciation of selenium can be carried out on a number of levels depending on the extent of chemical modification permitted (or required). Taking the example of selenium incorporated in protein, speciation analyses are capable of identifying the specific protein, identifying the selenium as being protein bound, showing the presence of selenomethionine protein residues, or just identifying selenomethionine. The thoroughness of the speciation is therefore governed by the type of analysis to be done and the availability of extraction procedures which are capable of quantitatively removing the species of interest from the sample matrix. It is rather unfortunate that those methods best suited to preserving the full speciation integrity are generally those which are the least efficient. It is frequently necessary to compromise and due attention must therefore be paid to the changes in speciation introduced by harsh extraction procedures.

In order to validate new procedures, standard reference materials are required which are characterized with respect to selenium content. Those which are currently available are only certified for total element concentrations and little information is available on the actual chemical form of selenium present in such samples. One exception to this is the work by Cutter [22]. In a study of the chemical forms of selenium in biogenic particles and sediments he employed an alkaline leach (1*M* sodium hydroxide, 4 hr in an ultrasonic bath) to release selenium species under mild conditions. This recovered 94% of ^{75}Se-labelled selenite and selenate that had been added to homogenized plankton. A 1*M* hydrochloric acid leach, however, removed less than 70% of the added selenium. In addition, the alkaline leach procedure did not convert selenocysteine into selenite. The concentrations of selenium found in some reference materials by Cutter [22] are shown in Table 11.1.

11.4 DETECTION

11.4.1 Atomic absorption

Atomic-absorption measurements of selenium almost exclusively employ the 196.0 nm line which, at the temperatures encountered in flame-in-tube atomizers, should have a line-width of about 0.3 pm due to Doppler broadening and at least 1 pm due to collisional broadening [26]. This compares with an estimated line-width of less than 0.35 pm for a selenium electrodeless discharge lamp, and true absorption is therefore believed to take place. The 196 nm selenium line is unfortunately far from ideal for most conventional instrumentation, owing to the poor sensitivity of normal photomultipliers at wavelengths shorter than 200 nm, and absorption by atmospheric and flame constituents. The performance of most spectrometers can be improved by replacing the photomultiplier by one offering extended response at below 200 nm, thereby improving detection limits through enhanced signal-to-noise characteristics. The choice of light-source is largely between the hollow-cathode lamp (HCL) and the electrodeless discharge lamp (EDL); lower detection limits are often obtained with an EDL.

Table 11.1 — Chemical speciation of selenium in reference materials and one freshwater sediment.[a] (Reproduced by permission, from G. A. Cutter, *Anal. Chem.*, 1985, **57**, 2951. Copyright 1985, American Chemical Society.)

Type	Total Se	Se(IV)	Se(VI)	Predicted[b] Se(IV)+(VI)
A. Biological materials				
Copepod (IAEA MA-A-I)	3.00±0.20	0.02±0.01 ($n=3$)[c]	0.02±0.01 ($n=3$)	—
Oyster tissue (NBS 1566)	2.60±0.30	<0.01 ($n=3$)	<0.01 ($n=3$)	—
B. Sediments				
River sediment (NBS 1645)	1.70±0.30	0.2±0.01 ($n=8$)	0.08±0.03 ($n=8$)	0.11±0.02 ($n=4$)
Estuarine sediment (NBS 1646)	0.43±0.02	0.001±0.0006 ($n=8$)	0.04±0.02 ($n=8$)	0.05±0.01 ($n=4$)
Hyco Reservoir sediment	6.45±0.66 ($n=4$)	0.21±0.03 ($n=4$)	0.51±0.09 ($n=4$)	0.63±0.04 ($n=3$)

[a] Concentrations in micrograms of selenium per gram.
[b] Predicted Se(IV)+(VI) equal to total selenium in the exchangeable + carbonate + iron/manganese oxide phase.
[c] n = The number of separate samples processed; each sample is determined in triplicate.

Whilst the direct introduction of gases and liquids into the conventional aspirator–burner systems of atomic-absorption spectrometers is appealing on the grounds of simplicity, it is not generally recommended; the volumes of most such inlet systems are generally large, leading to significant band-broadening and poor detection limits. Atom trapping can improve the sensitivity, but only by a factor of 3 unless slits are employed to view the region of high atom density close to the tube walls. This selective viewing can result in a further 5-fold increase in sensitivity [27].

The determination of selenium by atomic-absorption spectrometry has been reviewed by Verlinden *et al.* [28].

11.4.1.1 Gas-phase introduction

In recent years the preferred method of atomization of gas phase selenium species has become the heated silica atomizer. This can either be heated with a flame or, more controllably, an electric furnace held at between 700 and 900°C. In some applications an air–hydrogen flame is burnt inside the tube. The performance of silica atomizers depends on a number of factors, including cell design, condition and temperature, and carrier gas flow-rate. Dědina and Rubeška [26] and Dědina [29] have extensively studied the atomization of selenium hydride in cool, highly fuel-rich hydrogen–oxygen flames burning in T-shaped silica tubes, two examples of which are shown in Fig. 11.2.

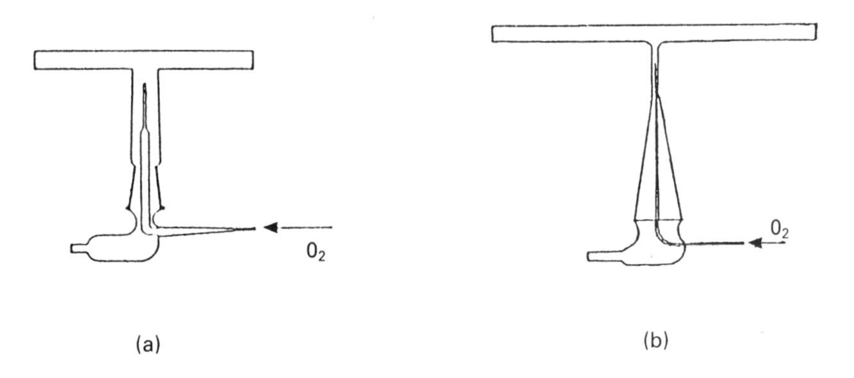

Fig. 11.2 — Examples of flame-in-tube atomizers. [Reproduced with permission. (a) From J. Dědina and I. Rubeška, *Spectrochim. Acta*, 1980, **35B**, 119; copyright 1980, Pergamon Press. (b) From J. Dědina, *Anal. Chem.*, 1982, **54**, 2097; copyright 1982, American Chemical Society.)

Three fused-silica tube atomizers were studied, differing in the dimensions of their atom cells. Cell A has an inner diameter of 7.8 mm and length of 14 mm, tube B was 3.2 mm in diameter and had a length of 152 mm. Tube C differed from tube B in that the distance between the oxygen inlet and the optical axis was increased from 10 mm to 100 mm. When a hydrogen flow of 6 l./min was employed, maximum response to selenium was obtained at oxygen flow-rates of 50 and 25 ml/min for tubes A and B respectively. Heating the tubes did not result in any increase in the response due to hydrogen selenide. Atomization was therefore attributed to interaction with H and OH radicals generated in the flame and not to thermal decomposition. Cell design C resulted in a diminished selenium response, possibly due to the recombination of selenium atoms before reaching the light-path.

Systems using heated T-pieces and no internal flame require oxygen for the formation of selenium atoms from H_2Se but enough for this purpose is normally present in the carrier gas (N_2) and in the sample solution. Atomization temperatures should be above 700°C and the residence time in the T-piece must be kept sufficiently long. If, however, more oxygen is added than is required for the combustion of the hydrogen generated by decomposition of the borohydride used as reductant, the absorption drops to zero. At low gas flow-rates dimerization of the selenium on the silica wall results in a drop in absorbance [30]. Hydrogen produced by the decomposition of sodium borohydride therefore has two effects. In addition to dilution of the selenium, the atomization of hydrogen selenide is decreased if excess of hydrogen is present [31].

In summary, hydrogen and oxygen react to give hydrogen radicals. These then react with the hydrogen selenide, leading to atomization.

$$H\cdot + SeH_2 \rightarrow SeH\cdot + H_2$$
$$H\cdot + SeH\cdot \rightarrow Se + H_2$$

This reaction is shifted to the left by the presence of excess of hydrogen, resulting in decreased sensitivity. Gas flows and atomizer cell design must be optimized to maximize atomization whilst preventing dimerization of selenium in the cell

$$2Se \rightleftharpoons Se_2$$

and dilution of the selenium with carrier.

There is an optimal period in the life of silica tube atomizers. New tubes can initially perform poorly and must be conditioned by use until the peak height for a given amount of selenium has increased to a constant value, or by treatment with 40% hydrofluoric acid for 15 min [32]. The end of the the useful working life of a silica tube atomizer is associated with devitrification of the quartz to β-cristobalite, which occurs over a period of time and affects both sensitivity and precision. Performance and lifetime can be improved by using sulphuric acid rather than hydrochloric acid in the hydride-generation step [33] but this may, in the case of selenium, lead to a reduction in sensitivity, owing to the absence of halide.

As an alternative to the silica atomizer it is possible to use a commercial graphite-furnace atomizer for the detection of selenium species separated by gas chromatography. Such systems can be employed to monitor selenium either continuously or in a pulsed atomization mode. In either case, furnace tube lifetimes are expected to be limited. Information on such techniques that is specific to the determination of selenium is, however, limited (see e.g. [34]).

11.4.1.2 Liquid-phase introduction

The atomization of material from liquid chromatography systems is rather more difficult than that for the corresponding gaseous systems, owing to the volume of liquid involved. This problem can be dealt with in two ways. Either the conventional nebulizer–burner system of flame spectrometers can be used or the column eluate can be periodically sampled. Neither approach is ideal, the former degrading resolution and having poor sensitivity. Periodic sampling of the eluate is generally employed with electrothermal atomization. This has been used in combination with HPLC separation techniques, but the non-continuous nature of the atomization step results in discontinuous output and consequentially potentially poor precision.

11.4.2 Flame emission

Selenium compounds in an air–hydrogen flame emit a broad band spectrum between 450 and 500 nm, due to Se_2, which can be measured by using conventional flame photometric detectors (FPD) fitted with a suitable filter. The selenium FPD response is weaker than that of sulphur, but sub-nanogram quantities can still be determined [35]. The band emission due to Se_2 and S_2 overlap, and compounds or mixtures of both elements mutually interfere. This is minimized in favour of selenium by using a fuel-rich flame but at the expense of the flame becoming easily extinguished by the solvent. As with the sulphur-responsive FPD, the selenium response is quenched by carbon-containing compounds but less severely. The performance of various methods of assessing the extent of these interferences has been investigated by Hancock *et al.* [36].

11.4.3 Plasma emission

Both the inductively coupled plasma (ICP) and microwave-induced plasma (MIP) have been employed in combination techniques for the detection of selenium. Plasma emission offers several potential advantages as a detection technique, not least of which being the simultaneous determination of a number of elements by using multichannel detection. The typical microwave-induced plasma system employs a 'Beenakker' TM_{010} cylindrical resonance cavity [37] operated with a 60–100 ml/min flow of helium and fed from a 2450 MHz microwave generator giving about 100 W of input power. The plasma is confined in a discharge tube made of refractory material such as fused silica or boron nitride.

Olsen *et al.* [38] have described systems using both low-pressure and atmospheric-pressure microwave-induced plasmas coupled to bonded-phase fused-silica capillary columns. The low-pressure system employed a 2450 MHz 1/4-wave Evenson-type cavity maintained at a pressure of less than 5 mmHg. The plasma was maintained in a 1 mm i.d. fused-silica tube and was protected from deposition of the analyte elements on the plasma tube, by the addition of a reactive scavenger gas such as hydrogen, oxygen or nitrogen. Plasma overload and contamination by bulk solvent were prevented by a bypass system. The atmospheric-pressure unit employed a 2450 MHz Beenakker cavity operated at up to 120 W. Four selenium emission lines, 196.026, 203.985, 206.279 and 207.479 nm, are available, the most intense being that at 206 nm.

11.5 GAS CHROMATOGRAPHY

Analyses based on gas chromatography of selenium compounds fall into two classes, those dealing with the natural volatile selenium compounds and those for the determination of inorganic selenium. The so-called inorganic forms of selenium require suitable derivatization to a volatile form prior to the chromatographic step and therefore rarely retain information on chemical form. Combination techniques of this type therefore frequently give multielement analyses.

11.5.1 Hydride procedures

11.5.1.1 Generation

The conversion of selenium into hydrogen selenide is widely used for the determination of total selenium and for obtaining information on inorganic selenium oxidation states. Although a variety of reductants such as Zn/HCl and $TiCl_3$/Mg/HCl have been employed for the generation of hydrogen selenide, in recent years attention has focused on the use of either tablet-form or dissolved sodium tetrahydroborate (borohydride). In acid solution Se(IV) is converted into H_2Se by the reaction

$$4H_2SeO_3+3BH_4^-+3H^+ \rightarrow 4H_2Se+3B(OH)_3+3H_2O$$

Despite the convenience of the tablets, some doubts have been raised over the efficiency with which hydride is generated in this way [39].

In aqueous solution, sodium tetrahydroborate undergoes slow hydrolysis, releasing gaseous hydrogen:

$$BH_4^-+H^++3H_2O \rightarrow B(OH)_3+4H_2$$

Solutions can be stable for several weeks if prepared with 0.1*M* potassium hydroxide. Unstabilized solutions must be prepared daily, but are generally less prone to contamination. Further improvements in stability can be achieved by filtration, sometimes with a concomitant reduction of reagent blanks.

Conditions recommended for the reduction vary. Quoted concentrations of sodium tetrahydroborate range typically from 0.01 to 80 mg/ml and the acid concentration ranges from 0.4 to 6*M*. Whilst selenite is reduced by sodium borohydride in 4*M* acid, selenate is not [39,40] and if it is to be determined must first be reduced to selenite by other means. A wide variety of mineral acids can be employed to maintain the low pH but in the absence of halides inefficient hydride generation can result. For this reason, hydrochloric acid is frequently employed. The addition of potassium iodide to the tetrahydroborate solution greatly affects the efficiency with which selenium(IV) is converted into hydrogen selenide [41] when the sample is acidified with suphuric acid. In the absence of iodide, at least 1 mg/ml tetrahydroborate solution is required for maximum conversion but with iodide present, 0.01 mg/ml tetrahydroborate is sufficient to achieve the same generation efficiency.

11.5.1.2 Interferences

All hydride methods are susceptible to interferences, and the selenium procedures are no exception. The problems originate both in the derivatization and detection stages but unfortunately little work has been done on this aspect of combination techniques for selenium determination. Much of the following information has therefore been drawn from the parent hydride procedures and must be carefully evaluated to assess its applicability to specific combination techniques.

Whilst there is general agreement that a number of elements interfere in the hydride methods, the effects of some elements are still in dispute. According to a variety of authors, and particularly Meyer *et al.* [42], interferences have been observed from the ions of the following elements: Ag, As, Au, Bi, Cd, Co, Cr, Cu, Fe, Ge, Hg, Mn, Ni, Pb, Pt, Sb, Sn, Te and V. According to other authors, however, Fe^{2+}, Fe^{3+}, Pb^{2+}, Sn^{2+} and Hg^{2+} do not interfere [29,43,44]. The magnitudes of the observed effect appear to be highly dependent on the exact nature of the apparatus and experimental conditions employed and should be evaluated for individual applications.

The mechanisms underlying the metal interferences are still very much in contention. Interferences can be divided into those originating in the general solution and those taking place in the gas phase. Those which have their effect in the generating solution can do so by several mechanisms. Some authors suggest that sodium borohydride reduces the metal ions to the metal, which then reacts with the hydrogen selenide. Precipitates are certainly formed under these circumstances but their colour and composition are often more characteristic of metal borides [43, 45]. Such precipitates might interfere in a number of ways, such as consumption of the reductant (thus reducing the rate of reduction of selenium so that the tetrahydroborate has decomposed before selenium reduction is complete), or causing decomposition of the hydrogen selenide, or presenting a surface on which the hydrogen selenide can be adsorbed. By adding the metals during and after acidification of the tetrahydroborate, Agterdenbos and Bax [41] were able to demonstrate that interferences from Co^{2+}, Ni^{2+} and Cu^{2+} arise at least in part from enhanced decomposition of the tetrahydroborate leading to reduced hydride formation.

The hydride-forming elements can interfere both in solution and in the gas phase. ^{75}Se tracer studies have shown that only As and Bi cause liquid-phase interferences. Sn, As, Sb, Bi and Te give gas-phase interferences, the magnitude of which depends on the atomization conditions. Fused-silica atomizer tubes employed in conventional hydride procedures have been observed to become coated with deposits that are believed to be Sb_2Se_3, $GeSe_2$, an As compound [44], Sn or Bi, depending on the interferent involved [42]. Once the fused-silica atomizer has been thus contaminated, subsequent analyses exhibit signal suppression. Gas-phase interferences at the atomization stage can result from an accelerated decay of the free analyte atom population or a decrease in hydrogen radical population [26,46].

Methods of overcoming interferences are sparse and many of the commonly employed masking agents such as EDTA are not effective with selenium. Copper decreases the selenium signal [43,46] but can be partially masked by the addition of potassium cyanide or use of cyanotrihydroborate instead of the tetrahydroborate [43]. Thiourea has also been found to be an effective masking agent for copper [47]. Nickel is another metal which severely depresses the selenium signal. The common masking agents 1,10-phenanthroline and thiosemicarbazide are ineffective in the selenium determination [43] but again, the situation can be improved by the use of cyanoborohydride or the addition of potassium cyanide before the analysis [43]. The addition of iodide to the tetrahydroborate solution results in a considerable decrease in the interferences caused by metal ions [41] but this may only be true if sulphuric acid is used to give the acidic reduction conditions.

Natural organic matter in groundwater can be successfully removed from samples for selenium analysis, whilst preserving the specification of the inorganic selenium, by column chromatography on Amberlite XAD-8 [48].

In the analysis of natural waters, one of the most significant interferences can arise from the presence of nitrite. Complete suppression of the selenium signal is caused by 2μ*M* nitrite, but this can be overcome by the addition of sulphanilamide, which is diazotized by the nitrite [49]. Nitrate does not interfere except in the presence of copper(II), but nitrogen oxides (and nitrous acid) generated by the use of nitric acid in digestion mixes can be a problem. This can, however, be overcome by the addition of sulphuric acid [43].

Further information on the determination of selenium by hydride methods can be found in the review by Gunn [50].

11.5.2 Organoselenium species and derivatives

11.5.2.1 Natural species

Several of the natually occurring selenium species, such as dimethylselenide, dimethyldiselenide and diethylselenide are sufficiently volatile for direct determination without chemical modification. Columns that have been employed for this purpose include 10% or 20% poly(metaphenyl ether) (PMPE 5 ring) on Chromosorb W (or WAW) [18, 51] and 16.5% silicone oil DC550 on 80–100 mesh Chromosorb W-AW-DMCS [21].

11.5.2.2 Inorganic selenium

When a hydride procedure is not to be used, gas chromatography of inorganic selenium is generally done after conversion of the selenium into selenite and reaction of this with a suitable 1,2-diamine to form a piazselenol. Various diamines have been employed for this purpose, including diaminonaphthalene [52], 1,2-diamino-4,5-dichlorobenzene [53] and 1,2-diaminonitrobenzenes [53,55,56]. The reaction between Se(IV) and 1,2-diamines is performed under acid conditions and can take place in the presence of both perchloric acid and nitric acid, which are commonly used in digestion procedures. Unfortunately, however, perchloric acid can adversely affect extraction of the piazselenol into toluene [57].

Alternative approaches to the derivatization of selenium include the formation of phenyl derivatives, and complexation with acetophenone or dithiocarbamate. Selenium can also be determined by gas chromatography after formation of its trimethylsilyl (TMSi) derivative $SeO_2(TMSi)_2$ [58].

11.5.3 Chromatography

AsH_3, SnH_4 and GeH_4 can be separated on a 3 ft Chromosorb 102 column, but the resolution of SnH_4 and H_2Se is poor [59, 60]. This problem has been circumvented by the separate determination of selenium [60].

A wide variety of columns have been employed for the determination of piazselenols. Amongst those reported as suitable are a 6 ft × $\frac{1}{8}$ in. glass column packed with 3% Silar 10C on silylated Chromosorb W [25], a 4 ft column containing 4%

SE30 on 30–60 mesh Chromosorb GHP [57], 7% OV225 on Supasorb AW-HMDS (100–120 mesh) [55], an SE30 column, and columns normally used for pesticide residue analysis — Dow 11, OV17, QF-1 and diethyleneglycol succinate [52].

11.5.4 Applications

11.5.4.1 GC–AAS

An extensive compilation of papers dealing with directly coupled gas chromatography–AAS systems has been published by Ebdon *et al.* [61].

Chau *et al.* [62] were amongst the first to couple a GC to a fused-silica tube atomizer AAS system. Their apparatus consisted of a cold trap connected in series with a packed column GC and AAS detector (Fig. 11.3).

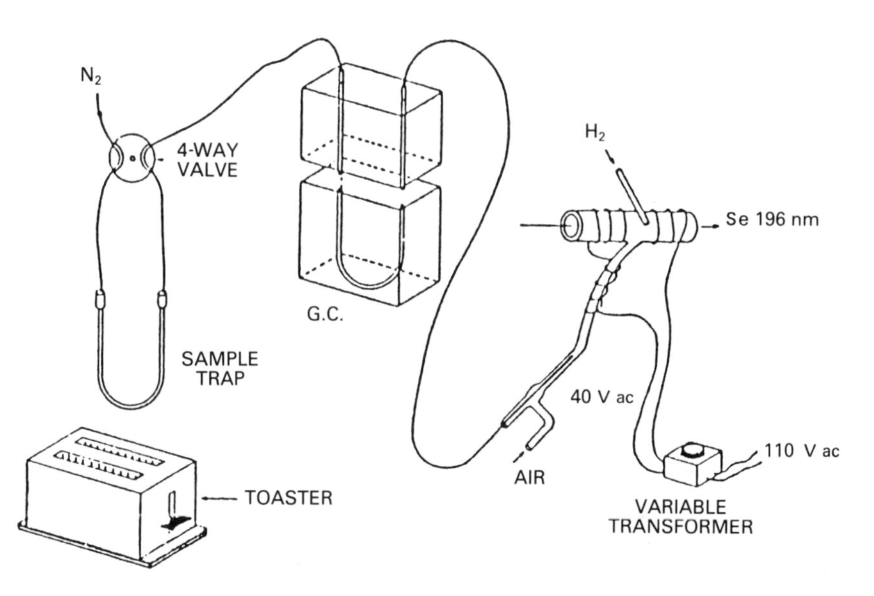

Fig. 11.3 — Schematic diagram of GC–AAS apparatus using cold trapping. (Reproduced with permission from Y. K. Chau, P. T. S. Wong and A. D. Goulden, *Anal. Chem.*, 1975, **47**, 2279. Copyright 1975, American Chemical Society.)

Methylselenium species were collected on a GC packing of 3% OV-1 on Chromosorb held at −80°C. The trap temperature was then raised to about 80°C and the sample components were swept onto the GC column (3% OV-1 on Chromosorb), which was temperature-programmed from 40 to 120°C (Fig. 11.4).

For about 10 ng of selenium the relative standard deviation was ~8% and the technique could detect 0.1 ng of Se with certainty.

Similar apparatus was used by Cutter [21], who converted a cold-trap hydride AAS system into one suitable for the detection of dimethylselenide and dimethyldiselenide by insertion of a gas chromatograph between the cold trap and the atomizer. To exclude air, the impinger tip had to be replaced by a fritted glass cylinder, and the

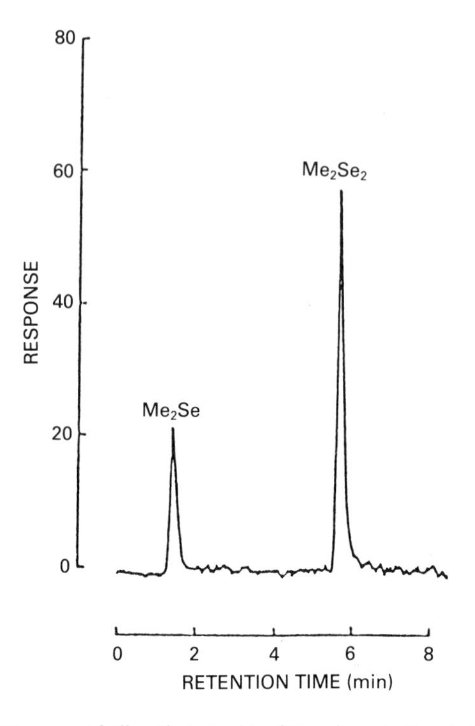

Fig. 11.4 — Chromatogram of dimethylselenide (10 ng) and dimethyldiselenide (16 ng) obtained with the apparatus shown in Fig. 11.3. (Reproduced with permission from Y. K. Chau, P. T. S. Wong and P. D. Goulden, *Anal. Chem.*, 1975, **47**, 2279. Copyright 1975, American Chemical Society.)

sample was introduced by injection through a septum. The dry-ice/2-propanol water-trap had also to be replaced by a calcium chloride drying tube.

Problems with heating interface lines have been minimized by Van Loon and Radziuk by construction of an integral GC–atomizer system. This consists of an electrically-heated fused-silica T-tube furnace heated to 900–1000°C and directly coupled to an integral, custom-built gas chromatograph (Fig. 11.5) [63].

This apparatus has been used in a study of the gases evolved by *Astrogalus racemosus.* Four selenium-containg compounds were discovered, two of which were identified as dimethylselenide and dimethyldiselenide (Fig. 11.6).

Jiang *et al.* [17,18] set up a simple GC-GFAAS combination for the measurement of dimethyldiselenide, dimethylselenide and diethylselenide. A 1 m nickel tubing transfer line connected the GC column to the graphite-furnace atomizer. Hydrogen was introduced into the gas stream at the beginning of the transfer line to increase detector selectivity. A typical chromatogram and calibration curve are shown in Fig. 11.7.

The apparatus was able to detect methylated selenium species (expressed as Se) in air down to 0.2 mg/m^3 [19] and was also employed to determine these compounds in the breath of mice which has been dosed with selenium [18].

In a study of the levels of selenium in air by GC–AAS, detection limits of

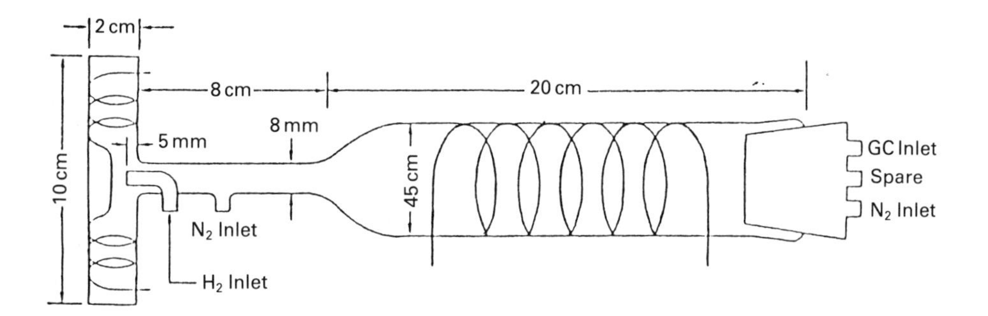

Fig. 11.5 — Combined gas chromatograph–atomizer unit for the study of alkylselenium species. (Reproduced with permission from B. Radziuk and J. Van Loon, *Sci. Tot. Environ.*, 1976, **6**, 251. Copyright 1976, Elsevier Science Publishers.)

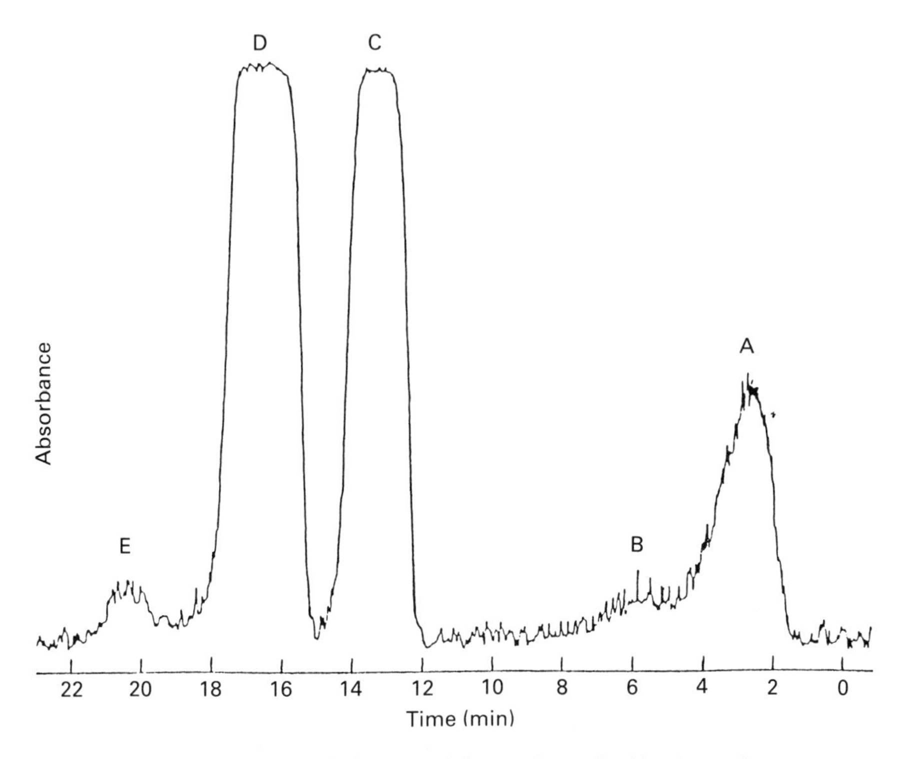

Fig. 11.6 — Chromatogram of selenium-containing species evolved by *Astragalus racemosus* (peak A=dimethylselenide, peak D = dimethyldiselenide, peaks B, C and E unidentified). (Reproduced with permission from B. Radziuk and J. Van Loon, *Sci. Tot. Environ.*, 1976, **6**, 251. Copyright 1976, Elsevier Science Publishers.)

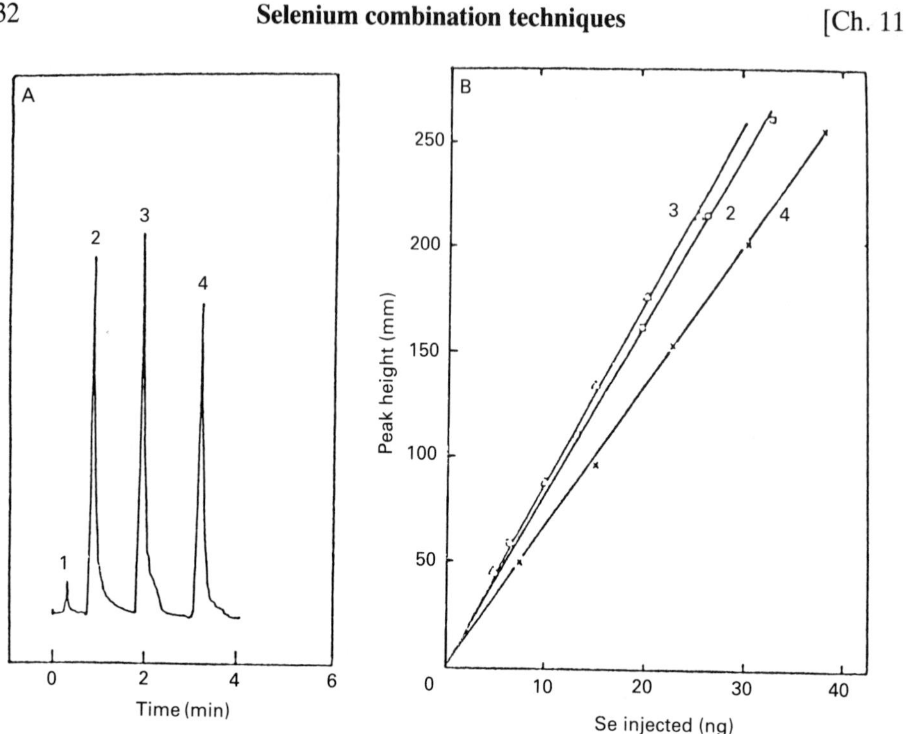

Fig. 11.7 — Chromatogram (A) and calibration curves (B) for standard mixtures of dimethylselenide (2), diethylselenide (3) and dimethyldiselenide (4). Peak 1 is due to the solvent. (Reproduced with permission from S. Jiang, W. DeJonghe and F. Adams, *Anal. Chim. Acta*, 1982, **136**, 183. Copyright 1982, Elsevier Science Publishers.)

0.2 ng/m^3 were obtained [17]. A chromatogram of material collected from air in the vicinity of a lake, showing the effect of adding hydrogen to the carrier gas, is shown in Fig. 11.8.

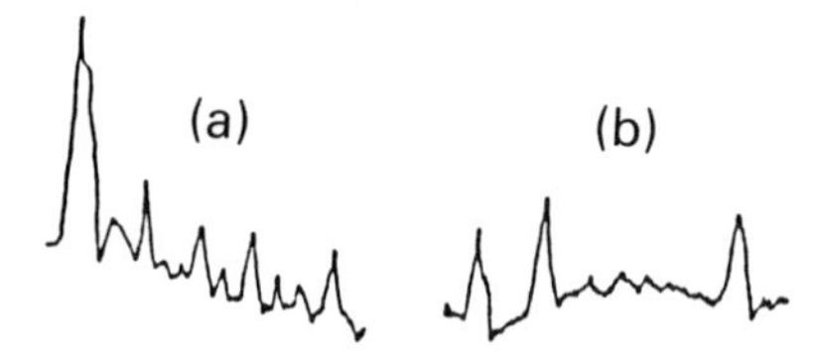

Fig. 11.8 — Chromatograms of an air sample collected from near a lake. (a) Without the addition of hydrogen to the carrier gas. (b) With 10% hydrogen added to the carrier gas. In (b) the first peak is due to residual molecular absorption; the second and third peaks are of dimethylselenide and dimethyldiselenide respectively. (Reproduced with permission from S. Jiang, W. DeJonghe and F. Adams, *Anal. Chim. Acta*, 1982, **136**, 183. Copyright 1982, Elsevier Science Publishers.)

The addition of hydrogen to the carrier gas was found to assist the efficiency of deuterium-lamp background-correction and increase atomization efficiency, resulting in a twofold improvement in sensitivity.

A system has been developed for the sequential determination of arsenic, selenium, germanium and tin as their hydrides by gas–solid chromatography with AAS detection. The hydrides are trapped at liquid-nitrogen temperatures before being released by warming (Fig. 11.9).

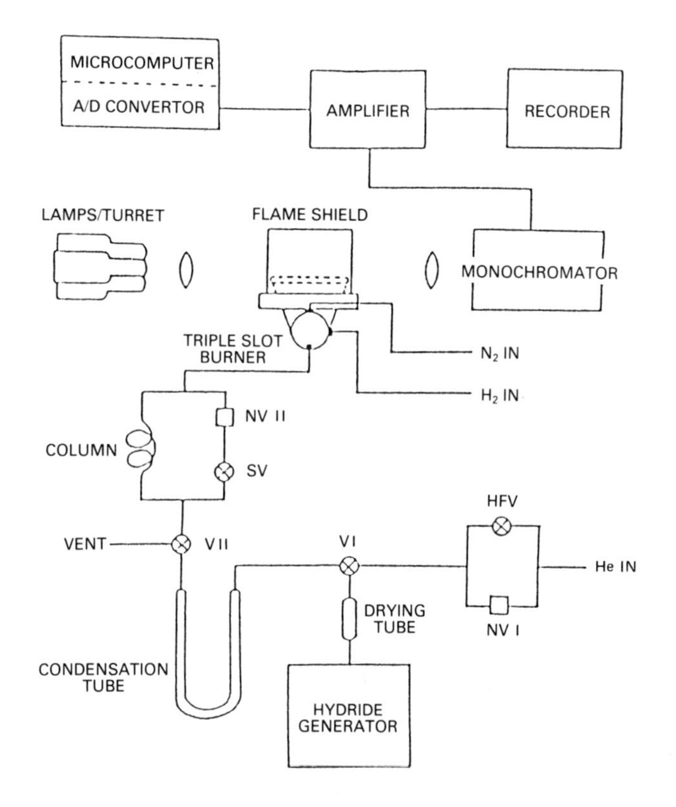

Fig. 11.9 — Schematic diagram of a gas chromatograph–AAS system for the sequential determination of arsenic, selenium, germanium and tin, using manual lamp selection. (NV = needle valve, V = 3-way valve, SV and HVF are 2-way valves). (Reproduced with permission from M. H. Hahn, K. J. Mulligan, M. E. Jackson and J. A. Caruso, *Anal. Chim. Acta*, 1980, **118**, 115. Copyright 1980, Elsevier Science Publishers.)

The gas flow is then split into two, one half passing to a Chromosorb 102 column for separation of arsine, stannane and germane, the other fraction going directly for selenium determination. This arrangement was necessary as it was not possible to resolve hydrogen selenide and stannane [60]. Detection limits for the system ranged from 3 to 13 ng/ml with relative standard deviations from 2 to 11%.

11.5.4.2 GC–atomic fluorescence

Atomic fluorescence offers the potential of multielement detection without the difficulties and expense involved with multiple wavelength selection. A four-channel non-dispersive atomic-fluorescence detector for GC has been developed by D'Ulivo and Papoff [64] for the determination of selenium, tin and lead species. The system uses a small hydrogen/argon flame atomizer and excitation is achieved with EDL sources (Fig. 11.10).

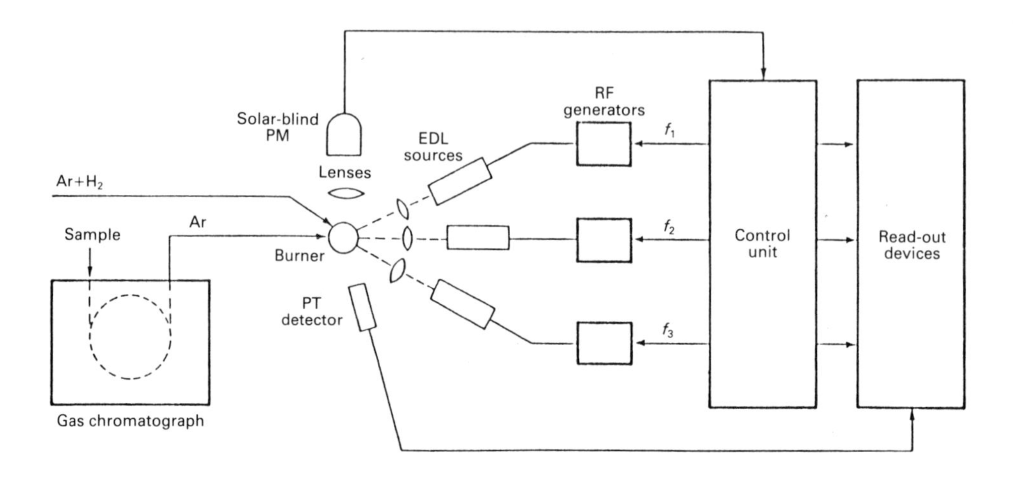

Fig. 11.10 — Schematic diagram of a simultaneous multielement non-dispersive atomic-fluorescence gas chromatographic detection system. (Reproduced with permission from A. D'Ulivo and P. Papoff, *J. At. Abs. Spectrom.*, 1986, **1**, 479. Copyright 1986, Royal Society of Chemistry.)

A variety of metal alkyls were separated on a 10 m fused-silica capillary column coated with a methylsilicone (Fig. 11.11).

Solvent signals were negligible except for those from solvents such as chlorinated solvents (which are difficult to combust) and carbon disulphide. By introducing organic compounds into the flame during the elution of analyte selenium species it was possible to show that co-elution could result in a significant quenching of the fluorescence signal.

11.5.4.3 GC–plasma emission

Total selenium has been determined by Talmi and Andren [57] by gas chromatography of piazselenols and MIP detection. The detection limit was 40 pg of Se, corresponding to 0.1 μg/l. for water and 15 ng/g for solid samples. Relative standard deviations were from 2.2 to 9.5%.

Fricke *et al.* [65] coupled a microwave-induced Ar/He plasma system (0.5 m monochromator) to a GC separation system (Chromosorb 102 column) to analyse for Ge, As, Se, Sn and Sb as their hydrides. A block diagram of the apparatus is shown in Fig. 11.12.

The performance of this system in the analysis of blood, flour and NBS orchard leaves has been reported [59]. The MIP is particularly well suited as a detector for capillary GC [66] but reports on the application of this technique to the determination of selenium-containing compounds are limited. Olsen *et al.* [38] have

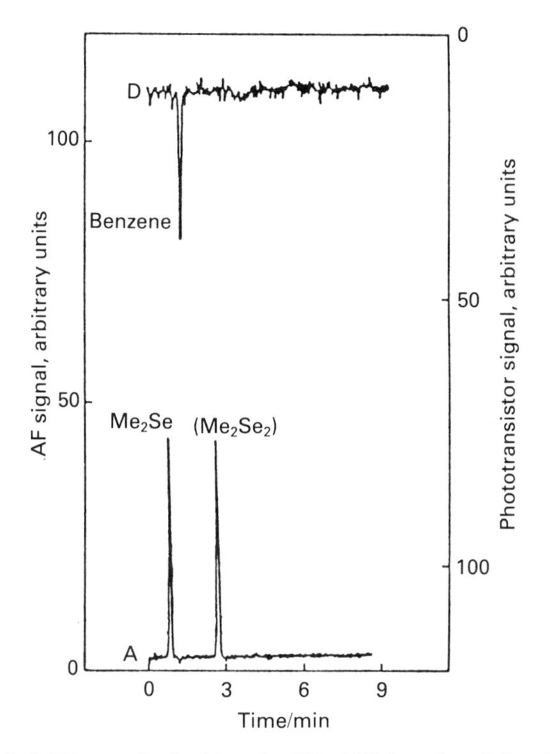

Fig. 11.11 — Typical GC trace obtained by using ND–AFS detection of dimethylselenide and dimethyldiselenide (0.3 ng of selenium in each component). (Reproduced with permission from A. D'Ulivo and P. Papoff, *J. At. Abs. Spectrom.*, 1986, **1**, 479. Copyright 1986, Royal Society of Chemistry.)

reported promising results for their capillary GC–MIP system but do not present results for selenium compounds.

Robbins and Caruso [67] demonstrated the application of a GC–MIP system for the determination of selenium in samples containing other hydride-forming elements. As well as separating the hydrides, the chromatographic system had the additional advantage of reducing background effects due to CO_2, HCl and water. Detection limits (concentration equivalent to twice the standard deviation of the blank signal) for selenium, in sequential and simultaneous modes, were 0.025 and 0.6 μg respectively. Precision was 5.5% for 1 μg of selenium measured sequentially and 27% for 4 μg of selenium measured simultaneously.

The performance of combination techniques involving gas chromatography is summarized in Table 11.2.

11.6 HIGH-PERFORMANCE LIQUID CHROMATOGRAPHY

The powerful separating ability of modern liquid chromatography techniques offers the potential of measurement of non-volatile or thermally unstable selenium species without recourse to derivatization procedures. Unfortunately, the coupling of HPLC

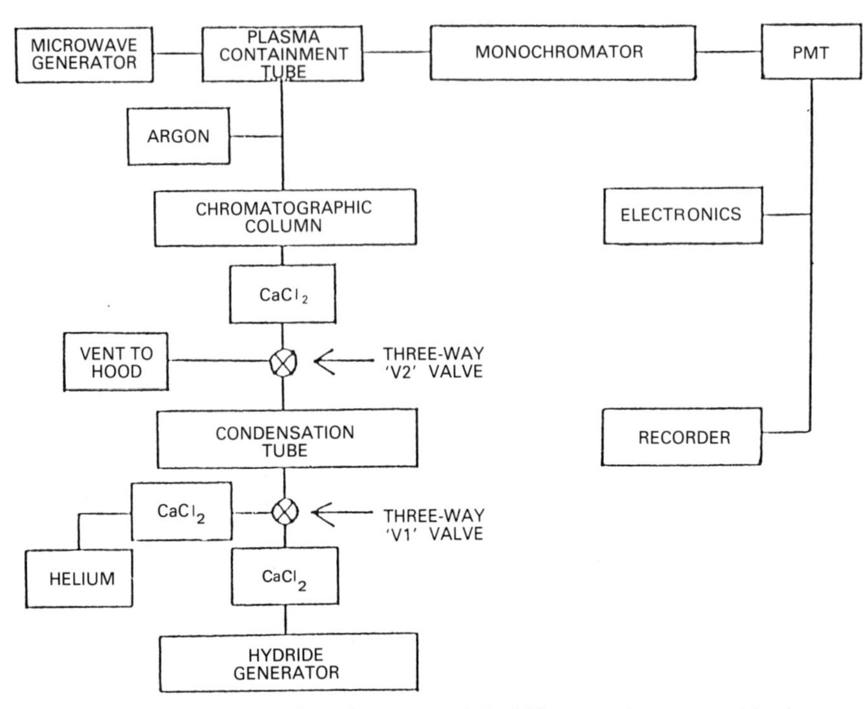

Fig. 11.12 — Block diagram of gas chromatograph–hydride generation system with microwave Ar/He plasma emission detection. (Reproduced with permission from F. L. Fricke, W. D. Robbins and J. A. Caruso, *J. Assoc. Off. Anal. Chem.*, 1978, **61**, 1118. Copyright 1978, Association of Official Analytical Chemists.)

Table 11.2 — Performance of gas chromatography combination techniques (NS = not stated).

Technique	Substance	Detection limit*	Relative standard deviation	Reference
Packed GC GFAAS	Me_2Se	7 ng	NS	[34]
GC–MIP	Piazselenol	40 pg	2.2–9.5%	[56]
Capillary GC ND–AFS	Me_2Se Me_2Se_2	10 pg	NS	[64]
Hydride GC–MIP	SeH_2	Sequential 0.04 μg Simultaneous 0.9 μg	5.5% (1 μg) 27% (4 μg)	[67]

to most forms of atomic spectrometry has proved to be more difficult than for the corresponding coupling of GC systems, owing to inefficient nebulization and the incompatibility of HPLC solvents with atomization systems. As a result, although the HPLC combination techniques offer great potential in the identification of a wide variety of organoselenium species, the potential has yet to be fully realized. Reports to date have been largely restricted to the use of HPLC-based combination techniques for the determination of total selenium or the common inorganic selenium anions.

11.6.1 Chromatography

The separation of selenite and selenate is commonly achieved by anion-exchange chromatography. With conventional HPLC instrumentation this can be done on a 25 cm strong anion-exchange column [Nucleosil–$NH(CH_3)_2$]. Selenite is eluted with 0.005*M* ammonium acetate/0.002*M* ammonium dihydrogen phosphate mixture adjusted to pH 4.6 with acetic acid. Ammonium dihydrogen phosphate (0.08*M*), adjusted to pH 6.9 with ammonia solution, is then used to elute the more strongly retained selenate [68]. Alternatively, ion-pair reversed-phase chromatography can be employed to separate selenite and selenate. For this, a water–methanol (90/10 v/v) mobile phase containing 5m*M* tetrabutylammonium hydroxide is employed as eluent, with an octadecylsilyl–silica reversed-phase column [69].

Ion-chromatography of selenite and selenate can be performed with a 50 × 3 mm anion precolumn (Dionex 30008) connected in series with a 150 × 3 mm anion separator column (Dionex 30589) and a 250 × 3 mm anion suppressor column (Dionex 30066). The 0.008*M* sodium carbonate mobile phase is delivered at 0.46 ml/min [70].

11.6.2 Applications

11.6.2.1 HPLC–AAS

For an extensive listing of papers dealing with directly coupled liquid chromatography–AAS systems the reader is referred to a review by Ebdon *et al.* [71].

The problem of interfacing liquid chromatography systems with AAS whilst preserving sensitivity, has largely been approached by the automated discrete sampling of the eluate into graphite-furnace atomizers. This approach, whilst not ideal, owing to the limited repetition rates dictated by the sampling and atomization systems, does provide a very useful approach to an otherwise difficult task.

Vickery *et al.* [72] have reported the design of a system to link liquid chromatography with flameless AAS. After passing through an ultraviolet and/or refractive-index detector, the eluate from the liquid chromatograph moves to an eight-port two-position sampling valve. Eluate is trapped in the sampling valve before being blown into the graphite furnace. A co-analyte (Ni^{2+}) was added to each aliquot of eluate to reduce the selenium volatility during the ashing stage and to enhance the selenium signal.

Ion-chromatography has been interfaced with Zeeman-corrected graphite-furnace AAS by Chakraborti *et al.* [70]. Selenate and selenite in natural waters were separated by using an anion precolumn in series with a 150 mm anion separator column. The eluate was passed through the conductivity detector and into a slider valve in the ion chromatograph–GFAAS system (Fig. 11.13).

The instrument produces a discontinuous selenium chromatogram (Fig. 11.14) and separates selenite and selenate from large excesses of other anions.

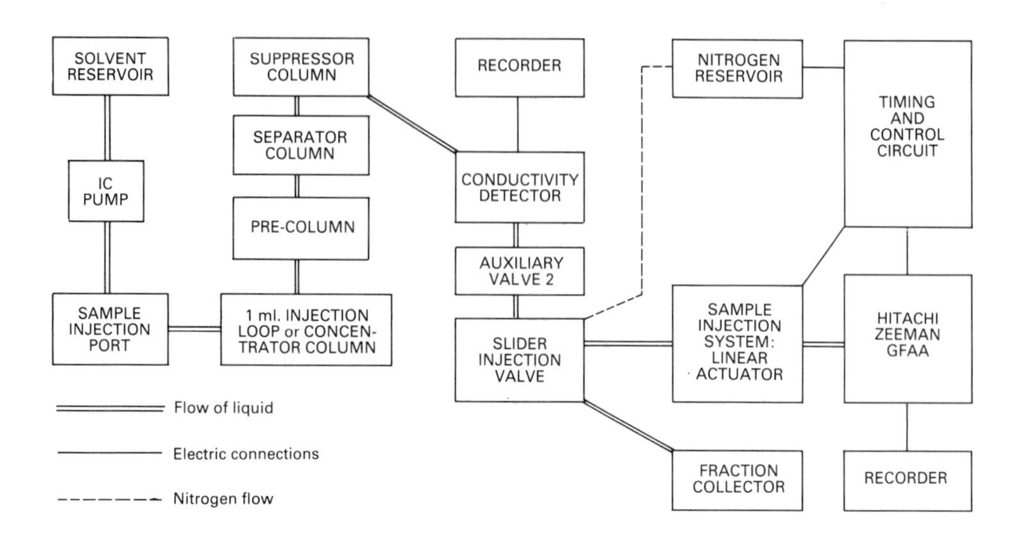

Fig. 11.13 — Block diagram of ion-chromatograph–Zeeman-corrected graphite-furnace AAS system. (Reproduced with permission from D. Chakraborti, D. C. J. Hillman, K. J. Irgolic and R. A. Zingaro, *J. Chromatog.*, 1982, **249**, 81. Copyright 1982, Elsevier Science Publishers.)

With the standard 1 ml injection loop the detection limit is 20 ng of selenium but 4 ml of anion-rich water can be preconcentrated to lower the detection limit to 5 ng. Distilled water samples can be preconcentrated to a greater extent and with 100-ml samples, 0.2 ng of selenium can be determined.

11.6.2.2 HPLC–AES

An HPLC system can be very simply coupled directly to the nebulizer of an inductively coupled plasma emission spectrometer with a short length of Teflon tubing. Such a system was employed by McCarthy *et al.* [68] for the determination of selenium(IV) and selenium(VI). The detection limits for selenium were 3.9 ng/sec, or, in absolute terms, 140 and 91 ng of Se(IV) and Se(VI) respectively. Reproducibility was 7%.

A similar approach was taken by Irgolic [73] who connected an HPLC system to a Meinhard concentric nebulizer of a simultaneous ICP emission spectrometer. The performance of the instrument was demonstrated on the separation of selenite, phosphate and a variety of arsenic species (Fig. 11.15).

A system has been described in which an ICP emission spectrometer was interfaced with a chromatography system comprising concentrator and separation columns. The apparatus separated selenium(IV) and (VI) and prereduction/hydride generation was carried out prior to detection [74].

Improved nebulization efficiency, resulting in better detection limits and retention of chromatographic resolution, may be possible by the use of a direct injection nebulizer. Such a system has been demonstrated for the detection of selenite and selenate separated by ion-pair reversed-phase HPLC [69].

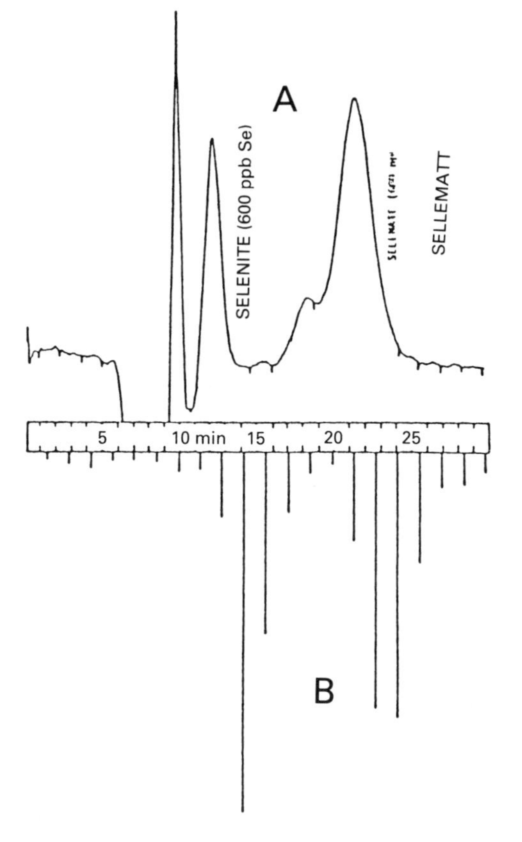

Fig. 11.14 — Chromatogram of selenite and selenate recorded with the apparatus shown in Fig. 11.13. (A) Conductivity detection. (B) GF–AAS detection of selenium. (Reproduced with permission from D. Chakraborti, D. C. J. Hillman, K. J. Irgolic and R. A. Zingaro, *J. Chromatog.*, 1982, **249**, 81. Copyright 1982, Elsevier Science Publishers.)

The performance of HPLC-based combination techniques is summarized in Table 11.3.

11.7 CONCLUSIONS

Though still in their early stages of development, the combination techniques are powerful tools for the study of selenium species in the environment. Gas chromatography–AAS systems are particularly well suited to the investigation of volatile selenium species, and are probably the most straightforward to apply to natural systems. Hydride-based systems, especially in combination with the emission techniques, are useful for multielement analyses but care must be taken to ensure that interferences are dealt with if results are to be reliable. Great potential is offered by the HPLC combination techniques for selenium speciation without derivatization,

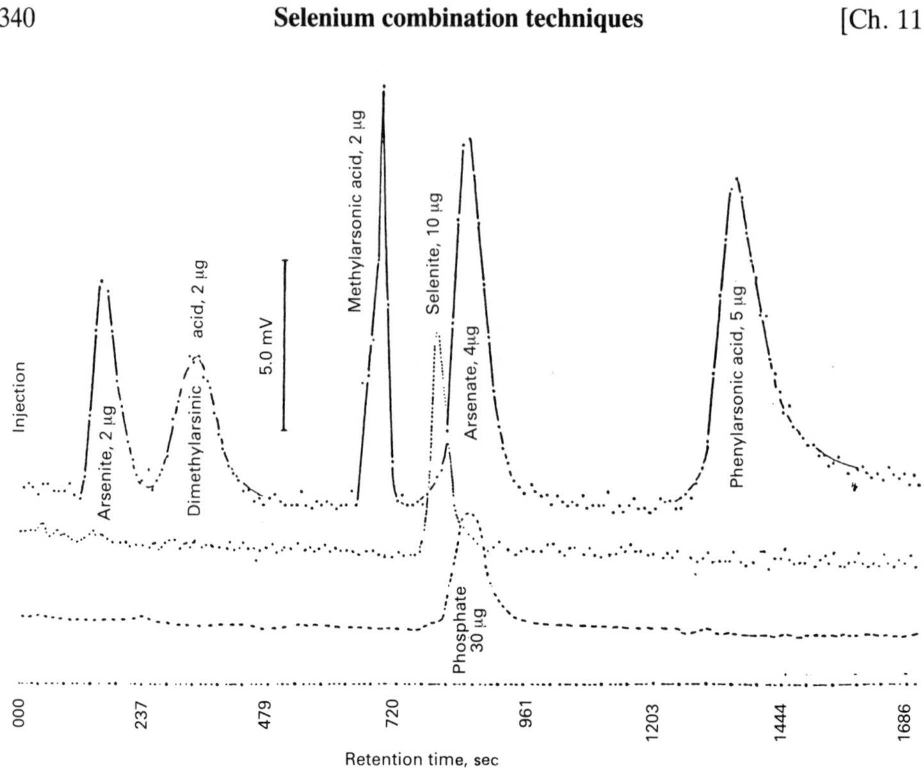

Fig. 11.15 — Chromatographic separation of arsenite, arsenate, methylarsonic acid, dimethylarsinic acid, phenylarsonic acid, selenite and phosphate by reversed-phase chromatography with multielement ICP–AES detection. (Reproduced with permission from K. J. Irgolic, R. A. Stockton, D. Chakraborti and W. Beyer, *Spectrochim. Acta*, 1983, **38B**, 437. Copyright 1983, Pergamon Press.)

Table 11.3 — Performance of liquid chromatography based combination techniques (NS = not stated).

Technique	Species determined	Detection limit*	Relative standard deviation	Reference
Ion-chromatography HG–AAS	Se(IV)	1 ng (2σ)	1.7% (100 ng)	[48]
Anion-exchange ICP–AES	Se Se(IV)	3.9 ng/sec (2σ) 140 ng	7% (2.8 µg Se)	[68]
HPLC–DIN–ICP–AES	Se	42–72 ng/g	NS	[69]
Anion-exchange ICP–AES	Se(IV) Se(VI)	1.6 ng/ml (3σ) 2.5 µg/ml	1.7% (10 ng/ml) 2.5% (10 ng/ml)	[74]

but it will be a few years before such systems reach the performance levels required for the study of trace levels of selenium in the environment.

REFERENCES

[1] K. Schwarz and C. M. Foltz, *J. Am. Chem. Soc.*, 1957, **79**, 3292.
[2] N. Wells, *N. Z. Geol. Geophys.*, 1967, **10**, 198.
[3] G. A. Fleming, *Soil Sci.*, 1962, **94**, 28.
[4] O. A. Beath, *Sci. News Lett.*, 1962, **81**, 254.
[5] Y. K. Chau, P. T. S. Wong, B. A. Silverberg, P. L. Luxon and G. A. Bengert, *Science*, 1976, **192**, 1130.
[6] D. C. Reamer and W. H. Zoller, *Science*, 1980, **208**, 500.
[7] G. A. Cutter and K. W. Bruland, *Limnol. Oceanog.*, 1984, **29** 1179.
[8] I. Rosenfeld and O. A. Beath, *Selenium*, Academic Press, New York, 1964.
[9] R. A. Zingaro and W. C. Cooper, *Selenium*, Van Nostrand Reinhold, New York, 1974.
[10] E. M. Bem, *Environ. Health Persp.*, 1981, **37**, 183.
[11] S. E. Raptis, G. Kaiser and G. Tölg, *Z. Anal. Chem.*, 1983, **316**, 105.
[12] J. D. Odom, *Struct. Bonding (Berlin)*, 1983, **54**, 1.
[13] L. Fishbein, *Int. J. Environ. Anal. Chem.*, 1984, **17**, 113.
[14] R. J. Shamberger, *Mutation Res.*, 1985, **154**, 29.
[15] D. C. Adriano, *Trace Elements in the Terrestrial Environment*, Springer-Verlag, New York, 1986.
[16] P. Craig, *Organometallic Compounds in the Environment*, Longmans, Harlow, 1986.
[17] S. Jiang, W. DeJonghe and F. Adams, *Anal. Chim. Acta*, 1982, **136**, 183.
[18] S. Jiang, H. Robberecht, F. Adams and D. Vanden Berghe, *Toxicol. Environ. Chem.*, 1983, **6**, 191.
[19] S. Jiang, H. Robberecht and F. Adams, *Atmos. Environ.*, 1983, **17**, 111.
[20] A. D. Shendrikar and P. W. West, *Anal. Chim. Acta*, 1975, **74**, 189.
[21] G. A. Cutter, *Anal. Chim. Acta*, 1978, **98**, 59.
[22] G. A. Cutter, *Anal. Chem.*, 1985, **57**, 2951.
[23] H. J. Robberecht, R. E. Van Grieken, P. A. Van Den Bosch, H. Deelstra and D. Vanden Berghe, *Talanta*, 1982, **29**, 1025.
[24] J. Agterdenbos, J. T. van Elteren, D. Bax and J. P. Heege, *Spectrochim. Acta*, 1986, **41B**,303.
[25] C. I. Measures and J. D. Burton, *Anal. Chim. Acta*, 1980, **120**, 177.
[26] J. Dědina and I. Rubeška, *Spectrochim Acta*, 1980, **35B**, 119.
[27] J. Khalighie, A. M. Ure and T. S. West, *Anal. Chim. Acta*, 1981, **131**, 27.
[28] M. Verlinden, H. Deelstra and E. Adriaenssens, *Talanta*, 1981, **28**, 637.
[29] J. Dědina, *Anal. Chem.*, 1982, **54**, 2097.
[30] J. Agterdenbos, J. P. M. van Noort, F. F. Peters and D. Bax, *Spectrochim. Acta*, 1986, **41B**, 283.
[31] D. Bax, F. F. Peters, J. P. M. van Noort and J. Agterdenbos, *Spectrochim. Acta*, 1986, **41B**, 275.
[32] B. Welz and M. Melcher, *Analyst*, 1983, **108**, 213.
[33] M. Verlinden, *Anal. Chim. Acta*, 1982, **140**, 229.
[34] G. E. Parris, W. R. Blair, and F. E. Brinckman, *Anal. Chem.*, 1977, **49**, 378.
[35] C. G. Flinn and W. A. Aue, *J. Chromatog.*, 1978, **153**, 49.
[36] J. R. Hancock, C. G. Flinn and W. A. Aue, *Anal. Chim. Acta*, 1980, **116**, 195.
[37] C. I. M. Beenakker, *Spectrochim. Acta*, 1976, **31B**, 483.
[38] K. B. Olsen, D. S. Sklarew and J. C. Evans, *Spectrochim. Acta*, 1985, **40**, 357.
[39] M. McDaniel, A. D. Shendrikar, K. D. Reiszner and P. W. West, *Anal. Chem.*, 1976, **48**, 2240.
[40] S. C. Apte and A. G. Howard, *J. Anal. At. Spectrom.*, 1986, **1**, 379.
[41] J. Agterdenbos and D. Bax, *Anal. Chim. Acta*, 1986, **188**, 127.
[42] C. H. Meyer, Ch. Hofer, G. Tölg, S. Raptis and G. Knapp, *Z. Anal. Cham.*, 1979, **296**, 337.
[43] R. M. Brown, Jr., R. C. Fry, J. L. Moyers, S. J. Northway, M. B. Denton and G. S. Wilson, *Anal. Chem.*, 1981, **53**, 1560.
[44] M. Verlinden and H. Deelstra, *Z. Anal. Chem.*, 1979, **296**, 253.
[45] R. Bye, *Talanta*, 1986, **33**, 705.
[46] H. Hayrynen, L. H. J. Lajunen and P. Peramaki, *Finn. Chem. Lett.*, 1985, **2**, 39.
[47] R. Bye, L. Engvik and W. Lund, *Anal. Chem.*, 1983, **55**, 2457.
[48] D. R. Roden and D. E. Tallman, *Anal. Chem.*, 1982, **54**, 307.
[49] G. A. Cutter, *Anal. Chim. Acta*, 1983, **149**, 391.
[50] A. M. Gunn, *Water Research Centre, Medmenham, Tech. Rep., TR169*, 1981.
[51] B. Radziuk and J. Van Loon, *Sci. Tot. Environ.* 1976, **6**, 251.
[52] J. W. Young and G. D. Christian, *Anal. Chim. Acta*, 1973, **65**, 127.
[53] T. Stijve and E. Cardinale, *J. Chromatogr.*, 1975, **109**, 239.

[54] Y. Shiomoishi and K. Toei, *Talanta*, 1970, **17**, 165.
[55] C. F. Poole, N. J. Evans and D. G. Wibberley, *J. Chromatog.*, 1977, **136**, 73.
[56] C. J. Cappon and J. C. Smith, *J. Anal. Toxicol.*, 1978, **2**, 114.
[57] Y. Talmi and A. W. Andren, *Anal. Chem.*, 1974, **46**, 2122.
[58] C. A. Burgett, *Anal. Lett.*, 1974, **7**, 799.
[59] W. B. Robbins, J. A. Caruso and F. L. Fricke, *Analyst*, 1979, **104**, 35.
[60] M. H. Hahn, K. J. Mulligan, M. E. Jackson and J. A. Caruso, *Anal. Chim. Acta*, 1980, **118**, 115.
[61] L. Ebdon, S. Hill and R. W. Ward, *Analyst*, 1986, **111**, 1113.
[62] Y. K. Chau, P. T. S. Wong and P. D. Goulden, *Anal. Chem.*, 1975, **47**, 2279.
[63] J. C. Van Loon and B. Radziuk, *Can. J. Spectrosc.*, 1976, **21**, 46.
[64] A. D'Ulivo and P. Papoff, *J. Anal. At. Spectrom.*, 1986, **1**, 479.
[65] F. L. Fricke, W. D. Robbins and J. A. Caruso, *J. Assoc. Off. Anal. Chem.*, 1978, **61**, 1118.
[66] K. J. Slatkavitz, P. C. Uden, L. D. Hoey and R. M. Barnes, *J. Chromatog.*, 1984, **302**, 277.
[67] W. B. Robbins and J. A. Caruso, *J. Chromatog. Sci.*, 1979, **17**, 360.
[68] J. P. McCarthy, J. A. Caruso and F. L. Fricke, *J. Chromatog. Sci.*, 1983, **21**, 389.
[69] K. E. LaFreniere, V. A. Fassel and D. E. Eckels, *Anal. Chem.*, 1987, **59**, 879.
[70] D. Chakraborti, D. C. J. Hillman, K. J. Irgolic and R. A. Zingaro, *J. Chromatog.*, 1982, **249**, 81.
[71] L. Ebdon, S. Hill and R. W. Ward, *Analyst*, 1986, **112**, 1.
[72] T. M. Vickrey, M. S. Buren and H. E. Howell, *Anal. Lett.*, 1978, **11**, 1075.
[73] K. J. Irgolic, R. A. Stockton, D. Chakraborti and W. Beyer, *Spectrochim. Acta*, 1983, **38B**, 437.
[74] F. Nakata, H. Sunahara, H. Matsuo and T. Kumamaru, *Bunseki Kagaku*, 1986, **35**, 439.

12

The determination of sulphur gases by gas chromatography and flame photometric detection

M. T. Shabbeer and Roy M. Harrison
Institute of Aerosol Science, University of Essex, Colchester CO4 3SQ, England

12.1 INTRODUCTION

This chapter reviews developments in the determination of different sulphur gases by GC/FPD (gas chromatography/flame photometric detector) techniques, and outlines the methods by which these measurements have been recently optimized. Since the first description of the FPD in 1962 [1], and its later development as a specific GC detector for sulphur and phosphorus compounds [2], many types of GC/FPD systems have been developed.

Considerable interest has arisen in accurately monitoring man-made as well as natural inputs of sulphur to the atmosphere in order to evaluate their contribution to the global sulphur cycle and the associated problems of 'acid rain'. Sulphur compounds are highly objectionable in many commercial products, owing to their toxicity and unpleasant odour, are a cause of metallic corrosion, reduce the effectiveness of many additives, and deactivate various catalysts in a variety of production processes. Table 12.1 shows sulphur compounds which have been identified by various GC/FPD techniques.

In most procedures sulphur gases are preconcentrated by trapping cryogenically and/or on adsorbents. This is in order to gain a detectable and reproducible signal, as ambient concentrations of sulphur are usually too low for direct detection (see Table 12.2). Special care must be taken, when sampling, to exclude co-trapped water and oxygen, and to avoid adsorptive losses of sulphur, by utilizing inert materials such as PTFE (Teflon). After trapping, samples are desorbed and injected into a GC column where different sulphur compounds are separated. Eluted compounds are combusted in the FPD and sulphur-containing compounds are detected by monitoring (with a photomultiplier tube) the light given off from the decay of an excited S_2^* molecule.

Table 12. 1 — Sulphur compounds identified by GC/FPD techniques.

Name	Formula	Molecular weight	Boiling point (°C)
Hydrogen sulphide	H_2S	34.08	− 60.7
Carbonyl sulphide (carbon oxysulphide)	COS	60.08	− 50.0
Sulphur dioxide	SO_2	64.06	− 10.0
Methyl mercaptan (methanethiol or ethylthiol)	CH_3SH	48.11	6.2
Dimethyl sulphide (DMS) (methylthiomethane)	CH_3SCH_3	62.13	37.3
Carbon disulphide	CS_2	76.14	46.2
Isopropyl mercaptan (2-propanethiol)	$(CH_3)_2CHSH$	76.17	52.5
tert-Butyl mercaptan (2-methyl-2-propanethiol)	$(CH_3)_3CSH$	90.19	64.2
n-Propyl mercaptan (1-propanethiol)	$CH_3CH_2CH_2SH$	76.17	67.0
Ethyl methyl sulphide	$CH_3CH_2SCH_3$	76.17	66.6
2-Butanethiol	$CH_3CH_2CHSHCH_3$	90.19	83.0
Thiophene	C_4H_4S	84.14	84.2
2-Methyl-1-propanethiol (isobutyl mercaptan)	$(CH_3)_2CHCH_2SDH$	90.19	88.7
Diethyl sulphide	$CH_3CH_2SCH_2CH_3$	90.19	92.1
1-Butanethiol (n-butyl mercaptan)	$CH_3CH_2CH_2CH_2SH$	90.19	98.4
2- Methyl-2-butanethiol	$CH_3CH_2C(SH)(CH_3)_2$	104.22	100.0
Dimethyl disulphide (DMDS) (methyldithiomethane)	CH_3SSCH_3	94.20	109.7
Diisopropyl sulphide	$CH_3(CH_2)_2S(CH_2)_2CH_3$	118.24	120.0
1-Pentanethiol	$CH_3(CH_2)_3CH_2SH$	104.22	126.6
1,2-Ethanedithiol	$HSCH_2CH_2SH$	94.20	146.0
Diethyldisulphide (ethyldithioethane)	$CH_3CH_2SSCH_2CH_3$	122.25	154.0
Benzenethiol	C_6H_6S	110.18	168. 7
Dimethylsulphoxide	$(CH_3)_2SO$	78.13	189.0
Methyl phenyl sulphide	$CH_3(C_6H_5)SH$	124.21	193.0

The following sections outline specific techniques which have been developed from this basic procedure in recent years.

12.2 SAMPLING TECHNIQUES

Quantitative measurements of atmospheric sulphur-containing gases at the sub-ppb level, (i.e. $<10^{-9}$ v/v), require an efficient sample preconcentration step. Sample enrichment techniques have included various types of chemically impregnated filters [3–9], as well as techniques which utilize metal-coated glass beads [10, 11]. However, cryogenic (cold-trap) and/or adsorbent enrichment techniques have become the most widespread and practical methods used for preconcentration.

12.2.1 Driers

A major problem encountered with cryogenic preconcentration is the freezing out of large volumes of atmospheric water vapour in the trap, which can seriously affect gas

Table 12.2 — Typical concentrations of sulphur gases found in the environment

Location	Sample	Compound	Range of concentrations (ppb)	References
Norsminde Fjord	Air above floating *Ulva lactuca*	H_2S	3.4–17	[24]
Various locations	Ambient air	COS	0.2–0.5	[20–22, 24]
General oceanic	Sea-water	DMS	3.2–32	[24, 30, 35]
AERE, Harwell	Ambient air	CS_2	0.19	[20]
Wallops Island, Va.	Ambient air	SO_2	<0.16	[26]
Atlantic Ocean from 50°N to 65°S	Sea-water	CS_2	0.22	[102]

chromatographic analysis, reduce the effectiveness of trapping (for example, by competing for adsorption sites in adsorbents), and increase the losses of some sulphur species [12]. This problem may be overcome by providing sequential traps at progressively lower temperatures, with the first trap designed to collect primarily water vapour [13]. The use of conventional desiccants to dry air samples, however, can lead to partial or complete loss of some sulphur species by adsorption. Nafion driers have been used as a method of removing water vapour without the loss of sulphur gases.

Foulger and Simmonds [14] introduced a Nafion drier consisting of 1 m lengths of Nafion tubing (Nafion is a copolymer of tetrafluoroethylene and fluorosulphonyl monomer, which acts as a perm-selective membrane), enclosed in a plastic outer container filled with desiccant, as shown in Fig. 12.1. This design has advantages in

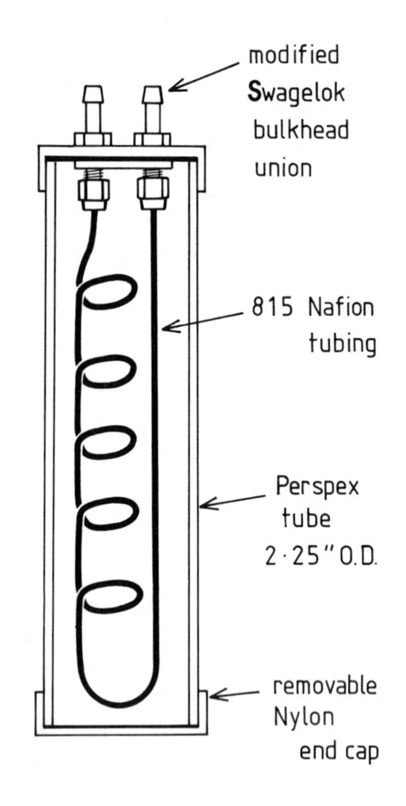

Fig. 12. 1 — Schematic of Nafion drier. Nominal dimensions 8 × 2.25 in.

that there is a low dead volume and a constant pressure drop compared with conventional bed desiccants. Water vapour is selectively removed by diffusion through the membrane, without any loss of sulphur compounds.

Carroll [15] constructed Nafion tubing driers for use in-line with sample loops. A countercurrent flow of ultrahigh-purity helium around thirteen strands of Nafion tubing, through which COS or H_2S was passed, produced no loss in up to 100 hr. Pleil *et al.* [16] give details for enhancing Nafion drier performance by heating while purging with a dry sample stream immediately before processing the gas sample of interest.

12.2.2 Cryogenic preconcentration

Various types of refrigerants are available for preconcentration by cold-trapping. Liquid nitrogen is often used, but may cause atmospheric oxygen to condense in the trap, and this routinely causes extinction of the FPD flame on desorption. This problem has been avoided by the use of liquid oxygen or argon [15, 17, 18], but liquid oxygen presents a serious combustion hazard. A less expensive alternative to argon may be the use of an isopentane slush; this is difficult to prepare, however, and cannot be stored easily. Another system has recently been developed at the University of East Anglia, in which the vapour from liquid nitrogen is used at − 150°C as the coolant, to prevent liquid oxygen from condensing out. A heating element is placed in a half-filled Dewar flask of liquid nitrogen, with a sample loop and temperature probe suspended in the vapour of the cold-trap system. The vapour temperature of − 150°C is maintained by a temperature-switching system which ensures that when the probe indicates an increase in temperature the liquid nitrogen is heated so that the vapour temperature is lowered (Dr. S. Turner, University of East Anglia, personal communication).

Several types of collecting devices have been used with cryogenic preconcentration. Rasmussen [19] used a $\frac{1}{8} \times 8$ in, FEP Teflon loop, packed with a 10 in. bed of 40–60 mesh glass beads immersed in liquid oxygen (− 183°C), to determine H_2S, CH_3SH, DMS and DMDS released from foliage.

Sandalls and Penkett [20] trapped trace quantities of COS and CS_2 in ambient air samples by condensing the air in an empty glass loop immersed in liquid nitrogen (− 196°C). High-purity helium, further purified by passage through a cold trap at − 196°C, was then bubbled through the trapped liquid air sample, still at − 196°C, to remove nitrogen and oxygen. The loop was then sealed with PTFE stopcocks and raised to room temperature, giving a sample concentrated by a factor of about 100, before GC/FPD analysis. The concentration of carbon tetrachloride in the collected samples and the atmosphere was simultaneously measured in order to find the factor by which the sample had been concentrated. In this technique measurements of COS showed similar individual values for each day of sampling. However, for CS_2, duplicate analyses of the same trapped sample showed substantially different values, and this was suggested as being the result of an increase of CS_2 in trapped samples stored overnight.

Carbonyl sulphide has also been preconcentrated in investigations by Maroulis *et al.* [21] and Torres *et al.* [22], and COS, H_2S and SO_2 have been determined by Goldberg *et al.* [17] in a similar procedure using GC/FPD techniques. Goldberg *et al.* concentrated sulphur gases by cold-trapping in sample loops constructed from 50 cm lengths of Teflon tubing (o.d. 3.2 mm, i.d. 1.6 mm), the centre sections of the trap being cooled to − 186°C with liquid argon [17].

Other workers [15,18,23] have also used empty Teflon loops to trap sulphur gases in air before GC/FPD. Carroll *et al.* [18] used such loops to measure H_2, COS, and CS_2 in air entering and leaving an emission flux chamber. These loops consisted of a

63.5 cm length of 0.32 cm o.d. Teflon tubing housed in a 48.3 cm length of aluminium tubing containing liquid argon.

Carroll [15] tested the efficiency of various trapping devices cooled in liquid nitrogen, oxygen and argon, and found that empty Teflon loops immersed in liquid argon gave the most reproducible enrichment surfaces. Lengths of empty Teflon tubing were found to have a much greater reliability, although the overall trapping efficiencies (collection and desorption) were only $72.4 \pm 3.9\%$ for COS and $62.6 \pm 2.0\%$ for H_2S. The trapping efficiencies were thought to be reduced by retention on the loop after desorption, as no evidence of breakthrough was found. Flow-rates ranging from approximately 10 to 250 ml/min through lengths of Teflon tubing ranging from 12 to 25 in. showed no significant differences in trapping behaviour. A rapid desorption by immersion in boiling water was also found to give sharp, well resolved, peaks, whereas sample decomposition and peak broadening were found with more prolonged heating. Samples could be safely stored in liquid oxygen or argon for up to 48 hr.

Jorgensen and Okholm-Hansen [24] cryogenically trapped air samples to determine emissions of H_2S, DMS, CH_3SH, COS and CS_2 from a Danish estuary. They trapped these compounds in a 20 cm long Pyrex U-tube, 5 mm o.d., deactivated by prior silanation with 5% dimethyldichlorosilane solution in toluene, and cooled in liquid nitrogen. A sample volume of 500–2000 ml of air was normally removed from an emission flux chamber by syringe and injected directly into the cold trap. Fifteen min after the trapping, the trap was mounted in a preparation system and the sample transferred to a second cold trap (3 mm o.d.) from which it was desorbed and injected into the GC/FPD system.

In controlled experiments with two traps in series, the trapping efficiency was found to be 100% in the first trap for all sulphur gases detected. Recovery efficiencies, detected by using internal sulphur standards introduced into the air stream from permeation tubes, were found to range from a mean of $55 \pm 3\%$ ($n = 17$) for H_2S to $98 \pm 4\%$ ($n = 8$) for DMS. Recoveries did not show a systematic variation with the total sample volume (200–2000 ml) or with the amount of sulphur trapped (50–500 pmole). For constant air flow and emission from permeation tubes, within the flux chamber, the measured emission rates showed relative standard deviations of 4–10%.

Farwell *et al.* [25] developed a GC/WCOT/FPD system compatible with cryogenic sampling. A U-shaped sample loop constructed from 6 mm o.d. Pyrex glass tubing packed with a 3–4 in. length of 50/80 mesh Pyrex glass beads was found to be the most suitable trapping device. The glass beads were held in the lower portion of the U-tube by two small plugs of Pyrex glass wool. These tubes were designed to fit into Dewar flasks filled with either liquid oxygen for sampling or with hot water for desorption into a cooled external glass-capillary trap and chromatographic system. Use of the cooled external capillary trap introduces a narrow band of sample into the head of the WCOT column for chromatographic separation, ensuring that the chromatographic resolution inherent with capillary WCOT columns is maintained.

Best results were obtained with traps that had internal surfaces coated with polysiloxane/SE-30 which gave almost quantitative results for COS, DMS, and CS_2. An apparently greater than 100% efficiency for DMDS was shown to be due to the conversion of some methyl mercaptan into DMDS during the cryogenic sampling, with consequently lowered recovery efficiency for methyl mercaptan. Only a 40% efficiency for H_2S was achieved, perhaps because of breakthrough losses, or a small, but finite, sublimation process and/or chemical oxidation to some non-volatile sulphur compounds retained in the trap after the desorption at 90°C.

12.2.3 Adsorptive preconcentration

Cryogenic techniques are particularly useful for trapping highly volatile substances, as oxidation or polymerization of the concentrated low molecular weight compounds is then minimal. However, these techniques have also been known to freeze out large amounts of water, and many attempts have been made to preconcentrate sulphur gases on solid adsorbents at ambient temperature, thereby simplifying the analytical techniques.

Black *et al.* [26] evaluated the effectiveness of trapping SO_2 and H_2S on various adsorbents. The most efficient adsorption tube consisted of a 13 cm length (0.8 g) of Molecular Sieve 5A packed into a 15 cm long, 4 mm i.d., 6 mm o.d. glass tube, sealed with plugs of silaned glass wool. Before use all adsorption tubes were conditioned by heating at 250°C for at least 12 hr with passage of nitrogen. Desorption efficiencies were between 83 and 87% for SO_2 and 75 and 82% for H_2S. Preliminary storage experiments showed that SO_2 was not lost from the trap after a two-week storage period, but H_2S was noticeably lost after one week.

Using this procedure, Carroll [15] found a 100% adsorption efficiency, with no breakthrough of sulphur gases for many hours. (The term ‘breakthrough’ denotes the presence of a test sulphur gas in the air stream leaving a trap, after an initial period of its absence in this air stream). However, low recoveries were obtained on desorption, when the traps were heated to 250°C. Tests conducted on Porapak QS in Teflon tubing also resulted in breakthrough at room temperature, and an attempt to utilize a Molecular Sieve/Tenax-GC trap, described by Steudler and Kijowski [27], did not achieve the adsorption/desorption efficiency reported by Black *et al.* [26] for a number of sulphur gases.

In the Steudler and Kijowski method [27], the adsorption tubes were 21.3 cm long, 6 mm o.d., 4 mm i.d. Pyrex glass tubes, packed with 9 cm (0.8 g) of 60/80 mesh Molecular Sieve 5A and 11.5 cm (0.2 g) of 60/80 mesh Tenax GC with silaned Pyrex wool end-plugs, activated before use, by heating at 325°C for at least 8 hr with passage of ultra high-purity nitrogen into the molecular sieve end. The traps were then cooled for 10 min, sealed, and stored in the dark at room temperature for up to 48 hr before use. The traps were used only twice, to avoid possible changes in adsorption characteristics with reactivation.

For sampling, the traps were covered with aluminium foil to reduce any photo-oxidation and change in temperature, and air was drawn through, (molecular sieve end first), at a flow-rate of 40–100 ml/min for 0.5–2 hr. They were then sealed with GC end caps and placed on dry-ice until they were returned to the laboratory and stored at −25°C. All samples were analysed within 20 hr.

The analysis consists of three steps (Fig. 12.2). First, the trap is connected to the desorption apparatus and helium is passed through it (Tenax end first) for 15 sec to purge air from the system. The Teflon loop is then immersed in liquid nitrogen and the trap is heated for 10 min, with passage of helium. Next, the helium flow is stopped and the loop connected to the injection valve so that the gas flow through the loop

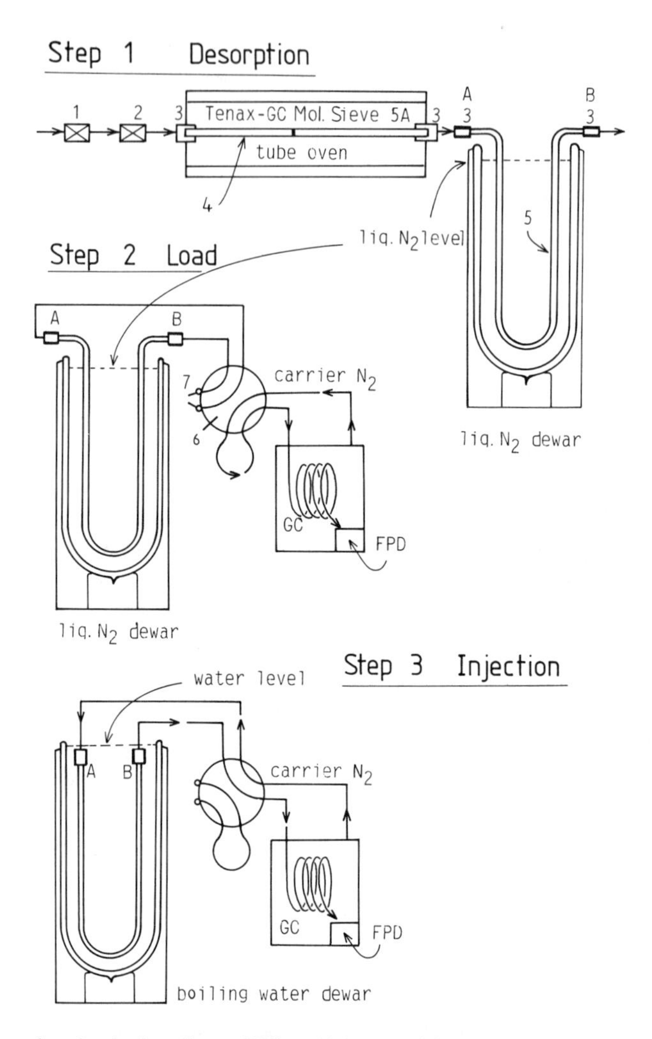

Fig. 12.2 — Procedure for the Steudler and Kijowski method: (1) toggle valve, (2) needle valve, (3) stainless-steel modified union, (4) trap, (5) Teflon loop, (6) 8-port valve, (7) plugs; A and B are ports of the loop. (Reproduced by permission from P. A. Steudler and W. Kijowski, *Anal. Chem.* 1984, **56**, 1432. Copyright 1984, American Chemical Society).

will be reversed. Finally the sulphur gases are volatilized from the Teflon loop by immersing it in boiling water, and injected into the GC through the switching valve for 3 min. The loop is then disconnected and purged with dry helium to remove any traces of moisture, before the next sample is analysed.

The capacity of the trap was tested and no breakthrough was detected during sampling for 2 hr at a flow-rate of 100 ml/min, for levels at least 10 times those observed in field experiments. From the detection limits and the sample loading tested it was calculated that the maximum loss of an individual sulphur compound during the 2 hr sample period was less than 4% for all the sulphur gases examined.

In the optimization study, use of longer desorption times and cryogenic trapping of the gases before injection into the GC was found to give much higher recoveries than direct thermal desorption did. Maximum recoveries were obtained with desorption for 10 min at 265°C at a helium flow-rate of 13.5 ± 0.5 ml/min with trapping in a 2.2 mm i.d. Teflon loop 41.5 cm long. These optimal conditions are a compromise between those optimal for recovery of individual sulphur gases, so that several sulphur compounds can be determined in the same sample. For example, higher recoveries of DMDS were achieved with higher flow-rates, but at the expense of lower recoveries of COS and H_2S.

Recoveries between 47 and 78% were found with humidified air standards (30 or 95% relative humidity) containing a mixture of the gases: the recovery of each gas was fairly constant over the humidity range. No loss of sulphur from the traps during a 24 hr storage period was found. No major interconversion of sulphur compounds was observed. The gases tested were COS, H_2S, CH_3SH, CS_2, $(CH_3)_2S$ and $(CH_3)_2S_2$.

A useful method for determining parts per trillion (ppt, 10^{-12} v/v) levels of atmospheric CS_2 has been developed by Bandy *et al.* [28] who used Carbosieve B as adsorbent and a GC/MS system. Traps were prepared by packing a 1-cm section of the centre portion of a glass tube (6 mm o.d., 2 mm i.d., 20 cm long) with 80/100 mesh Carbosieve B. Tests were made by adding ^{12}C and ^{34}S standards to ambient air and monitoring only ^{12}C and ^{34}S by GC/MS. A small oven surrounding the packed trap completely desorbed CS_2 from the trap in about 3 min at a temperature of 250°C, with a helium purge rate of 150 ml/min. The desorbed CS_2 was then retrapped cryogenically, volatilized with warm water and determined by GC/MS. Breakthrough volumes, defined as the sample volume at which 1% of the ^{12}C and ^{34}S had passed through the trap, were found to exceed 60 litres, at a trapping flow-rate of 11 ml/min, for all traps studied. Desorption efficiencies were 99% (± 13%) at the 95% confidence level.

12.2.4 Cryogenic/adsorptive preconcentration

Andreae and Barnard [29] used a combination of cryogenic and adsorptive trapping techniques to determine DMS in sea-water samples, and this was also used in several other investigations [30–33]. In this technique DMS is removed by purging the samples with helium, and collecting the volatile DMS on a column (cooled with liquid nitrogen) which acts as both a trap and a separating mechanism. After temperature programming to separate interfering compounds, DMS is detected with an FPD. A detection limit of 0.03 ng of sulphur (DMS) corresponding to a concentration of 0.3 ng/l. for a 100 ml sample, could be obtained.

Tangerman [34] has recently developed a GC/FPD system with a cryogenic/adsorptive preconcentration technique to determine sulphur compounds in human breath. Glass adsorption tubes (8 cm long, 6 mm o.d., 4 mm i.d.), filled with 200 mg of Tenax GC and immersed in liquid nitrogen, were used to trap the sulphur gases. The traps were designed to fit into an injection port of a gas chromatograph, where the adsorbed sulphur compounds could be desorbed at 200°C directly into the carrier gas stream. At − 196°C, no loss of sulphur volatiles was observed after concentration from 40 litres of air. Experiments showed that almost quantitative recoveries were obtained for the most volatile sulphur gases, after one week of storage in liquid nitrogen.

12.2.5 Other methods

Barnard *et al.* [30] used a GC/FPD system to determine DMS in marine air and improved upon previous preconcentration techniques [11]. In the trap used, air is drawn through an 8 mm i.d. silica tube filled with gold wool consisting of either 50 μm diameter gold wire or 154 μm thick gold foil. Reduced sulphur gases are concentrated on the gold surfaces by chemisorption, after SO_2 has been removed by a scrubber tube. Andreae *et al.* [35] recommend a scrubber consisting of 5% sodium carbonate on Anakrom C22 (40–50 mesh), which prevents the destruction of DMS on the gold wool. Before desorption 0.1 ml of ethyl iodide vapour is injected into the sampling tube, which converts H_2S and methyl mercaptan into their ethyl derivatives, leaving DMS unchanged. During a 5 min desoprtion period the compounds are released into helium carrier gas and retrapped in a capillary tube immersed in liquid nitrogen. They are then transferred to a separating column for subsequent detection. This system has been found to be 100% effective in removing DMS from a sample stream, with more than 90% of the DMS being recovered from the trap. The efficiency of the tubes declines only after several hundred uses.

Kagel and Farwell [36] have evaluated the use of metallic foils for preconcentrating sulphur-containing gases, which are then flash desorbed and detected with an FPD. Sub-ppb (parts per billion, 10^{-9} v/v) concentrations of H_2S, CH_3SH, DMS, DMDS, CS_2 and SiO_2 are drawn through a fluorocarbon resin cell containing a 30 μm × 7 mm × 0.025 mm metal foil. The preconcentrated sample species are thermally desorbed by a controlled pulse of current through the foil and swept from the fluorocarbon resin cell into the FPD with precleaned zero air. Sampling flow-rates, ambient temperature, sample humidity and common oxidants were examined for their effects on the collection efficiencies for these sulphur compounds on platinum and palladium foils. Analytical characteristics of this metal foil collection/flash desorption/flame photometric detector (MFC/FD/FPD) technique include a gas detectability of less than (0.05 ppb v/v), a response repeatability of at least 95% and field portability.

The application of the system is limited, however, to air samples known to contain only one sulphur gas or for which a total sulphur gas value is required. Preliminary results obtained with a Nafion-based sample-drying device to avoid the effects of high humidity on the MFC/FD collection efficiencies, seems promising for the application of this system to the measurement of total sulphur gas fluxes from biogenic sources.

12.3 ANALYTICAL TECHNIQUES

12.3.1 Column technology

Considerable research has been undertaken in order to optimize suitable materials for handling, transporting and separating mixtures of gaseous sulphur compounds by the gas chromatographic technique. The main difficulty has proved to be their highly reactive nature, which causes losses by adsorption and peak tailing.

The loss of gaseous sulphur compounds on various column materials and supports was evaluated by Koppe and Adams [37], using a microcoulometric titration cell with a minimum detection level of 0.05 ppm for SO_2 and 0.1 ppm for H_2S. A range of glass, Teflon and stainless-steel columns packed with a variety of solid supports and liquid phases was tested, but none was found that could measure SO_2 below 25 ppm of H_2S below 0.25 ppm. A stainless-steel column packed with 10% Triton X-305 on either Chromosorb DMCS (60–80 mesh) or Porapak Q was found to produce the most acceptable sample recovery and separation of H_2S, CH_3SH, SO_2, DMS and DMDS. Similar packings with Triton X-305 have been used by subsequent workers with improved column materials (e.g. Ronkainen *et al.* [38]) to separate several volatile sulphur compounds.

Considerable advances in the field were made by Stevens *et al.* [39] who determined ppb concentrations of volatile sulphur compounds in air by using specially developed column materials with PTFE components. Glass, stainless steel and conventional diatomite supports were found to adsorb gaseous sulphur compounds, whereas Teflon supports were nearly inert. Their column consisted of a 36 ft, 0.085 in. i.d. fluorinated ethylenepropylene (FEP) Teflon column packed with 40–60 mesh PTFE powder, flow-coated with approximately 10% poly(phenyl ether) and 0.5% phosphoric acid. After conditioning at 140°C for 6–8 hr, this column separated H_2S, SO_2, CH_3SH and DMS at 100°C at concentrations below 1 ppm.

Pecsar and Hartmann [40] give an intricate procedure for the preparation of this column, improving upon it and also showing that H_2S is resolved better by using air as a carrier gas. Although H_2S, SO_2, CH_3SH and DMS are separated, this column is relatively difficult to pack, often develops considerable back-pressure, especially if too tightly packed, and is unable to separate COS from H_2S or DMDS from DMS.

Banwart and Bremner [41] used a 1020 cm long 2.16 mm (i.d.) Teflon (FEP) column packed with 40–60 mesh Chromosorb T coated with 12% poly(phenyl ether) and 0.5% H_3PO_4 (*cf.* [39]), to separate a variety of sulphur gases. Better results were obtained with a column temperature of 100°C, instead of 50°C, with a hydrogen flow-rate of 75 ml/min, an oxygen flow of 20 ml/min and a nitrogen flow of 80 ml/min. This method resulted in the separation of SF_6, SO_2, CH_3SH, ethyl mercaptan, DMS, CS_2, isobutyl mercaptan, diethyl sulphide, n-butyl mercaptan, DMDS, and diethyl disulphide.

However, H_2S and CS_2, as well as ethyl methyl sulphide and n-propyl mercaptan, appeared as two composite peaks. Minimum concentrations of sulphur detectable were less than 0.2 μg/l, (i.e. <0.13 ppm v/v of sulphur gases in air). This column was successfully used to investigate the formation of volatile sulphur compounds from various soils (see [41–44]).

Bruner *et al.* [45] developed an alternative column for the separation of H_2S, SO_2, COS and CH_3SH at ppm and ppb concentrations in air. Selective inert supports were used in a short column, which produced much faster elution times than the columns developed by Stevens *et al.* and enabled use of a working pressure of less than 1 kg/cm^2. The column consisted of a 3 mm i.d., 1.25 m long Teflon column packed with Graphon, a graphitized carbon black support, treated with 0.5% phosphoric acid and 0.3% Dexsil 300. Graphon has a surface area of about 90 m^2/g and is deactivated by treatment with hydrogen at 1000°C. Peaks in H_2S, SO_2 and DMS were sharply defined at concentrations down to about 10 ppb. However, tailing problems were found with methyl mercaptan at concentrations below 100 ppm. At 80°C, good symmetrical peaks of methyl mercaptan were produced, at the expense of losing separation of H_2S and SO_2.

This column is commercially available as Carbopak BHT 100, and can separate

COS from H_2S, which was previously a problem. Bremner and Banwart [46] show that this column can also separate a variety of mercaptans, sulphides, and disulphides at a column temperature of 100°C with a carrier (nitrogen) flow-rate of 70 ml/min. This was used to confirm the identity of various compounds such as CS_2 as one of the gaseous products of microbial decomposition in soils. Jorgensen and Okholm-Hansen [24] also used a Carbopak BHT 100 Teflon column, to detect H_2S, COS, CH_3SH, DMS and CS_2 in studies on the emission of biogenic sulphur gases from a Danish estuary.

Deactivated silica gel columns have been used to determine atmospheric concentrations of sulphur at ppb levels and include commercially available compounds such as Deactigel, Chromosil 310 and Chromosil 330.

Acid-washed Deactigel was used by Thornsberry [47] to separate CO_2, COS, H_2S, CS_2 and SO_2 at concentrations above 50 ppm on a 2 ft long $\frac{1}{4}$ in. column at a temperature of 122°C with a helium carrier gas flow-rate of 55 ml/min. Losses of sulphur probably resulted from the use of conventional column materials such as glass, stainless steel or aluminium.

Silica gel columns have an advantage over other columns in that COS is eluted before H_2S, allowing the determination of trace concentrations of COS in the presence of H_2S. Banwart and Bremner [43] used a 30 cm, 2.16 mm i.d., FEP Teflon tube packed with 120–140 mesh Deactigel to establish the presence of COS in air. Deactigel was found to sorb H_2S, SO_2, mercaptans and alkyl sulphides, while allowing the separation and identification of trace amounts of COS, CS_2 and SF_6. Analyses of 500 μl samples of air treated with different amounts of COS and other compounds showed that Deactigel permits the identification of 1 ng of COS in the presence of 2000 ng of H_2S, mercaptans and alkyl sulphides.

Chromosil 310 has been found to sorb mercaptans and alkyl sulphides, and may be used for the separation and identification of SF_6, COS, H_2S, CS_2 and SO_2, at ppm or per cent concentrations [48]. Separations of ppm concentrations of COS, H_2S, CS_2 and SO_2 have been made on an 8 ft (6 ft packing) $\frac{1}{8}$ in. FEP Teflon column at a temperature of 50°C with a nitrogen carrier flow-rate of 20 ml/min. Per cent concentrations of COS, H_2S, SO_2 and CO_2 in air can also be separated by using a 6 ft long 4 mm i.d. glass column instead of Teflon, which improves the separation of sulphur gases from CO_2 and air.

Chromosil 330 has been used to separate ppb concentrations of COS, H_2S, SO_2, CS_2 and C_1–C_3 mercaptans on an 8 ft (6 ft packing) $\frac{1}{8}$ in. o.d. FEP Teflon column at 40°C, with a nitrogen carrier flow-rate of 20 ml/min [48]. If the temperature is raised to 65°C, higher molecular-weight mercaptans and alkyl sulphides, such as diethyl sulphide, butyl mercaptan and methyl disulphide, can be separated. Under these conditions, however, the pairs H_2S/COS, ethyl mercaptan/DMS and propyl mercaptan/ethyl methyl sulphide are not separated. Both Chromosil 310 and 330 are conditioned overnight at 70°C with dry nitrogen at a 20 ml/min flow-rate, and a maximum operating temperature of 70°C is recommended.

Steudler and Kijowski [27] separated ambient concentrations of COS, H_2S, CH_3SH, CS_2, DMS and DMDS in air by using a 10 ft long $\frac{1}{8}$ in. o.d. FEP Teflon column packed with Chromosil 330. This column was conditioned overnight at 70°C with a nitrogen flow of 35 ml/min. Separation of sulphur gases was achieved by temperature programming up to 100°C at a rate of 16°C/min, with an initial 2 min delay and a final 4 min hold. Similar columns have been used by other workers [49–51].

A versatile column for the determination of a number of sulphur gases is the Supelpak-S column [52, 53], which has several advantages over other columns:

(a) it has a very short length (76 cm, of which only 46 cm is effective) thereby producing less back-pressure;
(b) it can be temperature programmed up to 230°C;
(c) there is no liquid phase and hence no column 'bleed';
(d) it has a high resolving power;
(e) water and other highly polar molecules are quickly eluted and do not interfere with the separation of sulphur gases;
(f) it can separate H_2S from COS as well as fourteen other sulphur compounds which have a retention time of less than 10 min;
(g) there is no tailing after the SO_2 peak.

The column consists of specially treated Porapak QS, which is commercially available as Supelpak-S [53a]. Porapak QS is the trade name for a silaned porous polymer, ethylvinylbenzene cross-linked with divinylbenzene, which forms a uniform structure with a distinct pore size. This is jet-washed with chemically pure acetone for 5–10 min, then air-dried. An FEP Teflon tube 76 cm long, 0.32 cm o.d., is washed with acetone and packed with the treated Porapak QS to an effective length of 46 cm, with glass wool plugs on either side of the packing. The column is then conditioned overnight at about 230°C with helium (nitrogen can also be used) flowing at 30 ml/min. De Souza [53] analysed the performances of both untreated and acetone-washed Porapak QS. The treated Porapak QS column gave a far better separation and baseline with temperature programming, as shown in Fig. 12.3.

Using this column and a cryogenic trapping procedure, de Souza and Bhatia [54] succeeded in separating and measuring sulphur gases at concentrations between 100 and 200 ppb. Black *et al.* [26] used a Supelpak-S column to detect 0.1 ppb H_2S and 2 ppb SO_2 preconcentrated by adsorption trapping. More recently, Carroll [18] has used an acetone-washed Porapak Q column for the separation of COS, H_2S and CS_2.

A novel gas chromatographic technique has been developed by Farwell *et al.* [55] who used a deactivated high-resolution wall-coated open tubular (WCOT) column to separate H_2S, COS, CH_3SH, CS_2, DMS, DMDS and other organosulphur species at low and sub-ppb concentrations. This method has been successful in the field analysis of sulphur compound emissions from biogenic sources [25,26]. Chromatograms obtained from conventional glass WCOT columns showed varying degrees of peak broadening and tailing, whereas 30–38 m columns coated with OV-101 (SP-2100) and SE-54, (methylsilicone-coated glass capillary columns), produced high-resolution chromatographic separation of polar sulphur compounds.

Table 12.3 lists several types of column and the conditions under which they have been used to separate numerous volatile sulphur gases.

12.3.2 FPD optimization techniques

Several techniques have been used to optimize the response of the FPD since its initial application as a gas chromatographic detector [2]. The performance character-

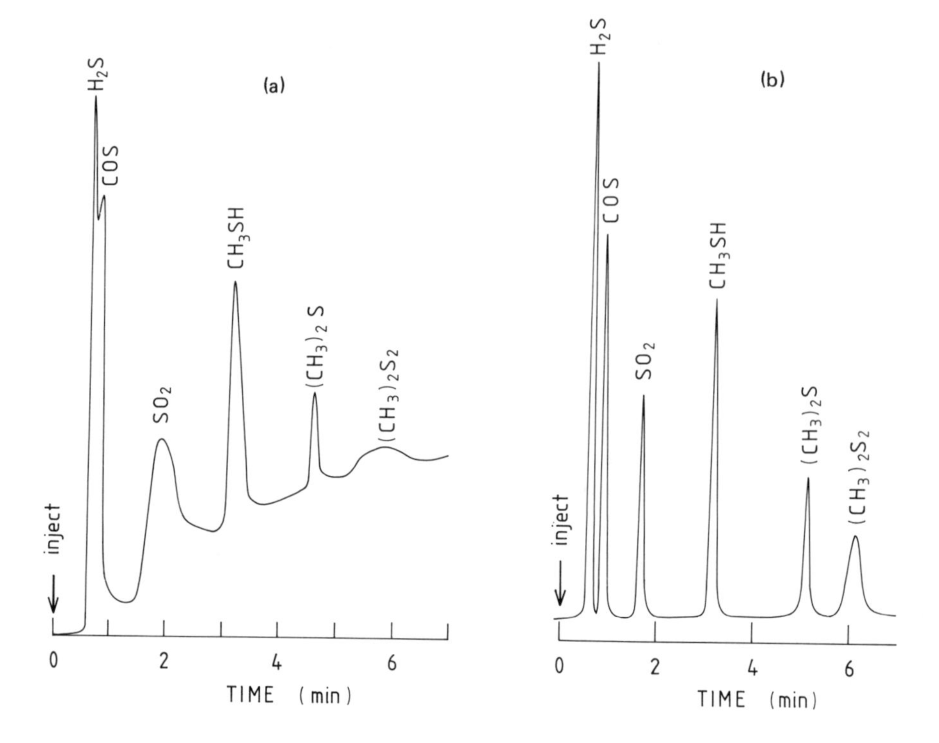

Fig. 12.3 — (a) A typical chromatogram obtained without pretreatment of Poropak QS. (b) a typical chromatogram obtained with Supelpak-S. [After 53].

istics and limitations of the FPD have been reviewed in several publications [58–66], as well as in Chapter 3 of this book. As a result, only a summary of some of the basic FPD optimization techniques is given here.

12.3.2.1 Detector design

Several commercial FPDs are currently available for GC/FPD application, each containing an enclosed H_2/O_2 burner, an optical filter and a photomultiplier tube (PMT) (see Fig. 3.1, p. 78). The column effluent, consisting of carrier gas and components of the injected sample, mixes with oxygen at the end of the column, and enters the burner. This is mixed with hydrogen and ignited inside a hollow tube which shields the flame from direct view by the PMT and mirror. When sulphur is present, emission occurs above the shielded flame, at the tip of the hollow tube, and this light is transmitted both directly and reflected by a mirror, through an optical filter having maximum transmittance at 394 nm, to the PMT (see Fig. 3.2, p. 79). All non-sulphur compounds burn inside the hollow burner tip, and any flame emission which occurs here is shielded from the flame. A potential of about 750 V is applied to the PMT, depending on the type used, and the output signal is amplified to generate a chromatogram.

Dual-flame detectors such as the Varian dual-flame FPD [67], the IBM dual-

Table 12.3 — Analytical columns and conditions useful for separation of sulphur gases.

Column and packing	Compounds separated	Carrier gas and flow-rate	Oven temperature	Detector	Detection limit	References
6–8 ft, 3/16 in. stainless steel, 10% Triton X-305 on Chromosorb DMCS, 60/80 mesh	H_2S, CH_3SH, SO_2, DMS, DMDS	He 50–170 ml/min	250°C, 70°C and various others	Microcoulometric titration cell	H_2S 0.25 ppm SO_2 25 ppm	[37]
2 ft, 1/4 in. o.d. glass; acid-washed Deactigel	CO_2, COS, H_2S, CS_2, SO_2	He 55 ml/min	122°C	TCD	> 50 ppm	[47]
36 ft, 0.085 in. i.d. Teflon; 10% PPE, H_3PO_4 on Teflon, 40/60 mesh	SO_2, H_2S, CH_3SH, DMS	N_2 100 ml/min	Unknown	FPD	2–10 ppb	[39]
36 ft 1/8 in. o.d. Teflon; 9% PPE, H_3PO_4 on Teflon T6, 40/60 mesh	SO_2, H_2S, CH_3SH, DMS	Air 80 ml/min	30°C	FPD	< 100 ng	[40]
6 m, 2 mm i.d. Teflon; 10% Triton X-30S, ± 0.5% H_3PO_4 on Chromosorb G (AW, DMCS, 70/80 mesh)	COS, H_2S, SO_2, CH_3SH, DMS, DMDS, CS_2	N_2 30 ml/min	3 min at 30°C then 20°C/min to 130°C hold	Melpar FPD	5–20 ng	[38]
34 ft, 0.055 in. i.d. Teflon; 12% PPE, 0.5% H_3PO_4 on Chromosorb T, 40/60 mesh	SO_2, H_2S, CS_2, COS, SF_6, CH_3SH	N_2 80 ml/min	100°C	Melpar FPD	1–4 ng	[46]
30 cm, 2.16 mm i.d. FEP Teflon; Deactigel, 120/140 mesh	COS, CS_2 SF_6	N_2 80 ml/min	50°C	Melpar FPD	1 ng	[43]
80 cm, 0.4 cm i.d. Teflon; graphitized carbon black, 40/60 mesh, 0.7% H_3PO_4, 0.7% XE-60 in CH_2Cl_2	SO_2, H_2S DMS, CH_3SH	N_2 100 ml/min	Room temperature	Tracor FPD	5–15 ppb	[57]
30–38 in. WCOT column; coated with OV-101 (SP-2100) and Se-54.	H_2S, COS CH_3SH, CS_2, DMS, DMDS	He 20–26 cm/sec	− 70°C to 100°C at 16°C/min	Hewlett-Packard Model 5713A FPD	sub-ppb	[55]
10 ft, 1/8 in. o.d. Teflon; Chromosil 330	COS, H_2S, CH_3SH CS_2, DMS, DMDS	N_2 35 ml/min	1°C to 100°C at 16°C min. Start 2 min hold and final 4 min hold	Hewlett-Packard Model 5730 FPD	8.8–20 pg S	[27]
18 in., 1/8 in. o.d. Teflon; acetone-washed Poropak QS, 80/100 mesh	H_2S, COS, SO_2, CH_3SH, DMS, DMDS + others	He 30 ml/min	30°C held for 1 min then to 210°C at 30°C/min	Model 5750 FPD	10 µg	[52–54]
2 m, 4 mm i.d. glass; 20% SE-30 on Chromosorb P, 60/80 mesh	H_2S, COS, CS_2, thiols, sulphides and disulphides	N_2 30 ml/min	30–100°C	Hewlett-Packard Model 906 FPD	ppt range	[34]

flame FPD and the DANI model 68/7 dual-flame FPD, utilize a different type of burner in which the column effluent is first burned in a small hydrogen-rich flame ($O_2/H_2 = 1:10$ v/v) housed within a larger tube carrying the analytical flame ($O_2:H_2 = 3:10$ v/v). Compounds entering the first flame are highly reduced (e.g. to H_2S, S_2, H_2O, CH_4, etc.), and this eliminates the competition for hydrogen atoms that normally occurs just after the initial oxidizing reactions in single-flame FPDs [66]. Several workers [67–70] have reported hydrocarbon quenching is dramatically reduced, although not eliminated, in such designs.

12.3.2.2 Flame chemistry and detector response

The basis for detecting sulphur-containing compounds by flame photometric detection is the measurement of a blue chemiluminescent emission from excited S_2 molecules (S_2^*) within a fuel-rich, hydrogen/air or hydrogen/oxygen flame, at temperatures of less than 1000°C.

Originally, S_2 molecules were thought to be the predominant source of chemiluminescence within such flames, but recent work suggests that they may not even be a dominant constituent [66]. Several sulphur species, such as H_2S, HS, S, S_2, SO and SO_2, may be formed in a hydrogen-rich flame, the concentrations of which are determined by the kinetics of the various reactions at each point in the flame. The types of sulphur species present will also depend on the ratio of hydrogen to oxygen in the gases entering the flame, which can change the relative concentrations of either H or OH radicals.

A common problem associated with the conventional single-flame FPD is extinction by the solvent, which is often exacerbated by incorrect mixtures of hydrogen with oxygen or air and too large an injection of solvent [63,67,68]. Most investigations give gas flow-rates and ratios most suitable for the maximum detector response to sulphur, with the least chance of extinction. Burgett and Green [71] claim to have eliminated this problem by using a hyperventilated oxygen flame.

The presence of hydrocarbons may also cause extinction and often reduces the response to sulphur compounds. The amount of light reaching a PMT may be affected by a gradual build-up of carbon deposited along the wall of the FPD glass window, from large amounts of hydrocarbon entering the detector. Moreover, additional components such as hydrocarbons or carbon (from COS or CS_2, for example) can produce negative or positive interferences with the chemiluminescence of S_2^*. Therefore, a reduction in the concentration of hydrocarbons reaching the FPD is highly desirable. Mowery and Benningfield [72] achieved this by using a 'heart-cut' technique in which a large proportion of the propylene gas under analysis was directed away from the detector by columns and valves. Patterson *et al.* [67] also report that both hydrocarbon quenching and solvent flame-extinction are minimized by using a Varian dual-flame FPD.

The chemiluminescence produced by sulphur-containing compounds in a hydrogen-rich flame consists of a series of roughly equally spaced emission bands between 350 and 460 nm, with the strongest emission occurring at 394 nm, as shown in Fig. 3.2 (p. 79). As such, most FPD optical filters have maximum transmittance at 394 nm in order to detect sulphur-containing compounds selectively. Most references suggest a selectivity of at least 10^4:1 [2].

Jones and Penkett [73] optimized the sensitivity of the FPD by increasing the amount of light entering the PMT. A filter was used which had a 48% transmission at 394 nm and a 35 nm bandwidth at half-height, instead of a narrow-band interference filter with a transmission of 43% at 394 nm and 9 nm bandwidth at half-height. This increased the signal to noise ratio by a factor of 1.4, which allowed the minimum detection limit for injection of 1 ml of a mixture of COS with nitrogen to be improved from 10 ppb to about 7 ppb.

The FPD response to sulphur is non-linear and is unique among common chromatographic detector responses. Several FPD response models have been suggested, (e.g. [8,35]), but all show that the response is proportional to $[SD]^n$ where n is predicted to be 2,because the rate-determining step of the reactions within the flame is the combination of two sulphur atoms to give excited S_2^*. The n values reported vary from approximately 1 to slightly greater than 2 [58,61,62,64,72–88]. This variation is mainly due to differences in flame conditions, which are altered by different gas flow-rates and oxidant to fuel ratios. To maintain any degree of analytical accuracy or precision, all flame parameters must be exactly adjusted and kept constant. In addition, the FPD responses (on a volume basis) for molecules containing one sulphur atom should be equivalent, but they are not, and the n-values have been found to be different for different types of sulphur-containing compounds, probably because of varying degrees of decomposition for each compound in fuel-rich flames [66]. Burnett *et al.* [81] have established relative response factors and experimental n-values for a number of important sulphur-containing compounds. For the most accurate work, the n-values should be experimentally established for the compounds under study and quantification based on calibration curves for individual compounds. Linearizing circuits, often available on many commercial FPDs, also assume that $n = 2$, and may therefore lead to considerable errors.

Different FPD designs are also known to have dissimilar response characteristics. McGaughey and Gangwal [82], for example, compared the performance characteristics of three commercially available FPDs, using the same support gases and conditions, and found different n-values for different detectors; for H_2S the n-values ranged from 1.81 to 2.00; for COS from 1.83 to 2.07; and for SO_2 from 1.82 to 2.10.

12.3.2.3 Sulphur doping

Crider and Slater [89] first reported that an intentional background of SO_2 improved the minimum detection limit (MDL) and linearity of an FPD sulphur-response. Other workers have since used an increase in the background level of sulphur [often using SO_2, COS or SF_6 as a make-up ('bleed') gas], to obtain the best possible MDLs [74,85,90]. This optimization technique utilizes the non-linear response of the detector as a means of increasing the signal for a given amount injected, the magnitude of the increase depending on the concentration of the gas mixture used. This is practical up to the point where the effect of a further increase in bleed gas level will be offset by an increase in noise and saturation, and a poorer signal to noise ratio.

Jones and Penkett [73] used a bleed gas mixture of 20 ppm COS in nitrogen, prepared in an aluminium vessel. The best results were obtained by introducing this gas between the end of the column and the detector at a rate of 2.2 ml/min, with a carrier gas flow-rate of 20 ml/min through the column, giving a background concentration of about 2 ppm of COS entering the detector. This increased the detector sensitivity at lower concentrations, and for a injection of 1 ml of a mixture of COS and nitrogen, the MDL was improved from 150 ppb to about 10 ppb (v/v).

Another advantage of using a bleed gas is the resultant linearization of the detector response to sulphur. It typically results in a Gaussian peak with a width at half-height that corresponds to the point where the mass-flow of the eluted analyte has reached half its maximum. This differs from the normal response obtained with the FPD, where the peak width should be measured at quarter peak height rather than at half peak height. Farwell and Barinaga [66] suggest that improved MDLs of 2–10 pg/sec can be obtained for sulphur by this doping method, but point out a disadvantage in that the linear range typically covers only 1–1.5 orders of magnitude (e.g. 2–30 pg/sec).

Bleed gases should not be introduced along with the carrier gas, because of possible changes in background sulphur concentrations during elution of sample components. Farwell and Barinaga [66] describe three doping techniques; first, a permeation tube containing a sulphur gas can be placed in the flow path of the detector supply-gas line, where the bleed gas concentration can be maintained by means of a controlled temperature and a constant gas flow-rate; second, the bleed gas can be premixed in one of the supply-gas cylinders and the concentration changed at constant flow-rate by adjusting a non-doped diluent with temperature-controlled devices; third, a pure sulphur gas source can be used, with the sulphur concentration controlled by temperature, pressure and control devices for highly restricted flow.

12.3.2.4 Other methods

The optimization of any FPD depends on several factors, such as flow-rates of oxygen, hydrogen, air, or carrier gas into the FPD, the type of sulphur compound being determined, detector and injector temperatures, and PMT voltage. Pecsar and Hartmann [40] suggested methods for obtaining maximum sensitivity from an FPD. The response, especially to H_2S, was found to vary with detector temperature, which was recommended to be maintained at an optimum of 110°C, to prevent the condensation of moisture and resultant flame instability found at below 100°C, and possible adsorption at above 130°C.

More critical factors in detector performance were found to be the flow-rates of the carrier and make-up gases. Omitting the detector air flow, which slightly reduced selectivity, gave an increased sensitivity to sulphur components. The nitrogen carrier gas flow-rate was recommended to be as high as possible to give minimum residence times in the column. A flow of 80 ml/min was found to keep the flame in a stable region of operation, and this determined the effect of variations of H_2 and O_2 flow rates.

This shows that it would be an advantage to keep as low a hydrogen flow as possible in order to maximize response, but flows below 70 ml/min produce flame instability and inadequate combustion. The oxygen flow is distinctly optimal between 20 and 25 ml/min. An optimum flow ratio of N_2:H_2:O_2 of 80:80:20 was therefore recommended, and slightly higher hydrogen rates and lower oxygen rates favoured more stable operation. Pecsar and Hartmann also attempted to optimize the signal to noise ratio of the PMT bias voltage; they chose to fix the voltage at 750 V but suggested optimizing the voltage for each phototube used.

12.3.3 Calibration techniques

Several methods have been developed for generating trace concentrations of sulphur gases for the calibration of sensitive analytical systems. Standard gaseous samples have been prepared by injections of known amounts of pure sulphur compounds (gas or liquid) into aluminium chambers [23], glass flasks [34,43], Tedlar bags [73], and several types of exponential dilution flasks [45,91,92], of known volume. Gas samples are then removed by syringe, diluted and analysed, by a set sampling procedure. Calibration curves are thus produced by plotting response against concentrations injected.

Another convenient method of preparing concentrations of sulphur gases in air in the range from ng/l. to μg/l., is by the use of permeation devices. These consist of a capsule, part of which is a permeable membrane (usually made of FEP Teflon), containing a liquefied gaseous substance hermetically sealed under its own pressure [93–96]. Such permeation tubes are commercially available in a wide variety of designs with many types of enclosed sulphur compounds, and give mass loss-rates ranging from a few ng/min to several μg/min. O'Keefe and Ortman [93] have shown that permeation rates are constant and reproducible when the temperature is controlled to within ± 0.1°C; the final permeation rate depends on dissolution of the gas in the Teflon, diffusion through the Teflon wall and evaporation from the outer surface of the device, the main driving force being the difference in partial pressure between the inner and outer walls of the tube [96]. Thus, the weight loss of a device at constant temperature over a known period of time constitutes the permeation rate. Once the rate is fixed by a tube design and set temperature, concentrations can be varied by controlling the flow of gas around the exterior of the tube.

Stevens *et al.* [84] first used permeation tubes as a source for calibrating a GC/FPD system for the determination of SO_2, CH_3SH and CS_2 in ambient air, and these have since become commonly used for calibration for a large range of sulphur gases, e.g. [15,18,24–33,36,40,50,57]. Bruner *et al.*, [57,97] used a combination of permeation tubes and an exponential dilution flask method to ensure accurate calibration for H_2S, SO_2 and CH_3SH measurements, while de Souza and Bhatia [54] used a similar technique to determine total reduced sulphur and SO_2 in the 0.1–100 ppb concentration range.

Continuing improvement in GC/FPD techniques and the lowering of detection limits, have meant that standards must now be produced that range from several ppb (v/v) to as low as 0.01 ppb (v/v). Such calibration standards should be kept out of contact with potentially adsorptive surfaces [25,98], be prepared in a gas matrix matching that of air samples, and span the expected concentration ranges of analytes in actual samples. Many workers have suggested or utilized single-stage or multi-stage dynamic dilution systems, and a number of single-stage permeation tube systems are commercially available.

Goldan *et al.* [99] have recently developed a three-stage dynamic dilution system for use with gravimetrically calibrated 'low loss' (≈ 100 ng/min) permeation tubes, in order to generate calibration standards below 0.1 ppb (v/v). Briefly, the permeation tube output is diluted with a mixture of nitrogen and oxygen (zero air) and CO_2 at normal atmospheric mixing ratios, with the gas stream passing through Teflon surfaces at all dilution levels. Water vapour may be admixed in the final stage to simulate relative humidities from near zero to approximately 90%. The system produces an equivalent one-stage diluent flow which can be varied between 10 and 100 l./min with a net actual consumption of 1.3 l./min. Accuracy of dilution is estimated to vary from ±2% to ±10% at 1000 l./min.

McTaggart *et al.* [100] have developed a system for the preparation and validation of sulphur gas standards at sulphur concentrations down to 0.01 μg/m^3. This system utilizes 'ultra low loss' permeation devices with emission rates (0.1–10 ng/min sulphur) determined by a non-gravimetric calibration procedure based on an MFC/FD/FPD technique (see Section 12.2.5) that can be standardized with aqueous sulphate solutions. The permeation tube output is diluted in a single-stage dynamic dilution system, either with compressed ambient air scrubbed to remove sulphur compounds [25], or by synthetic scrubbed ambient 'zero' air.

Farwell *et al.* [101] compared these two techniques for the calibration of H_2S, CH_3SH, CS_2 and SO_2 gas standards, using a GC/FPD measurement procedure for all four species and a fluorimetric measurement procedure for H_2S. Comparative measurements showed differences between the two techniques no greater than 12% for CS_2, SO_2 and CH_3SH, and 21% for H_2S in the 1–2 ppb (v/v) range.

REFERENCES

[1] H. Draegerwerk and B. Draeger, *West German Patent*, No. 113918, 26 July, 1962.
[2] S. S. Brody and J. E. Chaney, *J. Gas Chromatog.*, 1966, **4**, 42.
[3] D. F. Adams, W. L. Bamesberger and T. J. Robertson, *J. Air Pollut. Control Assoc.*, 1968, **18**, 145.
[4] T. Okita, *Atmos. Environ.*, 1970, **4**, 93.
[5] T. Okita, J. P. Lodge, Jr. and H. D. Axelrod, *Environ. Sci. Technol.*, 1971, **5**, 532.
[6] D. F. S. Natusch, H. B. Klonis, H. D. Axelrod, R. J. Teck and J. P. Lodge, *Anal. Chem.*, 1975, **47**, 2460.
[7] H. D. Axelrod and S. G. Hansen, *Anal. Chem.*, 1975, **47**, 2460.
[8] F. X. Meixner and W. Jaeschke, *Int. J. Environ. Anal. Chem.*, 1981, **10**, 51.
[9] B. W. Hermann and J. N. Seiber, *Anal. Chem.*, 1981, **53**, 1077.
[10] M. Buck, H. Ixfeld and H. Gies, *Schriftenr. Landesanst. Immissionsschutz Landes Nordrhein Westfalen*, 1978, **44**, 15.
[11] R. S. Braman, J. M. Ammans and J. L. Bricker, *Anal. Chem.*, 1978, **50**, 992.
[12] J. W. Russell and L. A. Shadoff, *J. Chromatog.*, 1977, **134**, 375.
[13] R. M. Harrison and R. Perry in *Handbook of Air Pollution Analysis*, 2nd Ed., R. M. Harrison and R. Perry (eds.), Chapter 6, Chapman & Hall, London, 1986..
[14] B. E. Foulger and P. G. Simmonds, *Anal. Chem.*, 1979, **51**, 1089.
[15] M. A. Carroll, *Ph.D. Thesis*, University of Massachusetts, 1983.
[16] J. D. Pleil, K. D. Oliver and W. A. McClenny, *J. Air Pollut. Control Assoc.*, 1987, **37**, 244.
[17] A. B. Goldberg, P. J. Maroulis, L. A. Wilmer and A. R. Bandy, *Atmos. Environ.*, 1981, **15**, 11.
[18] M. A. Carroll, L. E. Heidt, R. J. Cicerone and R. G. Prinn, *J. Atmos. Chem.*, 1986, **4**, 375.
[19] R. A. Rasmussen, *Tellus*, 1974, **26**, 254.
[20] F. J. Sandalls and S. A. Penkett, *Atmos. Environ.*, 1977, **11**, 197.
[21] P. J. Maroulis, A. L. Torres and A. R. Bandy, *Geophys. Res. Lett.*, 1977, **4**, 510.
[22] A. L. Torres, P. J. Maroulis, A. B. Goldberg and A. R. Bandy, *J. Geophys. Res.*, 1980, **85**, 7537.
[23] S. M. Turner and P. S. Liss, *J. Atmos. Chem.*, 1985, **2**, 223.
[24] B. B. Jorgensen and B. Okholm-Hansen, *Atmos. Environ.*, 1985, **19**, 1737.
[25] S. O. Farwell, S. J. Gluck, W. L. Bamesberger, T. M. Schutte and D. F. Adams, *Anal. Chem.*, 1979, **51**, 609.
[26] M. S. Black, R. P. Herbst and D. R. Hitchcock, *Anal. Chem.*, 1978, **50**, 848.
[27] P. A. Steudler and W. Kijowski, *Anal. Chem.*, 1984, **56**, 1432.
[28] A. R. Bandy, B. J. Tucker and P. J. Maroulis, *Anal. Chem.*, 1985, **57**, 1310.
[29] M. O. Andreae and W. R. Barnard, *Anal. Chem.*, 1983, **55**, 608.
[30] W. R. Barnard, M. O. Andreae and W. E. Watkins, *J. Geophys. Res.*, 1982, **87**, 8787.
[31] M. O. Andreae, W. R. Barnard and J. M. Ammons, in *Environmental Biogeochemistry*, R. Hallberg (ed.), pp. 167–177. Publishing House/FRN, Stockholm, 1983.
[32] W. R. Barnard, M. O. Andreae and R. L. Iverson, *Cont. Shelf. Res.*, 1984, **3**, 103.

[33] M. O. Andreae, *Limnol. Oceanog.*, 1985, **30**, 1208.
[34] A. Tangerman, *J. Chromatog.*, 1986, **366**, 205.
[35] M. O. Andreae, R. J. Ferek, F. Bermond, K. P. Byrd, R. T. Engstrom, S. Hardin, P. D. Houmere, F. LeMarrec, H. Raemdonck and R. B. Chatfield, *J. Geophys. Res.*, 1985, **90**, 12891.
[36] R. A. Kagel and S. O. Farwell, *Anal. Chem.*, 1986, **58**, 1197.
[37] R. K. Koppe and D. F. Adams, *Environ. Sci. Technol.*, 1967, **1**, 479.
[38] P. Ronkainen, J. Denslow and O. Leppänen, *J. Chromatog. Sci.*, 1973, **11**, 384.
[39] R. K. Stevens, J. D. Mulik, A. E. O'Keeffe and K. J. Krost, *Anal. Chem.*, 1971, **43**, 827.
[40] R. E. Pecsar and C. H. Hartmann, *J. Chromatog. Sci.*, 1973, **11**, 492.
[41] W. L. Banwart and J. M. Bremner, *Soil Biol. Biochem.*, 1974, **7**, 359.
[42] W. L. Banwart and J. M. Bremner, *Soil Biol. Biochem.*, 1975, **8**, 19.
[43] W. L. Banwart and J. M. Bremner, *Soil Biol. Biochem.*, 1974, **6**, 1132.
[44] J. M. Bremner and W. L. Banwart, *Soil Biol. Biochem.*, 1976, **8**, 79.
[45] F. Bruner, A. Liberti, M. Possanzini and I. Allegrini, *Anal. Chem.*, 1972, **44**, 2070.
[46] J. M. Bremner and W. L. Banwart, *Sulphur Inst. J.*, 1974, **10**, 66.
[47] W. L. Thornsberry, *Anal. Chem.*, 1971, **43**, 452.
[48] H. W. Grice, M. L. Yates and D. J. David, *J. Chromatog Sci.*, 1970, **8**, 90.
[49] P. A. Steudler and B. J. Peterson, *Nature*, 1984, **311**, 455.
[50] P. A. Steudler and B. J. Peterson, *Atmos. Environ.*, 1985, **19**, 1411.
[51] D. J. Cooper, W. Z. de Mello, W. J. Cooper, R. G. Zika, E. S. Saltzman, J. M. Prospero and D. L. Savoie, *Atmos. Environ.*, 1987, **21**, 7.
[52] T. L. C. de Souza, D. C. Lane and S. P. Bhatia, *Anal. Chem.*, 1975, **47**, 543.
[53a] *Analysis of Sulfur Gases, G. C. Bulletin* 722J, Supelco Inc., Belafonte, PA.
[53] T. L. C. de Souza, *J. Chromatog. Sci.*, 1984, **22**, 470.
[54] T. L. C. de Souza and S. P. Bhatia, *Anal. Chem.*, 1976, **48**, 2234.
[55] S. O. Farwell, A. E. Sherrard, M. R. Pack and D. F. Adams, *Soil Biol. Biochem.*, 1979, **11**, 411.
[56] D. F. Adams, S. O. Farwell, E. Robinson and M. R. Pack, EA-1516, Research Project 856-1, Sept. 1980.
[57] F. Bruner, P. Ciccioli and F. Di Nardo, *Anal. Chem.*, 1975, **47**, 1790.
[58] M. L. Selucky, *Chromatographia*, 1971, **4**, 425.
[59] D. F. S. Natusch and T. M. Thorpe, *Anal. Chem.*, 1973, **45**, 1184A.
[60] W. A. Aue, *J. Chromatog Sci.*, 1975, **13**, 329.
[61] J. Ševčík, *Detectors in Gas Chromatography*, pp. 145–64. Elsevier, Amsterdam, 1976.
[62] S. O. Farwell and R. A. Rasmussen, *J. Chromatog. Sci.*, 1976, **14**, 224.
[63] L. S. Ettre, *J. Chromatog. Sci.*, 1978, **16**, 396.
[64] S. O. Farwell, D. R. Gage and R. A. Kagel, *J. Chromatog. Sci.*, 1981, **19**, 358.
[65] H. V. Drushel, *J. C hromatog. Sci.*, 1983, **21**, 375.
[66] S. O. Farwell and C. J. Barinaga, *J. Chromatog. Sci.*, 1986, **24**, 483.
[67] P. L. Patterson, R. L. Howe and A. Abu-Shumays, *Anal. Chem.*, 1978, **50**, 339.
[68] P. L. Patterson, *Anal. Chem.*, 1978, **50**, 345.
[69] D. A. Ferguson and L. A. Luke, *Chromatographia*, 1979, **12**, 197.
[70] S. K. Gangwal and D. E. Wagoner, *J. Chromatog. Sci.*, 1979, **17**, 196.
[71] C. A. Burgett and L. E. Green, *J. Chromatog. Sci.*, 1974, **12**, 356.
[72] R. A. Mowery, Jr. and L. V. Benningfield, Jr., *Am. Lab.*, 1983, **15**, No. 5, 102.
[73] B. M. R. Jones and S. A. Penkett, *Internal Rept., Env. Med. Sci. Divn. AERE Harwell*, 1980.
[74] A. R. L. Moss, *Scan*, 1974, **4**, 5.
[75] J. M. Raccio and B. Welton, *Instrum. Res.*, 1985, **1**, 30.
[76] P. T. Gilbert, in *Analytical Flame Spectroscopy*, R. Mavrodineanu (ed.), pp. 281–87. Springer-Verlag, New York, 1970.
[77] A. I. Mizany, *J. Chromatog. Sci.*, 1970, **8**, 151.
[78] T. Sugiyama, Y. Suzuki and T. Takeuchi, *J. Chromatog.*, 1973, **77**, 309.
[79] T. Sugiyama, Y. Suzuki and T. Takeuchi, *J. Chromatog. Sci.*, 1973, **11**, 639.
[80] J. G. Eckhardt, M. B. Denton and J. L. Mayers, *J. Chromatog. Sci.*, 1975, **13**, 133.
[81] C. H. Burnett, D. F. Adams and S. O. Farwell, *J. Chromatog. Sci.*, 1978, **16**, 68.
[82] J. F. McGaughey and S. K. Gangwal, *Anal. Chem.*, 1980, **52**, 2079.
[83] C. E. Quincoces and M. G. González, *Chromatographia*, 1985, **20**, 371.
[84] R. K. Stevens, A. E. O'Keeffe and G. C. Ortman, *Environ. Sci. Technol.*, 1969, **3**, 652.
[85] T. J. Cardwell and P. J. Marriott, *J. Chromatog. Sci.*, 1982, **20**, 83.
[86] P. J. Marriott and T. J. Cardwell, *Chromatographia*, 1981, **14**, 279.
[87] C. D. Pearson, *J. Chromatog. Sci.*, 1976, **14**, 154.
[88] B. Wenzel and R. L. Aiken, *J. Chromatog. Sci.*, 1979, **17**, 503.
[89] W. L. Crider and R. W. Slater, Jr., *Anal. Chem.*, 1969, **41**, 531.

[90] C. M. Zehner and R. A. Simonaitis, *J. Chromatog. Sci.*, 1976, **14**, 348.
[91] M. D. Thomas and R. E. Amtower, *J. Air Pollut. Control Assoc.*, 1966, **16**, 618.
[92] H. P. Williams and J. D. Winefordner, *J. Gas Chromatog.*, 1966, **4**, 271.
[93] A. E. O'Keeffe and G. C. Ortman, *Anal. Chem.*, 1966, **38**, 760.
[94] F. P. Scaringelli, S. A. Frey and B. E. Saltzman, *Am. Ind. Hyg. Assoc. J.*, 1967, **28**, 260.
[95] F. P. Scaringelli, A. E. O'Keeffe, E. Rosenberg and J. P. Bell, *Anal. Chem.*, 1970, **40**, 871.
[96] O. P. Lucero, *Anal. Chem.*, 1971, **43**, 1744.
[97] F. Bruner, C. Canulli and M. Possanzini, *Anal. Chem.*, 1973, **45**, 1790.
[98] M. Thompson and M. Stanisavljevic, *Talanta*, 1980, **27**, 477.
[99] P. O. Goldan, W. C. Kuster and D. L. Albritton, *Atmos. Environ.*, 1986, **20**, 1203.
[100] D. L. MacTaggart, R. A. Kagel and S. O. Farwell, *J. Air Pollut. Control Assoc.*, 1987, **37**, 143.
[101] S. O. Farwell, R. A. Kagel, C. J. Barinaga, P. D. Goldan, W. C. Kuster, F. C. Fehsenfeld and D. L. Albritton, *Atmos. Environ.*, 1987, **21**, 1983.
[102] J. E. Lovelock, *Nature*, 1974, 248, 625.

Index